SECURITY AND PROTECTION IN INFORMATION PROCESSING SYSTEMS

IFIP – The International Federation for Information Processing

IFIP was founded in 1960 under the auspices of UNESCO, following the First World Computer Congress held in Paris the previous year. An umbrella organization for societies working in information processing, IFIP's aim is two-fold: to support information processing within its member countries and to encourage technology transfer to developing nations. As its mission statement clearly states,

> *IFIP's mission is to be the leading, truly international, apolitical organization which encourages and assists in the development, exploitation and application of information technology for the benefit of all people.*

IFIP is a non-profitmaking organization, run almost solely by 2500 volunteers. It operates through a number of technical committees, which organize events and publications. IFIP's events range from an international congress to local seminars, but the most important are:

- The IFIP World Computer Congress, held every second year;
- Open conferences;
- Working conferences.

The flagship event is the IFIP World Computer Congress, at which both invited and contributed papers are presented. Contributed papers are rigorously refereed and the rejection rate is high.

As with the Congress, participation in the open conferences is open to all and papers may be invited or submitted. Again, submitted papers are stringently refereed.

The working conferences are structured differently. They are usually run by a working group and attendance is small and by invitation only. Their purpose is to create an atmosphere conducive to innovation and development. Refereeing is less rigorous and papers are subjected to extensive group discussion.

Publications arising from IFIP events vary. The papers presented at the IFIP World Computer Congress and at open conferences are published as conference proceedings, while the results of the working conferences are often published as collections of selected and edited papers.

Any national society whose primary activity is in information may apply to become a full member of IFIP, although full membership is restricted to one society per country. Full members are entitled to vote at the annual General Assembly, National societies preferring a less committed involvement may apply for associate or corresponding membership. Associate members enjoy the same benefits as full members, but without voting rights. Corresponding members are not represented in IFIP bodies. Affiliated membership is open to non-national societies, and individual and honorary membership schemes are also offered.

SECURITY AND PROTECTION IN INFORMATION PROCESSING SYSTEMS

IFIP 18th World Computer Congress
TC11 19th International Information Security Conference
22–27 August 2004
Toulouse, France

Edited by

Yves Deswarte
LAAS-CNRS, France

Frédéric Cuppens
ENST-Bretagne, France

Sushil Jajodia
George Mason University, USA

Lingyu Wang
George Mason University, USA

Springer-Science+Business Media, B.V.

Electronic Services <http://www.wkap.nl>

Library of Congress Cataloging-in-Publication Data

A C.I.P. Catalogue record for this book is available from the Library of Congress.

Security and Protection in Information Processing Systems
Edited by Yves Deswarte, Frédéric Cuppens, Sushil Jajodia, and Lingyu Wang

DOI 10.1007/978-1-4020-8143-9

Printed on acid-free paper.
www.springer.com/mycopy

Contents

General Chair's Message

Security is probably the most critical factor for the development of the "Information Society". E-government, e-commerce, e-healthcare and all other e-activities present challenging security requirements that cannot be satisfied with current technology, except maybe if the citizens accept to waive their privacy, which is unacceptable ethically and socially. New progress is needed in security and privacy-preserving technologies.

On these foundations, the IFIP/Sec conference has been established from the eighties as one of the most important forum for presenting new scientific research results as well as best professional practice to improve the security of information systems. This balance between future technology improvements and day-to-day security management has contributed to better understanding between researchers, solution providers and practitioners, making this forum lively and fruitful.

The first IFIP/Sec Conference was held in Stockholm in May 1983. The IFIP Technical Committee on Security and Protection in Information Processing Systems, TC-11, was created in 1984, just before the second Sec conference. As we are celebrating its 20th anniversary, TC-11 has now seven active working groups on dedicated security topics. And even if the Sec conference remains the flagship event of TC-11, the working groups organize periodically workshops and working conferences on their own topics. This year, the 19th IFIP/Sec Conference embeds three of these workshops, on Information Security Management (ISM), on Information Security Education (ISE), and on Privacy and Anonymity in Networked and Distributed Systems (I-NetSec). These workshops had their own calls for papers, Program Committees and selection processes, and their papers are gathered in another volume of the IFIP/Sec 2004 Conference proceedings. The present volume contains all the papers that were selected for the main track of the conference.

In 2004, the IFIP/Sec conference is part of the IFIP World Computer Congress, in Toulouse. And security is also one of the main topics of other conferences and workshops that are hosted by the Congress, including CARDIS, the 6th Smart Card Research and Advanced Application Conference, I3E, the 4th Conference on e-Commerce, e-Business and e-Government, and FAST, the 2nd Workshop on Formal Aspects in Security and Trust. So there is no doubt that the 18th IFIP World Computer Congress will remain as a memorable event for many security professionals.

Yves Deswarte
General Chair

Program Chair's Message

SEC 2004 received 159 submissions, all of which were assigned to at least three program committee members or other external reviewers. At a one-day meeting of the program committee, all submissions were discussed and 35 papers were selected for presentation at the conference.

Several trends in computer security have become prominent since the beginning of the new century and thus received special room in the program: the proliferation of intrusions that exploit new vulnerabilities of computer infrastructure, the necessity to respond to every computer security incidents and the emergence of new security problems in home ad-hoc networks. The former trend convinced us to allocate two sessions on this topic, namely on ``Intrusion Detection" and ``Malicious Code Analysis". The second one is not directly reflected by research papers but we included a panel discussion on ``Meeting the Global Challenges of Security Incident Response". The latter is investigated in a session entitled ``Security Protocols and Home Security".

We gratefully acknowledge all authors who submitted papers for their efforts in continually enhancing the standards of this conference. It is also our pleasure to thank the members of the program committee and the external reviewers for their work and support.

Frédéric Cuppens and **Sushil Jajodia**
Program co-Chairs

IFIP/SEC2004 Conference Committees

Conference General Chair
Y. Deswarte, LAAS-CNRS, France

Program Committee co-Chairs
F. Cuppens, ENST-Bretagne, France
S. Jajodia, George Mason University, USA

Program Committee
H. Armstrong, Curtin University, Australia
W. Caelli, Queensland Univ. of Tech., Australia
H. Debar, France Telecom, France
B. De Decker, K.U. Leuven, Belgium
Y. Deswarte, LAAS-CNRS, France
M. Dupuy, SGDN/DCSSI/CERTA, France
M. El-Hadidi, Cairo University, Egypt
J. Eloff, Pretoria University, South Africa
C. Farkas, U. of South Carolina, USA
S. Fischer-Huebner, Karlstad Univ., Sweden
S. Furnell, University of Plymouth, UK
A. Gabillon, LIUPPA, France
D. Gritzalis, Athens U. of Economics, Greece
S. Katsikas, University of the Aegean, Greece
D. Khakhar, University of Lund, Sweden
K.-Y. Lam, PrivyLink, Singapore
W. List, Independent, UK
P. Liu, Penn State, USA

C. Majumdar, Jadavpur University, India
L. Mé, SupElec, France
J. Miettinen, Sonera, Finland
R. Molva, Eurecom, France
G. Papp, VRAM Communication, Hungary
H. Pohl, ISIS, University of Bonn, Germany
R. Posch, Graz Univ. of Technology, Austria
S. Qing, Chinese Academy of Sciences, China
K. Rannenberg, Goethe U. Frankfurt, Germany
I. Ray, Colorado State, USA
P. Samarati, University of Milan, Italy
R. Sasaki, Tokyo Denki University, Japan
R. von Solms, Port Elizabeth Tech., S. Africa
L. Strous, De Nederlandsche Bank, NL
V. Swarup, MITRE, USA
V. Varadharajan, Macquire Univ., Australia
J. Verschuren, TNO-EIB, The Netherlands
T. Virtanen, Helsinki U. of Technology, Finland
J. Vyskoc, VaF, Slovakia
M. Warren, Deakin University, Australia
T. Welzer, University of Maribor, Slovenia
G. Wenngren, Defence Research Agency, Sweden
L. Yngstrom, University of Stockholm, Sweden

Local Organizing Committee
Marie Dervillers, LAAS-CNRS

AN ABSTRACT REDUCTION MODEL FOR COMPUTER SECURITY RISK

Mohamed Hamdi, Noureddine Boudriga
Computer Network and Security Research Lab.
SUP'COM, University of 7th of November, Carthage, Tunisia
mmh@certification.tn, nab@supcom.rnu.tn

Abstract This paper presents an approach for decision making under security risks in a computer network environment. The proposed method relies on a many sorted algebraic signature and on a rewriting system. This latter is shown to be terminating and yielding a normal form, called the risk analysis equation, that models the cost-benefit balance. Furthermore, a gradual algebraic resolution of the risk analysis equation is described.

This formalism helps security analysts to automate the selection of the optimal security solutions that minimize the residual risk.

Keywords: Risk management, algebraic specifications, rewriting systems, risk analysis equation.

1. INTRODUCTION

AS the use of information systems and sophisticated communication infrastructures becomes widespread in modern enterprises, the growth of various attacks against those systems is inevitable. In spite of the important investments made by companies to secure their assets, they are continuing to be the target of harmful actions. To this end, security ought to be considered as a part of the management process. A variety of Risk Analysis (RA) methods have been proposed in this context [Alberts and Dorofee, 2002; Stonebumer et al., 2002; GoC, 1996; GAO, 1999] . They aim at evaluating accurately the loss resulting from potential attacks in order to make the appropriate decisions. Nonetheless, those approaches still have numerous shortcomings. In fact, most of them rely on qualitative reasoning which is, by nature, non precise despite of its ease. For instance, defining a four or five-step scale to assess the probability of success of an attack does not allow a suitable differentiation between threats. Moreover, existing RA approaches have not involved a considerable theoretical development. They introduce several simple concepts (e.g., RA matrix) that are intended to make their manual application simpler. However, in a system where hundreds (or even thousands) of assets, vulnerabilities, threats and decisions are considered, security management can not be addressed manually. Therefore, automated RA is a key concern which should be emphasized.

In this paper, we present a reduction system that permits to automate the reasoning made by security experts when performing RA. Our optimization framework consists of an algebraic rewriting system which yields the candidate security solutions and an algorithm for selecting the optimal countermeasure configuration through the use of an order relation on a lattice structure. The major merit of this work is that it allows to handle complicated situations where human intervention is not applicable. Furthermore, the algebraic framework includes a logic which offers the possibility to perform inferences and proofs. In addition, a formal proof of the efficiency of our approach is given through a demonstration of the termination of the rewriting system (using polynomial interpretations) and of the convergence of the algorithm. This shows that the system reaches the optimal security solutions independently from the situation.

The remaining part of the paper is organized as follows. Section 2 describes the RA signature representing the extension with a previous work [Hamdi and Boudriga, 2003]. Section 3 defines the rewriting system and studies its properties. Section 4 discusses the application of the concept of many-sorted algebras to RA. Section 5 proposes an algorithm to solve the RA equation. Section 6 concludes the paper.

2. THE RISK ANALYSIS SIGNATURE

2.1 Related work

Many-sorted signatures and first-order predicate logic have been previously used to model the RA problem [Hamdi and Boudriga, 2003]. Basically, a many-sorted signature Σ is characterized by a set S of sort names, an $S^\star \times S$-sorted set of operation names (denoted Ω), and an $S^\star$-sorted set of predicate symbols (denoted Π).

In [Hamdi and Boudriga, 2003], the authors have proposed the signature Σ_0 depicted in Table 1 to model the RA process. The richness of this algebraic signature allows the representation of attack scenarios (i.e., attacks performed on multiple steps). The introduction of this concept affects considerably the decision making process as the efficiency of a decision differs for two scenarios having two distinct semantic structures. This stems from the fact that the quantitative comparison attributes (e.g., probability of success, impact) of the main attack is computed using the attributes of the elementary attacks.

It can be demonstrated that security countermeasures can be viewed as pseudo-inverses of potential threats with respect to the composition law $\star$. More precisely, decisions aim at making the system recover from the effect of the various possible attacks. The major interest of this reasoning resides in

sig	$\Sigma_0=$	
	sorts	$asset, vuln, attack, decision$
	opns	$- \star - : attack \times attack \to attack$
		$1_a :\to attack$
		$(-,-) \bullet (-,-) : decision \times asset \times decision \times asset \to decision$ $\times asset$
		$a^\star :\to asset$
	preds	$ispresent : asset \times vuln$
		$exploits : attack \times vuln$
		$ispossible : attack \times asset$
		$isminimal : attack$
		$- \leq_a - : attack \times attack$
		$mitigates : decision \times attack$
		$(-,-) \succ_c (-,-) : decision \times asset \times decision \times asset$

Table 1. Many-sorted signature representing the RA process.

the fact that the decision making process, which is the essence of RA, is performed through the resolution of the equation $a \star d = 1_a$, where a models the possible attacks, d represents the potential countermeasures and 1_a is the neutral element of $\star$. Despite its interesting properties, this algebraic framework does not allow a rigorous resolution of the above equation since the operation $\star$ can be applied only for two attacks to build a scenario. For instance, $at_1 \star at_2$ expresses a scenario where at_1 occurs before at_2. Therefore, the expression $a \star d$ is mathematically incorrect since the second term (i.e., the decision d in this case) should be of sort $attack$. Hence, the signature mentioned above should be extended in order to allow a more accurate representation of the RA problem that relies on the same basic idea.

2.2 A more general framework

Basically, RA consists in studying the environment of the target system in order to predict the potential threats and to derive the corresponding security solutions. To support this reasoning, the signature presented in [Hamdi and Boudriga, 2003] should be enriched in order to allow a convenient modeling of both the environment and the potential events. As these events can be either destructive or preventive, a balance between the potential threats and a set of security countermeasures should be considered. To this purpose, we introduce the following definition of a RA signature.

DEFINITION 1 ***Risk analysis signature.*** *A RA signature is a many-sorted algebra $\Sigma = \langle S, \Omega, \Pi \rangle$ that verifies the following conditions:*

(1) $sysenv, action \in S$,

(2) $\oplus : sysenv \times sysenv \to action \in \Omega$,

(3) $\circledast : sysenv \times action \to action \in \Omega$,

(4) $\otimes$: $action \times action \rightarrow action \in \Omega$,
(5) $\leq_{act}$: $action \times action \in \Pi$.

The alert reader would have noticed that this definition is more general than Σ_0 in the sense that it just specifies kernel representing of basic risk management concepts. Two special sorts, $sysenv$ and $action$, have been considered to provide a formal view of the state of the analyzed system. Effectively, this state is characterized by the environment of the system (modeled by $sysenv$) and the actions that take it from one state to another (modeled by $action$). Furthermore, a special category of Σ-operations has been highlighted in this definition. They show the different methods that can be used to build actions. Three combinations to construct actions from the other sorts are provided. In fact, an action can be deduced purely from the system environment (using $\oplus$), or by taking into account a specific action (using $\circledast$). In addition, the combination of two actions can constitute an action (using $\otimes$). Finally, a predicate symbol $\leq_{act}$ has to be included in the signature so that actions can be compared.

A major advantage of this formal representation is that it translates perfectly the reasoning steps followed by the risk analyst which are: (1) the identification of the key components of the analyzed system, (2) the identification of the potential actions on the target system (i.e., attacks and countermeasures), (3) the selection of the best security decisions according to several preference criteria.

Table 2 gives an example of a RA signature. Four sorts are added to those mentioned in the above definition. Operations *ispresent*, *exploits*, *mitigates* and $\sqsubseteq$ give the environment information that the risk analyst should know at the first step. In fact, *ispresent* expresses the existence of a vulnerability inside a given resource while *exploits* tells whether a given vulnerability should be present to carry out a corresponding attack. On the other hand, *mitigates* means that a specified decision reduces, in a certain manner, the effect of a given attack. The last operation $\sqsubseteq$ states that an attack is a part of a more global attack scenario (i.e., the first attack is a sub-scenario of the second).

The second class of operators are related to the construction of attack scenarios. The composition law $\star$ is a precedence operator used to build composite attacks. The constant 1_a stands for the null attack which correspond to "*no action*" on the system.

3. THE REWRITING SYSTEM

Having built a RA signature, its properties must be correctly expressed in order to conduct an appropriate reasoning from RA point of view. Typically, this involves a set of axioms (or equations) and a corresponding inference system which derives all possible consequences from those axioms. In [Hamdi and

sig	$\sum$=	
	sorts	$asset, vuln, attack, decision, sysenv, action$
	opns	$ispresent : asset \times vuln \longrightarrow sysenv$
		$exploits : attack \times vuln \longrightarrow sysenv$
		$mitigates : decision \times attack \longrightarrow sysenv$
		$\sqsubseteq: attack \times attack \longrightarrow sysenv$
		$- \star - : attack \times attack \longrightarrow attack$
		$1_a :\longrightarrow attack$
		$\oplus : sysenv \times sysenv \longrightarrow action$
		$\circledast : sysenv \times action \longrightarrow action$
		$\otimes : action \times action \longrightarrow action$
		$1_{act} :\longrightarrow action$
		$\diamond : attack \times asset \longrightarrow action$
		$\circ : decision \times asset \longrightarrow action$
	preds	$\leq_{act}: action \times action$

Table 2. Example of a RA signature.

Boudriga, 2003], we used a first-order predicate logic and a classical Gentzen system to automate the deduction process. However, as this approach relies on replacement of equals by equals, it requires a special attention. In fact, knowing that $at \star 1_a = 1_a$ (i.e., 1_a is the neutral element on the set of attacks), the term at_1 in the equation $at_3 = at_1 \star at_2$ can be replaced by $at_1 \star 1_a$ leading to the equation $at_3 = at_1 \star 1_a \star at_2$. This does not conform with the risk analyst's objective which is to go through several deduction steps towards an equation representing the balance between attacks and security decisions. It turns out that rewriting systems (composed of directed equations) provide this possibility. In this section, we present a specific class of rewriting systems that is suitable for RA problems and we discuss its main properties.

3.1 Defining the rewriting rules

Term rewriting systems are widely used in the field of formal modeling. Their main feature is that they give the ability to automate the generation of canonical terms. Typically, a rewriting system $\langle T(\Sigma, \chi), \longrightarrow \rangle$ is a finite set of rewriting rules which are ordered pairs of terms denoted $\tau \longrightarrow \tau'$, where $\longrightarrow$ is a binary relation on $T(\Sigma, \chi)$ (the set of Σ-terms). A term t rewrites to a term t', denoted by $t \longrightarrow t'$, if there exists a rule $\tau \longrightarrow \tau'$ in $\langle T(\Sigma, \chi), \longrightarrow \rangle$, a position ω in t, a substitution σ, satisfying $t_{|\omega} = \sigma(\tau)$, such that $t' = t[\sigma(\tau')]_{\omega}$. A term t is called a normal form if there is not t' such that $t \longrightarrow t'$ (t is then denoted by $t \downarrow_{\langle T(\sum, \chi), \longrightarrow \rangle}$ in this case).

(ρ_1)	$ispresent(v, as) \oplus exploits(at, v) \longrightarrow at \diamond as$
(ρ_2)	$mitigates(d, at) \circledast at \diamond as \longrightarrow at \diamond as \otimes d \circ as$
(ρ_3)	$(at_1 \diamond as) \otimes (at_2 \diamond as) \longrightarrow (at_1 \star at_2) \diamond as$
(ρ_4)	$at_1 \sqsubseteq at2 \circledast (at_1 \diamond as \otimes at_2 \diamond as) \longrightarrow at_1 \diamond as$
(ρ_5)	$at \star 1_a \longrightarrow at$
(ρ_6)	$1_a \star at \longrightarrow at$
(ρ_7)	$act \otimes act \longrightarrow act$

Table 3. Example of a RA rewriting system.

In our context, we build a rewriting system $\rho = \langle T(\Sigma, \chi), \longrightarrow_\rho \rangle$ which simulates the reasoning of the human risk analyst. It allows the representation of the state of the analyzed system suitably for the decision selection process.

DEFINITION 2 ***Risk analysis rewriting system.*** *Let $\Sigma = \langle S, \Omega, \Pi \rangle$ be a RA signature. A RA rewriting system $\rho = \langle T(\Sigma, \chi), \longrightarrow_\rho \rangle$ is a set of rewriting rules which have one of the following forms:*

(i) Type I rules: $\tau \longrightarrow \tau'$ such that $\tau, \tau' \in T(\Sigma, \chi)$ and τ' is a sub-term of τ,

(ii) Type II rules: $\tau \longrightarrow \tau'$ such that $\tau, \tau' \in T(\Sigma, \chi)$, τ contains one function symbol and τ' does not contain any function symbol,

(iii) Type III rules: $\tau\varphi\tau' \longrightarrow \tau''\varphi'\tau'''$ such that $\tau, \tau', \tau'', \tau''' \in T(\Sigma, \chi)$ and $\varphi, \varphi' \in \{\oplus, \otimes, \circledast\}$.

This definition asserts that rewriting rules expressing RA reasoning are restricted to three types. This may seem constraining to the security specialist but we will show through some examples that RA rewriting systems are sufficiently general to allow an efficient representation of a practical context. For instance, the system ρ_0 given in Table 3 is a RA rewriting system. In fact, the reader can easily check that (ρ_4) is a type I rule ($at_1 \diamond as$ is a sub-term of $(at_1 \diamond as \otimes at_2 \diamond as) \otimes at_1 \sqsubseteq at2$) and that $\{(\rho_5), (\rho_6), (\rho_7)\}$ are type II rules. In addition, $\{(\rho_1), (\rho_2), (\rho_3)\}$ are type III rules.

The first rule shows how to state whether a threat is actually possible to perform on the system. In fact, this threat has to exploit a vulnerability which effectively exists in the corresponding asset.

The rule (ρ_2) is a heuristic that reduces the space where the security analyst searches for the optimal countermeasures. Instead of considering the whole set of security solutions, only those which potentially mitigate the effect of the possible attacks should be addressed. The two following rewriting rules are related to attacks scenarios. (ρ_3) states that if two attacks are possible to carry out on an asset, then so does their composition. On the other hand, (ρ_4) gives that if several attacks are considered as potential actions, then the system would focus on the one that includes the others (in the sense of the operator $\sqsubseteq$). Rules (ρ_5) and (ρ_6) simply mean that 1_a is the neutral element of the operation $\star$. The last rule permits to avoid redundancies when addressing potential attacks or candidate security solutions.

3.2 Termination and confluence of the rewriting system

The two main properties of a rewriting system are confluence and termination. A rewriting rule $\rightarrow$ is said to be confluent if it satisfies $^\star\leftarrow \circ \rightarrow^\star \subseteq \rightarrow^\star \circ^\star \leftarrow$, where $\leftarrow$ denotes the inverse of $\rightarrow$, and $\circ$ and * represent respectively the composition and the transitive closure of binary relations. On the other hand, termination means that there is not infinite chains of related elements $t_1 \rightarrow t_2 \rightarrow \ldots \rightarrow t_i \rightarrow \ldots$

Confluence means that the order of application of the rewriting rules to a given term is not important. In other terms, two normal forms obtained through applying two distinct sequences of rewriting rules to the same term must be equal. This property is not important in our case since the order that should be followed is known *a priori*. In fact, a typical scenario-based RA consists of the following steps: (1) threat identification, (2) attack scenario identification, (3) candidate security solutions identification, and (4) countermeasure selection. This situation is relatively easy to handle since operations are clearly ordered within each step. In other contexts where concurrent rewriting rules may be applied to achieve the same task, confluence would be seriously considered.

On the opposite, termination is an essential property in our context as it guarantees that the system reaches the normal forms, which is an major requirement from the RA point of view. In this paper, we adopt polynomial interpretations as a tool to demonstrate termination. This consists in assigning to each Σ-operation a polynomial verifying three basic requirements: (1) for each Σ-operation o of arity (i.e., number of variables) n, a polynomial of the same arity, denoted $[o](X_1, ..., X_n)$, is associated, (2) the variables $X_1, ..., X_n$ of any polynomial of arity n are supposed to be integers greater than or equal to a constant c, and (3) all polynomials should have non-negative integers coefficients.

This polynomial representation can be extended to terms by induction. For example, if $[\star](X_1, X_2) = X_1 + X_2$ and $[\sqsubseteq](X_1, X_2) = X_1.X_2$, then the interpretation of $at_1 \sqsubseteq (at_1 \star at_2)$ is expressed by $X_1.[\star](X_1, X_2)$, which is equal to $X_1.(X_1 + X_2)$. Thus, a polynomial function $[t](X_1, ..., X_n)$ can be mapped to every $t \in T(\sum, \chi)$. Moreover, an order relation on $T(\Sigma, \chi)$, denoted by $<_{[.]}$, can be set through the use of these interpretations such that $t <_{[.]} t'$ if, and only if $[t] < [t']$ for every two terms $t, t' \in T(\Sigma, \chi)$, where $<$ is the order relation on positive integers.

In [B. Cherifa and Lescanne, 1987], it has been shown that this order is stable by instantiation and that it can therefore be used to prove the termination of rewriting systems. In fact, to terminate, every rule $\tau \to \tau'$ of the rewriting system has to verify $[\tau] < [\tau']$. This method can be applied to RA rewriting systems.

THEOREM 3 *Let* $\Sigma = \langle S, \Omega, \Pi \rangle$ *be a RA signature and* ρ *be a RA rewriting system on* Σ *consisting only of type III rules.*

Let $\rhd$ *be a binary relation on* $T(\Sigma, \chi)$ *defined as follows:*

$\tau_1 \rhd \tau_2$ *iff.* $\exists \varphi, \varphi' \in \{\oplus, \otimes, \circledast\}$, $\tau_3, \tau_4 \in T(\Sigma, \chi)$ *st.* $(\tau_1 \varphi \tau_3 \to \tau_2 \varphi' \tau_4) \in \rho$.

If $\rhd$ *is non-reflexive, non-symmetric and transitive then the rewriting system* ρ *terminates.*

Proof. To the demonstration purpose, we associate the polynomial interpretation $[\tau_0] = \sum_{\{\tau | \tau_0 \rhd \tau\}} [\tau]$ to each term τ_0 appearing in type III rules. Since $\rhd$ is non-reflexive, non-symmetric and transitive, it can be concluded that $\{\tau \mid \tau_2 \rhd \tau\} \subset \{\tau \mid \tau_1 \rhd \tau\}$ if $\tau_1 \rhd \tau_2$. Which implies:

$$\text{if } \tau_1 \rhd \tau_2 \text{ then } [\tau_1] > [\tau_2]. \tag{1}$$

We assume now that the polynomial interpretation $[\varphi](X_1, X_2) = X_1 + X_2$ correspond to every function φ in $\{\otimes, \oplus, \circledast\}$. Thus, for every type III rewriting rule $\tau_1 \varphi \tau_2 \to \tau_3 \varphi' \tau_4$, we have to prove that $[\tau_1] + [\tau_2] > [\tau_3] + [\tau_4]$. This can be achieved for a rewriting rule $\tau_1 \varphi \tau_2 \to \tau_3 \varphi' \tau_4$, using Equation 1 and the fact that $\tau_1 \rhd \tau_3$,$\tau_2 \rhd \tau_4$.□

From the above theorem, it can be concluded that a system composed of type III rewriting rules terminates if it verifies:

If τ_1 *appears in the left term of a rewriting rule and* τ_2 *appears in the right term of the same rule, and if* τ_1*appears in the right term of an other rewriting rule, then* τ_2 *does not appear in the left term of the latter rule.*

EXAMPLE 4 ***Counter-example of termination****. The above theorem gives a sufficient condition for termination of type III rewriting rules. Obviously, if a specific system does not fulfill this condition, we can not conclude that it does not terminate. Nonetheless, we discuss the following example to show how this condition can affect the termination.*

Consider the RA rewriting system consisting of the two following type III rewriting rules:

$$\begin{cases} (\rho_1)\ mitigates(d, at) \circledast at \diamond as \to at \diamond as \otimes d \circ as, \\ (\rho_2)\ at \diamond as \circledast d \circ as \to mitigates(d, at). \end{cases}$$

It is clear that this system does not verify the assumption of the above theorem as (ρ_1) *gives that* $mitigates(d, at) \rhd d \circ as$ *and it comes from* (ρ_2) *that* $d \circ as \rhd mitigates(d, at)$. *In addition, the system does not terminate because the rewriting sequence* $\rho_1, \rho_2, \rho_1, \rho_2 ...$ *is infinite cycle.*

$[ispresent]\,(X_1,X_2) = X_1 + X_2,$	$[exploits]\,(X_1,X_2) = X_1 + X_2,$
$[mitigates]\,(X_1,X_2) = X_1 + X_2,$	$[\sqsubseteq]\,(X_1,X_2) = X_1.X_2,$
$[\star]\,(X_1,X_2) = X_1 + X_2,$	$[1_a] = 2,$
$[\oplus]\,(X_1,X_2) = X_1.X_2,$	$[\circledast]\,(X_1,X_2) = X_1.X_2 + X_2 + 2,$
$[\otimes]\,(X_1,X_2) = X_1 + X_2 + 2,$	$[1_{act}] = 2,$
$[\diamond]\,(X_1,X_2) = X_1.X_2,$	$[\circ]\,(X_1,X_2) = X_1.X_2.$

Figure 1. Polynomial interpretations corresponding to Σ-operations.

$$P_{\rho_1}(X_1,X_2,X_3) = X_1^2 + X_1.X_2 + X_1.X_3$$
$$P_{\rho_2}(X_1,X_2,X_3) = X_1.X_3.(X_2-1) + X_2^2.X_3$$
$$P_{\rho_3}(X_1,X_2,X_3) = 2,\ P_{\rho_4}(X_1) = X_1.X_3 + X_2.X_3 + 4$$
$$P_{\rho_5}(X) = 2,\ P_{\rho_6}(X) = 2,\ P_{\rho_7}(X) = X + 2$$

Figure 2. Polynomials corresponding to the rewriting system ρ_0.

THEOREM 5 *Let $\Sigma = \langle S, \Omega, \Pi \rangle$ be a RA signature and ρ be a RA rewriting system on Σ. If the sub-system composed of the type III rules of ρ verifies the condition of theorem 3, then ρ terminates.*

Particularly, the rewriting system ρ_0 terminates.

This theorem is a direct consequence of Theorem 3 since type I and type II rewriting rules terminate trivially (see [Goguen and Malcolm, 2000; Loeckx et al.] for more details). Concerning the system ρ_0, we use a set of polynomial interpretations which is different from the one mentioned in the above theorem. This allows a better illustration of the use of polynomial interpretations as a tool for proving termination of rewriting systems. Figure 1 associates an interpretation to each function of Σ.

The polynomials corresponding to the rules of our rewriting system are given in Figure 2. They have been computed using the induction principle discussed above. It is clear that all of these polynomials are positive if the variables are greater than 2 (i.e., $c = 2$). For the sake of simplicity, the computational steps to reach $(P_{\rho i})_{i\in\{1,..,7\}}$ are not given. We leave it to the reader to check them.

4. TOWARDS RISK ANALYSIS ALGEBRAS

4.1 From specification to algebra

A many-sorted algebra assigns a concrete aspect to a many-sorted signature by associating a set of data to each sort and a function to each operation. In other terms, if $\Sigma = \langle S, \Omega, \Pi \rangle$ is a many-sorted signature, a Σ-algebra α assigns: (1) a set $|\alpha|_s$ to each sort $s \in S$, called the carrier set of the sort s (as defined in [Loeckx et al.]), (2) a function $|f|_\alpha : |\alpha|_{s_1} \times \ldots \times |\alpha|_{s_k} \to |\alpha|_s$ to each operation $(f : s_1 \times ... \times s_k \rightarrow s) \in \Omega$, (3) a predicate $|p|_\alpha : |\alpha|_{s_1} \times ... \times |\alpha|_{s_k}$ to each predicate symbol $(p : s_1 \times ... \times s_k) \in \Pi$. In our case, we can state that if Σ is a RA signature then α simply represent a RA project consisting of a real analyzed system and of a knowledge database characterizing the risk analyst.

In this section, we show how a RA rewriting system can be used to perform deductions that translate the representation of the studied information system from a complex equation to a normal form. To this purpose, we propose a general reduction methodology composed of the following steps:

1 Model the state of the system through the use of functions having the form $|f|_\alpha : |\alpha|_{s_1} \times ... \times |\alpha|_{s_k} \rightarrow |sysenv|_s$.

2 Apply rewriting rules that combine terms of sort $sysenv$ and generate the related terms of sort $action$.

3 Apply rewriting rules that combine terms of sort $sysenv$ and terms of sort $action$ to generate terms of sort $action$.

4 Apply rewriting rules that combine terms of sort $action$ and generate terms of the same sort. These latter terms should be normal forms with respect to the used rewriting system.

This methodology allows the automated risk analyst to perform a progressive reasoning leading to a term representing the actions of interest (i.e., destructive and preventive actions). A key feature of this term is that it consists of two different parts. A static part corresponds to destructive actions. It is called *static* because the risk analyst can not act on them, he can only introduce other actions that reduce their effect. On the other hand, a variable part represents preventive actions. The RA decision maker should select some of these actions (i.e., a sub-part of the variable part) to provide the best security level for the analyzed networked system.

For instance, in the case of the rewriting system presented in the previous section, the normal form generated by our methodlogy is expressed by the following equation:

$$\left((|\otimes|_\alpha)_{i \leq n,\, j \leq n_i} (at_j \diamond as_i) \right) (|\otimes|_\alpha) \left((|\otimes|_\alpha)_{i \leq n,\, j \leq m_i} (d_j \circ as_i) \right). \tag{2}$$

This term will be used in Section 5 in order to search for the optimal countermeasures combination. For the seek of clarity, the subscript α will be removed from sorts and operations when no confusion can be made about the many-sorted RA algebra.

4.2 Illustrative example

The goal of this subsection is to illuminate the key concepts presented above at a sufficient level to ensure a fundamental understanding of their usage in practice. The first step is to build a many-sorted algebra α that conforms with the specification $\langle \Sigma_0, \rho_0 \rangle$. This means that a set $|\alpha|_s$ is associated to every sort s in the signature Σ_0.

We suppose that the analyzed system is composed of two assets, say as_1 and as_2 (i.e. $|\alpha|_{asset} = \{as_1, as_2\}$), that verify the hypotheses below:

(H_1) as_1 and as_2 are placed behind a packet filtering gateway that prevent connection establishment from the external network.

(H_2) The company has no security awareness program .

(H_3) The user working on as_1 introduced a strong root password.

(H_4) The user working on as_2 did put his name as root password for this machine.

(H_5) The FTP (File Transfer Protocol) port (21) is closed in both as_1 and as_2.

Furthermore, it is assumed that the basic knowledge of the RA system consists of the following attacks and vulnerabilities ((a_i) denote attacks and (v_i) stand for vulnerabilities):

(a_1) Execute remotely on the victim machine Netcat, a Unix utility which reads and writes data across network connections.

(a_2) Perform a SYN (synchronization) scan. This consists in sending a SYN packet to every port on the victim machine and waiting for responses. If the victim acknowledges the packet, then the port is open.

(a_3) Perform an ACK (acknowledgment) scan. This technique is more sophisticated as it permits to know if a port is open even it is protected through a packet filter.

(a_4) Get password files from the victim machine. For instance, in a Unix environment, the attacker gets the /etc/passwd and /etc/shadow.

(a_5) Perform a dictionary attack on the root password. This attack can be performed by testing all the combinations generated from a wordlist.

(a_6) Perform a buffer overflow on FTP server application to get remote root access to the victim machine.

(v_1) The user executes programs coming from non-trusted sources.

(v_2) The host responds to SYN scan. This vulnerability can be removed using a packet filtering firewall that blocks incoming connections.

(v_3) The host responds to ACK scan. Packet filtering firewalls are not efficient to overcome this weakness. Stateful inspection firewall, which monitor the state of TCP connections, can be used to cover this limit.

(v_4) Weak root password. This means that the password can be easily cracked (e.g., brute force, dictionary).

(v_5) FTP port is open. Many FTP server implementations contain security breaches, especially buffer overflow vulnerabilities.

In addition, two attacks scenarios, denoted a_7 and a_8 are considered. They are expressed by the following equation:

$$a_7 = a_1 \star a_2 \star a_4 \star a_5, \quad a_8 = a_1 \star a_3 \star a_4 \star a_5. \tag{3}$$

These scenarios correspond to two alternatives of the same main attack (password cracking), the first one relies on SYN scan while the second one is based on ACK scan. In other terms, $|\alpha|_{attack} = \{a_1, a_2, a_3, a_4, a_5, a_6, a_7, a_8\}$.

In addition to these sets, we use binary relations to represent operations and predicates of the signature $\sum$ in the algebra α. For instance, the following matrix models the relation which states whether a vulnerability is exploited by a given attack.

$$|exploits|_\alpha = \begin{bmatrix} 1 & 0 & 0 & 0 & 0 \\ 0 & 1 & 0 & 0 & 0 \\ 0 & 0 & 1 & 0 & 0 \\ 0 & 0 & 0 & 0 & 0 \\ 0 & 0 & 0 & 1 & 0 \\ 0 & 0 & 0 & 0 & 1 \\ - & - & - & - & - \\ - & - & - & - & - \end{bmatrix}.$$

If the value of $(|exploits|_\alpha)_{ij}$ equals 1 (respectively 0), this means that the attack a_i exploits (respectively does not exploit) the vulnerability v_j. The rows corresponding to a_7 and a_8 can not be filled because those attacks are composite scenarios.

In the following, we show how the oriented equations of ρ_0 can model efficiently two fundamental RA steps: vulnerability analysis and threat analysis.

Vulnerability identification

The conduction of the vulnerability identification process through the use of automated scanners and questionnaires, results in the following matrix (with regard to the hypotheses $(H_i)_{i\in\{1,..,5\}}$).

$$|ispresent|_\alpha = \begin{bmatrix} 1 & 1 \\ 0 & 0 \\ 1 & 1 \\ 0 & 1 \\ 0 & 0 \end{bmatrix}.$$

Threat identification

As it has been mentioned above, this step aims at identifying the attacks that are possible to carry out on the analyzed system. In fact, a direct application of the rewriting rule (ρ_1) states that this operator can be modeled by a binary relation which is the composition of $|ispresent|_\alpha$ and $|exploits|_\alpha$. Thus, $\otimes$ appears in α as a binary relation denoting the composition between relations. More formally, we have:

$$|\diamond|_\alpha = |exploits \otimes ispresent|_\alpha = |exploits|_\alpha \circ |ispresent|_\alpha .$$

Consequently, the matrix corresponding $|\diamond|_\alpha$ is the product of $|ispresent|_\alpha$ and $|exploits|_\alpha$. Nonetheless, it is worth noting that the two bottom rows can not be computed through the use of this product as attack scenarios are not considered in the rewriting rule (ρ_1). Hence, the rule (ρ_4) has been applied to compute $(|\diamond|_\alpha)_{(i,j)\in\{7,8\}\times\{1,2\}}$.

$$(|\diamond|_\alpha)_{(i,j)\in\{7,8\}\times\{1,2\}} = \begin{bmatrix} 0 & 0 \\ 0 & 1 \end{bmatrix}.$$

Candidate countermeasure selection

We suppose that the following actions are available to reduce the effect of potential attacks.

(d_1) Acquire a stateful inspection firewall.

(d_2) Configure the packet filtering firewall to close FTP connections.

(d_3) Elaborate a security training program for the employees.

(d_4) Change root password to guarantee more robustness.

It is obvious that only (d_1), (d_2) and (d_3) mitigate the attack scenario as they thwart respectively (at_3), (at_1) and (at_5). Therefore, according to the rewriting rule (ρ_2), these decisions constitute the set of candidate countermeasures for the resource as_2. Concerning as_1, it is not taken into account during this step because it was found that no attack scenario is possible to carry out against it. The normal form can then be represented by the following formula:

$$at_8 \, |\diamond|_\alpha \, as_2 \otimes d_1 \, |\circ|_\alpha \, as_2 \otimes d_3 \, |\circ|_\alpha \, as_2 \otimes d_4 \, |\circ|_\alpha \, as_2. \qquad (4)$$

5. SOLVING THE RISK ANALYSIS EQUATION

Having applied the rewriting rules according to the aforementioned order, we obtain an irreducible formula (see Equation 2). This formula expresses the potential destructive and preventive actions that correspond to the current state of the system. Since the aim of the security analyst is to thwart the attacks threatening the computer network, the countermeasure selection process can be thought of as the resolution of the equation:

$$\bigotimes_{i\leq n,\, j\leq n_i} (at_j \diamond as_i) \bigotimes \bigotimes_{i\leq n,\, j\leq m_i} (d_j \circ as_i) = 1_{act}, \qquad (5)$$

meaning that the chosen security solutions totally annihilate the effect of the possible threats. However, since perfect security is an utopia, a realistic objective would be to select the decisions that make the effect of the above formula closest to the one of 1_{act}.

The resolution of this problem can be viewed as the selection of an optimal subset of the set $\Delta = \bigcup_{i\leq n, j\leq m_i} \{(d_j \circ as_i)\}$ of candidate decisions. Elements

$(d_j \circ as_i)$ of Δ will be denoted by δ. Obviously, the set $\Delta^\star$ (set of partitions of Δ) has a lattice structure with respect to the inclusion binary relation $\subset$. The universal lower bound of this lattice is $\emptyset$ (the empty set) while the universal greater bound is Δ itself. The optimization process can be thought of as a traverse of the lattice $(\Delta^\star, \subset)$ from $\emptyset$ towards the direction of Δ. We propose the following algorithm to select the optimal set of decisions:

Algorithm decision_select
Begin
$\delta^\star := 1_{act}$;
$\Delta^{opt} = \emptyset$;

While $\left(\bigotimes_{i,j} (at_j \diamond as_i) \otimes \bigotimes_{\delta \in \Delta^{opt}} (\delta) \right) \leq_{act} \left(\bigotimes_{i,j} (at_j \diamond as_i) \otimes \bigotimes_{\delta \in \Delta^{opt} \cup \{\delta^\star\}} (\delta) \right)$

$\Delta^{opt} := \Delta^{opt} \cup \{\delta^\star\}$;
Find $\delta^\star \in Parent(\Delta^{opt})$ such that for every $\delta_0 \in Parent(\Delta^{opt})$

$\left(\bigotimes_{i,j} (at_j \diamond as_i) \otimes \bigotimes_{\delta \in \Delta^{opt} \cup \{\delta^\star\}} (\delta) \right) \leq_{act} \left(\bigotimes_{i,j} (at_j \diamond as_i) \otimes \bigotimes_{\delta \in \Delta^{opt} \cup \{\delta_0\}} (\delta) \right)$

EndWhile
End

where $Parent(X)$ represents the set $\{Y | Y \in \Delta^\star,\ X \subset Y \text{ and } |Y| = |X| + 1\}$ for every $X \in \Delta^\star$.

An informal reading of this algorithm shows that the optimal set of countermeasures ($\Delta^{opt} \subset \Delta^\star$) is reached through an iterative process which consists basically in adding, at each iteration, a security decision and evaluating its effect. At the beginning, the set Δ^{opt} is empty. Then, at each step, a solution $\delta^\star$ is selected from $Parent(\Delta^{opt})$ such that its combination with the elements of Δ^{opt} and the set of possible attacks be as close as possible to 1_{act} (with respect to $\leq_{act}$). The algorithm stops when the addition of any element $\delta \in \Delta \setminus \Delta^{opt}$ worsens the situation defined by Δ^{opt}.

THEOREM 6 *If the operation $\otimes$ is compatible with $\leq_{act}$ then the algorithm* **decision_select** *is $\leq_{act}$-optimal (i.e., every iteration gives a solution which is better than the solutions found at the previous iterations).*

Proof. We proceed to an *ab absurdo* proof. Supposing that the decisions δ_1 and δ_2 have been selected after the two first iterations of **decision_select** and that there exists $\delta_i, \delta_j \in \Delta \setminus \{\delta_1, \delta_2\}$ such that their combination is better than $\delta_1 \otimes \delta_2$:

$$\delta_i \otimes \delta_j \leq_{act} \delta_1 \otimes \delta_2. \tag{6}$$

As $\leq_{act}$ is compatible with $\otimes$ and $\delta_1 \leq_{act} \delta_i$, we can write:

$$\delta_1 \otimes \delta_j \leq_{act} \delta_i \otimes \delta_j. \tag{7}$$

Moreover, as δ_2 is the optimal decision at the second step:

$$\delta_1 \otimes \delta_2 \leq_{act} \delta_1 \otimes \delta_j \leq_{act} \delta_i \otimes \delta_j,$$

which conflicts with the main assumption expressed by Equation 6. □

Furthermore, the algorithm terminates because the search space at a given iteration is strictly included in the search space of the previous iteration. This means that the search space, which is finite, becomes more restricted across iterations.

A comprehensive example is given in the following to illustrate this algorithm. It is based on the concrete case studied in Section 4. The normal form reached at the end of that section (Equation 4) gives that the set of security decisions is equal to $\Delta = \{d_1 \circ as_2, d_3 \circ as_2, d_4 \circ as_2\} = \{\delta_1, \delta_2, \delta_3\}$. We propose to assess the cost and the benefit (in monetary terms) associated to each element $\delta \in \Delta$ and to state that $\delta \leq_{act} \delta'$ if, and only if, $(benefit - cost)_\delta \geq (benefit - cost)_{\delta'}$. We assume that the cost of δ_1, δ_2 and δ_3 are respectively equal to 1000, 400 and 30. On the other hand, the three candidate decisions have the same benefit, equal to 500, because they mitigate the same attack scenario.

Consequently, a direct application of the aforementioned algorithm gives that δ_3 is selected at the first step of **decision_select** because $(benefit - cost)_{\delta_1} = -500$, $(benefit - cost)_{\delta_2} = 100$ and $(benefit - cost)_{\delta_3} = 470$. At the second step, we evaluate the countermeasure combinations that belong to $Parent(\{\delta_3\})$ (i.e., $\{\delta_1, \delta_3\}$ and $\{\delta_2, \delta_3\}$). It is easy to check that $(benefit - cost)_{\delta_1 \otimes \delta_3} = -530$ and $(benefit - cost)_{\delta_2 \otimes \delta_3} = 70$. Therefore, no combination is selected at this level since both candidate decisions worsen the state reached at the previous iteration. Therefore, the final selected set of security decisions is $\{\delta_3\}$ according to the algorithm **decision_select**. In fact, the cost of δ_3 is acceptable while the stateful inspection firewall and the training are too expensive with respect to the attack that it prevents.

6. CONCLUSION

In this paper, we developed an algebraic decision making approach under security risks in a networked environment. Our method consists of two basic steps: a rewriting system that allows to identify the candidate countermeasures and a selection algorithm. We proved the termination of both of these steps meaning that the approach does not diverge.

Our work can be improved in the future by enriching the rewriting system (involving a study of its confluence) through the addition of a business-oriented reasoning.

References

Alberts, C.J. Dorofee, A.J. (2002).Managing Information Security Risks: the OCTAVE Approach, *Addison Wesley Professional*, ISBN: 0321118863.

Ben Cherifa, A. Lescanne, P. (1987). Termination of Rewriting Systems by Polynomial Interpretations and its Implementation, *Science of Computer Programming*, 9(2):137-160.

Claßen, I. Ehrig, H. Wolz, D. Algebraic Specification Techniques and Tools for Software Development: the ACT Approach, *AMAST Series in Computing (1)*, ISBN: 981-02-1227-5.

Goguen, J.A. Malcolm, G. (2000). Software Engineering with OBJ: Algebraix Specification in Action, *Kluwer Academic Publishers*, Boston, ISBN: 0-7923-7757-5.

Hamdi, M. Boudriga, N. (2003). Algebraic Specification of Network Security Risk Management, *First ACM Workshop on Formal Methods in Security Engineering*, Washington D.C.

Loeckx, J. Ehrich, H-D. Wolf, M. "Specification of Abstract Data Types," *Wiley and Teubner*, ISBN: 0-471-95067-X.

Stonebumer, G. Grogen, A. Fering, A. (2002). Risk Management Guide for Information Technology Systems, *National Institute for Standards and Technology*, Special Publication 800-30.

A Guide to Risk Management and Safeguard Selection for IT Systems, *Government of Canada, Communications Security Establishment*, January 1996.

Information Security Risk Assessment: Practices of Leading Organizations, *United States General Accounting Office*, GAO/AIMD-00-33, November 1999.

REMEDIATION GRAPHS FOR SECURITY PATCH MANAGEMENT

Vipin Swarup
The MITRE Corporation, 7515 Colshire Drive, McLean, VA 22102, USA
swarup@mitre.org

Abstract Attackers are becoming proficient at reverse engineering security patches and then creating worms or viruses that rapidly attack systems that have not applied the patches. Automated patch management systems are being developed to address this threat. A central function of a patch management system is to decide which patches to install on a computer and to determine the sequence of actions that will cause them to be installed.

In this paper, we introduce the notion of a *patch remediation graph* that depicts ways in which a system can be patched so that the system eventually reaches a desired, secure state. We present a language for specifying patch dependencies, patch conflicts, and user preferences for patch configurations. We then present efficient algorithms that construct and analyze remediation graphs from current computer configurations and various patch constraints. Our analysis algorithms use the graphs to compute maximal preferred configurations and sequences of patch actions that, if executed, will transition computers safely to those configurations.

Keywords: Security patch management, software configuration management, risk management, model checking.

1. INTRODUCTION

Security patches have become the primary technique for closing software vulnerabilities that are discovered after software systems have been deployed. Unfortunately, security patches also reveal details of the underlying vulnerabilities. Attackers are becoming proficient at using this information to create worms or viruses that rapidly attack systems that have not applied the patches.

Several patch management systems (e.g., Microsoft/SMS [9], Radia Patch Manager [10], etc.) have been developed to combat this threat. These systems include features from sophisticated software version management, software configuration management, and (post-deployment) software maintenance management systems [2, 5]. Software version, configuration, and maintenance management systems are designed to assist software developers in developing, composing, and maintaining large, evolving software systems. They provide features such as storage of multiple versions of a software package, retrieval of any identified version, tracking changes to a version, and notifying users affected by a change. Patch management systems extend this functionality by including sophisticated system inventory modules that determine the current state of a computer system, configuration management modules that compute patches that may be installed on a system, and secure patch deployment modules that deploy and install the patches in a secure manner [3, 4].

However, while patch management systems such as Microsoft/SMS provide numerous sophisticated capabilities, they still expect human administrators to make critical decisions regarding which patches to install. In particular, they only support coarse-grained patch management policies, e.g., vendors may mark certain patches as being critical, and administrators may designate certain patch sets as being reference configurations. However, no support is provided to represent the relative desirability of patches, or to help an administrator decide which patch sets to designate as preferred configurations, or for end-users to express their policy requirements regarding security patches (especially regarding patches which are not "recommended" or "critical"). For instance, a user may wish to specify that a certain application is critical and takes precedence over non-critical security patches which conflict with the application.

A second problem arises when a computer has not been maintained in an up-to-date patched state. In this case, when an administrator seeks to update the configuration, he must install numerous patches in order to take the computer to a recommended configuration. For instance, Sun Solaris 8 currently has 113 security patches of which 81 are in the Recommended Patch List. An administrator must first select some subset of these patches as the target configuration and the patch management system will then install the patches sequentially (the system will manage all dependencies and conflicts). However, since the patches are installed sequentially, the computer may pass through states that are insecure. This temporary vulnerability is important since the patch management system might fail or be suspended while the system is in the vulnerable state. Thus, the sequence of patch installations should be selected to minimize such vulnerable states, and the administrator should be informed of any potential vulnerabilities.

In summary, existing patch management systems do not answer two questions that are important to the security administrator of an individual computer: (1) What is the most desirable set of patches that should be installed on a computer? (2) What sequence of patch installation and removal actions will provide a safe transition path to the desired state? In this paper, we present an abstract model and practical algorithms that answer these questions for heterogeneous, distributed computers. We use a preference relation to model the relative desirability of patches. We define and build a patch remediation graph which permits us to analyze properties of patch installation sequences, e.g., whether a sequence includes vulnerable states. Note that since most subsets of patches are valid configuration states, the number of states and the number of possible patch installation sequences is enormous (i.e., the size of the remediation graph is exponential in the number of patches). Hence, we rely on model checking to manage the state explosion problem. We emphasize here that we do not address the practical problems of system inventory analysis, patch deployment and installation, etc. in this paper as these functions are present in current patch management products. Rather, we focus on the novel problem of remediation analysis as expressed by the previous two questions.

The remainder of this paper is organized as follows. Section 2 presents an abstract model for patch configuration management while Section 3 presents algorithms for patch remediation. Section 4 compares our approach with related work. Section 5 concludes this paper.

2. SECURITY PATCH MANAGEMENT: MODEL

Several models of software configuration management exist. Version models [2] label each distinct version of a software product and relate the different versions using labeling conventions or graphs. Change models model a software product in terms of a single base version and a series of changes (e.g., software patches). Hybrid models include attributes of both version models and change models. However, these models do not typically guide system administrators in selecting appropriate versions and changing a system's state from one version to another. In this section, we present an abstract security patch management model that permits a security administrator to reason about the desirability of possible patch configurations and to determine how to transition a system securely from one configuration to a preferable configuration.

2.1 Security Patch Management

We model the patch configuration state of a system in terms of a baseline configuration and a set of changes (e.g., security patches). We represent changes by a set of patches rather than a sequence of patch installation and removal actions; however, our model and algorithms can be generalized to handle patch sequences.

If $\mathcal{B}$ is a finite set of *baseline system configurations*, and $\mathcal{P}$ is a finite set of *security patches*, then $S = \mathcal{B} \times 2^{\mathcal{P}}$ is the set of all possible *patch configuration states*. If p is a patch and $s = \langle b, P \rangle$ is a state, then we will write $p \in s$ to denote that $p \in P$.

Let $\tau \subseteq S \times S$ be a *state transition relation*. τ represents the state change caused by installing or uninstalling a single security patch. Patch installation is often state-dependent, e.g., a patch may only be installed in states in which a specific version of an application is installed while certain other patches are not installed.

A *patch action graph* $G = \langle S, \tau \rangle$ is a finite directed graph whose vertices are the patch configuration states in S and whose directed edges are given by τ. We say that a state s' is *reachable* from state s if there is a directed path in G from s to s'.

Security patches typically have consistency constraints, e.g., two patches conflict with each other and must never be installed together in the same state, or one patch must always be installed together with another. We represent these in terms of a consistency predicate $V \subseteq S$, i.e., V defines the set of *consistent* states.

Due to the consistency constraints, it is not possible to reach the most secure state simply by installing all available patches. Rather, an administrator must make a choice regarding which subset of patches to install. We represent this choice using a *preference relation* $< \subset S \times S$ which is a quasi order over S, i.e., $<$ is irreflexive and transitive.

Informally, *patch remediation* is the process of transitioning a system from an initial state to a preferred, consistent state by performing a sequence of patch actions. We will present an algorithm to compute consistent states that are reachable from a given initial state and are maximal under $<$. We elaborate on this in the remainder of this section.

2.2 Patch Configuration States

There are several well-described inventory tools that collect information regarding the baseline software products installed on a computer as well as the patches that have been installed thus far. These tools are specific to a computer's operating system. For instance, tools for Windows-based systems may determine configurations by examining the

Windows Registry of each system. Tools for Unix-based systems may determine which applications are currently installed by using one of several techniques. One technique is to compute hashes of all executable files in certain directories and to compare those hashes with the standard hashes published by software vendors. Another technique is to look for patterns in audit records and to compare those patterns with standard patterns caused by the presence of an application.

Popular operating systems do not maintain adequate information to precisely define a specific patch configuration state. In particular, operating systems do not typically maintain the entire sequence of patch installs and uninstalls from the base version of an application. Rather, they only maintain a set of all patches currently installed on the computer. Hence, in this paper we model a computer's state as a set of patches, together with a baseline configuration that includes all the software products installed on the computer.

2.3 Patch State Consistency

Consistency constraints for patch states are properties that states must satisfy in order to be compliant with operational security policy. They can be defined in terms of the following three predicates on patch states:

(`critical` p) : This specifies that the patch p must be installed in all consistent states. That is, (`critical` p)(s) $=$ ($p \in s$). For instance, organizations typically mandate that all computers connected to the organizations' networks must have certain critical security patches installed. This can be expressed using `critical` constraints.

(p `conflicts-with` q) : This specifies that patches p and q cannot both be installed in any consistent state. That is, (p `conflicts-with` q)(s) $= (p \notin s) \vee (q \notin s)$. For instance, this can express the adverse interactions that a security patch may have with another patch or application, and indicates that they must not be installed together.

(p `requires` q) : This specifies that patch p is installed in a consistent state only if patch q is also installed. That is, (p `requires` q)(s) $=$ $(p \in s) \supset (q \in s)$. For instance, this can express the typical constraint that a set of patches must be installed together. This is so common that we introduce a derived construct to express it: (`coexists-with` $p_1 \ldots p_n$) is defined to mean that (p_i `requires` p_j) for all $1 \leq i, j \leq n$.

For instance, the following is an excerpt of the patch consistency constraint specification for Sun Solaris OS 2.3 (these patches have been superceded and are now obsolete):

(critical 101318)
(critical 101782)
(critical 102167)
(critical 103705)
(101318-32 conflicts-with 101524-01)
(101362-51 conflicts-with 101262-21)
(101782 requires 101318)
(102903 requires 101318)
(102932 requires 101318)
(101782 requires 101359)
(102167 requires 101359)
(103705 requires 101359)

A patch state s is consistent if it satisfies all the patch consistency constraints in the specification.

2.4 Patch State Transitions

A state transition represents the state change caused by installing or uninstalling a single security patch. Patches are specified as being applicable only to systems with a specific baseline configuration installed and a set of prerequisite patches already installed. Several patches also require the absence of conflicting patches. A state transition relation is specified as a set of rules. Each rule is a ground instantiation of one of the following rule templates with b^* being a sequence of baseline configurations, p^* and q^* being sequences of patches, and r being a patch. We write $r.p^*$ to denote the sequence whose first element is r and whose remaining elements are those in p^*.

(if (and (`baseline-in` b^*)
(`patches-present` p^*)
(`patches-absent` q^*))
(`install-patch` r))

(if (and (`baseline-in` b^*)
(`patches-present` r . p^*)
(`patches-absent` q^*))
(`remove-patch` r))

Each rule has a condition and a body. If the state satisfies the condition, then the rule is applicable and has the effect of either adding or removing the patch r from the state's patch set. The condition has three clauses. For a state $s = \langle b, P\rangle$ to satisfy the condition, its baseline configuration b must be in b^*, all patches in p^* must be in P, and the patches in q^* must not be in P. Note that the **`remove-patch`** rule also requires the patch r to be installed in the state. The state transition relation τ is given by the union of all specified rules.

2.5 Patch Preference Relation

A preference relation $<\subseteq S \times S$ is a quasi order (i.e., an irreflexive and transitive) relation between configuration states that represents the "desirability" of configurations. Preferences may be provided by software vendors (e.g., when a vendor recommends that certain patches be installed), by security administrators (e.g., when an administrator recommends that certain patches not be installed due to site-specific conflicts), or by end-users (e.g., when a user specifies that it is not desirable to install patches that conflict with mission-critical applications).

(`preferable` p^*): This returns a quasi order $<$ in which states with a larger set of patches in p^* are preferable to states with a smaller set. That is, if $s = \langle b, P\rangle$ and $s' = \langle b', P'\rangle$, then $s < s'$ if $(p^* \cap P) \subset (p^* \cap P')$. For instance, this can be used to specify recommended sets of patches as typically published by vendors. Note that if p^* consists of a single patch r, this returns a quasi order $<$ in which states with patch r are preferable to states without patch r.

(seq $<_1 <_2$): This returns a quasi order $<$ that first compares two states using $<_1$, and if they are incomparable then compares them using $<_2$. That is, $s < s'$ if either $(s <_1 s')$ or if (s and s' are incomparable under $<_1$ and $(s <_2 s')$). This construct is used to compose multiple quasi orders in order to construct a single preference relation.

2.6 Remediation

We can now define what we mean by patch remediation.

DEFINITION 1 *Let S be a set of patch configuration states, and let $\tau \subseteq S \times S$ be a patch transition relation. A* patch action graph $G = \langle S, \tau\rangle$ *is a finite directed graph whose vertices are the states in S and whose directed edges are given by τ.*

DEFINITION 2 *A* patch remediation posture *is a triple* $\langle S, V, \tau, < \rangle$ *where* $\langle S, \tau \rangle$ *is a patch action graph,* $V \subseteq S$ *is a set of consistent states, and* $<$ *is a preference relation over* S.

DEFINITION 3 *Let* $\langle S, V, \tau, < \rangle$ *be a posture. Then, the* patch remediation relation *for the posture is a relation* $R \subseteq S \times V$ *such that* $R(s, s')$ *holds if and only if* $s' \in V$ *(i.e., s' is a consistent state),* s' *is reachable from* s *in* $G = \langle S, \tau \rangle$*, and* $s < s'$.

Given an initial state s_0, a remediation related state s_f must be consistent, reachable from the initial state, and preferable to the initial state. However, a path from s_0 to s_f may take the system through intermediate states that are not consistent or that are not preferable to s_0. If an attack or fault causes patch remediation to abort in this intermediate state, then the system could be left in a final state that is less desirable (and possibly less secure) than the initial state. Thus, we define a special case of remediation where all intermediate states are preferable to the initial state, and where the system never moves from a consistent state to an inconsistent state.

DEFINITION 4 *Let* $\langle S, V, \tau, < \rangle$ *be a posture. Then, a* safe patch remediation relation *for the posture is a relation* $R \subseteq S \times V$ *such that* $R(s_0, s_n)$ *holds if and only if there exists a sequence of states* $s_0, \ldots, s_n$ *such that for all* $0 < i \leq n$, $\tau(s_{i-1}, s_i)$, $s_0 < s_i$, *and* $(s_{i-1} \in V) \supset (s_i \in V)$.

3. SECURITY PATCH MANAGEMENT: ALGORITHMS

We now present an algorithm for the computation of the remediation relation of a posture, for a given initial state. Let $\langle S, V, \tau, < \rangle$ be a posture, R be the induced remediation relation, and s_0 be an initial, possibly inconsistent, state. We want to compute all states s_n such that $R(s_0, s_n)$; we also want to compute the transition paths from s_0 to s_n. The states s_n are our goal states; they are consistent states that are reachable from s_0 and are preferable to s_0.

Since the transition relation can both add and delete patches and since it is sensitive to the absence and presence of patches, a simple state-space search for goal states will run into the state explosion problem in real-world systems. Hence, we consider an algorithm based on model checking. Model checking [1] is a technique for checking whether a formal model of a system satisfies a given property. A model, in our case, is a patch action graph. If a model does satisfy a specified property, then a model checker will return true; otherwise it will return a counterexample, i.e., a transition path that violates the property. Hence, we use a safety

```
GenerateRemediationGraph (S, V, τ, <, s0)
    p =  AG(¬(s ∈ V ∧ s0 < s))
    S_rg = modelCheck(S, τ, s0, p)
    τ_rg = τ ∩ (S_rg × S_rg)
    V_G = {s | s ∈ S_rg ∧ s ∈ V ∧ s0 < s}
    return (S_rg, τ_rg, s0, V_G).
```

Figure 1. Remediation Graph Generation Algorithm.

property of the form $\mathbf{AG}(\neg(s \in V \wedge s_0 < s))$. This expresses the property that it is not possible to reach a consistent state that is preferable to s_0. Model checking will return all states reachable from s_0 for which this property is false. That is, it will return all states s such that s is reachable from s_0 and a goal state is reachable from s; it will also return a transition path from s_0 to each s.

Sheyner et al [14] have described the use of model checking to construct attack graphs that represent all possible attacks on a networked system. We can adapt their algorithm to construct a remediation graph, i.e., a graph that depicts ways in which a system can be patched so that the system eventually reaches consistent states that are preferable to the initial state.

DEFINITION 5 *Let* $RP = \langle S, V, \tau, < \rangle$ *be a remediation posture and let* $s_0 \in S$ *be a state. Then, a* patch remediation graph *for the posture* RP *and initial state* s_0 *is a tuple* $RG = \langle S_{rg}, \tau_{rg}, s_0, V_G \rangle$ *where* $S_{rg} \subseteq S$ *is a set of states,* $\tau_{rg} \subseteq S_{rg} \times S_{rg}$ *is a state transition relation,* s_0 *is the initial state, and* $V_G \subseteq V$ *is a set of goal states.*

Goal states are defined to be all consistent states that are reachable from s_0 and are preferable to s_0. The algorithm (see Figure 1) first uses a model checker to compute the set S_{rg} of all states that are reachable from the initial state and that have a path to a goal state. It then restricts the patch action graph to the states in S_{rg}; the resulting graph is the remediation graph. As in [14], there are efficient BDD algorithms for all the operations used in this algorithm.

We can show that the patch remediation graph is exhaustive and succinct [14]:

LEMMA 6 1 *A sequence of patch actions takes a system from an initial state to a preferable, consistent state if and only if the sequence is a path in the system's remediation graph from the initial state to a goal state.*

```
ComputeMaximalGoalStates (V_G, <).
  for all s, s' ∈ V_G do
    if (s < s') then
      W = W ∪ {s}
  end
  V_MG = V_G − W
  return V_MG
```

Figure 2. Maximal Goal States Algorithm.

2 A state s (transition t) is in the remediation graph if and only if s (t) is in some path in the graph from s_0 to some goal state.

Once we have computed a remediation graph, we can use it to select a goal state and a transition path to that state. Note that a remediation graph may contain nodes that are not consistent states or that are not preferable to s_0. However, every node in a remediation graph has a directed path (possibly empty) to a goal state that is consistent, reachable from s_0, and preferable to s_0. Remediation graphs permit the efficient computation of *maximal* goal states s_n such that there are no other goal states that are preferable to the s_n. Figure 2 shows how to compute the set V_{MG} of goal states that are maximal under $<$.

The transition relation of the remediation graph provides the transition paths from the initial state to the maximal goal states. These can be used by any patch management system to deploy and install the selected patches. For instance, we have implemented similar algorithms within the Outpost System [8] that was developed at MITRE.

Note that a remediation graph can be analyzed further, e.g., to select safe transition paths that never traverse states that are less preferable than the initial state, or to determine the periods during remediation that a system may become vulnerable to attacks. Such analysis can help a security administrator reason about the implications of various defensive actions available to the administrator.

4. RELATED WORK

Several researchers have used model checkers to perform vulnerability analysis of computer networks. Ritchey and Ammann [13], Jha *et al.* [7, 6], and Sheyner *et al.* [14] use model checkers to generate attack paths (or attack graphs) by which an attacker can use known attacks to compromise a networked computer system. Our approach is based on this body of work; however, we apply those techniques to a completely different problem, namely the problem of security patch remediation. Further, we propose the new, novel concept of a remediation graph which represents all possible paths by which a defender can close vulnerabilities in a system.

Ramakrishnan and Sekar [11, 12] use a model checker to discover individual vulnerabilities on single hosts. While our approach is also host-based, we do not discover individual patches but rather use a model checker to discover patch installation sequences to goal states.

Patch management systems such as Microsoft/SMS [9] and Radia Patch Manager [10] include modules that compute patch installation sequences that satisfy patch dependencies and constraints. We extend this capability with a patch preference relation that models the relative desirability of patches and applications. Further, we also build patch remediation graphs that enable us to compute maximal goal configuration states and perform analyses such as finding safe patch installation sequences.

5. CONCLUSION

The main contribution of this paper is the automatic generation of patch remediation graphs. We presented language constructs for specifying patch consistency constraints, patch action rules, and user preferences for patch configurations. We showed how a model checker can be used to construct patch remediation graphs by generating counterexamples to a safety property. The graphs permit us to compute maximal goal states that represent the preferred final states for patch remediation. The graphs also give us the sequence of patch actions (patch installs and uninstalls) that would transition the systems to the desired final states. Finally, the graphs permit us to reason about properties of the transition paths, e.g., whether the system enters some vulnerable states during the remediation process.

Future work includes combining the model with a trust management model. With this, patch policy assertions (patch dependencies, constraints, preferences, etc.) would be represented as credentials. A delegation model would provide a flexible and powerful framework for handling patch management in a distributed and heterogeneous environment. We are also examining whether remediation graphs may be used to handle other security configuration management tasks.

References

[1] E. M. Clarke, O. Grumberg, and D. Peled. *Model Checking*. MIT Press, 2000.

[2] R. Conradi and B. Westfechtel. Version models for software configuration management. *ACM Computing Surveys*, 30(2), June 1998.

[3] P. Devanbu, M. Gertz, and S. Stubblebine. Security for automated, distributed configuration management. In *In Proceedings, ICSE 99 Workshop on Software Engineering over the Internet*, 1999.

[4] P. Devanbu and S. Stubblebine. Software engineering for security: a roadmap. In *The Future of Software Engineering*, volume Special Volume (ICSE 2000), 2000.

[5] R. S. Hall, D. Heimbigner, A. van der Hoek, and A. L. Wolf. An architecture for post-development configuration management in a wide-area network. In *Proc. of the 1997 International Conference on Distributed Configurable Systems*, pages 269–278, May 1997.

[6] Somesh Jha, Oleg Sheyner, and Jeannette M. Wing. Minimization and reliability analyses of attack graphs. Technical Report CMU-CS-02-109, School of Computer Science, Carnegie Mellon University, February 2002.

[7] Somesh Jha, Oleg Sheyner, and Jeannette M. Wing. Two formal analyses of attack graphs. In *Proceedings of the 2002 Computer Security Foundations Workshop*, pages 49–63, June 2002.

[8] Key, Proulx, Michnikov, Rubinovitz, Wittbold, and Wojcik. The Outpost system. In *MILCOM 2000 presentation*, 2000.

[9] Microsoft Corporation. *Patch Management Using Microsoft Systems Management Server 2003*, 2003.

[10] Novadigm. *Radia Patch Manager*, 2003. http://www.novadigm.com/ products/patchmanager.asp.

[11] C. R. Ramakrishnan and R. Sekar. Model-based vulnerability analysis of computer systems. In *Proceedings of the 2nd International Workshop on Verification, Model Checking and Abstract Interpretation*, September 1998.

[12] C. R. Ramakrishnan and R. Sekar. Model-based analysis of configuration vulnerabilities. *Journal of Computer Security*, 10(1-2):189–209, 2002.

[13] Ronald W. Ritchey and Paul Ammann. Using model checking to analyze network vulnerabilities. In *Proceedings of the IEEE Symposium on Security and Privacy*, pages 156–165, 2000.

[14] Oleg Sheyner, Joshua Haines, Somesh Jha, Richard Lippmann, and Jeannette M. Wing. Automated generation and analysis of attack graphs. In *Proceedings of the IEEE Symposium on Security and Privacy*, pages 254–265. IEEE Computer Society, 2002.

SECURITY MODELLING FOR RISK ANALYSIS

Lam-for Kwok[1] and Dennis Longley[2]
[1]*City University of Hong Kong, Department of Computer Science, Hong Kong, cslfkwok@cityu.edu.hk ;*[2]*Information Security Research Centre, Queensland University of Technology, Brisbane, Australia, d.longley@qut.edu.au*

Abstract: A security model to facilitate the recording and investigation of organisational security data is proposed; this model employs a directory structure for security entities and relationships. The model database with associated software may then be employed to develop and display organisational threat networks representing the risk environment of the organisational information processing and communication system. Thereafter the design of the defence systems may be facilitated by interactive procedures to determine appropriate countermeasure structures.

Key words: risk analysis, security standards, security models, security documentation, countermeasures, threat trees.

1. INTRODUCTION

The high cost of data collection for risk analysis projects has tended to favour high level methodologies focusing upon organizational structures; Baskerville[1] and Craft et. al[2] provide detailed accounts of such information security risk analysis methodologies. Nevertheless effective information security risk management for current highly complex systems depends upon a detailed knowledge of the system, and its environment. The Risk Data Repository (RDR)[3-7] was developed on the philosophy that risk analysis should be based upon a relatively simple and transparent analysis of comprehensive security data, rather than complex analyses of limited data. It aimed to store security data electronically so as to facilitate security studies and provide an effective means of developing and maintaining security documentation. When combined with a technique for representing the effectiveness of countermeasures[8], it could form the basis of a security officer's workbench.

Experience with the RDR indicated the requirement for a uniform method of representation for the various security entities and relationships. This paper explores the application of a security model based upon the RDR concepts and a directory structure for entity representation. An electronic security database and supporting software, of the type described in this paper, would not only alleviate the task of security documentation development, it would also significantly reduce the effort of initial data collection and subsequent updating. Moreover a common form of security documentation would greatly facilitate the importation of external security expertise, from vendors, consultants, standards bodies etc. It would also support processes to maintain system security when disparate systems are inter-networked or merged.

2. SECURITY MODEL

2.1 Security Documentation

Security documentation is an essential component of organisational information security because it provides a common source of information to the wide range of staff with information security responsibilities, and minimises the duplication of data collection effort. Such documentation should contain a broad range of information including inter alia business, management, administrative, operational and environmental contexts.

However, the cost of developing and maintaining such documentation is high, and it is suggested that such cost and effort may be reduced if conventional documentation were replaced by a security information database with supporting software. Such database / software would not only significantly reduce the effort of data recording and retrieval, it could also assist the security officer in the analysis of risk scenarios, design and maintenance of information security defence systems etc.

The effective cost of the database/ software design could be minimised if it were widely deployed, which implies the need of a common security model to guide the design of the database. This paper describes a security model for that purpose. The first stage in the model development lies in the classification of the various entities relevant to organisational security (see 2.2).

2.2 Classification of Entities

The classification of security entities in the proposed model was discussed in a previous paper by Fung et al[9] and was based upon a directory model[10]. The wide array of entities relevant to organisational security are grouped within classes, and uniquely identified, in the model. It is possible to generate a macro view of the system security by commencing the data collection with high level entities: major sites, large IT and communication systems, major applications and user groups etc. and then refine this view as detail of the component entities are recorded.

The proposed model uses but does not at this stage prescribe a directory structure. It simply suggests that an entity:

- has a unique identifier (an object identifier) indicating its position within its family grouping;
- has an arbitrary set of attributes defined by (Tag Value) tuples;
- can be linked with other entities, and such linkages are themselves entities of the model.

Description	OI	Description	OI	Description	OI
Systems	1	Hardware	1.1		
		Software	1.2		
		Platforms	1.3		
		Networks	1.4		
		Applications	1.5		
		Users	1.6		
		Assets	1.7		
Environment	2	Locations	2.1	Sites	2.1.1
		Services	2.2	Power	2.2.1
				Communications	2.2.2
				Water/Drainage	2.2.3
Security	3	Threats	3.1	Threat Types	3.1.1
				Threat Trees	3.1.2
		Defences	3.2	Countermeasures	3.2.1
				Threat Countermeasure Diagrams	3.2.2
Procedures	4	Internal	4.1	Policies & Procedures	4.1.1
				Standards	4.1.2
		External	4.2		
Relationships	5	Link Type 1	5.1		
		Link Type 2	5.2		
		Link Type 3	5.3		
		Link Type 4	5.4		

Figure 1. Directory Classification of Entities
(OI = Object Identifier is a unique numerical identifier for each entity)

A proposed directory system is illustrated in Fig. 1. Within this scheme every entity is given a unique identifier (termed object identifier OI) according to its position in the directory[9]; hence such an identifier may be allocated to an IT system, network, file server, room, class of users or a paragraph in the standard documentation. This classification scheme provides for groupings of entities within classes and hence facilitates a top down approach to the documentation.

2.3 Attributes, and Relationships between Entities

The classification scheme described above (see 2.2) provides a list of entities. Additional information about particular entities (e.g. the protocol of a network) can be given as an attribute of that entity (Tag = Protocol, Value = TCP/IP). Hence security relevant attributes may be allocated to entities as the need arises from the model.

Information processing systems involve inter-relationships between their various entities, e.g. workstations are connected to networks. These inter-relationships are themselves entities within the scheme (see Fig. 1). Relationships thus have object identifiers; a particular relationship is given a unique identifier that is itself, a child of a relationship type identifier. Hence if the model records that a given PC (with object identifier PC_OI) is *connected* to a specific LAN (LAN_OI) then this fact is recorded with a Relationship OI (Connect_OI.1) belonging to the Relationship family Connect (Connect_OI).

2.4 Developing the Model

The proposed scheme is capable of describing security scenarios of complex systems down to any required level of detail, but selecting the type of information to be collected, and the requisite level of detail, is no mean task. It is suggested that the process commences with a top level model, explores the risk environment at this macro level and use the results to guide the refinement of the model.

Hence the Systems Entity contains the class Platforms used to represent major organisational systems. A simple security macro model can then be developed by identifying the major security facets of a platform:

- its location, i.e. identifying the building or site hosting the platform and linking them with relationship of type – *located;*
- the classes of *assets* processed by the platform, i.e. linked to the platform with the relationship type – *processed by;*
- the external services required to ensure the availability of the platform.

Once a macro model has been developed the security features of the platform may then be explored, using the techniques described in the next sections (see 3 and 4). These studies using the interactive tools described below will then provide guidance on the next stage of the model environment.

3. SECURITY MODELLING

3.1 Overview

Information security officers are concerned with threat events having the potential to damage organisational information assets and systems. They need to predict and prioritise such threat events, in order to develop and maintain cost effective security measures.

A threat tree is such a tool for risk analysis, since it displays the potential set of outcomes for a given threat event. If all the potential threats are considered then a forest of threat trees can be developed, the upper level comprising the initial threats, the leaves the potential outcomes. The trees are not necessarily disjoint, since some of the tree nodes may be common to a number of trees, e.g. physical damage to computer equipment is likely to arise from fire, flood, malicious damage etc. Hence the collection of trees may actually represent a network, with intrinsic threats at the top and the undesirable outcomes at the bottom (see Fig. 6).

Having identified and prioritised the threat scenarios in terms of the probabilities and impact severities, the security officer has the task of selecting, installing, customising and maintaining appropriate security countermeasures. The role of a countermeasure may be considered to be that of cutting, or at least weakening, a branch of the threat network so as to inhibit one or more undesirable outcomes. Countermeasures themselves are physical devices, administrative procedures etc. that may be rendered ineffective or bypassed, and should often themselves be protected by supplementary countermeasures.

3.2 Threat Propagation

3.2.1 Threat Entity Relationships

The proposed security directory system recognises threats classified from top levels (see Fig. 2). Each of these threats may be assigned object identifiers (OI) according to their position in the hierarchy.

Threats become significant to an organisation when they impact upon a System or Environmental Entity, e.g. *Malicious Physical Damage to Finance LAN router, Accidental Physical Damage to a HQ Site Power Supply, Fire in I.T. Building*. Like Threats, each of these target System or Physical Environment entities is allocated an object identifier in the model.

Thus a threat event represents a threat impacting upon some specific entity, and can be represented by a link between the threat and the impacted entity OIs (see Fig. 3). The threat events are represented as nodes in the Relationship directory, under the class Threat_Entity: i.e.

- TE is a linkage (Threat OI, Entity OI).
- These linkages are of the class Threat Entity (TE) and each individual TE is allocated an OI in the model.

Threats (3.1.1)	Physical (3.1.1.1)	Natural (3.1.1.1.1)
		Malicious (3.1.1.1.2)
	Environmental (3.1.1.2)	Fire (3.1.1.2.1)
		Flood (3.1.1.2.2)
		Earthquake/ Volcanic Eruption (3.1.1.2.3)
		Severe Weather (3.1.1.2.4)
		Asteroid Impact (3.1.1.2.5)
	Personnel (3.1.1.3)	......
	Network (3.1.1.4)	....

Figure 2. Classification of Threats

3.2.2 TE – TE (TETE) Relationships

Threats propagate, for example:

- as fire through geographically related physical environments;
- by damaging essential services such as power supplies causing systems to shut down;
- by interactions among information processing/ communication systems.

A fire in a room may cause physical damage to equipment located there. In effect an *Incident Threat* impacting upon an *Incident Entity* may cause a *Target Threat* to impact upon a *Target Entity*. In the model this threat propagation is represented a link between two threat events or TEs (see Fig. 4):

- Incident TE: Link (IncTE_OI) between Incident Threat (IncT_OI) and Incident Entity (IncE_OI);
- Target TE : Link (TarTE_OI) between Target Threat (TarT_OI) and Target Entity (TarE_OI);
- TETE : Link (TETE_OI) between IncTE_OI and TarTE_OI.

Threat propagation is not inevitable, e.g. there are a number of factors determining if a given piece of equipment will be damaged by fire in a room: size of the room, flammability of materials in the room, fireproofing of the equipment etc. Hence a TETE may have an attribute termed Vulnerability Index (0 <= VI <=1) to indicate the probability of the threat propagation.

3.2.3 Threat Tree Development

Threat trees (see Fig. 5) may be developed automatically in the model. Hence there is a Threat Tree Development Algorithm:

1. Select the initial threat event (Root TE).
2. Find a TETE with Root TE as its Incident TE.
3. Get the corresponding Target TE (second component of TETE link)
4. Target TE is next node in Threat Tree.
5. Repeat 2 until there are no more candidate TETEs
6. Go to next child node in the Threat Tree and set this as Root TE
7. Repeat 1 until no more child nodes.

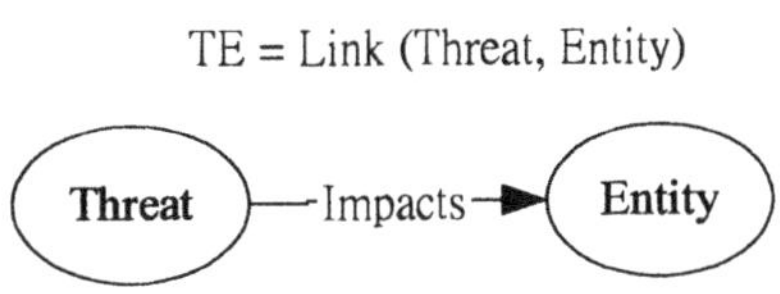

Figure 3. TEs Represent Threat Events

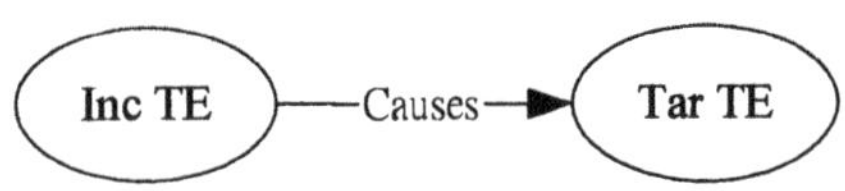

Figure 4. TETEs Represent Threat Event Propagation

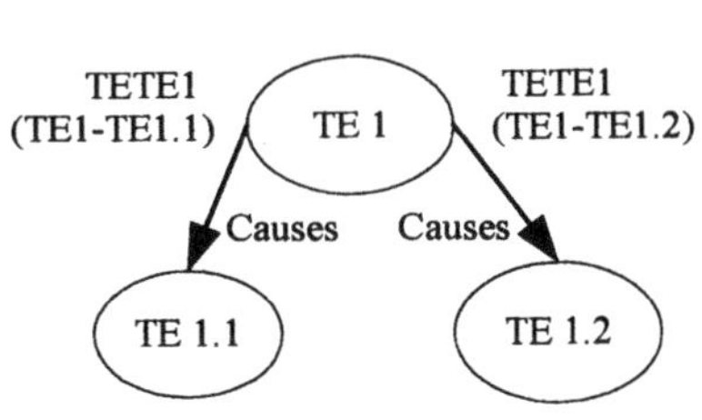

Figure 5. TETEs Provide the Links in a Threat Tree

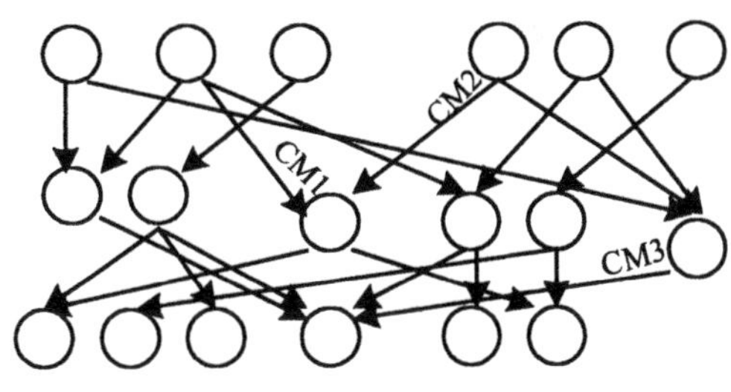

Figure 6. Threat Networks including Countermeasure

The Threat Tree nodes may be assigned Object Identifiers and hence the Threat Trees are themselves are represented within the model. The automatic development of threat trees described above, however, involves a major manual effort in the production of TEs and linkages between TEs i.e.

TETEs. These TETEs represent security knowledge on threat propagation; the potential number of TETEs could be in the billions for a reasonable sized organisation. Fortunately the model provides a solution to this problem, since the proposed directory approach organises security entities in classes. Multiple individual TETEs can thus be replaced by individual generic TETEs containing details of entity classes (see 3.2.4).

3.2.4 Generic TETEs

3.2.4.1 Incident and Target Entity Linking

The directory classification of security entities implies that simple relationships exist between common entities. Hence locations will have object identifiers that reflect the geographical relationships (see Fig. 7).

Physical Environent / Location / Site (OI = PELS)	HQ Site (PELS.1)	Admin Building (PELS1.1)	Ground Floor (PELS1.1.1)	Room 1 PELS (1.1.1.1)
				Room 2 PELS (1.1.1.2)
				.....
			1st Floor (PELS 1.1.2)	
		Manufacturing Building (PELS.1.2)		
	Branch Site (PELS.2)			

Figure 7. Object Identifiers for Locations

The generic TETE for fire propagation recognises that a fire at HQ Site can spread to the Admin Building, which in turn can spread to floors in that building and from the floors to rooms on those floors. Hence the generic TETE for fire propagation can be expressed in terms of wild card OIs indicating parent and child relationships of the Incident and Target Entities (i.e. sites, buildings, floors and rooms etc.)

In general the conditional relationships between the Incident and Target Entities may be more complex. For example a TETE may take the form:

Fire in a Physical Location causes Physical Damage to Hardware on condition that:- Hardware is Located in the Physical Location.

The TETE thus needs to store the Incident and Target Entities as wild card representing Physical Location and Hardware entities respectively. In addition the TETE stores the information regarding the required link between the Incident and Target Entities, as an attribute of the TETE. The use of object entities to identify Relationship classes facilitates the storage of this attribute information, e.g. *Located_In.* With more complex conditions between the TETE Incident and Target Entities, a Linkage Condition Table may be included as an attribute of the TETE.

A generic TETE may thus replace a multiple of specific TETEs, and such generic TETEs allows security knowledge to be directly imported to the model. The threat tree development now involves a search for specific Target Entities that:

- match a TETE Target Entity class, and
- satisfy the Incident – Target Entity relationships as specified in Linkage Condition Table.

3.2.4.2 Vulnerability Indices (VI).

The probability associated with a specific threat propagation will naturally depend upon the particular threats and entities, and their attributes; e.g. fire propagation will have a higher probability for wooden buildings. Hence VI Condition Tables may also be stored as TETE attributes, to facilitate the estimation of probability associated with a threat propagation between specific entities.

3.2.5 Role of Threat Networks

3.2.5.1 Development of Threat Networks.

Threat Trees have a single threat event (TE) root; these trees may contain duplicate nodes implying that such nodes have more than one parent. Combining trees with various threat event roots may also result in the combination of duplicate intermediate nodes. Hence the total set of threat scenarios will normally take the form of a network (see Fig. 6). Duplicate nodes are significant in the interpretation of Threat Networks and in the design of subsequent defence systems.

3.2.5.2 Outcome Probabilities.

It is extremely difficult to assign accurate probabilities to a plethora of potential information security events. The way forward, proposed in this model, is to allow for the inclusion of probabilities as transparently as possible, and to suggest the provision of interactive tools to facilitate the task of probability estimation, i.e. the estimator should at least be given the opportunity to explore the sensitivity of the outcome to the estimate.

A Threat Tree provides for a comparatively straightforward estimate of outcome probabilities based upon:

- the probability that the root threat event will occur; and
- the probability of threat propagation (VI) for each link between the root and the outcome.

The probability of an outcome, arising from a threat event, will decrease with the number of serial links in the path, and increase with the number of parallel paths. Given two parallel paths of different lengths, the outcome probability will tend to be more sensitive to the VIs of the links in the shorter path.

An interactive tool that displays the variation of outcome probabilities with variations in the VI of a selected link, would allow a security officer to focus on the links with the highest impact on outcomes and hence prioritise the effort in estimating probabilities.

3.2.5.3 Outcome Impacts.

An outcome in the Threat Networks (see Fig. 6) represents a threat to an organisational asset, e.g. loss of confidentiality of client account data. The previous sections have alluded to security officers prioritising their efforts towards outcomes with the largest impacts, but so far have not discussed the estimation of such outcome impacts. This task is onerous, particularly if there is a requirement to assign quantitative measures to such impacts.

The fundamental problem lies in the relationship between a technical event, such as compromise to a business dataset, and the subsequent impact upon the organisation. Security officers would normally experience significant difficulty in trying to obtain a host of such quantitative measures from business management, let alone maintaining the validity of such data as organisational environments evolve. This aspect of risk analysis was studied in the RDR development[6, 11]. The approach suggested was:

- Defer the allocation of the quantitative measure and assign textual statements to outcomes, e.g. *illicit disclosure of this asset could lead to major loss of client confidence.*
- Deduce impacts from information on impacts associated related assets.

The allocation of impact statement to assets can still prove to be an excessive task for organisations with large variety of datasets. Anderson[11] demonstrated that given knowledge of the inter-relationship of data items, it was possible to deduce impact statements for a set of data, from explicit impact statements given for a related dataset. The proposed security model facilitates such imputation of impacts.

3.2.6 Automatic and Interactive Threat Network Development

The concept of automatic Threat Network development sounds attractive but it depends upon the initial storage of a significant amount of general and local security knowledge. If the requisite generic TETEs are not stored in the model then the search will end prematurely and a potential threat path will be omitted, giving a false sense of security. Moreover, even if the requisite

TETEs are stored, the path will still end prematurely if local entities, or local relationships between those entities, are not included in the model database. As ever, the model results are only as good as the model itself.

However, the proposed directory structure does provide two ways forward. Firstly, given a common directory structure then generic TETEs may be imported into a local model. Hence a large organisation can effectively export its security expertise from head office to local branches. Moreover a study of the Incident and Target Entity types and conditional relationships for these imported TETEs provides a clue to the type of entity and relationship that should be included in the local model.

Secondly the proposed model provides guidance on the development of the local security database when Threat Tree Networks are developed in an automatic / interactive manner. Consider a node in the Threat Network with a premature end node in a path. An interactive session can report the class of Target Entity, and Incident – Target relationships for each TETE with an Incident Threat event matching the node; for example, the interactive session may suggest a check for equipment located in a room vulnerable to flood.

4. DEFENCE MEASURES

4.1 Overview

The previous section discussed the role of the model in providing a security officer with an insight into the organisational threat scenario. Threat Networks display the threat paths to organisational assets and Vulnerability Indices (VI)/ impact statements prioritise potential outcomes. The security officer then has the task of prioritising defence systems according to the system risk, expressed in terms of outcome impacts and probabilities.

The security officer is thus responsible for the design, implementation and operation of cost effective defence measures designed to minimise the identified system risk. This section describes how the proposed model may be used in this role.

4.2 Countermeasures

A study of Threat Networks can identify branches that have the effect of increasing the probability of undesirable outcomes. Countermeasures are employed to minimise the probability of a threat propagation leading to such unacceptable outcomes. Hence a study of Threat Network branches may assist in the optimum deployment of countermeasure (see Fig. 6)

Reducing the probability of a threat propagation is equivalent to reducing the effective VI of a TETE (see Fig. 8); e.g. if an attacker logs onto a computer system there is a moderate to high probability that a sensitive file stored on the system will be compromised with loss of confidentiality. A password access control system protecting the file will mitigate against such an event. Hence in Fig. 8,

- Incident TE: Attacker Accesses – File Server,
- Target TE: Loss of Confidentiality – Personnel File,
- TETE: (Incident TE , Target TE),
- TETE Condition: Personnel File STORED ON File Server
- Countermeasure: Password Access Control.

Countermeasures are included as Security Measures in the Directory scheme. An earlier publication on countermeasures[8] discussed a model highlighting the role and effectiveness of countermeasures. The effectiveness of any countermeasure depends upon its inherent components, and threats to those components can seriously affect the countermeasure performance. For example, a firewall is highly dependent upon the effectiveness of its rules. A logical attack may allow malicious packets to satisfy inadequately formulated firewall rules for access (i.e. bypassing the countermeasure), whilst unauthorised physical access to the firewall hardware could allow the attacker to modify the firewall rules, rendering the firewall ineffective.

Countermeasure structures (see Fig. 9) may thus be modelled in terms of:

- the incident TETE, i.e. the threat propagation to be countered;
- the components of the countermeasure, i.e. those aspects of the countermeasure that determine its effectiveness; and
- the residual TETEs, that could impact upon those components, and hence compromise the operation of the countermeasure.

The residual TETEs represent potential attacks on the countermeasure components, e.g. *Illicit Access to Firewall Equipment* CAUSES *Illicit Modification to Firewall Rules.*

The security officer has the role of protecting countermeasures against attacks on its components, and may therefore deploy supplementary countermeasures to ensure the effectiveness of the original countermeasure (see Fig. 10), e.g. an intrusion detection system may be employed to counter sophisticated attacks bypassing firewall rules. The supplementary countermeasures may also be implemented in the form of procedural security, e.g. an access control system may have a residual TETE: *Attacker Guesses Password Causes Attacker to Gain Access to System* and this could be countered by an organisational procedure mandating strong passwords.

4.3 Threat Countermeasure Diagrams (TCDs)

TCDs are used to record the complete structure of major and supplementary countermeasures (see Fig. 10). A TCD is a tree of countermeasure structures with the prime countermeasure as the root and the supplementary countermeasures as the child nodes. The TCD records the rationale of a countermeasure system. Given that the supplementary countermeasures will themselves have vulnerable components, and hence residual TETEs, it would appear that TCDs will grow indefinitely. The decision to employ supplementary countermeasures depends upon their role in the effectiveness of the root countermeasure and this aspect of TCDs is discussed in 4.5.

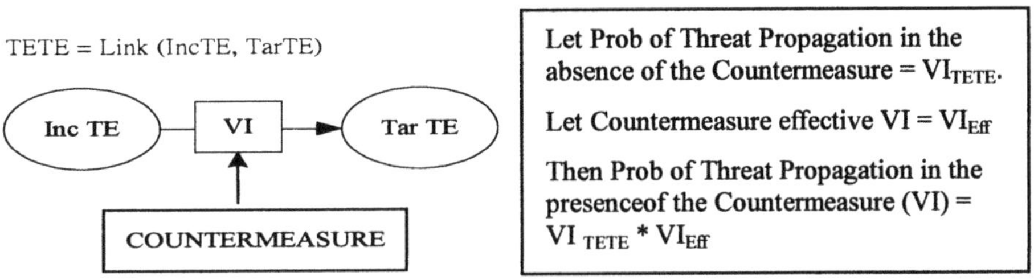

Figure 8. The Role of the Countermeasure is to reduce the Probability of the Threat Propagation, (VI) of a Threat Network Link (TETE)

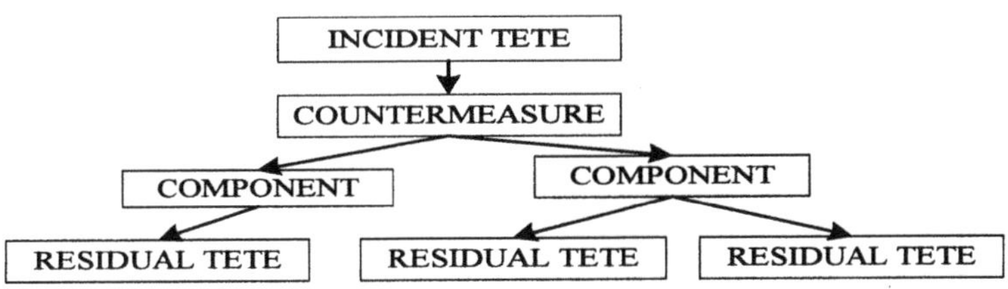

Figure 9. Countermeasure Structure

4.4 Design of Defence Systems

The countermeasure structures (see Fig. 9) and TCDs (see Fig. 10) fit well into the proposed directory structure. The various facets of the countermeasure (incident TETE, components and residual TETEs) may be allocated object identifiers under the countermeasure entity, and the nodes of a TCD may be treated in a similar manner to those of Threat Trees. These defence structures represent significant security knowledge. The countermeasure structure provides details of the countermeasure vulnerabilities; the corresponding TCDs give advice on the secure installation of the countermeasure. Hence generic countermeasure structures and corresponding TETEs may be imported into the model. The defence structures may be developed in a similar manner to the automatic development of Threat Trees.

The first stage in the development of a defence system lies in the construction of generic countermeasure structures (see Fig. 9). The security designer with detailed knowledge of the countermeasure determines the incident TETEs, essential components, residual TETEs etc. These incident and residual TETEs are developed as *generic* TETEs (see 3.2.4), i.e. the incident and target entities contained in the TETEs refer to a class of systems or physical environment entities. These constructs are imported into a local organisational information security model.

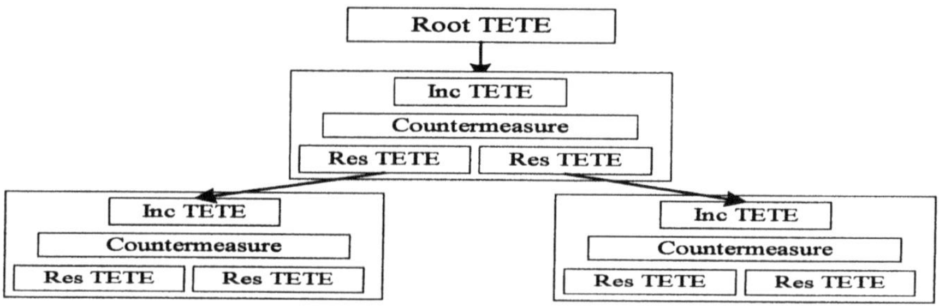

Figure 10. A Threat Countermeasure Diagram (TCD) displays Supplementary Countermeasures maintaining the effectiveness of the Root Countermeasure

A security officer now selects a link in a Threat Network for the placement of a countermeasure. This link represents a local TETE, since it contains details of the organisational actual entities, and now becomes the root TETE of a TCD. An automatic search amongst the generic countermeasure structures produces one or more countermeasures with incident generic TETEs matching the TCD root TETE. The security officer interactively selects one of the offered countermeasures and the first node of the TCD is inserted. The incident generic TETE is instantiated to the local entities, the residual TETEs are correspondingly instantiated.

The security officer may then proceed in a similar manner to the selection of supplementary countermeasures, countering the residual TETEs.

Alternatively complete generic TCDs may be imported and then instantiated by the entities of the root TETE. In effect the warnings and recommendations contained in security information brochures can be directly imported into the model.

There remains the problem of the depth of supplementary countermeasures. Currently these decisions are made, often implicitly, on the basis of experience and judgment. The model can assist with this aspect of countermeasure design by consideration of the VIs associated with the incident and residual TETEs in the total structure (see 4.5).

When countermeasure systems have been selected and implemented they become part of the total information security model and new equipment and procedures will be added. This raises the question, can the countermeasures introduce new threats, i.e. additional paths in the Threat Network? There is plenty of anecdotal evidence of water damage caused by fire extinguisher sprinklers. Thus the organisational Threat Network should be redeveloped on the updated model to test for such eventualities.

4.5 Defence Effectiveness

The role of a countermeasure is to reduce the intrinsic probability of threat propagation, i.e. reduce the VI of the incident TETE (see 4.2); the model allocates an effective VI (VI_{eff}) to the countermeasure (see Fig. 8). The countermeasure VI_{eff} depends upon the VIs of its residual TETEs. Consider a simple password system, and assume that the only means of rendering the system ineffective is to correctly guess a PIN. The residual TETE may take the form *Attacker Guesses PIN Causes Access System Compromise*. In this case the VI of the residual TETE (0.0001) is the countermeasure VI_{eff}. If there are *n* residual TETEs, then the worst scenario is that each potential attack on the countermeasure is simultaneously and independently undertaken; the countermeasure VI_{eff} is then given by

$$VI_{eff} = 1 - (1 - VI_1) * (1 - VI_2) *\ldots *\ldots (1 - VI_n)$$

If any residual TETE has a VI close to 1, then VI_{eff} is also close to 1 and the countermeasure is ineffectual. If a supplementary countermeasure guards against a high residual TETE then that residual TETE value is reduced by a factor equal to the VI_{eff} of the supplementary countermeasure. Tools may be developed to compute countermeasure effectiveness interactively based upon estimates of VIs of residual TETEs. A simple grading scheme of *high, neutral and low VI* can be easily incorporated to assist in the decisions on supplementary countermeasures.

5. CONCLUSIONS

The proposed model seeks to provide a methodology to assist in the problem of maintaining information security within complex organisational environments. It does not claim to be a simple solution to a complex problem. It does, however, claim to provide a methodology for information security defence design by focusing organisationally available skills and expertise to the local problem, making the most effective use of available system data, guiding the collection and recording of additional security data.

The basis of the methodology is a database of security entities with associated tools, used interactively to develop security insight of the modelled system, which in turn guides the collection and recording of additional security data so that the model increasingly reflects the organisational security environment. The model facilitates the distribution of security expertise and experience to operational environments. Security officers of a large organisation will now have common tools to capture local data, develop security models for local systems and use those models to improve the local security scenario.

To date experience with the model has been limited to the development of prototypes to test the concepts of Threat Tree and TCD development using the procedures described in the paper. The development of a user friendly package with graphical displays is due to commence shortly.

ACKNOWLEDGMENTS

The authors acknowledge the valuable help, advice and suggestions received from colleagues and students in the development of the ideas that lead to the production of the model; in particular, contributions from Prof W. Caelli, Dr A. Anderson, Dr A. Tickle, Ms P. Fung and Mr M. Branagan.

REFERENCES

1 R. Baskerville Information Systems Security Design Methods: Implications for Information Systems Development. *ACM Computer Surveys*, **25**(4) 1993, 375-414.

2 R. Craft, G. Wysss, R. Vandewart and D. Funkhouser. An Open Framework for Risk Management. *21st National Information Systems Security Conf.*, 6-9 Oct. 1998, Crystal City, Virginia, USA.

3 L. F. Kwok and D. Longley. Code of Practice: A Standard for Information Security Management. *Information Security in Research and Business, Proc. IFIP TC11 13th Int. Conf. on Information Security* (Editors: Yngstrom and Carlsen), IFIP Sec.'97, Copenhagen, Denmark, 14-16 May 1997, Chapman & Hall 1997, pp.78-90.

4 L. F. Kwok. A Hypertext Information Security Model for Organisations. *Information Management and Computer Security*, **5** (4) 1997, 138-148.

5 L. F. Kwok and D. Longley. Information Security Management and Modelling, *Information Management and Computer Security*, 7 (1) 1999, 30-39.

6 A. Anderson, L. F. Kwok and D. Longley. Security Modelling for Organisations. *Proc. 2nd ACM Conf. on Computer and Communications Security*, CCS'94, Fairfax, Virginia, USA, 2-4 Nov 1994, ACM Press 1994, pp.241-250.

7 L. F. Kwok and D. Longley.. A Security Officer's Workbench. *Computers and Security*, **15** (8) 1996, 695-700.

8 W. Caelli, D. Longley, and A. B. Tickle. A Methodology for Describing Information and Physical Security Architectures. *IT Security: The Need for International Cooperation, Proc. IFIP TC11 8th Int. Conf. on Information Security* (Editors Gable and Caelli), IFIP Sec.'92, Singapore, 27-29 May 1992, NY:Elsevier Science Publishers 1992, pp.277-296.

9 P. Fung, L. F. Kwok and D. Longley. Electronic Information Security Documentation. *ACSW Frontiers 2003* (Eds. Johnson, Montague and Steketee), Australasian Information Security Workshop (AISW2003), 4-7 Feb. 2003, Adelaide, Australia, pp25-31.

10 *The Directory, CCITT Rec. X.500-X.521* | ISO/IEC Standard 9594:1993.

11 A. Anderson. The Object Oriented Modelling of Information Systems SecurityRisk, *PhD thesis*, Queensland University of Technology 1997.

CONTRASTING MALICIOUS APPLETS BY MODIFYING THE JAVA VIRTUAL MACHINE

Vincenzo Ciaschini[1] and Roberto Gorrieri[2]

[1] *INFN CNAF*
Viale Berti Pichat 6/2, 40127 Bologna, Italy
vincenzo.ciaschini@cnaf.infn.it

[2] *Dipartimento di Scienze dell'Informazione*
Mura A. Zamboni 7, 40127 Bologna, Italy
gorrieri@cs.unibo.it

Abstract Java is the most popular language for web programming. However it suffers from some well-known denial-of-service attacks (e.g., obscuring the screen) due to the execution of malicious code that uses resources in an improper way. In this paper we present a new approach to alleviate these problems by patching the Java Virtual Machine, in order to force the needed checks on resources usage bounds directly at the level of the source code.

Keywords: Denial of Service, Mobility

1. INTRODUCTION

The Java[2] language, developed by Sun Microsystems, is nowadays an accepted standard in Web programming. Many applications are developed using it, and there are even more Java applets included in web pages from millions of sites. This is certainly due to its many desirable features like code mobility (supported by machine independancy at the bytecode level), simplicity of programming (due to similarities with C and C++), efficient execution, use of a virtual machine, and so on. Last but not least, Java is certainly a language for *secure* Web programming. As a matter of fact, Sun addressed from the very beginning and with great care the problem of stopping all potential system-breaking attacks due to code mobility, and it did so by adopting the *sandbox* [9, 4] approach. It consists of enclosing the applet in an isolated space from which it cannot escape unless the user gave it some special permissions. In this way, many limits are imposed on it and, assuming there is no bug in the implementation of the sandbox, most system attacks are effectively made impossible.

Unfortunately, Java still suffers from some security problems. Of course, not everything can be prohibited if we want to have a useful system, and this leaves the way open for downloaded malicious applets to start attacks more subtle but no less important, such as denial-of-service, antagonism and invasion of privacy. It is worth noting that these kinds of attack are simple to implement, generally ranging from 10 to 30 lines of code, and with no conceptual complexity in them, thus becoming a tempting instrument for malicious users.

In the next section, we survey the many approaches that have appeared in the recent literature to contrast malicious applets. This survey will end with the motivations for our current proposal, i.e., a monitor, called *Limiter* and implemented as a patch [1] for the Java Virtual Machine (JVM, for short)[4], that performs many checks on the Java source code before actually executing it. Section 3 describes Limiter; in particular, the general criteria that have guided its definition and the many different sources of attack that have been contrasted. Section 4 reports about the effectiveness of the approach and discusses performance issues. Finally, some remarks on possible improvements for future work conclude the paper.

2. ATTACKS AND DEFENSES

It is widely accepted (see, e.g., [6]) that there are four major categories of attacks:

1 *Attack applets.* They try to modify the system (e.g., the local filesystem) to allow, or at least facilitate, a successive intrusion. This kind of applet is the most dangerous one, as the whole system can be compromised and possibly destroyed. Sun has focused its efforts mainly on this class, and so the available defenses are strong.

2 *Privacy Invasion.* There are applets that try to gather information from the user or to impersonate him (for example by sending e-mails in his place). Some attacks in this class are easily implementable. The consequences of these attacks are usually moderate, but occasionally they may be very nasty (e.g., password cracking attack); the countermeasures Java offers in this respect are usually sufficient.

3 *Denial of Service.* These are applets that try to deny the legitimate user to access some services or resources that he could normally obtain from his local machine. For instance, allocating all of a system's memory, obscuring the screen, locking the keyboard and mouse, deny access to the internet, and so on. Implementing such attacks is usually not difficult, but contrasting them is hard and Java does not offer adequate countermeasures.

4 *Antagonism.* There are applets that antagonize or annoy the user in some way. Three examples are: continuously playing a sound, bringing up popup windows at unwanted moments, or even working in background to do something without the awareness of the user. At first sight, this last example of antagonism does not seem hostile at all: such an applet does not disturb the user or critically consume machine's resources, so that the user does not realize to be under attack; indeed, it is difficult to detect such applets.

Sun actively researched to stop only the first two classes of attacks, and with very good results. It is also important to cope with the other classes, because of the costs due to repair time and time offline. However, it must be said that the reason why Sun has not coped with the latter is that those classes of problems are not Java-specific, but on the contrary are common to all network applications.

While the problems presented by the latter two classes may seem trivial if they manifest themselves on a normal user workstation (a system reboot is generally enough to take care of them), they become of much greater gravity if the machine attacked is a server, because in this case a reboot may cause much greater problems: loss of data, network downtime, and so on. For more details on hostile applets, see[5–7].

2.1 Defensive Approaches

Many solutions have been proposed to deal with the problem, and here is a short review of the most representative ones.

1 *Disable Java.* There is no doubt that disabling Java would solve all these problems. However, it is a much too drastic decision. As a matter of fact, this way would also lock yourself out of all the advantages that the Java language has, such has a high portability of *compiled* code, and it can easily be said to do more harm than good. However, it should be noted these problems have in the past even pushed CERT to recommend disabling Java. An implementation of this solution, namely blocking Java directly at the firewall, has been tested in [10].

2 *Code Signing.* This is the solution implemented by the Java developers from Sun. Signing a piece of bytecode certifies its author, and on the base of this knowledge, extra rights and possibilities (like read or write access to the file system) may be given or denied to the bytecode itself. The problem with code signing is that it gives absolutely no guarantees on the behaviour of the signed code. It only offers a way to trace the code back to its author. Even this is not certain. In fact, if you need to sign some code, you need a certificate. There are essentially two ways to obtain one: you can make one yourself, or you can request one to a certification authority. In the first case it can be easily falsified, thus removing all the (limited) usefulness of code signing, while in the second case you have to pay a hefty sum, thus putting it out of reach for the non-profit programmers. Also there are certainly ways to fool the company by impersonating someone else. And there is always the possibility that the user's browser will not recognize your signature anyway.

3 *Code Inspecting.* This approach consists of examining the bytecode of every Java class before it is run, looking for suspicious contracts, and then deciding whether to allow execution or not. An example of this approach is Finjan's SurfinGate program [11], which attempts to discriminate code when it reaches the machine. The problem with this technique is that it permits only to screen for *known* attacks, and is generally useless against techniques like code obfuscation. Also, many successful attack techniques are very difficult to distinguish from normal code. For example, there are known attacks that consist simply of a sequence of type casting operations.

4 *Language Modification.* Many problems stem directly from some of the capabilities of the language, so it would seem natural to simply modify Java to remove or modify such capabilities. Examples of this are given in [6]. However, it is not a simple task. What is really executed is not the Java language, but its compiled version, called *bytecode.* The association between Java and bytecode is not particularly strong. For example it is possible to compile C code into bytecode. As a consequence, simply modifying Java would net you nothing, because the bytecode can be generated in many other ways. On the other hand, modifying the bytecode (for example by replacing it with abstract syntax trees, as proposed in [6] can be a much more rewarding task, but the price would be to remove compatibility with all the previous versions. Not a good thing

for a language that makes compatibility one of his most alluring features.

5 *Codebase filtering.* This strategy consists essentially of filtering out applets basing such a decision on the applet's codebase, and has been proposed and implemented in Princeton's JavaFilter[14]. The obvious problem is that it is rather easy to fool the controlling software into believing that the applet has a different origin than the real one.

6 *Sacrificial machine.* The basic idea of this approach is to setup a sacrificial machine with only minimum connections to the rest of the net, and execute all Java code on it [13]. While this approach can certainly be used to put a limit on the damage a malicious applet can wreak, it also foregoes the ability to use the Java language for distributed computing, and it feels like a step back. An implementation of this approach has been presented by Digitivity (now part of Citrix) [12].

7 *Monitors.* Finally, this approach consists of monitoring the applet behaviour to intercept and then to prohibit dangerous actions, obtaining the ability to stop even previously unknown attacks simply based on their effects. They can be implemented at various levels:

user level – At this level the monitor is nothing but another applet, obviously running *outside* of the sandbox, that runs all of the checks. The problem with this approach is that the applet can be interfered with by another misbehaving applet, and so you cannot even be sure that it is running. On the other hand, this approach is simple to implement and can work everywhere without any modification to existing systems. An example of this approach is [8].

browser level – At this level it is the browser itself that runs all the checks. The problem with this approach is that of a lack of uniformity, because every browser is independent of each other, and so this monitor should be reimplemented from scratch every time.

system level – At this level it is the JVM itself that runs the checks when the applet is running. An obvious advantage of this choice is that it is not possible to avoid the checks.

An obvious drawback to all these three approaches is that the added checks increase the execution time of each and every applet, whether they are malicious or not.

Other problems are shared by all these seven approaches. For example, it is possible to write applets capable of resurrecting themselves in case of death, or that simply refuse to die, and none of them can put a stop to it. We will approach again this issue in the conclusions.

2.2 Our Approach

The approach proposed in this paper is that of a patch to the JVM implementing a system-level monitor. The reason that led to this particular strategy is that by modifying the source code of the virtual machine, many weaknesses of all other defensive approaches can be circumvented.

For example, all the problems mentioned above can now be simply prevented by way of a few additional checks in the JVM, and the fact that they are implemented directly in the JVM itself serves to lessen their impact on performance.

There are other obvious advantages of this approach:

- It becomes impossible for applets to avoid the new checks the monitor implements, since they are done by the JVM on all running applets regardless of other factors, and this is a dramatic improvement on systems like [8], where the monitoring is done by an applet that can be shutdown or interfered with by other applets.
- This approach does not involve itself with an analysis of the code that it does execute, but instead simply takes step to limit or make impossible dangerous behaviour, with the result of catching also attacks that were unknown by the authors of the code.
- Finally, by directly modifying the JVM we sidestep the need to create patches for all existent browsers, since most of them allow the user to choose a specific JVM by using a plug-in system, like for example Mozilla does.

In light of all these factors, in the rest of this paper we describe the monitor, called Limiter, that we implemented.

3. THE MONITOR IMPLEMENTED IN THE JVM

3.1 General Criteria

Limiter was programmed with a few basic rules in mind:

1 *No changes to the class library.*

This rule (and partly the following one) is a direct consequence of the license under which the JVM source code is released. In fact, this license prohibits all changes to standard class library. Furthermore, the modified JVM, as configured out-of-the-box must be indistinguishable from the standard one.

2 *Conservative extensions*

It should be possible to configure the new VM in such a way that no difference can be found between it and the standard one. As the previous one, this rule is a consequence of the JVM source license.

3 *Changes to the semantic of the language should be limited to the minimum necessary.*

This rule is a consequence of the need to keep compatibility as high as possible between the modified version of the JVM and the original one, to keep the existence of the monitor as transparent as possible to all existing (non malicious) applets, and to avoid breaking them unless absolutely necessary.

4 *Easily identifiable modifications*

Every change to the original source code should be clearly highlighted as such in the source itself. This last rule is here as a way to distinguish original code from the new one, and is also required from the JVM source code license.

These rules are evidently straight-forward and are derived primarily from the JVM source license. Of particular interest are rules 2 and 3. They together mean that, while it was necessary to slightly alter the semantics of the language to implement checks on some operations, if those checks are disabled the semantics of the operations revert to the original ones. As a special note, those changes consist of adding the possibility of failure to operations that could not normally do so. This means that existing applets execute in exactly the same way in both the the original and modified virtual machines, except for the case in which the operations they attempt are going against the limits imposed by the user of the modified VM, and are in this case stopped. Exactly what should happen.

A few general observation are valid for all the changes that have been done to the system:

1 Applets are classified respecting to their codebase (i.e. the location from which they have been downloaded) and not their actual code (or hash function). This because this second choice is too strong. It can easily distinguish applets when there is no real reason to do so; for example, the presence of an additional unused variable is reason enough to distinguish them.

2 Also, no distinction is made among applets with the same codebase. This has been done because there is no use in putting restrictions on an applet when those same restrictions can easily be circumvented by having two or more instances of the same applet work together. For example, if a limit of 2M is put on memory usage per applet, to use 20M it would be sufficient to use 10 different instances of the same applet and having them work together.

3.2 Which problems does Limiter Contrast?

Here is a list, categorized by type and gravity, of the areas that have shown troublesome and that were checked in some way.

- **Denial of Service, grave.** This category includes those denial of service attacks that can potentially crash the system.

  - **Thread Explosion** It is possible for an applet to create an enormous number of threads, filling the process table and crashing the system. This happens because Java threads relies on the system ones in many implementations, most notably the one considered here.

    The solution chosen was to put a limit on number of threads in execution at the same time for each applet. In case an applet attempts to create more threads than it is allowed to, the operation fails as if for lack of memory, since this was the only case in which this operation could originally fail.

  - **Frame Explosion** Also, an applet can create a great number of frames in a very short time, thus overloading the event manager and effectively locking the user out of the keyboard and the mouse.

    There is also a limit on the maximum number of frames opened. In case an applet tries to created a bigger number of frames than it is allowed to, the operation fails with a `NullPointerException`, the only way the function could normally fail. It is to be noted that this limit applies not to a single applet (like all the others), but to all the applets at the same time, e.g. it is a limit on the *total* number of frames. This because:

    1 It is very difficult to deduce from a frame which applet is trying to create it, and

    2 The cause of problems is the number of frames opened, it does not matter what applet opened them.

- **Memory Exhaustion** An applet can also make unlimited use of available memory, possibly filling it and crashing the system.

  There is now a limit on the total amount of memory that a single applet can use. It is said *total* memory, because the limit applies to the whole amount of memory that an applet uses during its lifetime, whether it has been released it or not. This is because aside from calling `System.gc()` (a thing rarely done by an applet) there is no way for the applet to effectively release memory, or even to know that some memory has been released. In case memory is refused, the exception handling is the same that would occur if the requested memory were truly unavailable.

- **Denial of Service, Weak.** This category includes those denial of service attacks that are not strong enough to crash the whole system.

  - **Priority Juggling** As an applet can modify its own priority level without any limit, this means that the applet can be put it to `MAX_PRIORITY`, with the result that all other applets, that run at normal priority, are effectively almost never executed, bringing about an unacceptable degradation of the system.

    The solution chosen was to put a limit, configurable by the user, on the maximum priority of the threads of an applet, preventing it from stealing CPU cycles from other applets. In case an applet attempts to set a higher priority than it is allowed to, then the priority is silently lowered to the maximum acceptable value.

  - **Frame Resizing** An applet can create frames of unlimited size. This could be abused by creating a frame larger than the screen, and positioning it in such a way to completely cover it, making impossible for the user to see what is happening under it. It should be noted that this fact can also be abused by moving it in such a way that the borders (and the notice that it was really an applet who created it) are off-screen, and this would lead to the possibility of having an applet frame masquerading itself as a system frame, a typical invasion of privacy attack.

    A limit has been put on the maximum size of a frame. In case an applet attempts to create a bigger frame than it is allowed to or to resize it beyond acceptable values, the offending dimension is silently lowered to maximum acceptable value. It is to be noted that this limit is in effect only if the applet itself is trying to break it. The user can enlarge a frame without any limit.

  - **Thread Brokering** At last, it is possible for a thread to interfere with the execution of other threads, modifying their priority, calling `suspend()` or `resume()`, etc...

    As a consequence, now an applet can only execute these operations on its own threads. Trying to call these functions on threads belonging to another applet has no results.

- **Antagonism.** This category includes all those attacks that are meant to simply annoy the user.

  - **Juke-box** The ability to play sounds can be misused. In fact it is sufficient to continuously play an annoying sound to create problems to users.

    The solution chosen, configurable by the user, was to prohibit an applet from using sounds. This is done by fooling the system into believing that the sound device is perpetually occupied. This does have the unfortunate side effect of forbidding the use of sounds for the whole applet.

- **Invasion of Privacy.** This category includes those attacks that are meant to gather information from a user, or to masquerade as one, or to have him working from oneself.

  - **Mail Forging.** A dangerous characteristic of applets is that they can make internet connections. It is true that these connections can only be made to the originating host, but this is not sufficient to prevent attacks like mail forging, and is also a necessary part of work-for-me attacks.

    As a consequence, it is now possible to prevent *all* socket access for an applet, *except* for the accesses made by a class-loader. This makes impossible mail forging attacks and prevents work-for-me applets from returning the computed results. In case access is refused, the JVM is fooled into thinking that the `socket()` call returned an `EACCES` error, and the situation is treated accordingly to the already present code. The exception to this ban is necessary because an applet must in any case be allowed to completely load itself.

- **Miscellaneous.** This category include all those problems that, while not usually attacks by themselves, are commonly used as tools, aids or building blocks in creating them.
  - **Class loading** While applets cannot normally load external libraries, this ability is present in the JVM. Since external libraries cannot obviously be subjected to any check from the JVM, and the only thing that prevents an applet from using them is the security manager (that has been notoriously subject to bugs and attacks) this is an extremely dangerous features.

    The chosen solution chosen was to make it possible to prevent the loading of external libraries. In theory, this is the default behaviour and is enforced by the security manager, however in practice the security manager has often been circumvented, and this is the reason of the ban. In case the permission to load a library is refused, it acts as if it was not found. However, it is necessary for some libraries to be loaded for Java to work properly, and so libraries from the system paths can always be loaded regardless of this setting.
  - **Immortal Applets, Part 1.** One of the most consistently misused features of the Java Virtual Machine is the ability to catch the `ThreadDeath` exceptions. Because this is the exception raised when an applet is about to be terminated, catching it means preventing the death of the applet itself. Immortal applets are thus reborn.

    To prevent this, it is now possible to prohibit the catching of `ThreadDeath`. If the permission to catch this exception is refused, the system simply does not scan the class to look for `catch` or `finally` clauses, simply ignoring them. It is to be noted that system classes sometimes need to catch this exception, and so this check does not apply if the catching class is a system one.
  - **Immortal Applets, Part 2.** A similar problem can be presented by putting appropriate code in the `finalize()` method of the class. Since this method is called by the finalizer after the thread has been stopped but before memory is released, it could be used to resurrect a new thread at a certain moment after its end, when the user has already forgotten it.

    To solve this problem the finalizer thread, responsible of the resurrecting applets, has been subjected to the same limits as the applets, and it can so be configured to prevent the creation of new threads or the use of many resources, just as for applets, effectively putting an end to resurrecting applets.

It should be noted that we have been as conservative as possible: whenever an operation exceeds one of the configured limits, it is made to fail according to one of the already handled possibilities. For instance, when an applet tries to allocate memory above the configured limit, the operation fails as if there was an out of memory situation. Hence the "programmed" failures of Limiter occur in such a way that the class libraries are already prepared to handle. This has been done to limit the changes on the semantics of the JVM itself.

3.3 The Structure of the Original JVM

The version of the JVM we have modified is 1.2.2 for Linux. The source code itself is divided into three subdirectories, `src/`, `ext/` and `build/`. `build/` contains the makefiles and all the other files necessary to build the code, `ext/` contains the source for extensions like the support for internationalization; the actual source is inside the `src/` directory. This directory is also divided into two subdirectories, `share/` and `linux/`. The former includes all machine-independent code, while the latter has all linux-specific code. In fact, every component of the JVM is divided into two parts. The portable one into the `share/` directory and its subdirectories, the non portable one into `linux/` and its subdirectories. This also means that the two directories have (roughly) the same subdirectory structure.

The source of the JVM itself is written in C, C++ and Java. Java is used primarily in the system class library, while the JVM itself is divided into many subsystems (audio, graphics, threads, core, etc...) programmed in C and C++. Each one of these subsytems has its own subdirectory, either in the directory itself or in `native/`. The core of the JVM itself is included in the `javavm/` subdirectory. It is further divided into `export/`, `include/` and `runtime/`. `export/` includes the files available to all the JVM, `include/` includes files available only to this particular subsystem, and `runtime/` contains the source code itself. In these three directories we have placed the bulk of the modified (and all the newly added) files. A few other files have been modified into the `src/share/native/sun/audio/`, `src/share/native/sun/awt`, `src/linux/hpi/green_threads/src`, `src/linux/hpi/native_threads/src` and a few other subdirectories.

3.4 Implementation

Let us see, in short detail, exactly how Limiter has been implemented.

The first thing needed is a structure capable of holding the limits for a specific applet and the resources currently in use. This is the duty of the following structure, taken from **src/share/javavm/include/limiter.h**.

```
struct AppletInfo
{
        long int   memLimit;
        long int   memCurr;

        long int   thrLimit;
        long int   thrCurr;

        long int   priLimit;

        long int   frmX,frmY;

        long int   Flags;

        /* A single AppletInfo structure can be referenced in more than
         * one place. This field takes account of the various
         * references.
         */
        volatile int uses;

        int code;

        /* Thread specific info starts here... */
        int *ignoremem;
        int *memused;
};
```

The meaning of its fields should be easy to understand, with the possible exception of the fields **Flags**, and the last three. The first one is in effect a flip-flop set for limits that do not have an associated parameter, but are in effect a simple on/off decision, while the last three are used for internal safe-keeping and are not really relevant now.

This structure is normally memorized into an hash table in pairs (thread, structure). The algorithms used are defined into **src/share/javavm/runtime/hash.c**, and are pretty standard ones. The only thing that should be noted for further reference is that the access function is called **FindHash()** and its parameter is the thread of which we want to find the respective structure.

This structure can be accessed concurrently by many threads, and so a way is needed to assure its consistency. This is done by defining a monitor and requiring the code to enter it before accessing the structure. This is its implementation.

From **src/share/javavm/include/limiter.h**:

```
extern sys_mon_t *_ai_lock;
#define AI_LOCK_INIT() monitorRegister(_ai_lock, "AppletInfo_lock");
```

```
#define AI_LOCK(self)   sysMonitorEnter(self, _ai_lock)
#define AI_LOCKED(self) sysMonitorEntered(self, _ai_lock)
#define AI_UNLOCK(self) sysMonitorExit(self, _ai_lock)
```

and from **src/share/javavm/runtime/interpreter.c**:

```
/* CV */
    _ai_lock=(sys_mon_t *) sysMalloc(sysMonitorSizeof());
    if (_ai_lock == NULL) return FALSE;
    AI_LOCK_INIT();
/* CV */
```

The **/* CV */** comments are used to delimit the changes we have made in the original code.

These two fragments, together, define and initialize a monitor whose job is to regulate access to the AppletInfo structures, and to do so use the standard JVM mechanism.

Then, a particular set of files is modified for each check we added to the JVM. For example, the checks to prohibit an applet to interfere with the execution of another are implemented by modifying the **JVM_SuspendThread()**, **JVM_ResumeThread()**, **JVM_SetThreadPriority()** and **JVM_StopThread()** functions in **src/share/javavm/runtime/jvm.c** by having them call a new **limallowthread()** function in the new file **/src/share/javavm/runtime/limutil.c**. For more details, look directly in [1].

Let us now see the **JVM_SuspendThread()** function, the one called to cause a thread to be suspended.

```
JNIEXPORT void JNICALL
JVM_SuspendThread(JNIEnv *env, jobject this)
{
    Hjava_lang_Thread *p = (Hjava_lang_Thread *)DeRef(env, this);
/* CV */
    struct Classjava_lang_Thread *tid;
    ExecEnv *ee;

    tid = THREAD(p);

    ee = (ExecEnv *)ll2ptr(tid->eetop);
    if (limallowthread(NULL, ee)) {
/* CV */
        if (ll_nez(THREAD(p)->eetop)) {
            (void) threadSuspend(p);
        }
/* CV */
    }
/* CV */
}
```

As you can see, the modified function determines the thread that we want to interfere with, and then calls the **limallowthread()** function (detailed below) to see if the current thread (**NULL**) has permission to interfere with the **ee** thread. In case the permission is not granted, execution continues as if it had been granted, but nothing really occurs.

This is done in respect to the principles stated in subsection 3.1 to limit incompatibilities.

Similar changes are done to the three functions **JVM_ResumeThread()**, **JVM_SetThreadPriority()**, **JVM_StopThread()**.

Now, from **src/share/javavm/runtime/limutil.c**, a file which contains all functions directly regarding the new security checks, here is **limallowthread()**.

```
/*
 * This function is called to verify that thread 'this' is
 * allowed to mess with thread 'thread'. If 'this' is NULL,
 * it is intended to represent the current thread.
 *
 * return values are as usual.
 */
int limallowthread(ExecEnv *this, ExecEnv *thread)
{
        struct AppletInfo *aithis, *aithread;
        int res = 0;

        if (!this) this = SysThread2EE(sysThreadSelf());

        if (this == thread) return 1;

        ALLOCK(sysThreadSelf());
        aithis = FindHash(this);
        aithread = FindHash(thread);

        if ((aithis == aithread) || (aithis == NULL) ||
            (aithis->Flags & F_ALLOW_THREAD))
           res = 1;
        ALUNLOCK(sysThreadSelf());

        return res;
}
```

As can be seen, this function first determines which thread is calling it. This is done by looking at the **this** parameter, and in the case it is **NULL**, it is taken to mean the current thread.

Then, if the two threads (the one determined above and the one on which to work) are the same, permission is always granted.

Otherwise, inside a block of code protected by the monitor, the algorithm discovers the applets to which the threads belong, and if it discovers that the two belong to the same applet or that one belongs to a system thread, then permission is granted, otherwise the calling applet is first checked for the permission to interfere with other applets.

3.5 Configuration

A JVM can be configured to apply the changes shown above to applet with the use of two configuration files: **/etc/applist** and **$HOME/applis t**. The first one, **/etc/applist**, contains settings that are system-wide and apply to all users, whether they wish to make use of the new features or not, while **$HOME/applist** is a user specific one and can be used to modify the system-wide settings. Please do note, however, that the user settings cannot relax the system-wide ones, but can only made them even more stringent.

A useful feature of this files is that it is not necessary to specify the complete codebase of an applet, but it is possible to specify just part of the path or even of the hostname, thus permitting to specify settings that could be applied to whole sites and not just single pages.

4. EFFECTIVENESS AND PERFORMANCE

The proposed approach has shown that, if properly configured, our approach can stop many of the applets from [7]. In the cases of privacy invasion or work-for-me applets, Limiter does not stop them, but prevents them from sending back their results to the originator of the malicious applet.

The same is valid for the applets described in chapter 4 of [6], with the exception of the ones based on the Ackerman functions or analogues, against which this patch offers no protection.

In fact there are two features conspicuous by their absence. The first one is that Limiter has no reasonable way to limit the CPU usage of an applet, and this because we are at too low a level to obtain a meaningful value of CPU time. Indeed, at this level CPU usage can be measured with respect either to the whole system or to the JVM. In the latter case we are bound to obtain high values regardless of real usage, while in the former almost always the opposite holds true.

The second feature is that no attempt has been made to prevent the redefinition of the `stop()` method in an applet. This because there seem to be only two ways to do this: either by substituting the definition the user may have provided with a new one or by hiding the `stop()` method from the JVM. Both are unsatisfactory, because they will result in broken compatibility and changing code behind the programmer's back.

On a different matter it should be noted that the inclusion of this new tests has caused an inevitable hit on performance. However the approach of including them directly in the JVM has paid off again, and as it will be shown these losses are perfectly acceptable.

We executed some tests on a computer with 32MB of ram, and an AMD K6 200 processor and running the applets found in the `demo` directory of the JVM source.

The applets were run using the standalone applet runner distributed with JVM SDK. Execution times were measured by having the applet itself print the time when it started and when it ended and repeating the process several times in order to reduce measure errors. The functions of the applets tested ranged from sound execution, to drawing, to numer crunching. We measured the execution times of the applets on two different virtual machines: on a standard JVM, and on the patched JVM with all the checks turned on. We found that there was a mean degrade of 0.1%, which we judged inconsequential for normal use.

5. CONCLUSIONS

Our approach has proven to be rather effective in practice, stopping most mischievious applets right before they are allowed to cause damages, and doing so in such a way as to maintain maximum compatibility with all pre-existing code.

As it has been said in the previous section, however, there are two problems our approach does not solve. As a future work, it would be interesting to find a way to integrate these two functionalities in the monitor in a sensible way or, as an alternative, to use it in conjunction with another, external, monitor which could at least provide CPU usage monitoring. An example of such a monitor could be the Applet WatchDog, described in [8].

Another interesting idea, given the effectiveness of the approach, would be to try to implement it also on other architectures, like for example .NET[15] or similar.

References

[1] Vincenzo Ciaschini. The patch itself, http://www.cnaf.infn.it/~marotta/patch.html.

[2] J. Goslin, K. Arnold, The Java Programming Language, Addison-Wesley, Reading, MA 1996

[3] Mary Campione, Kathy Walrath, The Java tutorial, ftp://ftp.javasoft.com/docs/tutorial.tar.gz, May 2001

[4] T. Lindholm, F. Yellyn, The Java Virtual Machine Specification, ftp://ftp.javasoft.com/docs/specs/vmspec.html.tar.gz

[5] G. McGraw, E.W. Felten, Java Security. Hostile Applets, Holes and Antidotes, Wiley, 1997.

[6] G. McGraw, E.W. Felten, Securing Java, Wiley, 1999.

[7] M. LaDue, Hostile Applets Home Page, http://www.cigital.com/hostile-applets/index.html.

[8] M.F. Florio, R. Gorrieri, G. Marchetti, Coping with denial of service due to malicious Java applets, Computer Communications 23 (2000) 1645–1654.

[9] Li Gong, Inside Java 2 Platform Security, Addison-Wesley, Reading, MA, 1999.

[10] D. Martin, S. Rajagopalan, A.D. Rubin, Blocking Java Applets at the Firewall, Procs. Internet Society Symp. on Network and Distributed System Security (1997) 123-133.

[11] Finjan Software, www.finjan.com

[12] Citrix, www.citrix.com

[13] D. Malki, M. K. Reiter, A. D. Rubin, Secure Execution of Java Applets Using a Remote Playground, procs. IEEE Computer Society, Symposium on Security and Privacy, pages 40–51, 1998

[14] The Princeton Java Filter, http://www.cs.princeton.edu/sip/JavaFilter/

[15] David S. Platt, Introducing the Microsoft .NET Platform, Microsoft Press International, 2001

ANALYZING NETWORK MANAGEMENT EFFECTS WITH SPIN AND CTLA

Gerrit Rothmaier, Andre Pohl, Heiko Krumm
Materna GmbH, Dortmund; LS 4, FB Informatik, Universität Dortmund

Abstract: Since many security incidents of networked computing infrastructures arise from inadequate technical management actions, we aim at a method supporting the formal analysis of those implications which administration activities may have towards system security. We apply the specification language *cTLA* which supports the modular description of process systems and facilitates the construction of a modeling framework. The framework defines a generic modeling structure and provides re-usable model elements. Due to *cTLA*'s connection to the temporal logic of actions *TLA*, formal analysis can resort to symbolic reasoning. Supplementarily, automated analysis can be applied. We focus here on automated analysis. It is supported by translation of *cTLA* specifications into suitable model descriptions for the powerful model checking tool *SPIN*. We outline the utilized methods and tools, and report on the modeling and *SPIN*-based analysis of *IP-Hijacking*.

Key words: security incidents, network management, formal analysis, *SPIN*, *cTLA*, *TLA*

1. INTRODUCTION

The current practice of those computer networks which form the basis of the IT infrastructure for companies and institutions shows, that a considerable part of the occurring security incidents has to be ascribed to vulnerabilities which result from inadequate technical management. One can perceive that network and system administrators are often overstrained by the complexity of the system structure, by the frequency of dynamic changes, and by the high number of existing service interdependencies. Moreover, since management operations can potentially control all components of a network, interferences of attacks and inadequate management may result in substantial vulnerability propagation and a plethora of unforeseen threats. Therefore the security of IT-systems does not only depend on their suitable design, on the trustworthiness of their components and on the employment of suitable security services, but also on their proper technical management. Considering that background, our work focuses on the formal modeling and analysis of technical management effects in order to support clear descriptions of relevant attack procedures and management effects as well as a better understanding of the interrelations between management operations and attacks. Moreover, the work shall finally contribute to the discovery of new vulnerabilities and threats.

With respect to the type of modeling, we have to consider that management operations as well as attacks usually do not appear as atomic actions but rather as processes consisting of sequences of steps where each step performs a set of contiguous modifications of a system component. The interesting effects trace back to functional interactions between processes. Therefore we have to resort to a formal modeling technique for the functional aspects of concurrent process systems. The application of such modeling techniques for the analysis of security aspects is not new. In particular, they are applied in the context of protocol design for the verification of authentication protocols and in the context of trustworthy software system development for the verification of implementation code. In both fields the modeling regards rather narrow systems consisting of closely related elements which can mainly be represented in one major level of abstraction. Nevertheless, approaches of both fields report on the high complexity of models, on high costs and efforts for model design and analysis, and finally on the failure of automated analysis tools due to huge time and memory demands.

The application field of management operations and network-centered attacks imposes further requirements. Interesting scenarios comprise a broad spectrum of processes. Interactions can be relevant which concern operations on different abstraction levels. Therefore we attach special importance to the model specification technique. With respect to modeling it shall support the efficient description of broad models by means of composing re-usable specification modules. With respect to analysis, the defined models shall be concise and have a clear formal semantics. They shall be well-suited to the application of automated analysis tools, and – considering the limitations of these tools – they shall also support creative symbolic reasoning. Moreover, since most computer networks are very complex, there is a needs for partial modeling, i.e., one shall be able to focus the formal modeling on interesting parts and substructures of a network and the analysis of partial models shall supply clear contributions to the analysis of the network as a whole.

Our approach is based on the specification language *cTLA*, which supports the flexible and modular description of process systems [Her00]. In particular, in *cTLA* processes can describe the behavior of implementation components as well as general behavioral constraints (cf. [Vis88]). *cTLA*'s process composition has the character of superposition (cf. [Cha88, Kur93]), i.e. properties of subsystems are preserved in embedding systems. Therefore analysis results of partial models can be valid for systems as a whole. We already made good experiences with *cTLA* in the field of communication protocol verification where a formal modeling and verification framework was developed [Her00]. The framework consists of model architecture principles, re-usable specification modules, and theorems. It supports the efficient design of models as well as structured logical proofs of model properties. The semantics of *cTLA* specifications is directly defined by *TLA* formulas [Lam93] and the *TLA* methods for symbolic logical reasoning can be applied. Since *TLA* formulas refer to state transition systems, which is a well-understood and commonly used modeling technique, several suitable automated verification tools are available. Especially we translate *cTLA* specifications into the modeling language *Promela* and apply the corresponding model checker *SPIN* [Hol97], which is currently one of the most powerful and elaborated automated process system verification tools. We abstain from the direct use of *Promela* because it has a relatively complex C-language oriented semantics and does not directly support symbolic logical reasoning. Moreover, *Promela* specifications tend to be less abstract than *cTLA* specifications.

In the sequel we firstly refer to related work in the security analysis field. Then, after outlining the specification language *cTLA*, we describe the translation of *cTLA* specifications into *SPIN/Promela* model definitions. The next section explains the generic model structure, all models of our approach shall comply with. Thereafter, the example problem "*IP Hijacking*" is introduced. Its modeling and *cTLA*-based model specification are discussed. Finally, we report on the *SPIN*-based analysis of this model. Concluding remarks refer to the current state and planned developments. More details of the work are described in [Rot03].

2. RELATED WORK

There is a wide spectrum of work which uses formal modeling and analysis of execution and interaction processes for the verification of security properties. In the main, program verification techniques are applied to enhance the trustworthiness of software systems (e.g. [Bal00]) and protocol verification techniques to detect vulnerabilities of cryptographic protocols (e.g. [Mea95]). In both application fields, a variety of basic methods is employed covering classic logic and algebraic calculi (e.g. [Kaw03]), special calculi (e.g. [Bur89]), and process system modeling techniques (e.g. [Led99]). Many approaches consider tool support in order to facilitate the verification tasks and ensure the correctness of the reasoning. Most tools take advantage of existing more general analysis tools like Prolog, expert system shells, theorem provers, algebraic term rewriting systems, and reachability analysis based model checkers. Some powerful tools combine several analysis techniques [Mea96].

Formal model checking based analysis focusing on network management effects, network attacks, and vulnerability propagation is relatively new. [Ram98, Ram02] report on the analysis of computer systems for those vulnerabilities, which result from the combined behavior of system components. The underlying model is a composition of processes representing the components' and user's security-related behavior. The processes are specified in a special high level language and translated to Prolog. Mainly, simple security properties are modeled by labeling states safe or unsafe. A custom model checker built on top of a Prolog environment searches execution sequences which lead to unsafe states and correspond to vulnerability exploitations.

[Amm00] reports on checking networks of hosts. A network is statically modeled by a set of hosts, a set of vulnerabilities per host, an attacker access level per host, and a network connectivity matrix. The analysis investigates the possible combinations vulnerabilities and their stepwise propagation in the network. The model is implemented using the well-known model checker *SMV*. In [Noe02] this approach is extended and called topological vulnerability analysis. The model uses a multi-valued connectivity matrix which supports the representation of low level communication exploits.

3. CTLA

cTLA is based on TLA [Lam93]. In comparison with TLA it supplies explicit notions of processes, process types, and process composition [Her00]. Furthermore, there is a different look of *cTLA* specifications since canonical parts are not explicitly written down. A wide variety of data types can be defined, in particular those with finite (and preferably small) value sets in order to support the *SPIN*-based model checking.

3.1 TLA

TLA [Lam93] describes state transition systems by means of a so-called canonical formula. A state transition system STS ::= < S, S_0, T > is defined by a set of states S, a set of initial states $S_0 \subset S$, and a set of transitions $T \subset S \times S$. A canonical TLA formula f_{Sys} ::= $Init \wedge \Box\, [\, Next\,]_V \;\wedge FA$ directly addresses the state transition system model. V is a set of state variables (e.g. $V = \{x_1, x_2, \ldots, x_n\}$) spanning the state space S. *Init* denotes the *Init*-predicate defining the set of initial states S_0. *Next* denotes the next state predicate which defines the set of transitions T and is structured into a set of actions: *Next* ::= $act_1 \vee act_2 \vee act_3 \ldots \vee act_m$. The always-subformula $\Box\, [\, Next\,]_V$ means that each pair of two consecutive states <s_i, s_{i+1}> of each execution state sequence of *Sys* has to fulfil the formula: $Next \vee (x_1' = x_1 \wedge x_2' = x_2 \wedge \ldots \wedge x_n' = x_n)$, i.e. the state pair has to comply with one of the actions of *Next* or represent a so-called stuttering step where s_i equals s_{i+1}. *FA* specifies liveness properties by a conjunction of fairness assumptions *FA* ::= $\mathrm{WF}(act_j) \wedge \mathrm{WF}(act_k) \wedge \ldots \wedge \mathrm{SF}(act_p) \wedge \ldots \wedge \mathrm{SF}(act_q)$. A weak fairness assumption $\mathrm{WF}(act_j)$ assumes, that the action act_i has to be executed in situations, where the action is enabled and continuously will be enabled until its execution. A strong fairness assumption $\mathrm{SF}(act_j)$ assumes, that the action act_i has to be executed, if the action will be enabled again and again until its execution.

TLA facilitates formal verification. It supplies inference rules supporting the deduction of implications of the form $f_{Sys} \Rightarrow f_{Prop}$, where the formula f_{Sys} describes a system *Sys* in more detail and the formula f_{Prop} represents a more abstract property of *Sys*. The deduction of the implication formally verifies that the system *Sys* has the property *Prop*.

3.2 cTLA Simple Process Type

Each *cTLA* specification module describes a process type, an instance of which corresponds to a state transition system modeling a process of this type. A *simple process type* declaration defines a state transition system directly. As an example, the following type *Media* models the physical packet transfer of a local area network and is part of the "*IP Hijacking*" model:

```
PROCESS Media();
   VAR buf: ARRAY[MAXZONES] OF PacketBufT;
   INIT::= FORALL i:MAXZONES: [ buf[i].usd = FALSE ];
ACTIONS
   in( pkt: PacketT ) ::=
    fSrcToZone( pkt.scn, pkt.sci ) != UNKNOWN_ZONE
    AND buf[ fSrcToZone( pkt.scn, pkt.sci ) - 1 ].usd = FALSE
```

```
    AND buf[ fSrcToZone( pkt.scn, pkt.sci ) - 1 ].usd' = TRUE
    AND buf[ fSrcToZone( pkt.scn, pkt.sci ) - 1 ].pkt' = pkt;
  out( pkt: PacketT ) ::= ...
END
```

The state space of the state transition system is spanned by state variables. In *Media* the variable `buf` is an array of packet buffers. The set of initial states is defined by the predicate `INIT` which in the example claims that all buffer fields are flagged as unused. The actions directly constitute the next state relation. Action `in` applies to those state transitions which model that the Media accepts and transfers a sent packet `pkt`. Aktion `out` models that a packet `pkt` is delivered to a set of receiving nodes. Syntactically, actions are predicates over state variables (referring to the current state, e.g. `buf`), so-called primed state variables (referring to the successor state, e.g. `buf'`), and action parameters (e.g. `pkt`).

A subset of actions can be classified as so-called internal actions. An internal action defines a set of state transitions in exactly the same way as a normal action. The difference between both sorts of actions concerns the composition of systems from processes. When a process instance is employed as a component in a system, the internal actions of the process cannot be coupled with actions of other processes. Each internal action is accompanied by stuttering steps of all other system components.

With respect to liveness properties, it has to be mentioned, that currently fairness assumptions are not used since we concentrate on safety properties.

3.3 cTLA Process Composition Type

A process composition type describes a process system which is composed from process instances of imported types. It is exemplified by the following type *IpArpExample* which models a small local computer network and is part of the "*IP Hijacking*" model:

```
PROCESS IpArpExample();
CONTAINS
   med: Media();  bnA: IpArpNode( NA_ID, NA_MII );
                  bnB: IpArpNode( NB_ID, NB_MII );
                  bnC: IpArpNode( NC_ID, NC_MII );
ACTIONS    /* send system actions */
   snd_A( pkt: PacketT ) ::= bnA.snd( pkt ) AND med.in( pkt );
   snd_B( pkt: PacketT ) ::= bnB.snd( pkt ) AND med.in( pkt );
   snd_C( pkt: PacketT ) ::= bnC.snd( pkt ) AND med.in( pkt );
           /* receive system actions */
   rcv_A( pkt: PacketT ) ::= bnA.rcv( pkt ) AND med.out( pkt );
   rcv_B( pkt: PacketT ) ::= bnB.rcv( pkt ) AND med.out( pkt );
   rcv_C( pkt: PacketT ) ::= bnC.rcv( pkt ) AND med.out( pkt );
           /* broadcast receive system actions */
   rbc( pkt: PacketT ) ::=
```

```
        bnA.rbc( pkt ) AND bnB.rbc( pkt ) AND bnC.rbc( pkt ) AND med.out( pkt );
    /* implicitly added: internal actions of the nodes, e.g. rpcs, spcs */
  END
```

An *IpArpExample* network contains four process instances, one process `med` of the type *Media* which is described above, and three node processes `bnA`, `bnB`, and `bnC` of type *IpArpNode*. These processes are coupled according to the system actions (e.g. `snd_A`). The right side of a system action is a conjunction of process actions (e.g. `bnA.snd(pkt)`, `med.in(pkt)`) and lists those process actions which have to be performed jointly in order to realize a system action. Each process can contribute to one system action by at most one process action. If a process does not contribute to a system action, it performs a stuttering step. In the special *cTLA*-variant used here, process stuttering steps are not explicitly listed in the right side of system actions.

The state transition system which models an instance of a process composition type, is defined indirectly. Its state space is spanned by the vector of all state variables of all contained processes. Its *Init*-predicate is the conjunction of the *Init*-predicates of the contained processes, and its *Next*-predicate is the disjunction of the system actions. In this way the concurrent execution of processes is modeled according to interleaving semantics, while process interactions are modeled by joint actions where two or more processes simultaneously perform non-stuttering steps. System actions and process actions can be supplied with data parameters supporting the communication between processes.

The TLA formula describing a *cTLA* process system is equivalent to a conjunction of the process formulas and consequently a system implies its constituents. The process composition of *cTLA* has therefore the character of superposition (cf. [Cha88, Kur93]) which guarantees that a property fulfilled by a process or a subsystem is also a property of each system which contains this process or subsystem. In particular, superposition supports the so-called structured verification. In order to prove that a system *Sys* has a property *Prop* it is sufficient to find a subsystem of *Sys*, where *Prop* can be proven.

4. TRANSLATION TO SPIN/PROMELA

The model checker *SPIN* accepts models written in the *Promela* (Process Meta Language) specification language [Hol97]. *Promela*'s basic syntax resembles the *C* programming language. But most of the computational functions have been removed and constructs for modeling process systems have been added. These constructs are similar to Hoare's *CSP* (Communicating Sequential Processes) but more flexible. Process types can be declared and dynamically executed through instantiation. Communications between processes are modeled with special channels or through shared variables. Guarded statements can be used for synchronisation.

We developed a compiler, *CTLA2PC* (*cTLA* to *Promela* Compiler), for translating high level specifications written in *cTLA* into the more low level *Promela* language. *CTLA2PC* generates highly optimized *Promela* code and supports various switches. For example the `-optbitarrays` switch can decrease the state vector size considerably in comparison to a native *Promela* model (cf. Sec. 7.1).

The general steps of the translation process are as follows. First *CTLA2PC* builds a parse tree from the *cTLA* input. Using this tree the *compositional system* is then expanded to an equivalent *flat system*. The flat system representation is necessary because *Promela* specifications don't support systems of composed processes coupled through system actions. In the flat system only a single process with all actions exists. This process contains the system initialization (merged from all processes) as well.

The flat system is used for all further steps. Different back-ends can be used to generate code in a target language. A simple *cTLA* backend just outputs the flat system definition (switch `-ctla`). But for *Promela*, more transformations have to be done. Processes can't contain multiple actions in *Promela*. *CTLA2PC* therefore generates a non-deterministic selection with appropriate guards for each action. Furthermore, *Promela* doesn't support action parameters like *cTLA*. These parameters are implicitly existentially quantified. Hence *CTLA2PC* creates extra input generator processes in *Promela* and shared variables which replace the parameters. Each input generator uses a non-deterministic selection loop over all possible values of the parameter. The value selected is written to the parameter variable. An alternative approach would be to use *SPIN*'s channels instead of parameter variables for passing the value, but in our experiments *SPIN* needed more state space that way.

Many more points not mentioned here due to space reasons have to be taken into account during the compilation process. These include particularly the mapping of data type definitions, functions, and predicates.

5. GENERIC MODEL STRUCTURE

The goal of our modeling is to have a generic structure suitable for describing computer networks and applying automated verification techniques. It shall be abstract enough to hide low level details which would quickly enlarge the state space and make verification impossible. But it must allow specific models to be derived easily and augmented with more detailed behavior required by the example, e.g. *ARP* protocol handling. Thus we choose a middle level of abstraction which can be extended both for low level packet oriented modeling and high level service oriented processes.

The modeling shall be as natural as possible. From a bird's eye view, a real computer network consists of several nodes transmitting packets over media. Thus these are the basic elements of our modeling (cf. *Fig.* 1). A *Node* process models any object (host, printer, ...) that sends (action `snd`) or receives (action `rcv`) packets with an interface over media. The processing of packets takes a layered approach. Which layers (e.g. *Network Interface*, *IP/ARP*) are needed depends on the specific model, but the generic background is as follows. For both sending and receiving, each layer has its own packet buffer and appropriate actions for processing (`spcs`, `rpcs`). After a packet arrives at the layer the processing action can be activated. The processing of the packet may either terminate within this layer or the packet may be marked as "ready" to be passed to the next layer. Corresponding actions (`smov`, `rmov`) for moving "ready" packets between the layers have to be included in the specific model. These actions as well as the processing actions belong to the internal actions of the node.

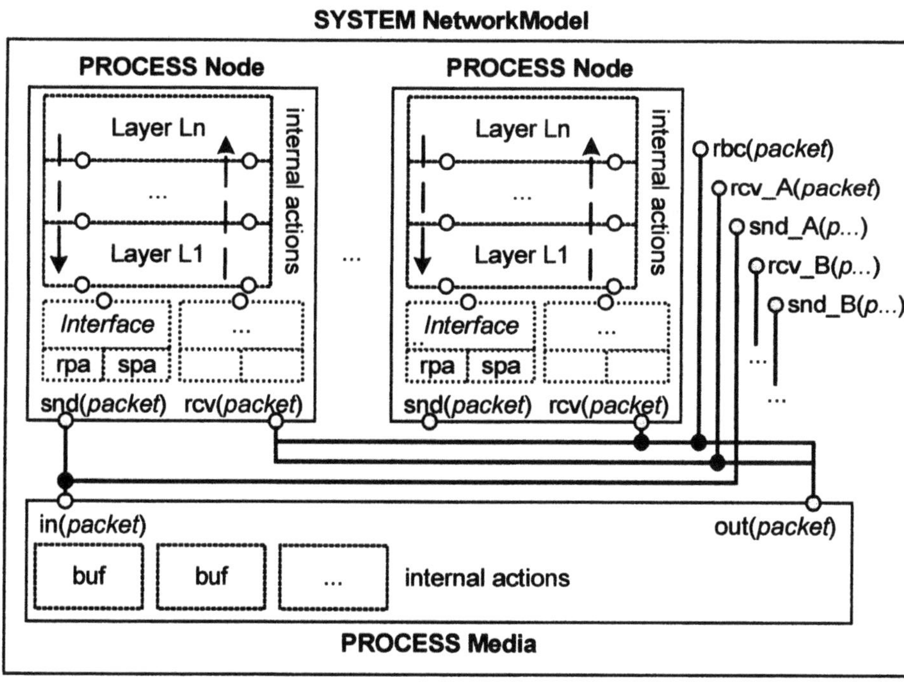

Figure 1: Generic model stucture

Because the generic model is independent of addressing schemes (e.g. Ethernet hardware addressing), no constraints on received packets (e.g. remove the packet if it isn't a broadcast packet and doesn't match the interface's address) are defined. These constraints are included in the low level packet processing of specific models. Similarly, stamping of packets with a specific source address (other than the model internal source node, source interface attributes) before sending has to be done there as well.

Each node has to have at least one interface, but may have multiple interfaces (e.g. a router node). Interfaces can be dynamically enabled or disabled (cf. *Unix ifconfig up/down*). Each interface has a buffer (variable `rpa`) to hold a packet after it has been received and before it is processed by the network stack on the node. Similarly each interface has a buffer (variable `spa`) to hold a packet that shall be transmitted to the media.

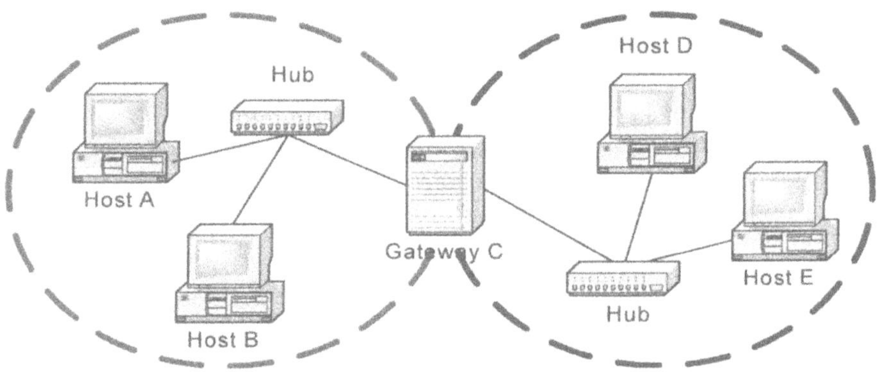

Figure 2: Example network

All interfaces which can directly reach other interfaces via the link layer (i.e. without using higher level, e.g. *IP*, routing functions on a node) belong to a *zone*. Because a node may have multiple interfaces, a node can be in multiple zones. A low level broadcast by a node with a selected interface will reach all others nodes with an interface in the zone. Zones are assumed to be static, i.e. a node may not be dynamically equipped with a new interface (but may enable or disable a previously existing interface).

The *Media* process is used for packet exchanges. For each zone *Media* buffers (variable `buf`) one packet currently in transit in the media. Only one packet can be in transit in any zone at one time. A node can only transmit a packet over an interface to the media (action `in`) if no packet is already in transit for the zone. Similarly a packet can only be received by a node over an interface from the media (action `out`) if a packet is currently in transit for the zone.

Within the generic model packets just contain two internal attributes for determining the zone of the packet. Without higher level routing a packet can't move between zones (by definition). For a specific model further attributes are typically added to packets.

The *System* process associates node and media instances. It defines system actions which couple actions of process instances. Each node *N* can send a packet to media (action `snd_`*N*, coupling node *N* and *Media*) or receive a unicast packet from *Media* (action `rcv_`*N*). Furthermore a broadcast packet may be received by all nodes in a zone (action `rbc`, coupling *Media* and all nodes of a zone).

6. EXAMPLE SYSTEM

The modeling structure described above provides a basis for concrete instances. Consider the example depicted in *Fig.* 2.

Two network zones are connected by a gateway. Zone 1 contains hosts *A*, *B* and the gateway *C*, zone 2 contains hosts *D*, *E* and the gateway *C*. Only the gateway has two interfaces. All hosts within a zone can directly reach each other without using the gateway. The hosts communicate over *IP*. We want to analyze the effects if an administrator (or ignorant user) hijacks an *IP* address (e.g. of the gateway) by changing the *IP* address of a host to an already used one.

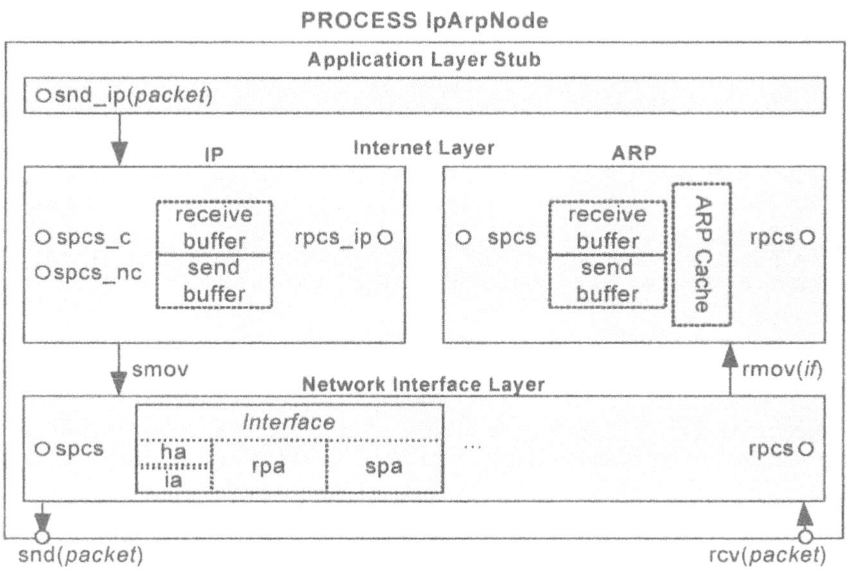

Figure 3: IpArpNode

For the modeling of this example we use the abstract node element as starting point. It has to be augmented with appropriate behavior and attributes for *IP* communication. As described in the generic model *IpArpNode*'s actions (cf. *Fig.* 3) are layered, in this case akin to the *TCP/IP* reference model [Com01]. The layers *Network Interface*, *Internet* (including *IP* and *ARP*) and a minimal *Application Layer Stub* are needed.

Submitted packets from the application (action `snd_ip`) are stored in a send buffer. Then the packet can be processed by the *IP* layer. Basically *IP* addresses have to be resolved to hardware addresses. If the *IP* address to hardware address assignment is not available within the cache, a broadcast *ARP* query is created (action `spcs_nc`), processed through the layers (especially *ARP*) and sent (action `snd`). After a matching *ARP* reply (or a query by the sought-after node) has been received and processed, the *ARP* cache is updated. Then the *IP* packet can be processed and finally sent. If the address was resolved before and is still in the *ARP* cache no *ARP* request is generated by the *ARP* layer (action `spcs_c`).

Received packets (action `rcv`) are stored in the interface's buffer and processed layer by layer as defined in the generic model. For this model, the low-level *Network Interface* layer checks the packet's hardware destination address against the interface's hardware address (if the interface is neither in promiscuous mode nor the packet is a broadcast packet). Invalid packets are thrown away immediately and don't reach the *Internet* layer. Similarly, the send packet processing (action `spcs`) has been extended to set the hardware source address of the packet to the hardware address of the interface.

Besides the layers and their actions each of *IpArpNode*'s interfaces is extended with a hardware (variable `ha`) and an *IP* address (variable `ia`). The initialization of *IpArpNode* has to be enhanced to include these as well. Furthermore the packet structure is augmented for the *Network Interface* layer with hardware source and destination address and frame type. Depending on the frame type, a packet has either *ARP* or *IP* attributes on *Internet* layer. *ARP* packet attributes include a query type and pairs of hardware and protocol addresses. *IP* packet attributes are the *IP* source and destination addresses.

To model the administrator, we introduce a new action for changing the *IP* address of a host to another address (action `chg_ip`). The system initialization is generated by *CTLA2PC* from the processes initialization automatically. So no further explicit system initialization action has to be added.

7. ANALYSIS

Most of the time model checking is a challenging task. Even small models quickly exceed given time and memory constraints. Marginal model additions can have a strong adverse impact on the performance of the verification tool. This effect is well-known as state space explosion.

7.1 Example System Optimizations

In order to enable automated analysis, we started with two major model optimizations. First, we realized that a smaller subsystem suffices to study the effects of the administrator's actions. The two zones are similar, so modeling just one zone, e.g. zone 1 with the hosts *A*, *B*, and *C*, satisfies our requirements. Due to the superposition property of *cTLA*'s process composition, the results can be carried over to embedding systems. Also, because all hosts are of the same type, we can assume that a certain host, namely *B*, changes its *IP* address to the address of host *C*.

Second, we apply an approach of the field of efficient protocol implementation in order to save on buffers and action steps. We refer to the *activity thread* approach which combines all those actions of several protocol layers into one integrated and non-interruptible execution thread, which processes the same stimulating packet request [Svo89]. Accordingly, we modify the host model by merging actions and eliminating interface buffers. As a result, just one processing action and working buffer for each direction (sending and receiving) remains. Furthermore we include the basic interface setup of the nodes (i.e. activation and assignment of original *IP* addresses) into the system initialization to save some steps.

To evaluate the possibility of checking the model we translated it to *Promela* using *CTLA2PC* and performed several *SPIN* runs. The state vector size was about 250 bytes. A *DFS* search depth of over one million was reached within a minute and *SPIN* quickly used up over 512 MB of memory. Approximate (supertrace and bit-state-hashing) *SPIN* verification modes prevented the memory overrun but did not give any results in a reasonable amount of time. One of these runs was cancelled after over 150 hours.

Hence we looked at the *cTLA* model again in order to cut down the state vector size further. In particular the action parameters were promising. *cTLA* action parameters correspond to existentially quantified value variables, and their translation to *Promela* introduces extra processes and state variables (cf. Sec. 4). We observed that for most of the action parameters value determining equalities exist, since many parameters of flat system actions serve as output parameters of constituting process actions. Therefore it is possible to replace these parameters with the corresponding symbolic output value. In turn this "paramodulated" version of the model was translated to *Promela*. Now the *SPIN* state vector size was shortened to about 210 bytes. Additionally we improved the code generation of *CTLA2PC* to consider specific oddities of *SPIN*. In particular, *SPIN*'s built-in handling of arrays of those variables which use only a few bits, requires much more state space than necessary. With the switch `-optbitarrays` *CTLA2PC* wraps such an array into an integer type following a generalized version of the bitvector approach suggested by [Ruy01] and appropriately maps all read and write accesses to array elements. This led to a further considerable reduction of the state vector to a size of 168 bytes.

7.2 Checking Assertions & Analyzing Trails

Based on this version we wanted to check the property that a node cannot receive non-broadcast *IP* packets destined for other nodes. We inserted the following assertion at the end of the non-broadcast receive action of each node:

```
assert( !(bnN_ifs[1-1].rpa.pkt.l2t==L2_IP &&
        BVGET(bnN_ifs[1 - 1].rpa.pkt.dat, 3, DI_IDA) != bnN_ifs[1 - 1].ia ));
```

It checks that a received *IP* packet's destination address equals the interface's *IP* address that received the packet. We had in mind, that through a change of the *IP* address of node *B* to the *IP* address of node *C* (*IP* hijacking) and a subsequent *ARP* query by *B* the entry for *C* in the *ARP* cache of the other nodes might be updated with *B*'s hardware address (*ARP* cache poisoning). This could lead to further packets intended for *C* to be received by *B* and violate the assertion. So we started a *SPIN* verification run. As expected the assertion was violated:

```
pan: assertion violated  !(((bnB_ifs.rpa.pkt.l2t==1) && (((bnB_ifs.rpa.pkt.dat>>(1*3)) &&
                          ((1<<3)-1))!=bnB_ifs.ia)))          (at depth 12)
pan: wrote ip-arp-example-verif-comp-flat-para-bitopt.promela.trail
(Spin Version 4.1.0 -- 6 December 2003)
   Warning: Search not completed + Using Breadth-First Search + Partial Order Reduction
   ...
State-vector 168 byte, depth reached 12, errors: 1
```

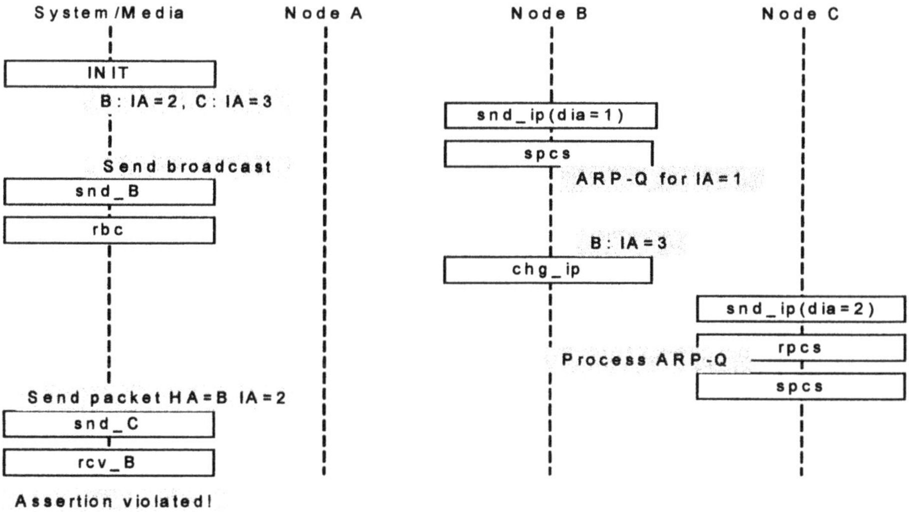

Figure 4: Violating path

But subsequent guided simulation with the created trail file revealed that the violation occurs in an unforeseen way (cf. *Fig.* 4). Node *B* changed its *IP* address (`chg_ip`) to 3 but has not yet updated the *ARP* cache of another node. Instead, a previously received (`rbc`) *ARP* query from node *B* still contains the old *IP* address and node *C* updates its *ARP* cache from this query (`rpcs`) just now. Thus an already buffered *IP* packet for the (now unused) *IP* address 2 can be processed (`spcs`) with the cached *IP* address to hardware address assignment. The packet is submitted to *Media* and received by node *B* (due to the matching hardware address). But the interface's *IP* address is different and this violates the assertion. A typical (non promiscuous) *IP* stack would throw away such a packet. Thus this demonstrates the simple fact that an *IP* change may lead to lost packets until all nodes in the subnet have updated their *ARP* caches correctly. This is especially relevant to hosts with static *ARP* configurations since their settings have to be maintained manually.

To analyze a scenario analogous to the originally intended situation another assertion can be used, which has to be included in the send action of node *A*:

```
assert( !(bnA_ifs[1 - 1].spa.pkt.l2t==L2_IP &&
        BVGET( bnA_ifs[1 - 1].spa.pkt.dat, 3, DI_IDA) ==
                bnC_ifs[1 - 1].ia && bnA_ifs[1 - 1].spa.pkt.dha != bnC_ifs[1 - 1].ha ));
```

The resulting trail file shows node *B* immediately changing its *IP* address and poisoning the *ARP* cache of the other nodes as described above. Then the assertion is violated by node *A* sending an *IP* packet intended for node *B* to node *C*.

8. CONCLUDING REMARKS

Technical management processes and their security implications constitute a new and practically relevant application field for formal modeling and analysis. It, however, has distinct demands. The modeling results in comprehensive models and shall therefore be supported by a modular and flexible specification technique, while on the other hand analysis tools shall be able to uncover unknown effects automatically. As shown by means of the *IP-Hijacking* example, the proposed combination of modeling guidelines, the *cTLA* specification language, and *SPIN*-based analysis is a feasible approach which can substantially contribute to improve the understanding of management effects.

Current work studies intertwined network and security service reconfiguration processes. Moreover we focus on an enhancement of the technique. The modeling framework will be completed by re-usable process type definition modules. The analysis of complex models will be supported by model architecture principles which are borrowed from the field of efficient communication protocol implementation. The provision of improved tool support will be based on the integration of *cTLA* and *SPIN* into an *Eclipse*-based development environment (cf. [Ecl02]).

We thank the anonymous reviewers for their valuable suggestions directing our future work towards methodic modeling abstractions for complex systems.

REFERENCES

[Amm00] P. Ammann, R. Ritchey: *Using Model Checking to Analyze Network Vulnerabilities*. IEEE Symposium on Security and Privacy, May 2000.

[Bal00] M. Balser, W. Reif et al.: *Formal System Development with KIV*. In: T. Maibaum (ed.), Fundamental Approaches to Software Engineering. Springer LNCS 1783, 2000.

[Bur89] M. Burrows, M. Abadi, R. Needham: *A Logic of Authentication*. In: Proceedings of the Royal Society, Volume 426, Number 1871, 1989, und in: William Stallings, Practical Cryptography for Data Internetworks, IEEE Computer Society Press, 1996.

[Com01] D. E. Comer, R. E. Droms: *Computer Networks And Internets*. Prentice Hall, 2001.

[Cha88] K.M. Chandy, J.Misra: *Parallel Program Design – A Foundation,* Addison-Wesley, Reading, 1988.

[Ecl02] Eclipse.org: *Eclipse Project FAQ*. URL: http://www.eclipse.org/eclipse/faq/eclipse-faq.html, 2002.

[Her00] P. Herrmann, H. Krumm: *A framework for modeling transfer protocols.* Computer Networks, 34(2000)317-337.

[Hol97] G.J. Holzmann: *The model checker spin,* IEEE Trans. Soft. Eng. 23 (1997) 279-295.

[Kaw03] K. Kawauchi, S. Kitazawa et al.: *A Vulnerability Assessment Tool Using First-Order Predicate Logic*. IPSJ SIGNotes Computer SECurity No.019 (2003)

[Kur93] R. Kurki-Suonio: *Hybrid models with fairness and distributed clocks*, in: Lecture Notes in Computer Science, Vol. 736 (Springer, New York, 1993) pp. 103–120.

[Lam93] L. Lamport: *The temporal logic of actions*, ACM Trans. Prog. Lang. and Sys. 16 (1994) 872-923.

[Led99] G. Leduc, O. Bonaventure, L. Leonard, E. Koerner, C. Pecheur: *Model-based verification of a security protocol for conditional access to services*. Formal Methods in System Design, Kluwer Academic Publishers, vol.14, no.2, March 1999 p.171-91.

[Mea95] C. Meadows: *Formal Verification of Cryptographic Protocols: A Survey*. In: Proc. Asiacrypt, Intern. Conf. on Theory and Application of Cryptology, LNCS, Springer-Verlag, 1995.

[Mea96] C. Meadows: *The {NRL} Protocol Analyzer: An Overview*. Journal of Logic Programming, 26, 2 (1996) pp. 113-131.

[Noe02] S. Noel, B. O' Berry, R. Ritchey: *Representing TCP/IP connectivity for topological analysis of network security*. Computer Society, IEEE (ed.), Proceedings of the 18th Annual Computer Security Applications Conference, Dec. 2002, p. 25-31.

[Ram02] C. Ramakrishnan, R. Sekar: *Model-Based Analysis of Configuration Vulnerabilities*. Journal of Computer Security, Vol. 10, Nr. 1, Jan. 2002, p. 189-209.

[Ram98] C. Ramakrishnan, R. Sekar: *Model-Based Vulnerability Analysis of Computer Systems*. Second International Workshop on Verification, Model Checking, and Abstract Interpretation, Pisa, Italy, Sep. 1998.

[Rot03] G. Rothmaier, H. Krumm: *Formal Modeling of Security Properties of Computer Networks*, Internal Technical Report, RvS Group, FB Informatik, University of Dortmund, URL: http://ls4-www.cs.uni-dortmund.de/RVS/P/ForSecMod.pdf, 2003.

[Ruy01] T. C. Ruys: *Towards Effective Model Checking*. PhD Thesis, Univ. Twente, 2001.

[Svo89] L. Svobodova: *Implementing OSI Systems*. IEEE Journal on Selected Areas in Communications 7 (7),(1989), pp.1115–1130.

[Vis88] C.A. Vissers, G. Scollo, M. van Sinderen: *Architecture and specification style in formal descriptions of distributed systems*, in: S. Agarwal, K. Sabnani (Eds.), Proceedings of the PSTV VIII, Elsevier, Amsterdam, 1988, pp. 189-204.

FORMAL REASONING OF VARIOUS CATEGORIES OF WIDELY EXPLOITED SECURITY VULNERABILITIES USING POINTER TAINTEDNESS SEMANTICS[1]

Shuo Chen, Karthik Pattabiraman, Zbigniew Kalbarczyk and Ravi K. Iyer
Center for Reliable and High-Performance Computing, Coordinated Science Laboratory, University of Illinois at Urbana-Champaign, 1308 W. Main Street, Urbana, IL 61801. {shuochen , pattabir, kalbar, iyer}@crhc.uiuc.edu

Abstract: This paper is motivated by a low level analysis of various categories of severe security vulnerabilities, which indicates that a common characteristic of many classes of vulnerabilities is pointer taintedness. A pointer is said to be tainted if a user input can directly or indirectly be used as a pointer value. In order to reason about pointer taintedness, a memory model is needed. The main contribution of this paper is the formal definition of a memory model using equational logic, which is used to reason about pointer taintedness. The reasoning is applied to several library functions to extract security preconditions, which must be satisfied to eliminate the possibility of pointer taintedness. The results show that pointer taintedness analysis can expose different classes of security vulnerabilities, such as format string, heap corruption and buffer overflow vulnerabilities, leading us to believe that pointer taintedness provides a unifying perspective for reasoning about security vulnerabilities.

Key words: Security Vulnerability, Static Analysis, Program Semantics, Equational Logic, Pointer Taintedness

[1] This work is supported in part by a grant from Motorola Inc. as part of Motorola Center for Communications, in part by MURI Grant N00014-01-1-0576, and in part by NSF CCR 00-86096 ITR.

1. INTRODUCTION

Programming flaws that result in security vulnerabilities are constantly discovered and exploited in real applications. A significant number of vulnerabilities are caused by improper use of library functions in programs. For example, omitting buffer size checking before calling string manipulation functions, such as *strcpy* and *strcat*, causes many buffer overflow vulnerabilities. Passing user input string as the format string in *printf*-like functions causes format string vulnerabilities. Heap corruption vulnerabilities are the result of invoking the *free* function with a pointer pointing to an overflowed buffer or a buffer that has not been allocated by the heap manager. Library functions are usually secure only under certain conditions, and therefore, formally extracting security conditions from library code can be a valuable aid in implementing applications free of security vulnerabilities.

We introduce the notion of pointer taintedness as a basis for reasoning about security vulnerabilities. The notion of pointer taintedness is based on the observation that the root cause of many reported vulnerabilities is due to the fact that a pointer value (including return address) can be derived directly or indirectly from user input. Since pointers are internal to applications, their values should be transparent to users. Thus a taintable pointer is a potential security vulnerability. By analyzing the application source code, the potential for pointers to be tainted can be determined and hence possible vulnerabilities can be identified.

Existing compiler-based techniques, such as CQUAL [7] and SPLINT [1], perform taintedness analysis by associating an attribute or a type qualifier with program symbols, i.e., variables, constants, arguments and return values. Although these techniques allow reasoning about taintedness of symbols, they cannot analyze the taintedness of pointers unless C statements explicitly perform the tainting. As we will show, there are many situations (e.g., format string vulnerabilities, heap corruptions and stack buffer overflows) where pointers become tainted without explicit assignment statements in the code. Pointer taintedness that leads to well known vulnerabilities usually occurs as a consequence of low level memory writes, typically hidden from the high level code. Hence a memory model is necessary to reason about pointer taintedness. This paper proposes a formalization to define and analyze pointer taintedness so as to uncover potential security vulnerabilities. We focus on commonly used library functions because many widely exploited vulnerabilities are caused by pointer taintedness occurring in library functions.

A memory model is formally defined in this paper using equational logic[2], which forms the basis of the semantics of pointer taintedness. A mechanical reasoning technique is applied to several library functions to examine the possibility of a pointer being tainted. The analysis process for each library function yields a set of formally specified preconditions that must be satisfied to eliminate the possibility of pointer taintedness. These pre-conditions either correspond to already known vulnerability scenarios (e.g., format string vulnerability and heap corruption) or indicate the possibility of function invocation scenarios that may expose new vulnerabilities.

2. RELATED WORK

Security vulnerabilities have been reported in many applications. The *Bugtraq* vulnerability list and *CERT* advisories maintain information about reported vulnerabilities. Security vulnerabilities can be modeled using finite state machines, by breaking them into multiple simple operations, each of which may be associated with one or more logical predicates [10].

Many static detection techniques have been developed based on the recognition of existing categories of security vulnerabilities. Techniques such as [1] and [2] can check security properties if vulnerability analysts are able to specify them as annotations in the code. Domain-specific techniques require less human effort, but each technique only detects a specific type of vulnerability. Static detection techniques are proposed to detect to buffer overflow vulnerabilities, e.g. [6]. Runtime mechanisms against security attacks have been introduced in [8] and [9]. Xu et. al. [5] recently proposed efficient approaches to randomize memory layout and to encode control flow information, which aims at defeating various attacks in a generic manner.

The notion of taintedness was first proposed in Perl language as a security feature. Inspired by this, static detection tools like SPLINT [1] and CQUAL [7] also use taintedness analysis to guarantee that user input data is never used as the format string argument in *printf*-like functions. In both these tools, taintedness is an attribute associated with C program symbols. A symbol gets tainted only if an explicit C statement passes a tainted value to it by assignment, argument passing or function return. As shown in the next section, in many real attacks, pointers are tainted without explicit C statements tainting program symbols. Since these tools do not have a memory model, they cannot determine whether an address is tainted and hence cannot reason about the underlying memory status. In [12], several

[2] An introduction to equational logic can be found in [11].

examples are provided where data can be tainted without being detected by SPLINT and CQUAL. The inability to reason about taintedness at the memory level is an inherent limitation of existing taintedness analysis tools.

3. POINTER TAINTEDNESS EXAMPLES

Our study indicates that many known security vulnerabilities, such as format string, heap corruption, buffer overflow and *Glibc glob* vulnerabilities, arise due to pointer taintedness. In these examples, the pointers become tainted without explicit C assignment statements. Due to space limitations, only the example of the format string vulnerability is discussed in detail. Other examples are described in the extended version of this paper [12]. In this section, we first give a high level description (Table 1) of the format string vulnerability, and show how it can be exploited. We then show that a memory model (Figure 1) is necessary for reasoning about pointer taintedness.

Table 1: Format String Vulnerability Illustration

```
  int Vfprintf (FILE *s, const char *format,
va_list ap)
 {     char * p;
       int count;
        p = format; ... ...
L1:    *(int *) ap = count;
        ... ...
 }

  int Printf (const char *format, ...)
 {   va_list arg;    ... ...
L2:  Vfprintf (stdout, format, arg);
    ... ...
  }

  int i,j;
  int main()
 { char buf[100];
```

```
      //This is how to call Printf correctly
      strcpy(buf,"hello");
      i=1234;
L3:   Printf("string=%s\ndata=%d\n%n",buf,i,&j);
      Printf("total output length=%d\n",j);

      //This is how format string vulnerability occurs
      scanf("%s",buf);
L4:   Printf(buf);
L5:   Printf("\ni=%d\n",i);
      }

Program Output:
O1:       string=hello
O2:       data=1234
O3:       total output length=23
O4:          1345147471234134519168 12345
O5:       i=31
```

The format string vulnerability is caused by incorrect invocation of *printf*-like functions (e.g., *printf, sprintf, snprintf, fprintf* and *syslog*). Table 1 gives examples of correct and incorrect invocations of *Printf()* (a simplified version of LibC *printf()* that we developed). This sample program produces five lines of output (shown in Table 1). Output lines O1 and O2 result from executing line L3 of the code: the string *buf* hosting "*hello*", and the integer *i* has the value 1234. In addition, line L3 uses the format directive *%n* to write the character count (i.e., the number of characters printed up to that point) into the address of the corresponding integer variable. In this case, the length of "*string=hello¶data=1234¶*" is 23, so *Printf()* writes 23 to the integer *j*.

This value is printed out in line O3 of the program output. The format string vulnerability is caused by incorrect invocation of *Printf* in line L4, which directly uses *buf* as the format string (the proper usage should be *Printf("%s", buf)*).

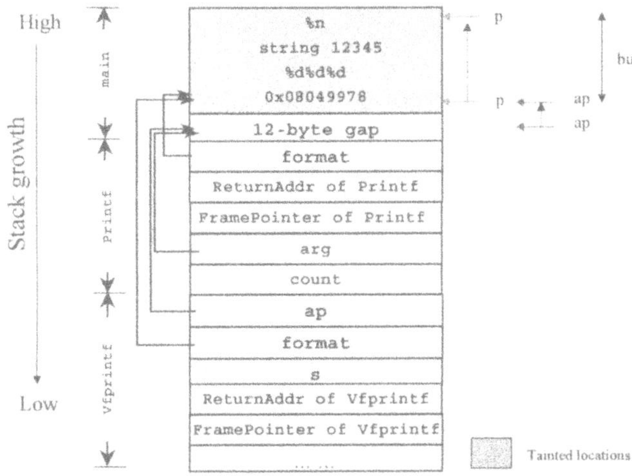

Figure 1: How to Overwrite the Global Integer *i*

We now show how an attacker can exploit this vulnerability. Let's assume that the attacker wants to corrupt an arbitrary memory location (e.g., the global integer *i*). In order to do this, he/she constructs an input string *buf* as given below (observe that the beginning of the input string corresponds to the address of global integer *i*):

\x78 \x99 \x04 \x08 %d %d %d '1' '2' '3' '4' '5' %n

The string is read by *scanf()* and passed to *Printf()*, which in turn, calls *Vfprintf()*. Just before Line L1 is executed, the stack layout is like the one in Figure 1. In *Vfprintf()*, there are two pointers: *p* is the pointer to sweep over the format string *buf* (from "\x78" to"%n"), and *ap* is the pointer to sweep over the arguments (starting from the 12-byte gap). The attacker deliberately embeds three "%d" directives in *buf* so that *ap* can consume the 12-byte gap and get to the word 0x08049978. A padding string "12345" follows the "%d" directives in order to adjust the character count. As we see in the program output line O4, the words in the 12-byte gap are printed as three integers followed by a padding string "12345". Eventually, when *p* arrives at the position of "%n" (i.e., the code line L1 is about to be executed), *ap* happens to arrive at the position of 0x08049978. Line L1 writes the character count *count* to the location pointed by **ap*. In this case, since the content in the location pointed by **ap* is the address of the integer *i*, the character count 31 is written to *i*. Note that this attack can overwrite any memory location, including locations containing return addresses or the global offset table of an application, which can result in the execution of the attacker's code.

We can view the above vulnerability as a consequence of pointer taintedness. In the above code (Table 1), the string *buf* is obtained from user input and is hence tainted (as indicated in Figure 1 as a grey area). When the pointer *ap* sweeps over the stack and points to *buf*, **ap* becomes tainted. *ap* is then dereferenced in Line L1, and the tainted value of **ap* is the target address of the write operation. This can lead to the corruption of an arbitrary memory location. Thus we see that pointer taintedness is the root cause of this vulnerability. Note that the pointer **ap* gets tainted because *ap* moves into the tainted memory locations, and there is no explicit assignment of a tainted value to **ap* in the C code. Hence a memory model is necessary to reason about the taintedness of **ap*. The next section defines the formal semantics of pointer taintedness using a memory model, and Section 5 shows how the semantics can be used to reason about security vulnerabilities in library functions.

4. SEMANTICS FOR POINTER TAINTEDNESS

Starting with the programming semantics of Goguen and Malcolm [3], this section proposes a formal semantics to reason about pointer taintedness in programs. The semantics proposed in [3] defines instructions, variables and expressions. We extend this semantics to include memory locations and addresses. Using the memory model, the notion of taintedness is incorporated into the semantics.

We define tainted data as: (1) data coming from input devices (e.g., by *scanf(), fscanf(), recv(), recvfrom()*), or (2) data copied or arithmetically calculated from tainted data. A tainted pointer is a pointer whose value (semantically equivalent to "data") is tainted. This definition can be formalized in equational logic using the Maude tool [4], which we used to reason about pointer taintedness.

In the semantics defined in [3], a *Store* represents the current state of all program variables. We extend this definition of a *Store* to be a snapshot of the entire memory state at a point in the program execution. The execution of a program instruction is defined as a function taking two arguments, a *Store* and an instruction, and producing another *Store*. There are two attributes associated with every memory location of a *Store*: *content* and *taintedness*. Accordingly, two operations, *fetch* and *location-taintedness*, are formally defined. The *fetch* operation Ftch(S,I) gives the content of the address I in store S; the *location-taintedness* operation LocT(S,I) returns a

Boolean value indicating whether the content of the specified address is tainted.

There is no notion of "variable" in this semantics. Any variable in a C program is mapped to a memory location addressed by the integer with the same name as the program variable. For example, the C program variable foo is mapped as a memory location addressed by the integer foo. We define the ^ operator to dereference an integer, i.e., to fetch the location addressed by the integer. Note that the address (a.k.a., the left value) of the C program variable foo is represented by the integer foo in the semantics; and the content (a.k.a., the right value) of the C program variable foo is represented by (^ foo). The expressions in the semantics are arithmetic operators (e.g., +, - and *) concatenating integers and integer dereferences. For example, expression 200+(^foo) represents "200 plus the content of the C program variable foo". Expression 200+foo represents "200 plus the address of the C program variable foo".

We define two operations – *evaluation* and *expression-taintedness* – for expressions, based on the *fetch* and *location-taintedness* operations. The *evaluation* operation Eval(S,E) gives the result of evaluating the expression E under store S; the *expression-taintedness* operation ExpT(S,E) indicates whether expression E contains any data from a tainted location, e.g., ExpT(S,(^foo)+2) indicates whether the expression (^foo)+2 contains any data from a tainted location, which is equivalent to checking whether the memory location addressed by foo is tainted. *Thus pointer taintedness is defined as a dereference of a tainted expression.*

Table 2 lists a set of axioms for the *evaluation* and *expression-taintedness* operations, and gives examples of applying the equations.

Table 2: Axioms of *Evaluation* and *Expression-Taintedness* Operations

	Axioms	Examples
1	Eval(S,I) = I	Eval(S,5) = 5
2	Eval(S, ^ E1) = Ftch(S, Eval(S,E1))	Eval(S, ^ foo) = Ftch(S,Eval(S,foo)) = Ftch(S,foo)
3	Eval(S, - E1) = - Eval(S, E1)	Eval(S, - 30) = - Eval(S, 30) = - 30
4	Eval(S, E1 – E2) = Eval(S, E1) - Eval(S, E2)	Eval(S, 3-2) = Eval(S, 3) – Eval(S, 2) = 3–2 = 1
5	Eval(S, E1 + E2) = Eval(S, E1) + Eval(S, E2)	Eval(S, 3+2)=Eval(S, 3)+Eval(S,2) = 5
6	Eval(S, E1 * E2) = Eval(S, E1) * Eval(S, E2)	Eval(S, 3*2)=Eval(S, 3)*Eval(S, 2) = 6
7	ExpT (S,I) = false	ExpT(S,5) = false
8	ExpT(S, ^ E1) = LocT(S,Eval(S,E1))	ExpT(S,^foo) =LocT(S,Eval(S,foo)) =LocT(S,foo)
9	ExpT(S, - E1) = ExpT(S,E1)	ExpT(S, -5) = ExpT(S, 5)= false
A	ExpT(S,E1-E2)=ExpT(S,E1) or ExpT(S,E2)	ExpT(S, (^foo)–2) = ExpT(S, (^foo)) or ExpT(S, 2) = LocT(S,foo) or false = LocT(S,foo)
B	ExpT(S,E1+ E2)=ExpT(S,E1)or ExpT(S,E2)	ExpT(S, (^foo)+2) = … = LocT(S,foo)
C	ExpT(S,E1*E2)=ExpT(S,E1) or ExpT(S,E2)	ExpT(S, (^foo)*2) = … = LocT(S,foo)

Lines 1-6 define how to evaluate an expression under store S. For example, line 1 indicates that the evaluation result of a constant I under store S is the constant I. Line 2 indicates that Eval(S,^E1) can be computed by first evaluating E1 under S, then applying *fetch* operation on the evaluation result. The semantics of arithmetic operations are defined in Lines 3-6.

Lines 7-C define the *expression-taintedness* operator. Note that the relationship between the *expression-taintedness* and *location-taintedness* operators is similar to the relationship between the *evaluation* and *fetch* operators. Line 7 indicates that an integer constant is not a tainted expression. Line 8 indicates that determining whether the expression ^E1 is tainted is equivalent to checking whether the location addressed by the evaluation result of E1 is tainted. Line B indicates that the expression E1+E2 is tainted if either E1 or E2 is tainted. The example of Line C shows that the expression (^foo) + 2 is tainted if and only if the location pointed to by foo is tainted, according to the equation in Lines 7 and 8.

Table 3: Semantics of Statements

Statement	**Semantics**
mov [E1] <- E2	Move the evaluation result of the expression E2 to the memory location addressed by the evaluation result of the expression E1
if T then P1 else P2 fi	If the condition T is true, Execute P1 otherwise execute P2
while T do P od	If the condition T is true, execute P, repeat until T is false

Table 3 gives the informal semantics of a subset of the supported statements. Their formal semantics are similar to the specifications given in [3], and are sufficient to analyze a wide variety of program constructs in the C language. However, they are not sufficient to faithfully model all C statements. For example, the program counter has not been defined in the semantics. So certain C statements, such as goto, break, continue, return and exit cannot be modeled, but it is relatively easy to extend the semantics for these also.

Formal specifications of statements other than the mov statement are fairly straightforward. Axioms defining mov statement semantics are shown in Table 4. The goal is to define the *fetch* (Ftch) and *location-taintedness* (LocT) operations after applying a mov instruction on store S.

Table 4: Equations Defining mov Statement Semantics

1	Ftch((S ; mov [E1] <- E2),X1) = Eval(S,E2) if (Eval(S,E1) is X1) .
2	Ftch((S ; mov [E1] <- E2),X1) = Ftch(S,X1) if not (Eval(S,E1) is X1) .
3	LocT((S ; mov [E1] <- E2),X1) = ExpT(S,E2) if (Eval(S,E1) is X1) .
4	LocT((S ; mov [E1] <- E2),X1) = LocT(S,X1) if not (Eval(S,E1) is X1) .

The semicolon operator in our notation represents the execution of an instruction on a store, which results in a new store. For example, (S;mov[E1]<-E2) is the store after executing mov[E1]<-E2 on store S. Line 1 indicates that if the expression E1 evaluates to X1 under store S, then when fetching the location X1 after executing the instruction mov[E1]<-E2, we get the evaluation result of E2 under store S. Line 2 indicates that if the expression E1 does not evaluate to X1 under S, then when fetching the location X1 after executing the instruction mov[E1]<-E2, we still get the content in the location X1 under S (i.e., before executing the instruction mov[E1]<-E2). Similarly, the LocT operation is defined for mov statement in Lines 3 and 4.

5. FORMAL REASONING ON POINTER TAINTEDNESS VIOLATIONS

This section presents pointer taintedness analysis for three common library functions based on the defined semantics, and extracts their associated security preconditions. The analysis identifies several known vulnerabilities, such as format string, buffer overflow and heap corruption vulnerabilities, thereby showing that pointer taintedness based reasoning is able to unify different kinds of vulnerabilities.

Our experience suggests that statements needing critical examination for pointer taintedness are typically indirect writes, where a pointer points to a target address to be written, e.g., the pointer p in *p = foo and memcpy(p,foo,10). Checking indirect write statements is important because these statements can result in two types of pointer taintedness violations. For example, in the statement *p = foo, (1) if the value of p is tainted, then data foo can be written to any memory location; (2) if p is a pointer to a buffer but points to a location outside the buffer, then the statement *p = foo can taint the memory location p points to, which may be a location of a return address, a function frame pointer or another pointer.

5.1 Analysis of *strcpy()*

A simple but interesting example is *strcpy()*, which copies a NULL-terminated source string to a destination buffer. The string manipulation functions, including *strcpy()*, *strcat()* and *sprintf()*, are known to cause a significant number of buffer overflow vulnerabilities. Our formal reasoning extracts security preconditions from the implementation of *strcpy()*. The source code of *strcpy()* and its formal representation are given in Table 5.

As the only indirect write operations in the source code are in Line 1 and Line 2, it is enough to prove the three theorems listed in Table 6. We assume that the NULL-terminator (i.e., the character '\0') of the source string src is at the location (src + srclen), and that the size of the buffer dst is dstsize. Theorem NV1 ensures that before Line L1, the content of the variable dst is not tainted; Theorem NV2 ensures that for any address A outside the buffer dst (from dst to dst+dstsize), the taintedness of location A after Line L1 is same as the taintedness of location A before Line L0. NV3 is similar to NV1, but proves the property for the memory state before Line L2 is executed.

Table 5: Source Code and Formal Semantics of *strcpy()*

Source code	Formal semantics
`char * strcpy (char * dst, char * src) {`	
`      char * res;`	
`0:    res =dst;`	`L0: mov [res] <- (^ dst) ;`
`      while (*src!=0) {`	`    while (~((^ ^ src) is 0)) do`
`1:     *dst=*src;`	`L1:     mov [^ dst] <- (^ ^ src) ;`
`        dst++;`	`        mov [dst] <- (^dst + 1) ;`
`        src++;`	`        mov [src] <- (^ src + 1)`
`      }`	`    od ;`
`2:    *dst=0;`	`L2: mov [^ dst] <- 0 .`
`      return res; }`	

Table 6: Theorems to Prove for Function *strcpy()*

```
Theorem NV1: If S1 is the store before Line L1, then
                    LocT(S1,dst) = false
Theorem NV2: If S0 is the store before Line L0, and S2 is the store after Line L1,
   ∀ A ∈ INT •   A < Eval(S0, ^dst)  or Eval(S0, ^dst+dstsize) ≤ A  =>
                    LocT(S2,A) = LocT(S0, A)
Theorem NV3: If S3 is the store before Line L2, then
                    LocT(S3,dst) = false
```

Table 7 gives a set of security preconditions extracted in the process of proving the theorems. Among the four preconditions, Condition 4 is known because of the large number of buffer overflow vulnerabilities caused by string manipulation. This condition has already been documented on Linux MAN page of *strcpy*. Condition 2 indicates the scenario of overlap between src and dst and is examined further in Section 6.2. Violation of Condition 3 may occur when a program miscalculates the location of a stack buffer, causing the function frame of *strcpy()* to be covered by the buffer and is discussed in Section 6.1.

Table 7: Sufficient Conditions to Ensure the Validity of Theorems NV1 – NV3

1. Initially, the location of dst is not tainted.
2. The buffers src and dst do not overlap in such a way that the buffer dst covers the NULL-terminator of the src string.
3. The buffer dst does not cover the function frame of *strcpy()*, which consists of the locations &dst, &src and &res.
4. srclen < dstsize

5.2 Analysis of *Free()*

We implemented a binary buddy heap management system including function *Malloc()* and *Free()*. The memory block to be freed is pointed to by pointer FreedBlock. The binary buddy heap management algorithm requires the deallocated memory block FreedBlock to be merged with its buddy block if the buddy block is also free. The pointer BuddyBlock points to the buddy block. FreedBlock and BuddyBlock are structs of type HEAP_BLOCK as shown in Table 8. The Size field indicates the size of the memory chunk. The busy field indicates whether the memory chunk is free. Fields Fwd and Bak are pointers to maintain a doubly-link list of free memory chunks.

Table 8: Indirect Write Statements in *Free()* Source Code

```
typedef struct _HEAP_BLOCK {
    int         Size;       // The size of the block.
    int         Busy;       // Is this block busy?
    struct _HEAP_BLOCK * Fwd,* Bak; // List to the free blocks of the same size
} HEAP_BLOCK;
```

There are three lines in the *Free()* function where indirect write operations are performed. Six pointers are involved in the operations, including FreedBlock, BuddyBlock, FreedBlock->Fwd, FreedBlock->Bak, BuddyBlock->Fwd and BuddyBlock->Bak. Table 9 states the theorem to be proved for conditions guaranteeing that pointers are not tainted. It is assumed that the offset of the Fwd field in the HEAP_BLOCK structure is 2, and that the offset of the Bak field is 3. Theorem NV1 ensures that none of the six pointers is tainted before executing any indirect writes.

Table 9: Theorems to Prove for Function *Free()*

Theorem NV1: If S is the store before executing the indirect writes, then	
ExpT(S, (^ FreedBlock)) = false and	//FreedBlock is not tainted
ExpT(S, (^BuddyBlock)) = false and	//BuddyBlock is not tainted
ExpT(S, ^((^ FreedBlock) + 2)) = false and	//FreedBlock->Fwd is not tainted
ExpT(S, ^((^ FreedBlock) + 3)) = false and	//FreedBlock->Bak is not tainted
ExpT(S, ^((^ BuddyBlock) + 2)) = false and	//BuddyBlock->Fwd is not tainted
ExpT(S, ^((^ BuddyBlock) + 3)) = false	//BuddyBlock->Bak is not tainted

The process of proving the theorems extracted a set of formally specified conditions that guarantee the validity of Theorem NV1.

Table **10** describes the conditions. The function *Free()* is safe to be called when the caller function can guarantee these conditions. Violations of condition 1 are unlikely to occur, and the same is true for condition 7. Violations of condition 6 cause the classic double-free errors, and violations of condition 3 and 4 lead to the popular heap buffer overflow vulnerability. An example illustrating violation of condition 2 is presented in Section 6.3.

Table 10: Sufficient Conditions to Ensure the Validity of Theorem NV1

1. The memory range of the heap and the memory range of the current function frame do not overlap.
2. Immediately before *Free()* is called, FreedBlock points to a location on the heap.
3. Immediately before *Free()* is called, the Fwd and Bak links of the block of FreedBlock are not tainted.
4. All free-chunk double-linked lists are within the heap range, i.e., no Fwd or Bak links points to any location outside the heap.
5. No Fwd or Bak pointers in any free-chunk double-linked list are tainted.
6. Before *Free()* is called, FreedBlock is not linked in any free-chunk list.
7. If BuddyBlock is free, then BuddyBlock is linked in a free-chunk double-linked list.

5.3 Analysis of *Printf()*

We implemented a function *Printf()*, similar to the LibC function *printf()*, except that *Printf()* calls its child function *Vfprintf()*, which is a simplified version of LibC function *vfprintf()*. *Vfprintf()* implements the format directives *%%*, *%d*, *%s* and *%n*. The total length of *Vfprintf()* is 55 lines. Pointer p is used to sweep over the format string format. The argument list is swept over by pointer ap.

There are only two lines of indirect write operations in the function (Table 11). Line L1 is to get the last digit of data and save it in the nth position of the buffer buf. Line L2 is to assign the character count to the memory location pointed by the current argument. Note that ap is the argument list pointer pointing to the current argument.

Corresponding to the two indirect write operations, we need to prove the theorems given in Table 12. Theorem NV1A ensures that before executing code in Line L1, the memory location containing the variable n is not tainted. Theorem NV1B ensures that after Line L1, the memory location buf+10 is not tainted. Theorem NV1A and NV1B can be easily proved by the theorem prover. Theorem NV2 ensures that before Line L2, the expression (^ ^ ap) is not tainted, i.e., the memory location pointed by (^ ap) is not tainted, i.e., the memory location pointed by the content of variable ap is not tainted. The preconditions extracted in the process of proving Theorem NV2 are given in Table 13.

Table 11: Indirect Write Statements in *Vfprintf()* Source Code

```
L1:  buf[n]=data%10+'0';
L2:  *(int*)ap = count ;
```

Table 12: Theorems Need to Prove for *Vfprintf()*

Theorem NV1A: If S1 is the store before executing Line L1, then ExpT(S1,(^ n)) = false

Theorem NV1B: If S2 is the store after executing Line L1, then LocT(S2,(buf + 10)) = false
Theorem NV2: If S3 is the store before executing Line L2, then ExpT(S3, (^ ^ ap)) = false

Table 13: Sufficient Conditions to Ensure the Validity of Theorem NV2

1. ap never points to any location within the current function frame.
2. *ap never points to the location of variable ap, i.e., *ap != &ap.
3. Suppose the memory segment that ap sweeps over is called *ap_activitiy_range*, no locations within *ap_activitiy_range* are tainted before *Vfprintf()* is called.
4. *ap never points to any location within *ap_activitiy_range*.

The four conditions form a set of sufficient conditions, which if satisfied, guarantee that there is no pointer taintedness situation in the analyzed version of *Vfprintf()*. Format string vulnerabilities do not satisfy condition 3 (Table 13). As illustrated in Figure 1, the tainted data (word 0x08049978) is located in the activity range of ap, i.e., ap points to this data. For the other three conditions, we are currently unaware of any existing applications violating them.

6. EXAMPLES ILLUSTRATING VIOLATIONS OF LIBRARY FUNCTIONS' PRECONDITIONS

In the previous section, we have given a significant number of preconditions for common library functions. Not all of them are likely to occur in real application code. In this section, we give possible scenarios (constructed examples) in which some of the preconditions detailed in the previous section are violated, and explain how an attacker can exploit them. To the best of our knowledge, these vulnerabilities have not been reported in any real application or described in the literature.

6.1 Example of *strcpy()* violation – condition 3

Condition 3 in Table 7 for *strcpy()* states that the buffer *dst* does not cover the function frame of *strcpy()*, which consists of *dst, src* and *res.* Otherwise it is possible to overwrite the stack frame of *strcpy()* and modify the address of the *dst* string. Since *strcpy()* can write to the location *(*dst),* this can be used to write to any memory location, including function pointers, and hence transfer control to malicious code.

Consider the code sample in Table 14a, in which *buf* and *input* are allocated on the stack in the function frame of *foo()*. The string *input* is obtained from the user and passed as the *src* argument of *strcpy()*. The *dst* argument of *strcpy()* is *buf + index*, where *index* is computed by subtracting the length of *input* from the end of the buffer *buf*. After *strcpy()* is called, the stack frame looks as shown in Table 14b. Assume that the attacker enters an input string longer than 20 bytes as input. Since the input buffer has a size of 100 bytes, this may not cause buffer overflow. However, this makes the value of *index* computed to become negative, which in turn makes *dst* point to a stack location before *buf* and in the function frame of *strcpy()* (thereby violating the pre-condition). In the above example, setting the *index* to (-16) makes *dst* point to the location of itself on the stack. The *strcpy()* code then writes to the location of *(*dst)*, thereby overwriting *dst* itself. Subsequent writes to *(*dst)* can then modify the contents of the location pointed to by this new value of *dst*. This allows the attacker to write any value to any memory location, including sensitive locations such as function pointers.

The functionality of the code shown in Table 14a is to push data to the end of a buffer. We believe it is possible that applications require such a functionality. For example, a program may need to copy data at the end of a buffer and prefix headers in front the data. The pointer arithmetic shown in Table 14a is an efficient means of implementing such an operation, so we argue that the sample code demonstrates a possible scenario in real applications.

Table 14: Violation of condition 3 of *strcpy()*

a) Sample Code

```
void foo( ) {
  int index;
  char input[100];
  char buf[20];
  scanf("%s", input );
  index = 20 - strlen( input );
  strcpy( buf + index , input );
}
```

b) Stack Status

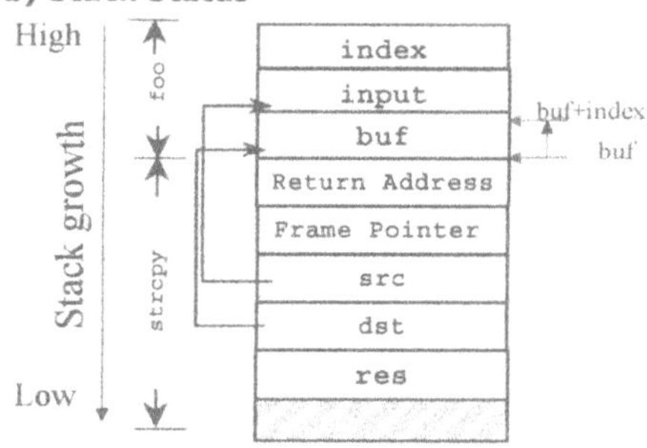

6.2 Example of *strcpy()* violation – condition 2

In Table 7, condition 2 of *strcpy()* states that *src* and *dst* do not overlap in such a way that *dst* covers the null-terminator of *src*, otherwise the null-terminator of *src* string gets overwritten and the program can go into an infinite loop. This can happen in two ways: by a buffer overflow error or by an inadvertent free error, as illustrated in Table 15a and Table 15b, respectively.

Table 15: Examples of *strcpy()* condition 2 violations

a) Buffer Overflow Error	b) Inadverent Free Error
char* src = malloc(20);	src = malloc(40);
char* dst = malloc(20);	snprintf(src, 30, "some string of 30 or more characters");
sprintf(src,"string with > 20 characters");	free(src);
strcpy(dst, src);	foo = malloc (10);
	dst = malloc(20);
	strcpy(dst, src);

In the first piece of code (Table 15a), two buffers are allocated on the heap and one of them is overflowed. This buffer is then passed as the *src* argument to *strcpy()*, and the other one as the *dst* argument. Upon running this code multiple times[3], we found that the memory manager consistently allocated nearly consecutive, successive memory addresses to *src* and *dst* respectively. As a result, when the *src* buffer is overflowed, its contents spill into the *dst* buffer, causing it to overlap with the *src* string and cover the null-terminator, leading to a violation of the pre-condition.

The *src* and *dst* arguments can also overlap if the destination buffer is allocated from some portion of the source buffer. This situation is illustrated in Table 15b. Here *src* is first allocated on the heap and then freed, which returns the *src* buffer to the free pool. When *malloc()* requests are made subsequently for *foo* and *dst*, the memory manager reuses the block most recently returned to it, namely the *src* buffer, for allocating the buffers *foo* and *dst*. When *strcpy()* is called, *src* and *dst* overlap in such a way that *dst* covers the null terminator of *src*, which is a violation of the pre-condition. In real codes, this can happen as a result of using a buffer that is freed on an infrequently executed path and may not be uncovered during testing.

6.3 Example of *free()* violation – condition 2

Condition 2 of the *free()* function in

Table **10** states that the pointer passed to *free()* must be within the heap range. This arises from the fact that the *free()* function itself does not perform this check. When a block is freed, the *free()* function checks for an integer value at the beginning of the block, which represents the size of the block to be freed. If it finds such an integer, it does the free, irrespective of whether the block is on the heap or not.

[3] We tried it with glibc on x86-linux and Sun Solaris platforms. Our results indicate that this is not an OS or platform specific phenomenon, but a feature of glibc.

Consider what happens when a local buffer on the stack is passed to the *free()* function in Table 16. In this code, the local array *buf* of function *foo()* is passed to the function *print_str()*, which checks if the length of the string passed to it is more than the value *n* specified by the user, and if so, frees the buffer. The pointer p which is freed by *print_str()* is aliased to *buf*, which is allocated on the stack in the function frame of *foo()*, leading to a violation of the pre-condition. Since this happens only when the user enters a string of more than 50 characters, it may not be uncovered while testing. In this example, the integer *i*, which is a local variable of *foo()* is present on the stack at the beginning of the block *buf*. The *free()* function assumes that this is the size of the buffer *buf* and attempts to deallocate a block of that size. Since the user also supplies this value, it is possible to free a block of any arbitrary size on the stack, and overwrite the contents of any memory location.

Table 16: Violation of condition 2 of *free()*

void foo() {
    char buf[100];
    int i;
    scanf("%d", &i);
    scanf("%s", buf);
    print_str(buf , 50); }
	void print_str(char* p, int n) {
	    if (strlen(p) > n) {
	        free(p);
	        return;
	    }
	    printf(stdout, "%s", p); }

7. CONCLUSIONS AND FUTURE DIRECTIONS

This paper is motivated by a low level analysis of various security vulnerabilities, which indicates that the common cause of the vulnerabilities is pointer taintedness. To reason about pointer taintedness, a memory model is needed. The main contribution of this paper is the equational definition of a memory model, which associates the *taintedness* attribute with memory locations rather than with program symbols. Pointer taintedness analysis is applied on several C library functions to extract security preconditions. The results show that pointer taintedness analysis can expose different classes of security vulnerabilities, leading us to believe that pointer taintedness provides a unifying perspective for reasoning about security vulnerabilities. We plan to extend this work in a number of ways: (1) Reduce the amount of human intervention in the theorem proving tasks; (2) Define semantics for other C statements, such as goto, break and continue to make the technique more widely applicable; (3) Incorporate this technique into compiler-based static checking tools to extend to large applications.

ACKNOWLEDGMENTS:

We thank Jose Meseguer for his insightful suggestions on the project, and Tammi O'Neill for her careful reading of an early draft of this manuscript.

REFERENCES

[1] D. Evans and D. Larochelle. Improving Security Using Extensible Lightweight Static Analysis. In IEEE Software, Jan/Feb 2002

[2] B. Chess. Improving Computer Security Using Extended Static Checking. IEEE Symposium on Security and Privacy 2002

[3] J. A. Goguen and G. Malcolm. Algebraic Semantics of Imperative Programs. MIT Press, 1996, ISBN 0-262-07172-X

[4] M. Clavel, F. Durán, S. Eker, P. Lincoln, N. Martí-Oliet, J. Meseguer and C. Talcott *The Maude 2.0 System.* In Proc. Rewriting Techniques and Applications, 2003, 2003.

[5] J. Xu, Z. Kalbarczyk and R. K. Iyer. Transparent Runtime Randomization for Security. To appear in Proc. Symposium on Reliable and Distributed Systems, 2003.

[6] D. Wagner, J. Foster, E. Brewer, and A. Aiken. A First Step Towards Automated Detection of Buffer Overrun Vulnerabilities. Network and Distributed System Security Symposium (NDSS2000).

[7] U. Shankar, K. Talwar, J. Foster, and D. Wagner. Detecting Format String Vulnerabilities With Type Qualifiers. 10th USENIX Security Symposium, 2001.

[8] C. Cowan, C. Pu, D. Maier, et al. Automatic Detection and Prevention of Buffer-Overflow Attacks. 7th USENIX Security Symposium, San Antonio, TX, January 1998.

[9] A. Baratloo, T. Tsai, N. Singh, Transparent Run-Time Defense Against Stack Smashing Attacks, Proc. USENIX Annual Technical Conference, June 2000.

[10] S. Chen, Z. Kalbarczyk, J. Xu, R. K. Iyer. "A Data-Driven Finite State Machine Model for Analyzing Security Vulnerabilities". in IEEE International Conf. on Dependable Systems and Networks, 2003.

[11] Introduction to equational logic. http://www.cs.cornell.edu/Info/People/gries/Logic/Equational.html

[12] S. Chen, K. Pattabiraman, Z. Kalbarczyk, R. K. Iyer. Formal Reasoning of Various Categories of Widely Exploited Security Vulnerabilities By Pointer Taintedness Semantics (Full Version). http://www.crhc.uiuc.edu/~shuochen/pointer_taintedness

MEETING THE GLOBAL CHALLENGES OF SECURITY INCIDENT RESPONSE

Vijay Masurkar,[1] Simone Fischer-Hübner,[2] and Morton Swimmer[3]

[1]*Sun Microsystems, Inc., Burlington, MA, USA;* [2]*Karlstad University, Karlstad, Sweden;*

[3]*IBM Research Laboratories, Zurich, Switzerland*

Abstract: Responding to computer security incidents has become a critical function within an information technology program of any enterprise. These incidents are threats not only to computing equipment but also to the stability of establishments, such as small to large governments or utilities serving large populations. New types of security-related incidents emerge frequently and massive activities take place across the globe to mitigate the violations of security policies or recommended security practices. In spite of the concerted efforts from many organizations, complete solutions are still lagging behind. Proactive or predictive research and planning activities are on the rise in many industrialized nations. However, they seem to fall short on coping up with an unexpected global incident before incurring a substantial damage. What can the global security-aware organizations and communities do? This paper is intended to help set the stage for a panel discussion to be chaired by the first author with the members of the IFIP WG9.6/11.7, "IT Misuse and the Law".

Keywords: Security, Incident, Attack, Response, Global, Risk, Vulnerabilities, Legislation

1 INTRODUCTION

Since the Internet Worm of 1988 (Spafford, 1988), the world has seen quite an acceleration of security attacks as the Internet proliferated globally. New types of security-related incidents emerge on a frequent basis and massive activities take place across the globe to mitigate the violations of security policies or recommended security practices. Most local and national governments recognize the need to protect their infrastructures and citizens (West-Brown et al, 1999). In addition, many also are aware that it is not enough to address such incidents only at their local level. Besides technological challenges, the socioeconomic and legal challenges are crossing national boundaries. Let us take look at the current state of security incidents and response capabilities, and briefly review the factors that help prevent or mitigate risks arising from these incidents. This review is intended to highlight the points for starting the proposed panel discussion.

2 CURRENT STATE: SECURITY INCIDENT RESPONSE CAPABILITIES

There are many known security threats to computing or networked resources. Perhaps the most common are port scanning, denial-of-service (DoS), IP address spoofing and sniffing, and injection of viruses, trojan horses or worms. With readily available tools and information on the Internet, finding specific vulnerabilities for a system is not very hard. Publicly available vulnerability assessment tools such as Nessus (2004) can be used to quickly identify where these threats are likely to gain a foothold and cause problems. Hackers also use port scanning tools such as Nmap (2004) to search through various network devices looking for ports that are open or enabled. They can compare this information to known, published vulnerabilities to see if they can gain access to the targeted devices. A person with a malicious intent can find information about vulnerabilities from one of the several organizational sites such as CVE (2004), Common Vulnerabilities and Exposures, or IAVA (2004), Information Assurance Vulnerability Alerts, that provide full disclosure. The hacker can then trace a vulnerability number from CVE to a source code clip on the public website such SecurityFocus (SecurityFocus, 2004) or via a general Internet search. He or she can then find corresponding detailed instructions on the exploit on the SANS (2004) or some other public website. This can all be done in minutes.

Most industry experts admit that a system or an environment of computer systems and networks that is hundred percent hacker-proof is yet to be invented. And, for the foreseeable future, IT users can best protect their assets by managing the evolving vulnerabilities that continue to threaten their environments (Masurkar, 2004). In spite of the the concerted efforts from many organizations, complete solutions are lagging behind. Traditionally, malicious code and backdoors (such as *rootkits*) and enabling vulnerabilities due to errors or negligence in system administration tasks caused exposures. In recent years, hackers have exploited multiple vulnerabilities at a time and have successfully launched multi-vector viruses.

These pose *blended threats*. It started with *Nimda* and *Code Red*, then *Slammer* and others, and, recently, *Blaster,* that caused widespread damage.

With the added advantage of knowing an organization's internal workings, insiders commit crimes that have been known to be on the rise for some time. This ability of malicious internal users to gain additional privileges on any system is not uncommon. Unfortunately, it is often overlooked by IT administrators and managers. What happens when an authorized user, such as an employee or a contractor working on the internal premises, turns malicious? She or he will gain access to another user's account through various possible means; via trusted relationship, poor password management, sessions left unlocked, to name a few. Once on a system, malicious users can use buffer overflow and other forms of attacks by compiling and executing exploit code available on the Internet to gain root privileges. Then the next step is to use a sniffer, install backdoors, maintain and gain additional access privileges using rootkits, and launch subsequent attacks.

Masurkar (2003) has compiled a list of worldwide organizations fighting security incidents. FIRST (2003) and CERTs from many countries continue to progress, yet struggle to coordinate the information regarding past incidents, trends and models to deal with modus operandi of attackers. To a vast majority of small to medium enterprises, the benefits don't seem as immediate. As against this, larger, multi-national enterprises can take advantage of CERTs. But these enterprises pose more ways to be vulnerable than smaller ones because of the multitudes of access methods into their intranets, DMZ servers, partner networks or Internet ASP sites.

As data centers have grown larger and more complex, upgrading hardware and software and managing security vulnerabilities has become a common tactic that regularly borders on strategy. What we mean is that due to the lack of any longer term solution IT organizations have a notion that, for the foreseeable future, it is an effective way to mitigate business risk. The same IT organizations, however, would not deny that it is an increasingly resource intensive tactic. Some vendors recommend patch management strategies (Radhakrishnan, 2003) to mitigate risks to critical assets by classifying business environments. Certainly, clients can benefit from this in deploying change management. But this doesn't solve the problem of segregation and analysis of massive amounts of security event data. Beyond segregation, the data also needs to be correlated so business decisions can be made to mitigate any perceived risk.

To help reduce cost and complexity, Managed Security Services providers (MSSP) sell user-friendly scalability management solutions (Outsourcing MSS, 2004) called *Managed Security Services* (MSS). More and more organisations are turning to MSSPs for a range of security services to reduce costs and to access skilled staff whose full-time job is security.

MSS-based vulnerability scanning matured in 2003, to be followed by intrusion detection in 2004, security monitoring and response in 2004, and authentication and administration by 2004 and 2005 (King, 2001). But the downside to all of this is that keeping up with the newer security devices and appliances and continuously developing attack signatures are daunting tasks for these vendors. Secondly, any organization relying on IT needs solid information security policies and practices. This is a challenge for many. Outsourcing components of information security should be evaluated as a solution, but the business must always retain responsibility, - thus underlining the importance of understanding business and regulatory implications of outsourcing security. Also, users examining managed security services should realize that multiple providers may be warranted, depending on the breadth of function outsourced. Lastly, in spite of the significant strides in recent years, customer-vendor trust (a factor that can be an issue for any service provider) remains a significant hurdle in delivering managed security services. This can slow down the fight against security incidents made affordable to some willing enterprises by the MSS business model.

There is a serious lack of training, awareness and standards across many IT organizations. Even if there is training provided in certain few ones, practitioners and technologists lack authority to make policy decisions. Aside from the global issue of *"digital divide"* that centers around the many poor and developing countries (such as in Africa) who are short on keeping up with the IT advancement, this lack of training, standards and authority has had a counter-effect on the fight against security attacks. It has not only encouraged negligence on the part of personnel who maintain information systems, but has promoted ad hoc approaches to incident response and change management. There are a few noteworthy efforts that address this issue. OCTAVE (2003) is one example. CERT-USA has introduced OCTAVE and its training, but OCTAVE as well as other such methodologies have yet to take hold in the industry.

3 FUTURE GLOBAL THREATS AND CHALLENGES

The global threats of the future are not simply from the hackers but also could come from international terrorists. Coordinated attacks are likely to be planned that may target power grids, nuclear reactors, water purification systems and other critical infrastructures that are the foundations of today's established societies. Secondly, the speed of the spreading of the attacks, such as viruses, will increase significantly. Dealing with high speed networks that are carriers of fast spreading viruses and worms, faster and massive scale coordinated attacks could be used as a weapon to bring quick destruction. The Distributed DoS (DDoS) attacks prevented some of the biggest Internet vendors from functioning few years ago. That certainly sent a warning that, in the near future, the Internet could be a strong catalyst for even a bigger orchestrated crime.

Distributed intelligent, wearable (on-person) devices containing advanced sensing and communication capabilities hold promise for many societal benefits. But as these intelligent devices take hold, they will give rise to opportunities for miscreants to reenter, "camp out" or inject malware in ways unlike ever before (Duquenoy and Masurkar, 2004). Their ubiquity, mobility and ability to distribute information pose new challenges in the area of security, privacy and ethics. Extreme care is needed in designing them to prevent the malicious activities.

For many years to come, if we have to address Internet terrorism, increasing and highly efficient cooperation and communication between incident response teams will have to remain as a high priority for worldwide organizations such as United Nations that promote cooperation among nations. FIRST is currently carrying that responsibility, yet some aspects need progress. In particular, it is known that key incident information is not always conveyed effectively between response teams such as CERTs and incident response standards are lagging.

To a great extent, the information society can progress when IT security issues are adequately addressed. IT Security is one of the *eEurope 2005* initiative's six policy priorities. For supporting eEurope 2005 security policies, the European Network and Information Security Agency (2004) was set up to enhance the capability of the Community, the Member States and consequently the businesses to prevent, address and respond to computer security problems. It seeks to achieve an IT security culture within Europe. The tasks of ENISA will focus on advising the European Commission and the Member States on information security and in their dialogue with industry to address security-related problems, promote risk assessments research, deliver training and collaborate between public and private industries. How will ENISA maintain cooperation with the existing CERTs in some of the advanced European countries? And, with FIRST and CERTs across the world?

Survivability, in this context, is the ability of a computing system to provide essential services in the presence of attacks and failures, and recover full services in a timely manner. Some research efforts are focused on establishing new methods for risk assessment and by developing engineering technologies for analysis and design of survivable systems as at CERT-USA (Ellison et al, 2002) . In spite of difficulties in funding, furthering the research efforts in threat and risk assessment is important. One group of researchers (Rathmell et al, 2000) claims that a deeper understanding of the potential threat actors, combining technical and psycho-social aspects, may increase the probability of developing effective attack indicators. Thereby improving prospects for substantive warning before the threat materializes into a damaging incident. Such approaches require compilation of multi-source data (which needs to be up-to-date) and establishment of a monitoring regime limited to certain hacker groups and cyber-terrorists.

For any enterprise, meeting the legal and regulatory compliance is a necessary goal. To comply with the legislation and, when necessary, to make use of law to bring justice to computer criminals are difficult challenges. Obviously, legal activities can consume critical human and financial resources. Citizens' rights are sometimes overlooked when new laws come into play; so this is a serious consideration for the governments. In USA, for example, following the "*9/11*" disaster, the USA Patriot Act (2001) enacted by the President gives full authority to government agencies to investigate individuals without search warrants. Canada has also passed legislation following the events of 9/11; for example, Bill C-36, the Anti-terrorism Act, which received Royal Assent (2001). These laws can certainly help to swiftly investigate computer criminal activity, but are also seen by many citizens as the violation of their fundamental right to privacy.

Consider also that legal and technical personnel need to understand and be aware of each other's needs. Governments around the globe are continuously introducing stricter laws to protect the enterprise as well its clients and end users. So, here, the challenges lie in bringing the awareness and practical training of these laws and their implications to the incident response teams. Are there any simple answers to questions in any computer or Internet criminal case? For example: How to prosecute international computer criminals and bring them to justice?

4 ADEQUACIES OF LEGISLATIONS & REGULATIONS

Most industrial countries have computer crime legislation in place that are criminalising cyber attacks such as hacking, denial of service attacks or malware attacks. The question as to how much the computer criminal law can actually help to combat IT Misuse has been much debated (Brunnstein et al. 1995). It has also been a question of high interest for the IFIP working group 9.6/11.7, *"IT Misuse and the Law"*.

Computer criminal law could, alongside its criminalisation function, has perhaps the more important objective of acting as a deterrent. Besides, it can stand as an ethical marker and can have an educational function. However, it must be noted that, in general, criminal law is the strongest sanction available to states and carries considerable stigmatising effect. Hence, we should be careful not to extend this powerful approach to cases where its use is unnecessary or inappropriate (Wasik, 1994). Secondly, hacking and malware-related crimes are often committed by young people that have social problems. They strive for attention and will hardly be deterred by legislation.

Problems in the prosecution of computer crimes arise because criminal acts are often either not detected or not reported by organisations which are concerned about the organisation's reputation. Furthermore, prosecutors (investigation officers, judges, and attorneys) are often lacking specific technical knowledge. Besides, detailed digital evidence or traces are often missing. In relation to the number of malware attacks, there have been only a few successful prosecutions of malware-related crimes. However, if there is no effective computer law enforcement in place that finally leads to punishment because just the computer crime laws by themselves can hardly have much of a deterrence and crime prevention function. Most computer crime legislations are implementing a general approach to regulate malware-related crime in which regulations are formulated to cater to a broad spectrum of criminal acts, including malware as a subset (e.g. offences against integrity of computer data and systems, computer sabotage).

Ten years ago, it had been discussed whether we needed new or improved laws to deal more aggressively with the virus threat. Some countries had introduced new regulations to apply specifically to various aspects of virus-related criminal acts, such as the generating, writing, promoting, offering, making available and circulating of viruses (e.g. Art. 144.2 Swiss Criminal Code). This specific approach to malware-related crime was enacted to fight the publishing of virus code or the selling of virus construction sets. However, today, malware is most effectively spread as worms, mass mailing viruses or via spam, and, consequently, distribution of virus code via publishing or selling plays an insignificant role.

Today, cyber attacks are usually trans-national computer crimes and an international cooperation and harmonisation of computer criminal law and regulations are necessary. The Cyber Crime Convention of the Council of Europe [CoE 2001], which was opened for signature in November 2001, is an important international instrument for combating trans-border cyber crimes. The CoE Conventions deals with cyber crimes directed against information and communication system, computer fraud and computer-forgery, copyright-related offences and content-related offences. It provides legal instruments for international cooperation and mutual assistance for fighting cyber crime.

On April 5, 2002, hackers accessed a California State payroll database containing sensitive personal financial information belonging to approximately 260,000 state employees. Legal and regulatory developments since the proposal of *SB 1386* became effective in July 2003 (California Senate Bill 1386, 2002). One of the primary motivations for many companies is avoiding exposure to the uncapped civil action authorized by the law: *"Any customer injured by a violation of this title may institute a civil action to recover damages."* The law's intent is to alert people, whose personal data has been compromised to the danger of identity theft, in a timely way. But this is not enough. First, there needs to be a legislation at the U.S. federal level, and, secondly, other states need to follow California's lead.

There are more U.S. laws to comply with; HIPAA - Health Insurance Accountability and Portability Act (1996), GLBA - the Gramm-Leach-Bliley Act (1999), Sarbanes-Oxley Act (2002), to name a few prominent ones. These laws require enterprises to protect sensitive data such as customer records. Without compliance, companies may face heavy fine, bad publicity and other legal ramifications. According to these regulations, companies must identify, authorize and track user activities. Numerous other related laws exist; for example, USA Patriot Act of October 2001. Then there are Basel II, German KonTrag and others to discipline financial and risk management. They all must be kept in perspective and updated ,- which is a challenge for the law makers. But, also, this legislative burden is seen as overwhelming by most profit-oriented enterprises. So, the question comes up: how can governments promote legislation in the industry in a cooperative way?

An internationally harmonised approach is needed for fighting spam, which has become a major privacy and security problem. The EU has implemented a restrictive opt-in approach for regulating spam in the EU Directive on privacy in the electronic communication sector (EU-Directive, 2002), which disallows to send unsolicited emails to users unless they have given their prior consents or are in a customer relationship with the senders. However, under the U.S. Law, "CAN-SPAM Act" (2003), no prior permission is required for sending commercial messages as long as the receiver has the chance to "opt-out" of receiving future message from the same sender. It has been criticised that the U.S. legislation is not going far enough and could even make matters worse by approving spam that follows certain rules. Since approximately eighty percent of spam is believed to be sourced from North America, it is difficult to successfully fight spam in Europe if the U.S. legislation is following a weaker approach. The EU has called on the OECD (2004) to set up international efforts to fight spam. How can global security-aware communities help OECD to aggressively pursue this issue?

5 ORGANIZATIONAL DEFICIENCIES

Even though an organization may view a particular attack as its own problem, it is far more likely to be a piece of a larger puzzle. The attack in question is probably not aimed at that one particular (which is taken to mean corporation, academic institute or any other form of organization) but at achieving a certain goal, for instance enhancing the attacker's monetary situation. Phishing spam is an example where neither viewing the attack from the point of view of the account holder nor the monetary institute helps. Instead the entire picture needs to be taken into consideration. Global terrorism is an analogy here; the global threat cannot be dealt locally.

It was obvious right from the first major Internet incident, the Internet Worm, that coordinated incident response was needed. Therefore, CERTs were created at a per-organization and sometimes at country level to respond. To enhance the cooperation between these localized organizations, the FIRST was also founded of which many of the CERTs are members (FIRST, 2003). These help establish vital person-to-person contacts that can be useful in an emergency (Brownlee et al, 1998). However, is it enough? With so many fires to put out, shouldn't some of the communication be handled automatically and with automatic responses as was done in the Digital Immune System (Kephart et al, 1998)? For this to happen, the trust between the individuals would have to be extended to the mechanisms put in place. There may be legal issues to contend with too. On a technical level, even though the core protocols of the Internet are fairly robust, will that be enough to facilitate communication while the same network is under attack? Perhaps a back-channel network between CERTs would be on order that would allow automated alerts and responses.

The larger an organization's network grows, the more unmanageable it becomes. Even when the core network devices are well understood, the client and server machines are not. When an alert is registered for, say, an IDS system, claiming it has logged an attack package from one machine to another, it is not at all clear if the targeted the machine was vulnerable to the attack or not. Was the appropriate operating system, patch level or server software installed and active that displayed the exploitable vulnerability? Most exploit attacks are very finicky about the exact configuration of the target. Of course, the attack may not even have reached the target because it was filtered or dropped. So, the alert does not carry as much weight as one would like to think.

A good system administrator will want to pursue any likely alert, though. The next problem that faces him or her is tracking down the source and target machines in question. In the case of network devices, there exist usually plans of deployment that tell where these are geographically located and who owns them. Because these are often closed devices, with one set of firmware (as opposed to an OS with various servers installed), it is easier to keep track of the software state they are in.

Servers are a bit more difficult. Server ownership may be more diverse, and their locations may not be confined to certain locations. A workgroup server may be set up in someone's office, for instance. The software state of these machines are also more complex to track, although usually servers are treated more conservatively than client machines.

Client machines, which are globally ubiquitous, are, of course, a nightmare. Apparently, to make administration easier, the DHCP protocol is used to assign IP addresses on demand. This is even the method of choice for fixed-location desktops in many organizations. Laptops are, of course, mobile by nature. They are taken into *'hostile'* environments and then back into the organization, possibly carrying malware of some sort. The IP address tells us little and the MAC address not much more as a laptop user may spontaneously decide to use a different network adapter. And, then there are the wireless roaming laptops that are allowed to keep their IP address even as they change location. So, tracking down a misbehaving laptop is not easy.

This organizational deficiency in terms of lack of self-awareness is problematic when trying to do risk analysis. In principle, we must update our exposure snapshot at every mutation of the network. That is an unlikely proposition. Then we would like to know how the current set of known vulnerabilities impact our given system and update this every time a new vulnerability becomes known. Also quite inconceivable.

Despite improvement in network management tools, and perhaps the recently more conservative spending on IT infrastructure, network management challenges have grown still faster. Instead of congealing on a limited set of architectures and software (which would be problematic for other reasons), the current trend is diversification, catalyzed by Java and Linux.

6 TECHNICAL EXPERTISE

Diversification also brings with it the problem of finding experts in these diverse fields who also understand security. Accreditation is of dubious benefit as it implies the field is stable enough to lay down in stone and test against. What is probably more important is a certain hacker mindset; however, with adult ethics packed in. But this can't be tested in any consistent way.

Size matters in converse ways. If an organization is too small, the few security experts may not have the resources between them to tackle all of the incident problems successfully. When an organization becomes too large, the resources tend to get distributed which does not facilitate close cooperation. One possible solution is to outsource part of the security duties to a third party (such as the earlier mentioned MSS) that is both centralized and diversified enough to provide the qualified resources. Such specialized companies may also be then better able to attract (and retain) people of the right mindset to tackle difficult security problems. The danger is, of course, that they may not be able to pay enough attention to each individual organization when a global incident breaks that threatens all its clients. We could, however, argue that seeing the global picture may be beneficial to efficiency.

Harking back to the risk analysis discussion before, the consequences of a configuration change are probably not understood well by those making the decision. Even a security expert may have to research the consequences if he or she were asked beforehand. Security education must reach those that enable or implement features because very often simple changes can avoid a big exposure without much loss of functionality.

7 EDUCATION, RESEARCH, ADVANCED TRAINING AND INFORMATION RESOURCES

For the last decade, with the growing demand for secure systems and products, many universities have started to build up IT security research groups and to offer IT security courses or degree programs, particularly in the United States, Europe and in some specific advanced institutes in several parts of the world such as India and Israel. But how many of the current IT security curricula help develop expert knowledge needed for computer security incident analysis and response covering such areas as operating systems internals, intrusion detection systems, reverse engineering methods, computer forensics and methods of computer emergency response?

Europe is proposing a research effort to build a "Trusted Computing" platform, reminiscent of the TCPA (2004) initiative. Concerns have been raised, however, that an EU institution in this area could promote security by obscurity to the advantage of a few players.

In USA, there are several agencies of the federal government (e.g. NIST-National Institute of Standards and Technology) that are involved in research and creating guidelines related to the preventive as well as reactive side of security incidents. It is not clear how the federal guidelines will be adopted in the industry even though members of some of the major vendors partake in the development of the guidelines. Some of the agencies also fund universities to perform research or develop security education; for example, CERT-USA at Carnegie Mellon University or Purdue University. Among many efforts to inspire research in academia, there is the initiative to create centers of academic excellence by the National Science Foundation in several universities for education in information assurance. In spite of this valiant effort, one can bring up an argument that currently, in the industry, matching jobs for such education may not exist in plenty in the near term, and that can be an issue for new recruits into the program or for graduating students.

Many training courses are offered to equip budding security experts. For those who already have some experience, security conferences offer a peek into potential problems and their defenses, and there are also specialized training events such as the Black Hat (2004) briefings. Little is known as to ultimately how successful such training actually is. With the right disposition, we can at least expect the trainee to know how to identify the right tools and go through the correct motions to handle an attack. Security incident response probably lends itself better to the apprenticeship model of training than the school house model because mindset cannot be taught by books.

It is also mandatory to stay up-to-date with the issues. There are mailing lists and information repositories for this purpose. BugTraq (2004), for instance, is a mailing-list where security risks are announced. Most people involved with security incident response keep track of such lists and a lot of important information is distributed through them. Vendors offer their own security advisories, but these only concern their own products. Security companies often have their own advisories, compiled from the vendors' advisories, other sources and their own research. Finally, some of the hacker groups publish their findings in different ways. Information glut is therefore a problem.

To distill the vast amount of information, various security bulletins are available that list the most recent threats. Some of these have the benefit of being translated into the native language of the reader so that the information can be absorbed more easily. Unfortunately, the publication cycle of such bulletins does not move as quickly as Internet time does, and the bulletins are often of little use by the time they reach the incident responder. A critical update may have been tragically delayed and an attack successfully launched.

The trend in updates is definitely going towards automatic updates in which a daemon periodically checks for new updates and installs them automatically. A lot of trust has to be placed in the creators of the patches to avoid mis-patching or introducing unwanted features. In the past, this trust has been occasionally misplaced (Knight, 2002). If the patch requires the restarting of software or the machine, the process may be inapplicable to mission-critical servers. The update distribution system should scale well and remain immune to the loss of one central server (as happened with a recent worm, which attacked the machine hosting the patches that prevented its spread). However, on the whole, it is a worthy idea. Especially, for those vulnerable and difficult-to-locate client machines.

The security reports also have other problems. First of all, it is hard to tell how reliable the information is, as often the text is kept intentionally vague. It would be useful to have a ranking scheme (perhaps a la Amazon) whereby others can elevate or depress its importance. Even an important report on a Linux flaw will be irrelevant to a Solaris user. Often it gets more subtler than even that. Each report has to be studied carefully and then compared with the existing installation base to determine the risk. There have been bugs in code that related to a very large population of machines across architectures because they share the same design or code (NISCC, 2004). This would be made easier if the reports were well structured or at least well written. Of course, the intentional vagueness does not help, but there are circumstances when a report still needs to be used while there is no real fix or it is felt that hackers may profit from a detailed description. Often a more detailed write-up follows a more widespread disclosure of the problem. Even then, it is obvious that natural language is not the most precise form of expression for such information. Sometimes, the wording in ambiguous, or the objects not accurately described. And, often, write-ups are done by non-native speakers and much is lost in translation.

Some publications have more structure to them. CVE dictionary (which is not a database) brings some structure into vulnerability reporting and defines some of the terms. However, the description text is still free-form. In the field of alerts, the IETF (2004) has defined the IDMEF, the Intrusion Detection Message Exchange Format (Curry et al, 2004), and it is already much richer in vocabulary. And yet, much is still open to interpretation limiting the usefulness of it. Beyond an appropriate syntax, clear semantics are required to accurately describe a vulnerability or an alert. Building elaborate ontologies of all aspects of security seems the right path to take. However, it will be a very long and hard one and will probably never be perfect (Undercoffer et al, 2002; Debar et al, 1999).

Setting aside the complexities of building such an ontological framework for security incident communication, the holy grail for the responder will be the automatic risk assessment to his or her situation. Given all the possible problems that need dealing with, which one needs most attention?

8 PREPARATORY REQUIREMENTS

Contingency measures must be practiced or else they will not be effective. A backup is useless if it can't be restored. And often this means restored quickly. But before a machine can be restored, the effects of the attack should be analyzed with forensic tools so that all aspects of the attack can be understood. Since all this has to be done under time pressure, the users of the tools must be trained.

Containment methods must be practiced and their impact on daily operations assessed. Can a single rogue machine be isolated by the incident response team quickly enough to prevent it from affecting other machines? Can network traffic be choked to prevent a worm from spreading too fast? In particular, worms need to be responded to in a matter of seconds to prevent widespread outbreaks.

Penetration testing is a useful method for probing one's defenses (Palmer, 2001). By putting oneself into the shoes of an attacker, the network takes on a different light and many faults become apparent. The question is, should this testing be done by an external team or internally? Certainly, an external team is likely to have more experience in the tools and tricks of the trade. However, the experience of trying the attacks is an important educational experience for the incident responders, and this is not something a lengthy report can replicate. Of course, once the penetration testing has been performed, one must ask if there are any lessons to be learned. Sometimes, the incident response team is not the same that sets the security policy and many problems remain unaddressed. However, if such a testing was performed internally, at least the experience will facilitate in maintaining the right mindset for dealing with real attacks.

9 CONCLUSION

In the present state, it is inexpensive, quick and easy to launch Internet or intranet attacks. However, for taking actions to effectively respond and prevent the security incidents the road ahead is tough. We need to harden our infrastructures, significantly improve global communication, cooperation and awareness, collaboratively fight hacking and terrorism, and perform cutting edge research such as in survivability engineering and threat assessment, and in setting standards.

The world's critical infrastructures and the government and business operations that rely on them are at serious risk. All nations should bear the responsibility to improve Internet security and help coordinate effective international global response to IT security incidents and events. In order to be successful, we need to ensure participation and cooperation among commercial enterprises, governments, law enforcement, security research organizations, and practitioners who can help further the art and deployment of incident response and prevention.

This paper summarized the current state as well as pointed out some key interdependent ideas that can help improve the overall capabilities.

However, as we pointed out, there are many open questions. The panel members intend to present these as only the starting points for discussion.

10 ACKNOWLEDGEMENTS

The authors wish to thank the following individuals for joining them in the discussion panel: Kai Rannenberg, Goethe University Frankfurt, Germany; Albin Zuccato, Karlstad University, Sweden; Gunnar Wenngren, Swedish Defense Research Agency, Sweden.

REFERENCES

Black Hat Briefings website, 2004, http://www.blackhat.org/

Brownlee, N., Guttman, E., "Expectations for Computer Security Incident Response", IETF RFC 2350, June 1988.

Brunnstein, K., Fischer-Hübner, S., "How far can the Criminal Law help to control IT Misuse?", International Yearbook of Law Computers and Technology, Volume 9, 1995.

BugTraq, http://www.securityfocus.com/

California Senate Bill 1386, 2002, http://info.sen.ca.gov/pub/01-02/bill/sen/sb_1351-1400/sb_1386_bill_20020926_chaptered.html

CAN-SPAM Act of 2003, "Controlling the Assault of Non-Solicited Pornography and Marketing Act of 2003", or the "CAN-SPAM Act of 2003", http://www.spamlaws.com/federal/108s877enrolled.pdf .

CoE - Council of Europe Cyber Crime Convention, 2001, http:// conventions.coe.int/Treaty/en/Treaties/Html/185.htm.

CERT - Computer Emergency Response Team (2004), http://www.cert. org/

CVE - Common Vulnerabilities and Exposures (2004), http://cve.mitre. org/

Canadian Government's Bill C-36, http://www.parl.gc.ca/37/2/parlbus/ chambus/house/bills/summaries/C17-e.pdf

Curry,D., Debar, H., Feinstein, B., "The Intrusion Detection Message Exchange Format", IETF, 2004 http://www.ietf.org/internet-drafts/draft-ietf-idwg-idmef-xml-11.txt

Debar, H., "Towards a taxonomy of intrusion-detection systems", Computer Networks, 1999.

Duquenoy, P., Masurkar, V., "Surrounded by Intelligence", in Risks and Challenges of the Networked Society. Proceedings of 2nd IFIP Summer School, 2003. (Forthcoming)

Ellison, R., Lipson, H., Linger, R., Mead, N., Moore, A., "Foundations for Survivable Systems Engineering", CrossTalk, The Journal of Defense Software Engineering, July, 2002.

ENISA, European Network and Information Security Agency, http://www.ffii.org/proj/enisa/index.en.html

European Commission's (EU) Directive on Data Protection (1999). http://europa.eu.int/comm/internal_market/privacy/index_en.htm

EU Directive 2002 - DIRECTIVE 2002/58/EC OF THE EUROPEAN PARLIAMENT AND OF THE COUNCIL of 12 July 2002 concerning the processing of personal data and the protection of privacy in the electronic communications sector (Directive on privacy and electronic communications).

FIRST Operational Framework, http://www.first.org/about/op_frame/ op_frame. 20030627.html, 2003.

GLBA – Gramm-Leach-Bliley Act (1999). http://www.ftc.gov/privacy /glbact/

HIPAA - Health Insurance Portability and Accountability Act (1996). http://www.hhs.gov/ocr/hipaa/privacy.html

Information Assurance Vulnerability Alerts (2004). http://www.iava.org/

IETF – Internet Engineering task Force, http://www.ietf.org/

IFIP Working Group, - IT Misuse and the Law, http://

Kephart, J., Sorkin, G., Swimmer, M., White, S., "Artificial Immune Systems and Their Applications", Springer, 1998.

King, C. "META Report: Are Managed Security Services Ready for Prime Time?" INT Media Group. July 13, 2001, http://itmanagement.earthweb.com/secu/article/ 0,,11953_801181,00.html.

Knight, W., "Microsoft's anti-piracy plans spark controversy", New Scientist, July, 2002.

Masurkar, V., "Responding to Customer's Security Incidents, Part 3: Following Up", Sun Blueprints Online, September, 2003, http://www.sun.com/solutions/blueprints/0903/817-3733.pdf

Masurkar, V., "On Developing a Methodology for Managing Security Vulnerabilities", Information Institute's 3rd Security Conference, April, 2004.

Nessus Remote Security Scanner, 2004, http://www.nessus.org/

NISCC, 2004, "Vulnerability issues in TCP", http://www.uniras.gov .uk/vuls/2004/236929/index.htm

Nmap Security Scanner, 2004.,http://www.insecure.org/nmap/

OCTAVE Methodoloy (2003). http://www.cert.org/octave/

OECD- Organization for Economic Cooperation and Development, http://www.oecd.org/

Outsourcing Managed Security Sources, http://www.cert.org/security-improvement/ modules/omss/b.html

Palmer, C., "Ethical Hacking", IBM Systems Journal, 2001.

Radhakrishnan, R. (2003). A Patch Management Strategy for the Solaris Operating Environment, Sun BluePrints Online, http://www.sun.com/ blueprints, January 2003.

Rathmell, A., Dorschner, J., Knights, M., Watkins, L., "Early Warning & Threat Assessment for Information Assurance", RAID 2000 Proceedings, October, 2000.

SANS Website, 2004, http://www.sans.org/

Sarbanes-Oxley Act, 2002, http://www.sec.gov/spotlight/sarbanes-oxley.htm

SecurityFocus Website (2004), http://www.securityfocus.com/

Spafford, E., "The Internet worm program: An analysis", CSD-TR-823, Purdue University, 1988.

TCPA - Trusted Computing Platform Alliance, 2004, http:// www.trustedcomputing.org/home/

Undercoffer, J., Pinkston, J., "Modeling Computer Attacks: A Target-Centric Ontology for Intrusion Detection", RAID 2002 Conference Proceedings, October, 2002.

USA Patriot Act of 2001, http://www.epic.org/privacy/terrorism/ hr3162.html

Wasik, M., "Foreword in Security and Control of IT in Society", edited by Sizer, R., Yngström, L., Kaspersen, H., and Fischer-Hübner, S., North-Holland, 1994.

West-Brown, M., Kossakowski, K., "International Infrastructure for Global Security Incident response", CERT-USA Report, SEI, CMU, June, 1999.

SECURITY IN GLOBALLY DISTRIBUTED INDUSTRIAL INFORMATION SYSTEMS

Petri Saloma, Ronja Addams-Moring, and Teemupekka Virtanen
Telecommunications and Multimedia Laboratory, Helsinki University of Technology

Abstract Today's industrial market is global. Manufacturing and measuring products are geographically distributed all over the world. Furthermore, the products are sophisticated and include know-how, often in the form of software. Thus the manufacturer is often the only party that can provide the maintenance services. In this paper, we focus on single intelligent and resource limited devices, which need to be remotely monitored or controlled. With the global Internet, satellite and telephone networks the access itself becomes possible. However, security is a challenge, because the remote devices' limited resources are pitted against threats from the Internet. We discuss the assets of and threats against a globally distributed industrial information system. To protect the assets, three possible security architectures, Centralized connections, Layered architecture and Integration of centralized and layered architecture, are proposed. Of these three, the integrated architecture meets best the requirements of a globally distributed industrial information system.

Keywords: Network security, Security policy, Information flow

1. INTRODUCTION

Nowadays, the manufacturer of a car is the only one, who has the know-how needed to repair any non-trivial problems in the car. The situation in industry is becoming quite similar – only the manufacturer of a device has the know-how needed for its maintenance. The devices are getting more sophisticated and they are distributed globally. Furthermore, industrial companies are themselves global and cooperate with other companies (suppliers of devices, maintenance services, Internet Service Providers). (figure 1)

The device manufacturers must often provide maintenance services remotely, because repair personnel cannot travel to every single device sold worldwide. Further, the device's remote control and monitoring capabilities might be valuable for the device owner.

The focus of this paper is firstly on single devices which are crucial parts of the industrial systems they belong to and which need to be remotely controlled or monitored. Such remote control and monitoring has become possible through the development in Information Technology (figure 1).

In this paper we aim to find one or more useful secure architecture concepts for how integrated single devices could be remotely monitored and/or controlled. We study three alternatives: the centralized architecture, the multi-zoned architecture, and last, we conceptually combine the two into a hybrid architecture.

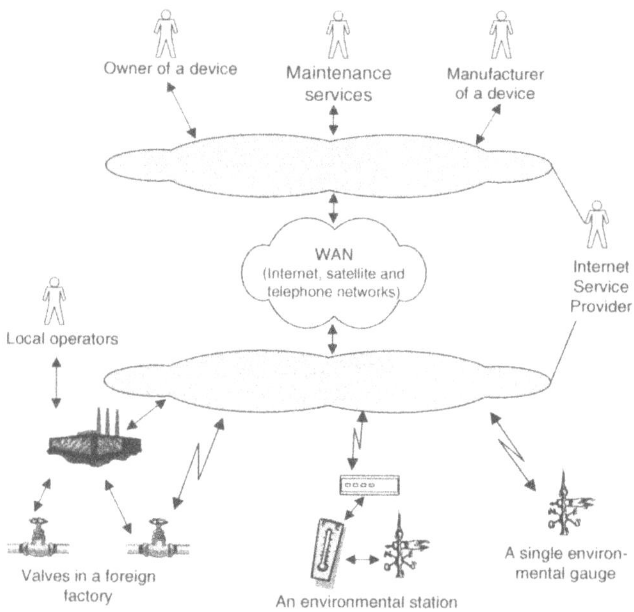

Figure 1. Parties in a globally distributed industrial information system, GDIIS

2. ASSETS AND THREATS IN A GDIIS

Because remote devices are connected to an open and insecure network, the security and trustworthiness of a GDIIS are imperative. From the device owner's viewpoint, information and services are the main assets of the system. Services can be divided into plain information query (e.g. a product recipe), monitoring (e.g. process temperatures) and controlling services (e.g. parameter setting or direct control of the device).

The remote device itself as well as the process and the system which the remote device is part of must be protected. Compromising even one of those can lead to process or product flaws, to the device breaking or even to a plant explosion.

From the manufacturer's viewpoint, the software can be more valuable than the device itself. Hence, the software should never end up in a competitor's hands. It might also be necessary to update the software remotely.

Security threats against these assets are basically the same as in any system which is connected to an open and insecure network. Eavesdropping (including man-in-the-middle attacks), spoofing and Denial of Service (DOS) attacks are all such network related threats that might lead to the compromising of the remote device's security. The assets can be considered secure as long as the confidentiality, integrity and availability of the GDIIS are preserved.

Confidentiality

Most often, the information and software of a GDIIS is confidential (e.g. a product recipe). But, this may not always be the case. The remote device's data might provide no information without the know-how to interpret the values (e.g. configuration parameters) or have only little value for the attacker (e.g. process temperature).

Integrity

Generally, the integrity of information, services and software must always be ensured in the GDIIS. This holds especially for software and services where every bit is significant.

However, some level of integrity compromisation can be acceptable, because e.g. monitoring information often has much redundancy and is not necessarily security critical. However, this does not always hold even in monitoring data, because sometimes every change needs to be traceable and non-repudiable. One example is the U.S. National Lightning Detection Network (NLDN) used to locate lightning strikes (Vaisala,). The NLDN is used by insurance companies to expose insurance fraud.

Also, laws or authority regulations may set strict bounds for the integrity of information, e.g. US. Food and Drug Administration (FDA) has regulations for food and drug manufacturing practices (including information systems and information traceability). Hence, the software and services of the GDIIS must always be malware free, whereas some integrity flaws might be accepted in the information.

Availability

Services and information are often time dependent, and hence availability is emphasized. There is no point in a service which is not available, or in information which cannot be accessed. A no-access signal could also indicate a broken device and a costly serviceman might be sent to the station without real reason. Therefore, even services in an environmental monitoring station must always be accessible when needed.

QoS (Quality of Service) is not a core security property, but has some security related features and can be included in the availability category. In industry, it is important to be able to predict the behavior of the system and instead of asking "How high QoS level on average can be provided?", we ask "Which QoS level can be guaranteed?".

3. SECURITY MECHANISMS

Security mechanisms are needed to protect the assets. In addition to technical security mechanisms, physical and psycho-social security mechanisms should be taken into account as well. Examples of physical security mechanisms are passage control and physical separation of a web server and the company's intranet. Psycho-social security mechanisms try to solve the problems associated with human beings, e.g. own personnel. Also, laws and regulations are psycho-social security mechanisms and they try to point out to the attacker that the benefits gained are less than the possible costs (consequences) (Naedele,).

General security guidelines

A system which has very little security is typically in great danger of being compromised. In contrast, a system with excessive security can become impossible to use. Therefore, at the minimum, the risks of the system must be evaluated. Further on, the pros and cons of used security mechanisms must be evaluated, and when the disadvantages are greater than the benefits, the mechanism should be left out.

Security might seem to have only a minor role in a remote device, because it does not directly hold any confidential components. However, even then the risks must be evaluated, because firstly, the components might include a backdoor to the company's intranet and secondly, sooner or later new components are added to the system or the existing components are updated. The integration of those new or updated components can lead to a security vulnerability, even if the new components are as secure alone as the existing system.

Authentication and authorization

Naedle states that also the connection itself (and not only the user) should be authenticated, and that authorization can be seen as action authentication. Hence, authentication can be divided into three: 1) *connection authentication* i.e. "Is this user permitted to connect the remote device at all?", 2) *user authentication* i.e. "Is this user permitted to use the remote application at all?"

and 3) *action authentication* i.e. "Is this user permitted to execute the action in question?" (Naedele,).

Connection authentication. Connection authentication specifies who in general is allowed to make a connection, i.e. to continue to the user authentication step. In TCP/IP networks, connections only from/to certain IP-addresses or ports can be permitted. Also, only certain applications might have permission to access the network / to be used over the network. (Naedele,) However, the spoofing of an IP address is quite easy and hence IP-based blocking gives only light security.

In a GDIIS, system specific properties can be used to drop unauthorized connection attempts. The message flows are often deterministic: size, frequencies and sequences of the messages are known beforehand. Furthermore, if the communication is not urgent in timing, the connection can be restricted to certain time windows. It might be also possible to give only to certain groups direct access to the remote machine, while others must communicate through a less valuable web server. (Naedele,)

User authentication. The management of usernames and passwords can be challenging in a GDIIS, and hence instead of user-based authentication, role-based authentication should often be used. Without role-based authentication, always when a new user is added to the system, every remote devices' password files must be updated. In the role-based authentication, the username and password must be sent to the new user.

Action authentication (authorization). Action authentication is controlling which actions and action sequences a user is allowed to do. This is often tightly related with services, software and the underlying operating system. Therefore, action authentication is often handled by those. Application level firewalls are counted as action authentication, too. They can be used to manage and monitor application level communication. (Naedele,)

Intrusion detection

Virus detection, network traffic monitoring and authentication failure detection are all needed parts of a GDIIS's intrusion detection. In addition to those, system files and file system integrity as a whole can be checked against unauthorized change (Naedele,).

Encryption

The NIST (National Institute of Standards and Technology) recommends that 128-bit protection should be used to achieve relatively long lasting security (up to the year 2036) (Vanstone, 2003). In symmetric encryption this means moving from popular 3DES (Data Encryption Standard, regarded to provide 112 bit security (Vanstone, 2003)) to AES (Advanced Encryption Standard, where security level increases as the key size increases (Anderson, 2001, p. 93)).

128-bit security is more challenging in traditional asymmetric algorithms (e.g. Rivest, Adi Shamir (RSA) and Digital Signature Algorithm (DSA)). According to (Vanstone, 2003), to provide 128-bit security, those algorithms need 3072 bit keys, and to provide 256-bit security a 15360 bit key is needed. However, those keys are too heavy for the remote devices. The problem gets even further complicated as the greater key size leads to even greater increase in computational cost (Vanstone, 2003).

In contrast, with elliptic encryption, the key size is only double to the security level. Therefore, the elliptic curve-algorithms have definite advantages especially in those devices which are limited in computational power, storage space, bandwidth or power consumption. (Vanstone, 2003)

A deeper study on PKI, Elliptic Curve Cryptography, and Digital Signatures is found in (Caelli et al., 1999).

4. THREE SECURITY ARCHITECTURES FOR THE GDIIS

In the previous chapters we discussed the assets and threats of a GDIIS as well as security mechanisms to protected the assets. In this chapter, we examine three security architectures for the GDIIS. In the first architecture, all connections are centralized in order to ease management and use of security mechanisms. In the second architecture, the GDIIS is divided into zones, and instead of relying on only one solution, also security is distributed. In the third architecture, the centralized and the layered (zone) architectures are integrated.

Centralized connections

In the centralized connection architecture, all connections from the human interface to the remote devices go through a server, a stepping board. Behind the stepping board there can be many remote devices or device groups which are geographically apart from each other. Hence, the centralized connection architecture consists of three parts *human interface*, *remote device* and *stepping board.* (figure 2)

Human interface. We can place the human interface authentication and authorization functionalities in the stepping board. In this way, all user passwords can be managed at the same place (at the stepping board). The stepping board makes it also possible to use sophisticated and heavier security mechanisms

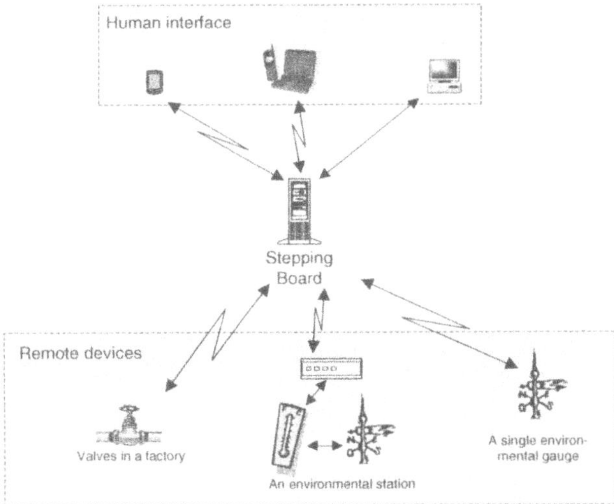

Figure 2. Centralized connection architecture

between the human interface and the stepping board, such as smart cards or AES encryption. Installing the counterpart of the smart card-system to the every remote device is often unpractical because of economic and management reasons, whereas AES is often too heavy for the remote devices.

The human interface devices can be anything from a mobile phone to a desktop computer. The capabilities of these (e.g. display) demand different servicing. In addition to content specific servicing, also security mechanisms may differ. For a mobile phone, we could use lighter encryption and allow the user to access only monitoring services. For a desktop computer, we could use stronger encryption and also enable control services. Because of needed resources for such a service, we cannot really place those in the remote device.

Remote device. We can utilize proprietary security mechanisms between the stepping board and remote devices, which means, that stronger security is achieved by obscurity. However, proprietary security mechanisms mean here more taking advantage of known and deterministic communication than building a really new mechanism. The use of proprietary solutions directly between the human interface and the remote devices is harder, because of the heterogeneity and changeability of the human interface devices. Further, we can also take the advantage of the fact, that the stepping board is trustworthy from the remote device's viewpoint. (Even if the human interface was properly authenticated, it could always try to do unauthorized things whereas the stepping board most likely would not).

Stepping board. Because all communication goes through one computer, we can utilize centralized IDS in the stepping board. The network traffic can be scanned for viruses and other anomalies. We can place a sophisticated firewall in the stepping board to protect remote devices from DOS attacks and port scanning. Also, logs and other information from all remote devices can be integrated at the stepping board to get the general picture of the whole system.

When the stepping board can act as a proxy server between the remote devices and the human interface, we can prohibit direct access to the remote devices. Sometimes it might even be possible to make the connection between the remote devices and the stepping board only one-way. A remote device only pushes its data to the stepping board. If we can ensure the one-way communication (e.g. by hardware), monitoring of very critical remote devices can be enabled, too.

Generally, the security mechanisms in the stepping board can be much more sophisticated and heavier and hence give stronger security. However, this does not mean that we can leave out all security mechanisms in the remote devices. The remote devices must still have their security mechanisms, but those can be lighter.

Layered architecture

A common architecture to protect the assets is to divide the system into trusted and untrusted zones and place a wall between them (e.g. intranet from the Internet). Traditionally, the separation of the zones is achieved by a firewall (e.g. a stepping board) which monitors the connections to and from the trusted zone. However, this type of *hard perimeter* solution has severe problems.

Firstly, the wall must have doors to and from the trusted zone. Because of complex programs, the wall has always weak points and unknown doors. Those might (will) be used by an attacker. (Naedele,)

Secondly, building the solution on only one principle is dangerous: if that principle fails, the whole solution fails. Once the attacker gets inside, the system is without security and everything is already lost. Hence, the wall should be infinitely strong to last for an infinitely long time. (Naedele,)

Lastly, technology is developing all the time, which gives attackers better penetration tools. (Naedele,)

Accordingly, instead of dividing the GDIIS into two zones, we should divide it into several security zones. This type of *defense-in-depth* architecture is used e.g. in automation systems of nuclear power plants. In this way, the core services and assets can be placed into the innermost zone, which then has several security zones around it. Basically, the system is in safe as far as the formula $P > D + R$ holds, where P is the time which is needed to break the security, D is the delay until the attack is detected and R is the time until a reaction against the attack has been completed (Schwartau, 2001).

When there are many zones, the administrator has better capabilities to fight the attacker. The compromisation of the first zone can raise alarms. When actions are taken early enough, the attack can be stopped.

To utilize the *defence-in-depth* architecture in the GDIIS, we can divide it into four zones: inter-zone, intra-zone, remote device -zone and application-zone. The inter-zone is an open and insecure global network. The intra-zone is open and insecure as well, but is usually national and managed by a trusted party (e.g. a ISP's GSM network). The remote devices lie in the remote device -zone. The information and the core services of the GDIIS are in the application-zone.

Inter-zone. In the GDIIS, the inter-zone is usually the Internet and can be described with the adjectives untrustworthy, globally reachable and heterogeneous. On the other hand, very light security, openness and globality are some of the Internet's strengths: it keeps the core protocols and routing simple and effective and connects almost everyone to almost anywhere. However, this unsecureness is a problem for the GDIIS.

To make the Internet more secure does not automatically mean heavy encryption or disposal of openness and globality. We could place some light and sophisticated intrusion detection systems in the backbone network. This could be, for example, the detection of the newest viruses and simple DOS detection systems. However, our goal is not to make the Internet 100% secure, but to detect the most common attacks by means of simple and light security mechanisms.

Placing security mechanisms in the backbone network of the Internet would cost. The costs could be charged from all Internet users thorough ISPs, and because there are millions of Internet users, the cost per user would not be high, especially when compared to the benefits. An other possibility is to build secure tunnels over the Internet and sell those tunnels for additional cost. From industry's viewpoint, either one would be acceptable as long as the costs are kept reasonable.

Intra-zone. The intra-zone can be defined as a network hosted by one ISP. Usually, the intra-zone is an open network, such as the Internet, but we can often restrict access to certain user groups and use more sophisticated security mechanisms. In the GDIIS, the intra-zone is often a GSM/GPRS network, but satellite networks are also used. Because of the popularity of GPRS network, we study the intra-zone from GPRS network's viewpoint.

Brief architecture of GPRS-network is found in figure 3 (Chang, 2002). A more detailed description and special security threats in GSM network can be found in (Rautpalo, 2000) and (Chang, 2002).

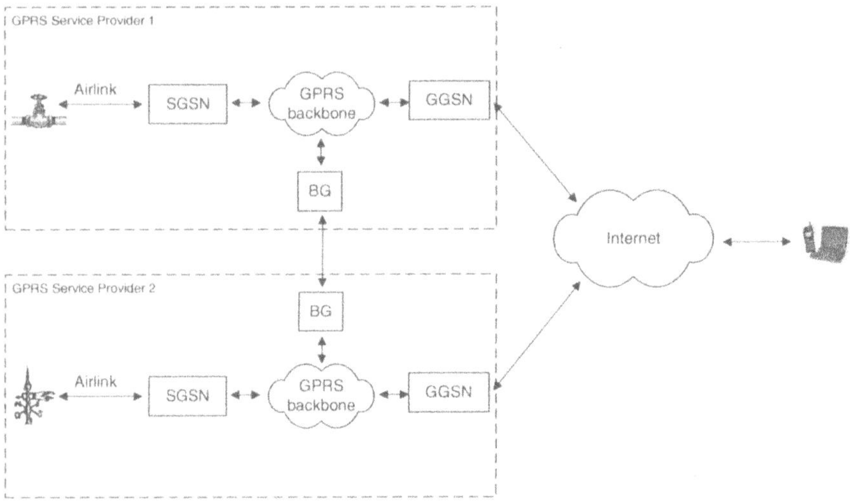

Figure 3. Arcihtecture of a GPRS network

SGSN Servicing GPRS Support Node - A gateway between a GPRS and a GSM network

BG Border Gateway - A gateway between two GPRS networks

GGSN Gateway GPRS Support Node - A gateway between a GPRS-net -work and the Internet

In chapter 3, we divided authentication into three categories: connection, user and action authentication. The last of these is tightly related with the service, hence it is not appropriate in the intra-zone.

We can place some connection authentication mechanisms in GGSN (Gateway GPRS Support Node). When it is known beforehand from which IP-address(es) certain network blocks or individual devices are accessed, we can drop unauthorized connection attempts. However, if there are public services accessible from any IP address, it is hard to place IP-based blocking in GGSN. In Border Gateway (BG), IP-based blocking is easier, because the address spaces of both networks is known.

For user authentication, GPRS utilizes a A3 GSM authentication algorithm. A unique subscriber authentication key is stored in a physical Subscriber Identity Module (SIM). Hence, to gain access to the GPRS network, a user must have an authorized SIM card. Compromisation of a SIM card is hard. Another means of gaining access to the GPRS network is to steal an authorized card. Thus, as long as the SIM card cannot be stolen, authentication in GPRS is sufficient for most purposes. (Chang, 2002) In some cases, we could utilize GSM A3 user authentication as such in the GDIIS.

Dealing effectively with viruses is today's challenge and industry welcomes all well-working IDS services. It has been seen that traditional anti-virus software in a mail gateway decreases virus problems dramatically. We could adopt the scanning technique in the other traffic as well.

In addition to virus detection, we could prevent port scanning of the remote devices in the GGSN. Today's port scanning methods may last months and therefore need logs over long time periods. Further on, DOS is one of the biggest threats against availability. Even heavy servers are in great danger, all the more the lightweight remote devices.

After the authentication procedure, the connection from the remote device to SGSN (Servicing GPRS Support Node) is encrypted with the GPRS-A5 algorithm (Chang, 2002). According to (ETSI,), the compromisation of that is hard even with massive computing resources. An other advantage is, that this encryption algorithm suits well the remote devices, because it is designed for resource limited mobile phones. However, the utilization of this algorithms for other purposes, (e.g. end-to-end encryption) is not possible, because the algorithm is not public.

In the GPRS backbone from SGNS to GGSN (or BG) special GPRS Tunneling Protocol (GTP) is used. GTP has no security mechanisms. However, the GTP is not totally insecure, because the communication occurs over a private network (Chang, 2002). Additionally, the ISP can encrypt the information in the backbone. Hence, the encryption in the GPRS network is enough for most purposes as long as the ISP's own personnel cannot access the information.

Earlier described NAT is often a default in today's GPRS networks, because of the lack of IPv4 addresses (Rautpalo, 2000). NAT protects the remote devices against direct access from the Internet, but does not protect against unauthorized access from inside the GPRS network. However, the problem of NAT is that it prevents all direct access to the remote devices (also the authorized one). Therefore, NAT is also a problem, because the connection is often initiated from the Internet. However, we can use other techniques (e.g. SMS) to tell the remote device to initiate the connection.

NAT makes it also difficult to use some end-to-end encryption mechanisms. For example, IPSec packets will be dropped, because the integrity check fails (address changes in the header). One solution for that is to encapsulate IPsec packets in UDP, which can be used without the integrity check. The drawback of UDP is that the additional encapsulation and possible key management slows down the setup and the connection itself. (Rautpalo, 2000).

Remote device-zone. The remote device -zone is the last lock before the core services. The surrounding zones can prevent some attacks from happening, but not all. Actually, we must still be prepared to face any attack in the remote device -zone. The task of ensuring security becomes nearly impossible when talking about e.g. a single pump which should be directly controlled from the Internet. Luckily, the situation is seldom like that.

In all cases, we must authenticate connections and users and need some encryption mechanisms. Also, we should place some level of malicious code detection in the remote device -zone. We can utilize the time window technique such that the network interface of the remote device is turned off, when the service are not needed. In general, because the remote device is near the core services, we can and should utilize the known properties of the communication in the security mechanisms.

We should also protect the security mechanisms themselves. We can easily disable unnecessary services. This saves us from many surprises. We can also download the core configuration parameters regularly from a secure server. Furthermore, we can utilize component based solutions (e.g. Java Embedded Server or Open Service Gateway Initiative), where all services are separate modules on top of operating system. This makes it possible to add, remove and update the services easily and safely.

To conclude, it might be impossible to reach the desired security level in the remote device -zone. Then we should firstly utilize every security mechanism, which the limited resources permit. Secondly the level of services should be reconciled with the achieved security level.

Application-zone. The application-zone is the most critical part of the system. However, there is not much we can do at this level. Here, the emphasis is particular on action authentication: who is allowed to perform which actions. This can be an inherent part of the underlying operating system, can be handled by the application, or both. In any case, we need action authentication, because by breaking the remote device -zone the attacker would automatically have full control of the device.

In the application -zone, service specific properties are known and can be used to increase security. We can utilize those e.g. to drop unauthorized connection attempts or to detect anomalies in communication. Also, if possible, we do not send the messages in self-explaining format, such as xml. To conclude, appropriate mechanisms in the application-zone are rare and utilize often service specific properties.

Integration of centralized and layered architecture

Some problems exist in the both above proposed architectures. The centralized architecture is like a hard perimeter solution. In the layered architecture, the biggest problem is that the whole path from the remote device to the human interface is still assumed to be insecure. Here, the remote device often becomes the bottleneck and hence reduces the end-to-end security mechanisms to be light.

Most presumably the biggest threats come from the inter-zone. Hence, if we can place the stepping board in the border of the inter- and intra-zone, we can use lighter security mechanisms in the inter-zone and heavier ones in the intra-zone. That is, we utilize some light and proprietary security mechanisms between the remote device and the stepping board and the heavier ones between the stepping board and the human interface. Therefore, the challenges in the inter-zone (human interface - stepping board) are solvable by standard security mechanisms.

In the intra-zone, perhaps the best location for the stepping board is at the border of the GPRS network and the Internet (figure 4). In the stepping board, we can utilize security mechanisms of GGSN in more suitable way for the GDIIS. Encryption of GPRS can be enough, but some level of connection, user and action authentication is still needed, because the connection attempts can always come from the inside of the intra-zone. Because only the stepping board is allowed to connect to the remote device, the stepping board-remote device -authentication can be simple and strong at the same time. This fact also simplifies also the connection authentication (all other attempts can be dropped except the stepping board).

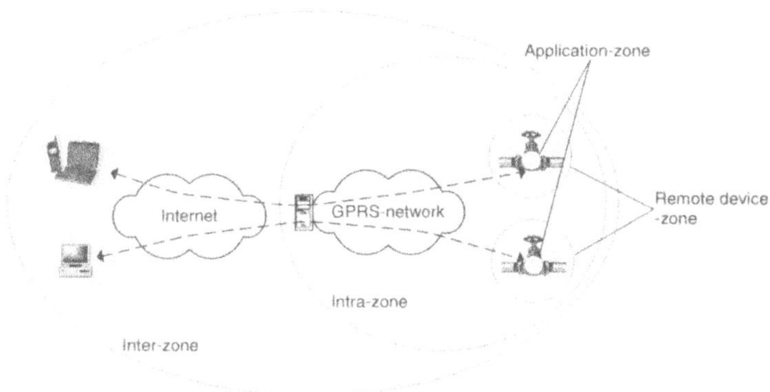

Figure 4. Integrated connection architecture

When the stepping board is in an ISP's network, the ISP can provide the stepping board as part of the service. Also, the ISP can join and analyze the information from both the stepping board and GGSN. But, even if it were possible to integrate the stepping board and GGSN, it should not be done, because the GGSN serves all of the ISP's customers. Therefore, the integration of the stepping board and the GGSN would reduce the benefits gained.

5. DISCUSSION

We have now examined three different architectures. The biggest advantages of *the centralized architecture* is its centralized security and the positive consequences of it. Management of passwords becomes easier and centralized IDS is possible. Between the stepping board and a remote device, we can utilize lightweight and proprietary security mechanisms, whereas heavier ones can be placed between a human interface and the stepping board. Also, human interface specific servicing is possible (e.g. controlling services only for PC). Further, with the stepping board, we might be able to ensure one-wayness of the connection, which makes it possible to monitor also very critical remote devices.

However, the stepping board is a hard perimeter solution, i.e. it rests only on one security principle. Also some of the advantages might be only theoretical. Firstly, the network between the stepping board and the remote devices can be so insecure that everything has to be built twice. Secondly, it can be impossible to place the stepping board close enough to the remote devices to really protect them, because we should have enough remote devices for a stepping board to make the architecture economically viable.

With *the layered architecture*, we can reduce the problems of hard perimeter solution. In the layered architecture, the security mechanisms in the limited remote devices can be lighter. A layered architecture makes it also possible to have different parties to manage different zones. From the industrial viewpoint, it would be nice to be able to buy all other zones except the remote device and the application -zone. Then, industry could concentrate on its core competence without having to study new security mechanisms.

A drawback of the layered architecture is that when we examine it in detail, it might not seem to protect the remote devices more effectively than the centralized architecture. We must still be prepared to face actually any threat in the remote device -zone, because the intra-zone is also a public network. Therefore, the security of the intra-zone plays a big role. With only little security in the intra-zone, we might be forced to reduce the level of services of the GDIIS (e.g. from control to monitoring).

We can partially solve the problems of the hard perimeter solution (centralized architecture) and the limited end-to-end security mechanisms (layered architecture) by *integration of the two architectures*. The layered architecture automatically removes the problem of hard perimeter solution. Further, with the stepping board we can place stronger security mechanisms in the inter-zone without loading the remote devices excessively.

That said, even integration leaves some problems. The topology of GPRS networks is not necessarily such that the stepping board really protects the remote devices. Also, the remote devices cannot be distributed globally, because the GPRS networks are mostly national. Furthermore, the stepping board means (almost) necessarily longer paths between a remote devices and a human interface.

If the ISP can (in addition to the intra-zone), manage by itself or through a third party, the stepping board and the inter-zone, the industry could outsource the management of the inter- and intra-zones to the ISP. The drawback of this approach is that the ISP must be trustworthy. However, we must not trust the ISP blindly. The ISP should strictly restrict of who its personnel and how is allowed to administrate the stepping board. All changes should be accountable and non-repudiable and the ISP's system must be audited by a trusted third party.

Table 1 shows, which security mechanism should be applied in which zone. From the industry viewpoint, the intra-zone should include all computational power and deep security knowledge demanding security mechanisms, which are complemented by the mechanisms in the remote device-zone.

Security mechanism	**Inter-zone**	**Intra-zone** (stepping board)	**Remote device -zone**	**Application -zone**
Connection authentication		X	X	
User authentication		X	X	
Action authentication			(X)	X
Intrusion detection	(X)	X	X	(X)
Encryption		X	X	(X)
Security mechanism protection		X	X	

Table 1. Appropriate security mechanisms in each zone.

6. CONCLUSIONS

Development in Information Technology has made it possible to move from centralized systems to distributed ones. Today's industrial devices, e.g. valves, pumps and rainfall gauges, have remote control and monitoring capabilities. However, one of the biggest hindrances for further development is security. There are assets to be protected, i.e. information, services and remote devices themselves, as well as software. Threats coming from the Internet are real and challenging. Because remote devices have limited resources, standard security mechanisms cannot be applied as such.

When we compare the communication of a GDIIS and the Internet, we find that the properties of the GDIIS's communication are often more deterministic.

Among others, we might be able to predict the content, frequency and paths of the messages. We can and should utilize those to increase the security, especially in the near of the core services (the remote device-zone and the application-zone).

We studied three possible architectures for the GDIIS, from which the integration of the centralized and the layered architecture seems to have the most advantages. Security mechanisms at the border of the inter zone (Internet) and intra zone (GPRS) should be such that the inner zone's mechanisms can be lighter to save remote devices' limited resources. Hence, the integrated architecture applies the general principles of defense in depth. Even if the outer zones have very strong and sophisticated protection, those will eventually fail and security in the inner zones is needed. The inner-zone security protects also against threats coming from inside the system.

To conclude, the security in a globally distributed industrial information system is a challenge. But, instead of trying to place security in one place, we should distribute the security also. We should see the distributed system as a whole and have security mechanisms as a integrated parts of different zones, not as add-on features.

Acknowledgments

We would like to thank Teemu Tommila for his patient guidance and Elina Valtonen for sharing her (unpublished) evaluation of earlier work in this area.

References

Anderson, Ross (2001). *Security Engineering*. John Wiley & Sons, USA.

Caelli, William J., Dawson, Edward P., and Rea, Scott A. (1999). PKI, elliptic curve cryptography, and digital signatures. *Computers & Security*, 18:47–66.

Chang, Dung (2002). Security along the path through gprs towards 3g mobile telephone network data services, version 1.3. SANS Institute 2002, As part of the Information Security Reading Room, referenced 27.9.2003.

ETSI, European Telecommunications Standards Institute. GSM calls even more secure - thanks to new A5/3 algorithm. referenced 17.10.2003.

Naedele, Martin. IT security for safety-critical automation systems. ABB.

Rautpalo, Jussi (2000). GPRS security - secure remote connections over GPRS. Technical University of Helsinki, Department of Computer Science, referenced 27.9.2003.

Schwartau, Winn (2001). Network security it's about time: An offer for a metric. *Network Security*, 2001:11–13.

Vaisala. National lightning detection network. referenced 24.10.2003.

Vanstone, S.A. (2003). Next generation security for wireless: elliptic curve cryptography. *Computers & Security*, 22:412–415.

A CASE FOR INFORMATION OWNERSHIP IN ERP SYSTEMS

Prof. S.H. von Solms, M.P. Hertenberger
Rand Afrikaans University, Johannesburg, South Africa

Abstract: This study investigates the lack of information ownership in current Enterprise Resource Planning (ERP) software systems. The purpose is to show how difficult, time consuming and costly the implementation of security within such systems is. The focus is on the investigation of security implementations within well-known ERP software packages such as SAP R/3 and Oracle E-Business Suite. The results of the study indicate that central administration, control and management of security within the ERP systems under investigation weaken security. It was concluded that central administration of security should be replaced by a model that distributes the responsibility for security to so-called information owners. Such individuals hold the responsibility for processes and profitability within an organization. Thus, they are best suited to decide who has access to their data and how their data may be used. Information ownership, coupled with tight controls can significantly enhance information security within an ERP system.

Key words: database security; security policy; misuse detection; authentication; information flow.

1. INTRODUCTION

Enterprise Resource Planning software systems are in use by many different organizations and businesses worldwide. To facilitate the understanding of the reader, the term ERP is defined here as any software system that has been designed to support and automate the business processes of medium and large businesses.

Due to the fact that the ERP system in use by an organization contains critical business data, it is essential that such information be protected from unauthorized access. Unauthorized access to the data within the ERP system's database must be prevented, especially since a large percentage of fraud takes place within the organization[1].

To protect all data within the ERP system's database, data security has traditionally been implemented by providing a security infrastructure or subsystem within the ERP system. Such a security subsystem generally allows an administrator to define profiles and roles for each user accessing the system. Once the user has received a logon name and a password, the system permits access only to the areas permitted by the roles allocated to the user master record. If required, the user can be restricted to certain screen forms and be prevented from entering certain values in various fields.

2. STUDY RESULTS

The approach introduced above is implemented in most modern and current ERP software packages. During the course of this study, four different ERP software packages were investigated to determine the method used to implement a security subsystem. The following ERP software packages were investigated:

1. SAP R/3 by SAP AG[2];
2. Oracle E-Business Suite by Oracle Corporation[3];
3. Navision Attain by Microsoft Corporation[4];
4. Navision Axapta by Microsoft Corporation[5]

The above products were selected based on market penetration and target segment. This permitted various approaches and technologies to be compared.

The primary conclusion of the investigation into the security implementation and provision of the above products was that all rely on a traditional, centralized approach to information security. Though specific implementation differences do exist, most follow the route of requiring the creation of profiles, roles or permissions[6]. The profiles, roles and permissions determine various actions that may be completed within the system. Possible actions may be the generation of a report, the entry of a sales order or the execution of a program, for example. Once these actions have been translated into profiles, roles or permissions, the security administrator assigns them to individual user master records. The user is presented with a user name and password. Once a logon action has been completed, the user is restricted to performing only those tasks that are permitted by the roles allocated to the user master record.

3. THE CENTRALIZED APPROACH TO SECURITY

The approach provided by the software products under investigation relies on one or more administrators to determine and define the security requirements centrally. In other words, one or more administrators are tasked with the creation and generation of roles and profiles that are allocated to user master records. Once these roles and profiles have been created, the user to whom they are allocated is able to complete only a certain subset of possible tasks within the system.

In contrast to legacy systems, ERP systems provide functionality specific to a user's requirements directly to that user's desktop. However, security is still configured centrally. This is a very traditional approach that mirrors administration in legacy software systems and environments. The results of the investigation into various ERP software packages concluded that the centralized approach to security is not ideal. Particularly, the practical implementation of security within ERP environments is often error-prone, time consuming and very costly when completed in the centralized fashion.

A diagrammatic representation of the traditional centralized approach to security within ERP environments is presented in Figure 1.

A number of the problems associated with centralized security implementations are discussed below.

3.1 Error-prone configuration of the security subsystem

The central approach to a security implementation within an ERP environment is faced with numerous challenges. As an ERP software package supports all facets of the organization or business, numerous functional areas have to be covered by the software package. For example, a large manufacturing organization may require financial, sales and material management functionality. For each of these functional areas, very specific objects, data types and processes have to be configured within the ERP software package, to support compliance with the organization's strategy and operation. As each such functional area is under the supervision of skilled and experienced employees and members of the organization, a very specific pool of knowledge is present.

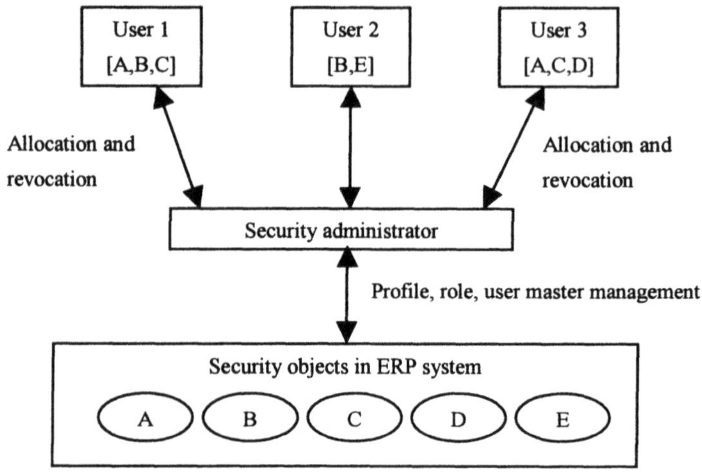

Figure 1. Centralized security within an ERP environment

To assume that a central system or security administrator has the ability to understand all nuances and specifics of each such functional area is often incorrect. Instead, the security administrator must gather information from each area of the business. Once all these details have been gathered, the security administrator is able to translate the requirements of each business area into the appropriate roles and profiles within the ERP system. In many cases, the security administrator has to select objects manually to create the appropriate access authorization for the user. It should be clear that such a process is often completed with a number of errors and omissions.

3.2 Time consuming and costly configuration of the security subsystem

ERP systems are generally installed in organizations with a large number of users. The creation of profiles and roles for a user population in excess of 200 users becomes a very complex and administration-intensive operation. As mentioned in the paragraph above, the central creation and maintenance of user master records, roles and profiles assumes knowledge of the specific functional area for which the security and access authorization settings are to be made.

Though all ERP systems that were investigated during the course of this study provide a means of creating profiles and roles in an accelerated fashion, very little support is provided to ensure accurate creation of security objects. This means that the security administrator, together with a knowledgeable member of the organization, has to spend a significant time testing the validity of the created access rights of each user. This testing cycle often involves an entire project team and is often completed during the testing of business process mapping within the ERP system. Unfortunately, testing is often performed with only a single user type in mind. This means that users with fewer access permissions may be denied access to certain functions that are required by them. In other cases, users holding more access authorizations may be able to access parts of the system that should not be accessible to them. Due to the fact that the testing process requires manual intervention, the requirement for additional project resources often increases costs. If incorrect decisions were made during the creation of access authorizations within the system, substantial amounts of work may have to be repeated.

3.3 Lack of change management and documentation support

The points made in the paragraphs above involve time, cost and knowledge constraints that are apparent due to the centralized nature of the security implementation within ERP software environments. A further increase in time and cost can be considered when attempting to document any of the security subsystem configuration and settings. As is customary in large-scale software implementation projects, the need for both system and end-user documentation is critical. During the development and integration phase of the new system, some form of change management is also required.

Most of the ERP software packages selected for investigation during the course of this study support change management and documentation for the mapping of the business processes to distinct process flows within the ERP system. Hence, the documentation of business process mapping to the configuration of the system is supported in most cases. In a similar fashion, a change management module is available to track changes made to key objects and elements within the system. Unfortunately, none of the systems selected for this study provide any form of change management or documentation support for the security subsystem. It is not known why no change management system has been implemented for the security subsystems in the ERP systems investigated during the course of this study. The lack of such a feature is considered to be a grave omission, however[6].

The lack of change management may be considered very serious. Without any form of change management documentation, workflow or control, all changes made within the system to effect changes to user master records and access permissions cannot be traced. Though the systems under investigation record the name of the user who effected the last change, there is no way of tracking the change and ensuring it complies with security requirements set out within the organization. Once again, the central nature of security object administration is a cause for concern. Security administrators are often under pressure to provide access to certain functions at short notice. Paper-based change control systems are difficult to maintain and tracking of changes becomes almost impossible.

4. DECENTRALIZING THE APPROACH TO SECURITY

Within traditional ERP environments, the centralized approach to implementing access control and access restrictions enables one or more security administrators to create and maintain profiles, roles and user master records. As has been mentioned above, this approach suffers from a number of problems, most notably that the security administrator cannot and usually does not understand the complexities of the actual business processes within the organization and how these have been mapped to the functionality of the selected ERP software package. To combat this problem and to promote more rigid and adequate security within an ERP environment, it is necessary to deal with complexity within the system as a whole. Figure 2 below indicates the changes from the centralized approach. It is important to notice that the security administrator still performs a management and auditing role, but is no longer concerned with the detail of the individual user requirements.

The decentralized approach suggested in Figure 2 ensures that the technical complexity associated with the creation and allocation of security objects is removed. This is achieved by presenting the identified information owners with an interface layer. The interface layer hides technical complexity from the information owners and presents them with only the necessary data relevant to their area of the system and business. As the control of the overall security subsystem must rest with one or more individuals, the security administrator has the usual, full access to all security objects at the most detailed, technical level. This permits the security administrator to tailor and allocate objects for the specific use of the information owners. Once the information owners are satisfied that all required security objects are accessible to them, the require security objects can be tailored and allocated to users within the business sphere of the relevant information owner. In a more advanced model, the security administrator may not be able to allocate any security objects to any users, but rather prepare security objects for use by the information owners. This addition to the decentralized model would ensure a higher degree of separation and ensure that no access to the system could be gained through the security administrator.

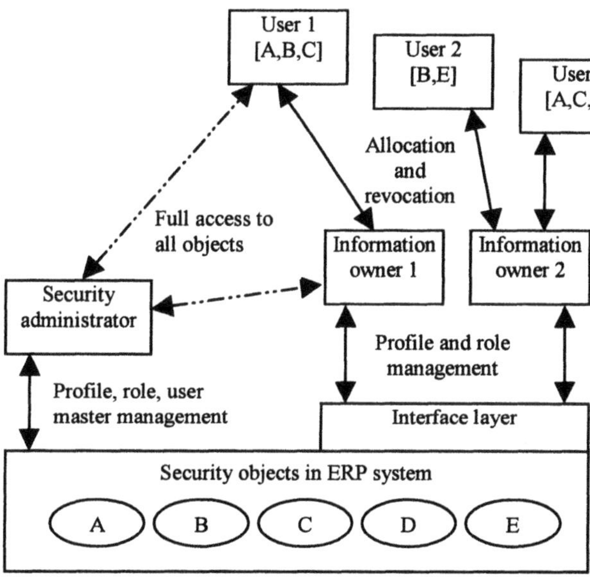

Figure 2. Decentralized security within an ERP environment

4.1 Dealing with complexity

ERP software packages provide integrated functionality across an entire enterprise. Generally, such software packages may be deployed in various industries, across various disciplines and across various countries. This makes ERP systems rather complex in their architecture and feature set.

Due to the complexity of modern ERP software packages, the ability to fully understand all functionality offered by the software package is virtually impossible. Specialist areas of expertise are thus created. The creation and maintenance of access restrictions and their related objects within the system may be considered such a specialist area of expertise. The ability to create and maintain a security infrastructure for areas of the business that are merely described and documented by knowledgeable members of the organization should not be considered adequate. To state this problem in a different way: the security administrator has the technical knowledge to create and maintain security related objects within the system, but is unable to fully grasp the complexities of the business processes and all related nuances. Conversely, the business owner has the necessary knowledge relating to the business processes, risks and areas of concern, but does not possess the technical knowledge to create and maintain the necessary security settings and objects within the system.

The problem of complexity is further exacerbated by the fact that the communication from the business owner to the technical security administrator may be unclear or misunderstood by either party. This results in a business owner and a technical security administrator who both believe that adequate security and access restrictions have been placed within the system. Any miscommunication or mistakes may be detected only once fraud has occurred or an audit on the system has taken place. It should be clear that such a situation is not acceptable.

A proposed method of dealing with this complexity is to permit each business owner to become responsible for the setting up of access restrictions within the system. However, the business owner should be able to perform such activities only within the area that the business owner is responsible for. In providing this feature, the level of security could be increased dramatically and the timeframe for its implementation reduced significantly. The reason for this is that the business owner is aware of all required access restrictions and security requirements within his sphere of responsibility.

The above should indicate that complexity can be dealt with by permitting the business owner the means to ensure adequate access restrictions to all objects and data within his sphere of responsibility. However, the topic of responsibility should be touched upon briefly.

4.2 Improving responsibility and accountability

The complexity of creating various roles, profiles and user master records within an ERP system can be eased by permitting each business owner to take over the function of allocating and creating security objects within the ERP system. An important side-effect of this concept is that of increased responsibility and accountability.

To elaborate on this concept, consider the traditional approach to implementing access restrictions within an ERP environment: a business owner determines who requires a certain type of access to complete certain tasks. Often, such requirements are documented by the business owner in a spreadsheet, indicating required functions per user. This set of documentation is passed on to the security administrator, who determines the requirements and configures appropriate security objects within the system. Such security objects are often roles or profiles. Once these are created, the security administrator allocates them to the appropriate users' master records within the ERP system. Some unit testing may be performed.

Though the above procedure seems to be acceptable, it should be considered that the business owner is generally not aware of the technical details of the security objects allocated to the users. This may have serious implications if the business owner is to be held accountable for an action taken by one of the users who have been allocated one or more of these security objects. If one of the users is able to perform a task or function that results in fraud, for example, the business owner may be held responsible for allocating an inappropriate function to that user. However, the security administrator may have made a mistake or added the extra function to cater for functionality required elsewhere. Clearly, the business owner cannot be held accountable for a mistake made by the security administrator. It is vitally important therefore, to ensure that the correct access authorizations are allocated to the correct users and entities within the system. In an ideal model, the accountability for the allocation of access rights to the system and its objects should rest with only one person. This person should be the identified information owner for a section of the business.

By providing the ability of creating and maintaining security objects to the business owner, accountability and responsibility is more visible within the organization. The occurrence of fraud due to the inappropriate allocation of a function or transaction to a certain user can now be traced directly to a single source.

4.3 Moving towards information ownership

The concepts discussed briefly above contribute most towards creating a decentralized security infrastructure within an ERP environment. Although numerous other concepts can be mentioned, these are beyond the scope of this study. Reduction of complexity and increased accountability and responsibility are considered the most important elements for the move towards increased information security with ERP systems. The concept of permitting individuals within the business to manage and maintain their own information security is termed information ownership. This concept will be expanded upon in the remainder of the paper.

4.4 Decentralizing without the need for more technical knowledge

It is necessary to stress the importance of reducing the need for more technical knowledge. The paragraphs above have dealt with the requirement to reduce complexity and to increase accountability. This cannot be accomplished if the security administrator's task is simply handed over to the individual business owners.

The business owners should not have to be taught the technical implementation details of creating roles, profiles and user master records. Instead, the technical complexity should remain with the security administrators, who maintain a support function. The creation of roles and profiles, together with the allocation of these to user master records should take place in a user-friendly environment. Ideally, the business owner should simply allocate functions and access requirements to a user by means of "point and click". If required, a more complex access requirement could be created by the security administrator for allocation by the business owner.

The need to make the security subsystem user-friendly and less technical for the business owners cannot be overstated. In fact, this can be considered the most important aspect for supporting the concept of information ownership.

5. INFORMATION OWNERSHIP

Though the term information ownership has been in use for some time, the concept of information ownership has not moved into the sphere of ERP software systems. The sudden popularity of ERP software packages stemmed primarily from their ability to integrate all data within the organization, to deliver real-time results and reporting and to make specific functionality available to the user at the desktop level. Unlike various business functionality and automation of business processes, the implementation of security is still considered a centralized function.

In stark contrast to the adaptability and flexibility of being able to configure the ERP software package to the needs of the business, the configuration of security related objects is completed by technical staff. Mention has already been made of the complexity of ERP software packages[7]. This complexity is necessary for the software package to be adaptable to different industries and legal requirements. The ERP software packages investigated during the course of this study provide full support for the configuration of the software to adapt it to support various business processes. Similar functionality for the configuration of security objects is missing. In fact, all security-related objects are generally grouped together and are not easily distinguished from one another. The ability to document the necessary access restrictions and security objects is hampered by very technical naming conventions. In ERP systems targeted at organizations with a smaller user population, the configuration of the security subsystem is often fairly trivial, offering the security administrator very little flexibility. It is clear that concept of information ownership is not supported by the ERP software packages investigated during the course of this study.

5.1 Supporting information ownership

The support for information ownership within the ERP software packages investigated during the course of this study is not natively possible. In other words, extensive changes to the user interface and possibly data structures would be necessary. The primary reason for this is the architecture employed by the software vendors to support their centralized security subsystems.

All the ERP systems under investigation employ a repository within which all information is stored. This pertains to data specific to the software system, customer information and security information. To enable the user master records to be filled with the necessary access and authorizations, various security objects need to be configured by a security administrator. Though the method and implementation differs from one vendor to another, the procedure can be unified and described as follows: each object within the ERP system has a certain access entry associated with it. To gain access to such an object, the user master record must contain an entry that corresponds to the access entry. Within the code of the ERP software, certain routines determine whether or not the user is permitted to access certain information or whether a function may be executed. If the entry within the user master record matches with the required entry, the user is permitted to continue.

The user master record is stored in one or more tables. All access entries required by the objects within the system are also stored in tables. Due to the fact that an ERP system spans many different functional areas, also known as modules, numerous access entries exist. General ledger, sales, materials management and asset management are examples of common modules within an ERP system. The centralized nature of the security subsystem permits the security administrator to allocate as many different access entries as required. This is regardless of the functional area under consideration. The inability of an ERP system to natively support information ownership can be summed up as follows: there is no support for determining what access entries the current user is permitted to allocate. In the centralized model favoured by the ERP software vendors, the information owner would be able to allocate all access rights to all modules within the system. The reason for this is that the system is unable to determine which area of the system the information owner is responsible for. Clearly, a mechanism is required to enforce such validation.

5.2 Validating information owners

The validation of information owners is important to support information ownership. For information ownership to be viable, the system must allow only certain users to be able to allocate certain tasks and functions. If this is not the case, the architecture reverts back to a centralized security configuration where a single user is able to allocate all access authorizations across all modules of the system.

To validate information owners, a simple check has to be performed by the security subsystem before any access entries for objects may be allocated. Thus, the system has to ensure that the information owner is in fact an information owner and that the information owner is indeed authorized to allocate the security objects in question. As multiple information owners exist within the system, these two checks would have to be performed by the system. In effect, this becomes a double authorization check within the code of the ERP system, but is in effect only when the information owner is busy with the creation of roles and profiles. To ensure correct operation, the system should be capable of displaying only those objects pertaining to the module or area within which the information owner is supposed to be working.

Equally important is the scope of users to which the information owner may allocate roles and profiles. In a similar fashion to the way in which the system should restrict access to only the access entries available for allocation to roles and profiles, so too should access be restricted to only the users reporting to the information owner. The information owner may not allocate access authorization to users outside the scope of his sphere of responsibility.

5.3 Shared responsibility

Once the information owner is able to allocate only certain access rights to certain users, the concept of information ownership is assured. However, a potential pitfall is the case where various users may require access to functions or information that exist within the sphere of another information owner. As an example, a user creating a sales order may require access to determine the credit worthiness of the customer. In this example, two information owners may have to be involved: the information owner for the sales module and the information owner for the financial module. For this to be possible, a mechanism must be in place to permit the information owner for the sales module to allocate a certain function or task to a user in the financial module. Similarly, a user within the financial module may require access to a function or task within the sales module.

The mechanism to achieve this is required due to the integration of functionality and data within an ERP system. A number of tasks may require some form of cross-module integration within the ERP system. The simplest solution to the problem of shared responsibilities is to permit each information owner to expose basic functions to other information owners. Within the ERP system's security subsystem, such a function could be possible by permitting an information owner to create basic access rights for common functions that may be required by other users in the system. If a function needs to be executed by a user under the responsibility of another information owner, for which the access rights have not been created, a formal request would have to be made. It would be up to the information owner to determine whether or not such access would be granted. In this way, the responsibility and accountability of the information owner would be maintained and assured.

6. INFORMATION OWNERSHIP

From the discussion above, the advantages of the approach to security within ERP software systems should be clear. This is especially so when compared to the traditional centralized approach favoured by current ERP software vendors.

In summary, the primary advantages of the information ownership approach to enhancing security within ERP software systems are:
1. a reduction of complexity;
2. the ability to increase responsibility and accountability within the organization;
3. a faster implementation time by providing decentralized access to security objects;
4. to improve the quality of the security configuration as a whole;

7. ERP SYSTEM SUPPORT FOR INFORMATION OWNERSHIP

It has been stated in a previous paragraph that existing ERP software packages cannot simply be retrofitted to support the information ownership concept. The primary reason is the requirement for a different architecture to support the information owner and the mechanism of allocating access rights to users assigned to the information owner's group.

In addition to the requirement for an extended architecture for the support of the decentralization of the security objects, an integrated change management and documentation module is required. Ideally, the change management and documentation module should be integrated with a workflow engine that permits automatic routing of tasks and requests from one information owner to another. A detailed discussion of these aspects is beyond the scope of this paper. A complete and ideal model for a security subsystem that supports all aspects of information ownership to enhance security within ERP software packages is under development[6].

7.1 Corporate governance in support of information ownership

A strong case for the information ownership concept is the current implementation of various legal requirements in different countries of the world. In the United States of America, the Sarbanes-Oxley act has become increasingly important. In Europe and the United Kingdom, the Basel 2 Accord and Turnbull Report respectively are gaining increasing importance[8]. Though many of the world's largest and most influential organizations make use of ERP technology, very basic support for the segregation of duties issues required by legislation is present in existing ERP systems. Segregation of duties is seen to be one of the most important aspects to prevent fraud and heighten security[9].

8. CONCLUSION

This paper has briefly described some of the problems encountered when implementing security within existing ERP software packages. An alternative has been proposed, that permits the formerly centralized security architecture to be decentralized. The concept of information ownership was briefly explained, and its importance highlighted. By supporting information ownership, the security of ERP software packages can be increased and improved.

REFERENCES

1. Joseph R. Dervaes, Internal Fraud and Controls, Washington Finance Officer's Association, 48th Annual Conference, 19 September 2004
2. SAP AG, SAP R/3 Online Help CD-ROM, 2003
3. Oracle Corporation, Oracle Applications System Administrator's Guide, Release 11i, Volume 1, 2002
4. Microsoft Business Solutions, Installation and System Management Manual: Navision Attain, Navision, 2002
5. Microsoft Corporation, Axapta System Administration Guide, Microsoft Business Solutions, 2002
6. M. Hertenberger, Prof. S.H. von Solms, Ph.D. study in progress: "Security in ERP environments", Rand Afrikaans University, Johannesburg, South Africa, 2004
7. K. Vuppula. BW security approaches, http://www.intelligenterp.com/feature/2002/12/0212feat1_1.shtml, 2002
8. P. Manchester, Financial Times, 12 November 2003
9. Elizabeth M. Ready, Emerging Fraud Trends, State of Vermont, 2003

INTERACTIVE ACCESS CONTROL FOR WEB SERVICES*

Hristo Koshutanski and Fabio Massacci
Dip. di Informatica e Telecomunicazioni - Univ. di Trento
via Sommarive 14 - 38050 Povo di Trento (ITALY)
{hristo, massacci}@dit.unitn.it

Abstract Business Processes for Web Services (BPEL4WS) are the new paradigms for lightweight enterprise integration. They cross organizational boundaries and are provided by entities that see each other just as business partners. Web services require shift in the access control mechanism: from identity-based access control to trust management and negotiation, but this is not enough for cross organizational business processes. For many businesses no partner may guess a priori what kind of credentials will be sent by clients and clients may not know a priori which credentials are required for completing a business process.

We propose a logical framework for reasoning about access control for BPEL4WS and a BPEL4WS based implementation using Collaxa server. Our model is based on interaction and exchange of requests for supplying or declining missing credentials. We identify the formal reasoning services (deduction, abduction, consistency checking) that characterise the problem and discuss their implementation.

Keywords: Web Services; Business Processes; Credential-Based Systems; Interactive Access Control; Internet Computing; Logics for Access Control

1. INTRODUCTION

In the past millennium the development of middleware marked influenced the IT sector efforts to integrate distributed resources of a corporation. The new century has seen the rise of a new concept: virtual enterprises, the result of the outsourcing trend of the last 10 years in the IT sector.

*This work is partially funded by the IST programme of the EU Commission FET under the IST-2001-37004 WASP project and by the FIRB programme of MIUR under the RBNE0195K5 ASTRO Project and RBAU01P5SS Project.

Conceptually, a virtual enterprise is born when a business process is not longer closed within the boundary of a single corporation. It is composed by partners that offer their services on the web and lightly integrate their efforts into one (hopefully coherent) process.

To support the process of lightweight integration of partners' resources, a number of specifications and standards have emerged. SOAP and Web Services Description Language[1] (WSDL) help organisations in exposing their basic functionalities as Web Services. Business Process Execution Language[2] (BPEL4WS) and Electronic Business XML initiative[3] (ebXML) describe the behavior of complex business processes.

Intuitively, business processes are hierarchical graphs where each composite node represents an orchestration activity and primitive nodes are Web Services interfaces described in WSDL.

Considering the nature of a virtual enterprise – orchestration and choreography of WS, global and local business processes, complex business transactions – the picture gets complicated. Distributed processes, in a virtual enterprise, become more dynamic, allowing new partners and services to be selected at runtime.

The scenario offered by business processes for web services is particularly challenging for the definition of its security features. It has aspects of trust management systems and aspects of workflow security management.

From the trust management systems (see e.g. [18, 8, 15]) it takes the credential-based view: a (web) service is offered on its own and the decision to grant or deny access can only be made on the basis of the credentials sent by the client. In contrast with trust management system, we have a process and thus a notion of assignment of permissions to credentials that requires to look beyond the single access decision.

From workflow access control systems (see e.g. [2, 3, 10, 11]) we borrow all classical problems such as dynamic assignment of roles to users, dynamic separation of duties, and assignment of permissions to users according the least privilege principles. In contrast with workflow security management schemes, a business process for web services crosses organizational boundaries and is provided by entities that see each other as partners and nothing else. We have something even more loosely coupled than federated databases.

Also, we can no longer assume that an enterprise will assign tasks and roles to users (its employees) in a way that makes the completion of

---

[1] WSDL–http://www.w3.org/TR/wsdl
[2] BPEL4WS–http://www-106.ibm.com/developerworks/webservices/library/ws-bpel
[3] ebXML Business Process Spec. – www.ebxml.org/specs/ebBPSS.pdf

the workflow possible w.r.t. its security constraints. The reason is that such enterprise no longer exists. So, it must be possible for a user to communicate missing credentials.

In this paper we propose a logical framework for reasoning about access control for business processes for web services. We identify the different reasoning tasks (deduction, abduction, consistency checking) that characterize the problem and clarify the problems of temporal evolution of the logical model.

2. SYSTEM ARCHITECTURE

In this section we sketch the architecture of a system for distributed access control for Web and Business Processes that we have implemented. We refer to [12] for additional information on the rationale behind the architecture. At the time of writing we have done an initial prototype including the main entities of the system given below.

PolicyEvaluator makes endpoint decisions on access control. All partners involved in a business process are likely to be as different entities, each represented by a PolicyEvaluator. It encapsulates the partner's specific authorization policy, and presents it as a service using standardized WS interface (e.g., WSDL).

PolicyOrchestrator is an entity responsible for the workflow level access and release control. It decides which are the partners that are involved in the requested service and on the basis of some orchestration security policies combines the corresponding PolicyEvaluators in a form of a business process that is suitable for execution by the AuthorizationServer.

AuthorizationServer is responsible for *locating*, *executing*, and *managing* all needed PolicyEvaluators, and returning an appropriate result to the ApplicationServer. Also it is responsible for managing all the *interactions* with the Client.

At the application level, the architecture does not envisage the typical exchange of messages in access control system: "data" level (credentials, policies, requests, objects, etc.) that must be interpreted by the recipients. We can exchange messages at "source code" level and in particular at the level of business process description. Partners exchanges "mobile" processes (namely BPEL files) passing from one entity to another indicating themselves what the recipient has to do.

The mobility of authorization processes has a number of advantages. First of all, a server simply needs an off-the-shelf interpreter for business processes for a quick implementation. Second, we have more flexibility for describing the process leading to an access control decision.

To say few words on the implementation, Collaxa[4] is used as a main BPEL manager (on the AuthorizationServer side) for executing and managing all policy composition processes returned by the PolicyOrchestrator, as well as, for the implementation of the of the AuthorizationServer itself.

The AuthorizationServer itself is a BPEL process deployed under Collaxa that internally deploys the policy process returned by the PolicyOrchestrator as an internal web service and then also internally executes it. The advantage in this case is that if the AuthorizationServer is requested to get an access decision for a service that has already been asked for it and there is no change in the workflow policy then the AuthorizationServer *does not* deploy the service's policy process again but just (internally) executes it. In that way we speed up the access decision time.

PolicyOrchestrator in the current prototype is just a mapping between a service resource and its workflow policy process. We assume that the the process is already created by some GUI (e.g., could be used any BPEL visual tool generator that actually connects all involved partners' PEs in a BPEL process) and is available to the orchestrator.

PolicyEvaluator is another key point in our system. In its core, it is a Java module that acts as a wrapper for the DLV[5] system and implements our interactive algorithm for stateless WS described in Section 6.

3. THE FORMAL FRAMEWORK

Our formal model for reasoning on access control is based variants of Datalog with the stable model semantics and combines in a novel way a number of features

- logic for trust management by Li et al. [15];
- logic for workflow access control by Bertino et al. [3];
- logic for release and access control by Bonatti and Samarati [4].

We consider the view of a single partner since we cannot assume sharing of policies between partners. In [12] it is explained how the entire process can be orchestrated by using "mobile" business processes, while keeping each partner policy decision process as a black-box.

[4]Collaxa BPEL Server – www.collaxa.com
[5]DLV System – www.dlvsystem.com

In our framework each partner has a *security policy for access control* $\mathcal{P}_{\mathcal{A}}$ and a *security policy for disclosure control* $\mathcal{P}_{\mathcal{D}}$, whose syntax will be defined later in section 5.

The policy for access control is used for making decision about usage of all web services offered by the partner. We will use abduction to infer the missing credentials from the access policy and the credentials already presented by the user. The disclosure policy is used (as the name implies) for controlling disclosure of credentials. Basically, we ask the client only the missing credentials that are disclosable according to $\mathcal{P}_{\mathcal{D}}$.

To execute a service of the fragment of a business process under the control of the partner the user will submit a set of *presented credentials* $\mathcal{C}_{\mathcal{P}}$, a set of *declined credentials* $\mathcal{C}_{\mathcal{N}}$ and a *service request* r. We assume that $\mathcal{C}_{\mathcal{P}}$ and $\mathcal{C}_{\mathcal{N}}$ are disjoint.

For the syntax we build upon [3, 4, 15]. We have three disjoint sets of constants: one for users identifiers denoted by $\mathsf{User}\!:\!U$; one for roles denoted by $\mathsf{Role}\!:\!R$; and one for services denoted by $\mathsf{WebServ}\!:\!S$.

The predicates can be divided into three classes: predicates for assignments of users to roles and services (Fig. 1a), predicates for credentials (Fig. 1b), and predicates describing the current status of the system. The last class of predicates keeps track on the main activities done by users and services, such as: a predicate specifying successful activation of services by users; a predicate for successful completion of services; its dual one for abortion; predicates indicating granting a service to a user and, the opposite one, denial user's access to a service.

Furthermore, for some additional workflow constraints we need to have some meta-level predicates that specify how many statements are true. We use here a notation borrowed from Niemela **smodels** system, but we are substantially using the **count** predicates defined by Das [6]:

$n \leq \{X.\,Pr\}$ where n is a positive integer, X is a set of variables, and Pr is a predicate, so that intuitively $n \leq \{X.\,Pr\}$ is true in a model if at least n instances of the grounding of X variables in Pr are satisfied by the model. The $\{X.\,Pr\} \leq n$ is the dual predicate.

We assume additional comparison predicates (for instance for equality or inequalities) or some additional monadic predicates for instance to qualify service, users, or keys for credentials.

4. LOGIC PROGRAMMING BACKGROUND

Normal logic programs [1] are sets of *rules* of the form:

$$A \leftarrow B_1, \ldots, B_n,\ not\, C_1, \ldots,\ not\, C_m \qquad (1)$$

$\mathsf{Role}:R_i \succ \mathsf{Role}:R_j$ when role $\mathsf{Role}:R_i$ dominates role $\mathsf{Role}:R_j$.

$\mathsf{Role}:R_i \succ_{\mathsf{WebServ}:S} \mathsf{Role}:R_j$ when role $\mathsf{Role}:R_i$ dominates, just for service $\mathsf{WebServ}:S$, the role $\mathsf{Role}:R_j$.

assign $(P, \mathsf{WebServ}:S)$ when an access to the service $\mathsf{WebServ}:S$ is granted to P. P can be either a $\mathsf{Role}:R$ or $\mathsf{User}:U$.

forced $(P, \mathsf{WebServ}:S)$ when access the service $\mathsf{WebServ}:S$ must be forced to P. Principal P can be either a $\mathsf{Role}:R$ or $\mathsf{User}:U$.

(a) Predicates for assignments to Roles and Services

declaration $(\mathsf{User}:U)$ it is a statement by the $\mathsf{User}:U$ for its identity.

credential $(\mathsf{User}:U, \mathsf{Role}:R)$ when $\mathsf{User}:U$ has a credential activating $\mathsf{Role}:R$.

credentialTask $(\mathsf{User}:U, \mathsf{WebServ}:S)$ when $\mathsf{User}:U$ has the right to access $\mathsf{WebServ}:S$.

(b) Predicates for Credentials

Figure 1. Predicates used in the model

where A, B_i and C_i are (possibly ground) predicates among those described in Section 3. A is called the *head* of the rule, each B_i is called a *positive literal* and each *not* C_j is a *negative literal*, whereas the conjunction of the B_i and *not* C_j is called the *body* of the rule. If the body is empty the rule is called a *fact*. A normal logic program is a set of rules.

In our framework, we also need *constraints* that are rules with an empty head.

$$\leftarrow \quad B_1, \ldots, B_n, \ not\, C_1, \ldots, \ not\, C_m \tag{2}$$

One of the most prominent semantics for normal logic programs is the *stable model semantics* proposed by Gelfond and Lifschitz [9] (see also [1] for an introduction). The intuition is to interpret the rules of a program P as constraints on a solution set S (a set of ground atoms) for the program itself. So, if S is a set of atoms, rule (1) is a constraint on S stating that if all B_i are in S and none of C_j are in it, then A must be in S. A constraint (2) is used to rule out from the set of acceptable models the situation in which B_i are true and all C_j are false is not acceptable.

We now consider ground rules, i.e. rules where atoms do not contain variables.

DEFINITION 1 *The* reduct P^S *of a ground logic program P with respect to a set of atoms S is the definite program obtained from P by deleting:*

1 *each rule that has a negative literal* $not\, C$ *in its body with* $C \in S$;

2 *each negative literal in the bodies of the remaining rules.*

The reduct P^S is a definite logic program. Let $M(P^S) = M_{P^S}$ be the semantics of the definite logic program P^S, i.e. its minimal model.

DEFINITION 2 *A set of atoms* S *is a stable model of a normal logic program* P *iff* $S = M(P^S)$.

A program can have none, one or many stable models. The definition of stable models captures the two key properties of solution sets of logic programs.

1 Stable models are minimal: a proper subset of a stable model is not a stable model.

2 Stable models are grounded: each atom in a stable model has a justification in terms of the program, i.e. it is derivable from the reduct of the program with respect to the model.

Though this definition of stable models in terms of fix points is non-constructive there are constructive definitions [1] and systems [17, 14] that can cope with ground programs having tens of thousands of rules.

Logic programs with variables can be given a semantics in terms of stable models.

DEFINITION 3 *The stable models of a normal logic program* P *with variables are those of its ground instantiation* P_H *with respect to its Herbrand universe*[6].

If logic programs are function free, then an upper bound on the number of instantiations is rc^v, where r is the number of rules, c the number of the constants, and v the upper bound on the number of distinct variables in each rule.

DEFINITION 4 (LOGICAL CONSEQUENCE AND CONSISTENCY) *Let* P *be a logic program and* L *be a (positive or negative) ground literal.* L *is a* logical consequence *of* P ($P \models L$) *if* L *is true in every stable model of* P. P *is* consistent ($P \not\models \bot$) *if there is a stable model for* P.

DEFINITION 5 (ABDUCTION) *Let* P *be a logic program,* H *a set of predicates (called hypothesis, or abducibles),* L *a (positive or negative) ground*

[6] Essentially, we take all constants and functions appearing in the program and combine them in all possible ways. This yields the Herbrand universe. Those terms are then used to replace variables in all possible ways thus building its ground instantiation.

literal, and $\prec$ *a p.o. over subsets of* H*, the* cautious solution *of the abduction problem is a set of ground atoms* E *such that*

1 E *is a set ground instances of predicates in* H*,*

2 $P \cup E \models L$

3 $P \cup E \not\models \bot$

4 any set $E' \prec E$ *does not satisfy all conditions above*

Traditional p.o.s are subset containment or set cardinality. Other solutions are possible with orderings over predicates.

5. THE LOGICAL MODEL

In this section we define the semantics of our logical model and give formal definitions of the security policies introduced in Section 3.

DEFINITION 6 *An* access control policy $\mathcal{P_A}$ *is a logic program over the predicates defined in Section 3 in which (i) no credential and no execution atom can occur in the head of a rule, (ii) role hierarchy atoms occur as facts, (iii) for every rule containing a head A which is the (possibly ground instance of) predicate* $\mathsf{forced}\,(P, \mathsf{WebServ}\!:\!S)$ *there is the (possibly ground instance of) rule* $\mathsf{assign}\,(P, \mathsf{WebServ}\!:\!S) \leftarrow \mathsf{forced}\,(P, \mathsf{WebServ}\!:\!S)$.

An access request is a ground instance of an $\mathsf{assign}\,(\mathsf{User}\!:\!U, \mathsf{WebServ}\!:\!S)$ *predicate.*

The request r *is a security consequence of a policy* $\mathcal{P_A}$ *if (i)* $\mathcal{P_A}$ *is logically consistent and (ii)* r *is a logical consequence of* $\mathcal{P_A}$*.*

In contrast to the proposal by Bertino et al. [3] for workflows we don't need any special rule for determining which services cannot be executed and which services must be executed by a specific user or role. The $\mathsf{forced}\,(,)$ predicate and the constraints guarantee the same result.

EXAMPLE 7 *Consider a security policy in which having a credential for the role accountant is incompatible with the assignment of any role manager, and that the execution of a service phoneCall from user billG requires that the service answer must be executed by anybody having the role* $headOfStaff$*. The following rules guarantees the desired behavior:*

$$\leftarrow \mathsf{credential}\,(\mathsf{User}\!:\!U, \mathsf{Role}\!:\!accountant), \mathsf{assign}\,(\mathsf{User}\!:\!U, \mathsf{Role}\!:\!manager).$$
$$\mathsf{forced}\,(\mathsf{Role}\!:\!headOfStaff, \mathsf{WebServ}\!:\!answer) \leftarrow \mathsf{running}\,(\mathsf{User}\!:\!billG, \mathsf{WebServ}\!:\!call, \mathsf{number}\!:\!N). \quad \blacksquare$$

EXAMPLE 8 *Consider an e-stock portal where we have roles associated to services as follows: role* eSeller *– for selling shares and bonds on the*

floor; role eBuyer *– for buying shares and bonds; role* eAdvisor *– used by accredited consultants to sell their advice to other customers of the portal. Then examine the case where one could send the* eAdvisor *credential to the service publishing advisories and suggest to sell shares, and at the same time the* eBuyer *credential to the service hosting bids.*
In such situations we can define separation of duty rules:

$customer(\mathsf{eSeller}) \leftarrow .$
$customer(\mathsf{eBuyer}) \leftarrow .$
$\leftarrow \mathsf{assign}\ (\mathsf{User}: U, \mathsf{Role}: R_1), customer(R_1), \mathsf{assign}\ (\mathsf{User}: U, \mathsf{Role}: \mathsf{eAdvisor}).$

The access control rule on reviewing selling bids is the following:

$$\mathsf{assign}\ (\mathsf{User}: U, \mathsf{WebServ}: S) \leftarrow \mathsf{credential}\ (\mathsf{User}: U, \mathsf{Role}: R), \mathsf{assign}\ (\mathsf{Role}: R, \mathsf{WebServ}: S).$$
$$\mathsf{assign}\ (\mathsf{Role}: R, \mathsf{WebServ}: reviewSell) \leftarrow \mathsf{Role}: R \succ \mathsf{Role}: \mathsf{eSeller}. \blacksquare$$

As mentioned, we will use the disclosure policy $\mathcal{P_D}$ to decide which missing credentials are to be asked from the client.

DEFINITION 9 *A* disclosure policy $\mathcal{P_D}$ *is a logic program in which no role hierarchy atom and no execution atom can occur in the head of a rule.*

DEFINITION 10 *A credential c is disclosable if it is a logical consequence of the disclosure policy $\mathcal{P_D}$ and presented credentials $\mathcal{C_P}$ ($\mathcal{P_D} \cup \mathcal{C_P} \models c$).*

EXAMPLE 11 *Considering again the access policy in Example 8. A possible (part of) the disclosure policy $\mathcal{P_D}$ could be:*

$\mathsf{credential}\ (\mathsf{User}: U, \mathsf{Role}: eUser) \leftarrow \mathsf{declaration}\ (\mathsf{User}: U).$
$\mathsf{credential}\ (\mathsf{User}: U, \mathsf{Role}: eSeller) \leftarrow \mathsf{credential}\ (\mathsf{User}: U, \mathsf{Role}: eUser).$
$\mathsf{credential}\ (\mathsf{User}: U, \mathsf{Role}: eSellerVIP) \leftarrow \mathsf{credential}\ (\mathsf{User}: U, \mathsf{Role}: eSeller).$

The second rule says: to reveal the need for a eSeller *credential there should be already a credential attesting the client as a valid user of the system together with a declaration of its identity.* ■

So, the request $\mathsf{assign}\ (\mathsf{User}: fm, \mathsf{WebServ}: reviewSell)$ together with $\mathsf{credential}\ (\mathsf{User}: fm, \mathsf{Role}: eUser)$ and $\mathsf{declaration}\ (\mathsf{User}: fm)$ will yield a counter request – $\mathsf{credential}\ (\mathsf{User}: fm, \mathsf{Role}: eSeller)$ – specifying the need for additional privileges necessitated to get the service.

Note that the need for a credential attesting the role *eSellerVIP*, disclosed together with *eSeller*, should not be considered as a potential output by the system because the "intuition" says that *eSeller* is enough.

REMARK 1 *The choice of the partial order has a major impact in presence of complex role hierarchies. The "intuitive" behavior of the abduction algorithm for the extraction of the minimal set of security credentials is not guaranteed by the straightforward interpretation of H (abducibles) as the set of credentials and by the set cardinality or set containment orderings.*

Consider the following program:

$$\begin{aligned}&\mathsf{Role}{:}r_2 \succ \mathsf{Role}{:}r_1 \leftarrow .\\ &\mathsf{assign}\,(\mathsf{User}{:}U, \mathsf{WebServ}{:}ws) \leftarrow \mathsf{credential}\,(\mathsf{User}{:}U, \mathsf{Role}{:}R),\\ &\qquad\qquad\qquad\qquad\qquad\qquad\quad \mathsf{Role}{:}R \succ \mathsf{Role}{:}r_1.\end{aligned}$$

Request $\mathsf{assign}\,(\mathsf{User}{:}fm, \mathsf{WebServ}{:}ws)$ has two $\subseteq$-minimal solutions:

$$\{\mathsf{credential}\,(\mathsf{User}{:}fm, \mathsf{Role}{:}r_1)\}\,, \{\mathsf{credential}\,(\mathsf{User}{:}fm, \mathsf{Role}{:}r_2)\}$$

Yet, our intuition is that the first should be the minimal one.

So, we need a more sophisticated partial order. For example, if $E \preceq E'$ is such that for all credentials $c \in E$ there is a credential $c' \in E'$ where $c = c'$, we can revise it so that $E \prec E'$ if $c \in E$ there is a credential $c' \in E'$ where c' is identical to c except that it contains a role R' that dominates the corresponding role R in c. This p.o. generates the "intuitive" behavior of the abduction algorithm.

Another alternative, currently implemented in out prototype, is to include extra information to credentials in the hypotheses (abducibles), specifying the position of a role in the role-lattice hierarchy. Then it is easy to select the set(s) with the lowest role-position values. After having obtained the missing credentials, we drop this extra information from the set that is to be sent back to the client.

DEFINITION 12 (FAIR ACCESS) *Let $\mathcal{P_A}$ be an access control policy, let $\mathcal{C_D}$ be the set of ground instances of credentials occurring in $\mathcal{P_A}$, and let $\prec$ be a p.o. over subsets of $\mathcal{C_D}$. The policy $\mathcal{P_A}$ guarantees $\prec$-*fair access *if for any ground request r that is an instance of a head of a rule in $\mathcal{P_A}$ there exists a set $\mathcal{C_M} \subseteq \mathcal{C_D}$ that is a solution of the abduction problem.*

DEFINITION 13 (FAIR INTERACTION) *Let $\mathcal{P_A}$ and $\mathcal{P_D}$ be, respectively, an access and disclosure control policies, and let $\mathcal{C_D}$ be the set of ground instances of credentials occurring in $\mathcal{P_A}$, and let $\prec$ be a p.o. over subsets of $\mathcal{C_D}$. The policies guarantee $\prec$-*fair interaction *w.r.t. a set of initial credentials $\mathcal{C_I}$ if (i) $\mathcal{P_A}$ guarantees $\prec$-fair access and (ii) for any solution of the abduction problem $\mathcal{C_M} \subseteq \mathcal{C_D}$ and any credential $c \in \mathcal{C_M}$ if it $\mathcal{P_D} \cup \mathcal{C_I} \models c$. If the set $\mathcal{C_I}$ only contains declarations then the disclosure is* unlimited.

The above process does not take into account the progressive disclosure of credentials in the interactive process.

REMARK 2 *It is possible to define a process of trust negotiation along the lines of Yu et al. [19] if at each interaction step we ask only for the credentials that are entailed by a 1-step deduction over $\mathcal{P}_\mathcal{D}$ and $\mathcal{C}_\mathcal{P}$. In this case, the interaction policy must be a* monotonic *logic program.*

6. REASONING

In this section we show how the various notions that we have seen so far can be combined into a complete authorization mechanism. PolicyEvaluator receives the request r, processes it according to the access control algorithm and eventually takes a decision. A decision may have involved interactions and so we also keep track of the current set of active credentials and the history of the requests made by the client.

Since the client must collect all relevant credentials (if required) for getting access to a service, one could borrow mechanisms for discovering distributed credentials' chains from [16, 5].

Once again it is worth noting that this view is partial as we only focus on the knowledge of one single partner: there is no authorization domain crossing partnerships.

To allow for an easier grasp of the problem, we start with a basic framework shown in Figure 2. This approach is the cornerstone of most logical formalizations [7].

---

1 verify that the request is a logical consequence of the credentials, namely $\mathcal{P}_\mathcal{A} \cup \mathcal{C}_\mathcal{P} \models r$

2 if the check succeeds then grant access else deny access

Figure 2. Traditional Access Control

A number of works has deemed such blunt denials unsatisfactory and therefore it has been proposed by Bonatti and Samarati [4] and Yu et al. [19] to send back to the client some of the rules that are necessary to gain additional access (see Figure 3). In their work it is revised to allow for the flow of rules and information to users.

Since the systems proposed by both Bonatti and Samarati [4] and Yu et al. [19] are flat, in p the client will find all missing credentials to continue the process until r is granted.

In many cases, this is neither sufficient nor desirable. For instance, if the policy is not flat, it has constraints on the credentials that can

1 verify that the request is a logical consequence of the credentials, namely $\mathcal{P}_{\mathcal{A}} \cup \mathcal{C}_{\mathcal{P}} \models r$

2 if the check succeeds then access is granted, otherwise select some rule $r \leftarrow p \in \texttt{PartialEvaluation}(\mathcal{P}_{\mathcal{A}} \cup \mathcal{C}_{\mathcal{P}})$ and send the rule back to the client

Figure 3. Disclosable Access Control

be presented at the same time (e.g., separation of duties) or a more complex role structure is used, these systems would not be complete. Also repeated queries allow for the disclosure of the entire policy, which might well be undesirable[7].

Another point in our formal model, worth discussing here, is the way we address the disjunctive information in the partner's disclosure policy (e.g., "present either a VISA or a Mastercard or an American Express card"). In presence of such disjunctive information an arbitrary disjunct will be selected and on the next interaction step (if the client has declined the credential) the abduction algorithm is informed that the previous solution was not accepted. We approach this by discarding the set of declined credentials from the set of newly computed disclosable credentials. In this case the abduction algorithm does not consider the declined credentials, from the last step, in the next interaction step.

Our interactive access control solution for Web Services is shown in Figure 4.

This is all we need for business processes made up by *stateless web services*, in which all decisions are taken on the basis of the current input set of credentials, and which envisaged to be the large majority.

This type of decision is characteristic of most logical approaches to access control [15, 3, 4]: we only look at the policy, the request and the set of credentials. The failure of the access control process at step *3c* (Fig. 4) may be due to the presence of badly designed constraints for separation of duties such that no possible set of credentials can unlock the service r. In same cases this might also be a feature of the systems.

[7]In the negotiation process of Yu et al. [19] rules are only disclosed when all preliminary credentials have been already sent by the client. Still this is unsatisfactory because we may well want to tell a user all credentials we may possibly ask him, but not how we are going to evaluate them.

1 extract from the client's input the set of presented credentials $\mathcal{C_P}$ and the set of declined credentials $\mathcal{C_N}$

2 verify that the request is a logical consequence of the credentials, namely $\mathcal{P_A} \cup \mathcal{C_P} \models r$

3 if the check succeeds then access is granted, otherwise

(a) compute the set of *disclosable credentials* $\mathcal{C_D}$ as $\mathcal{C_D} = \{c \mid c \text{ credential that } \mathcal{P_D} \cup \mathcal{C_P} \models c\} \setminus \mathcal{C_N}$

(b) use abduction to find a minimal set of missing credentials $\mathcal{C_M} \subseteq \mathcal{C_D}$ such that both $\mathcal{P_A} \cup \mathcal{C_P} \cup \mathcal{C_M} \models r$ and $\mathcal{P_A} \cup \mathcal{C_P} \cup \mathcal{C_M} \not\models \bot$

(c) if no such set exists then $\bot$ is sent back to the user,

(d) otherwise communicate the set of missing credentials $\mathcal{C_M}$ back to the client and iterate the process.

Figure 4. Interactive Access Control for Stateless WS

7. IMPLEMENTATION OF THE LOGICAL MODEL

For the implementation of the algorithm in Figure 4 we used DLV (a disjunctive datalog system with negations and constraints) as a core engine for the basic functionalities of deduction and abduction. The disjunctive datalog front-end (the default one) is used for deductive computations while the diagnosis front-end is used for abductive computations. We refer to Section 5 for definitions of deduction and abduction.

What follows is a step-by-step description of the implementation employing the DLV system:

1 Extract from the client's input the two sets of credentials $\mathcal{C_P}$ and $\mathcal{C_N}$, transform them to predicates suitable for the underlying formal model (ref. Fig. 1) and store them in temporary files;

2 Use the DLV's disjunctive datalog front-end. Specify as input the partner's access policy, the two sets from step 1 and the service request r marked as a query over the models computed by DLV. The output of this step are those models of the access policy in which r is true.

3 If it exists a model in step 2 that satisfies r then grant, otherwise:

(a) use again the DLV's front-end as input partner's disclosure policy $\mathcal{P_D}$ together with presented credentials $\mathcal{C_P}$. In this case DLV computes all models of $\mathcal{P_D}$ that are disclosable by $\mathcal{C_P}$.

Then from the computed models we remove all credentials that belongs to $\mathcal{C}_{\mathcal{N}}$.

(b) find a model, out of the ones in step 3a, for which it exists a subset satisfying the abductive computation described below. Use the abductive diagnosis (subset minimal) front-end with the following input: $\mathcal{P}_{\mathcal{A}}$, $\mathcal{C}_{\mathcal{P}}$, the set of credentials from the model being checked stored in a temporary file with special extension .hyp (called hypotheses or abducibles) and the service request r also stored in a temporary file with extension .obs (observations). The output of such computation are all possible subsets of the hypotheses that satisfy the observations. In that way we find all possible missing sets of credentials satisfying r. Then we filter them, first against role-minimality criterion, and then against set cardinality criterion. The former filters those sets with lowest possible role-position values and the latter filters the ones with minimal cardinality.

(c) if no such set exists reject otherwise send the missing set back to the client.

REMARK 3 *The sequence, the two criteria,* set cardinality *and* role minimality *makes sense in different contexts. The sequence* role minimality/set cardinality, *tries to keep the minimal set as lower in the role hierarchy as possible, i.e. selects those sets that have a larger number of not so powerful roles. The other alternative,* set cardinality/role minimality, *selects those sets with fewer roles but with higher privileges.*

The latter may be useful if getting or transmitting credentials is expensive (e.g., in a mobile setting).

8. STATEFUL BUSINESS PROCESSES

If the authorization decisions of business processes are stateful, and the corresponding workflow of the partners has constraints on the execution of future services on the basis of past services this solution is not adequate enough. For instance in the workflow example described by Atluri and Bertino [3, pag.67] a branch manager of a bank clearing a cheque cannot be the same member of staff who has emitted the cheque. The problems are the following:

- the request may be inconsistent with some role that the user has taken up in the past;
- the new set of credential may be inconsistent with requirements such as separation of duties;

- in contrast to intra-enterprise workflow systems [3], the partner offering the web service has no way to assign to the client the right set of credentials for consistency future request.

So, this means that we must have some roll-back procedure by which, if the user has by chance sent the "wrong" credentials, he has some revocation mechanism to drop them. A preliminary solution has been described in [13].

9. CONCLUSIONS

In this paper we proposed a logical framework for reasoning about access control for stateless business processes for web services. Our formal model for reasoning on access control is based on variants of Datalog with the stable model semantics and combines in a novel way a number of features: the logic for trust management by Li et al. [15]; the logic for workflow access control by Bertino et al. [3]; the logic for controlling the release of information by Bonatti and Samarati [4].

We identified the different reasoning tasks (deduction, abduction, consistency checking) that characterize the problem and clarify the problems of temporal evolution of the logical model.

Future work is in the direction of more effective trust negotiation for stateful business processes.

References

[1] APT, K. Logic programming. In *Handbook of Theoretical Computer Science*, J. van Leeuwen, Ed. Elsevier, 1990.

[2] ATLURI, V., CHUN, S. A., AND MAZZOLENI, P. A Chinese wall security model for decentralized workflow systems. In *Proceedings of the 8th ACM conference on Computer and Communications Security* (2001), ACM Press, pp. 48–57.

[3] BERTINO, E., FERRARI, E., AND ATLURI, V. The specification and enforcement of authorization constraints in workflow management systems. *ACM Transactions on Information and System Security (TISSEC) 2*, 1 (1999), 65–104.

[4] BONATTI, P., AND SAMARATI, P. A unified framework for regulating access and information release on the web. *Journal of Computer Security 10*, 3 (2002), 241–272.

[5] CLARKE, D., ELIEN, J.-E., ELLISON, C., FREDETTE, M., MORCOS, A., AND RIVEST, R. L. Certificate chain discovery in SPKI/SDSI. *Journal of Computer Security 9*, 4 (2001), 285–322.

[6] DAS, S. *Deductive Databases and Logic Programming.* Addison-Wesley, Reading, MA, 1992.

[7] DE CAPITANI DI VIMERCATI, S., AND SAMARATI, P. Access control: Policies, models, and mechanism. In *Foundations of Security Analysis and Design - Tutorial Lectures*, R. Focardi and F. Gorrieri, Eds., vol. 2171 of *LNCS*. Springer Verlag Press, 2001.

[8] Ellison, C., Frantz, B., Lampson, B., Rivest, R., Thomas, B. M., and Ylonen, T. *SPKI Certificate Theory*, September 1999. IETF RFC 2693.

[9] Gelfond, M., and Lifschitz, V. The stable model semantics for logic programming. In *Proceedings of the Fifth International Conference on Logic Programming (ICLP'88)* (1988), R. Kowalski and K. Bowen, Eds., MIT-Press, pp. 1070–1080.

[10] Georgakopoulos, D., Hornick, M. F., and Sheth, A. P. An overview of workflow management: From process modeling to workflow automation infrastructure. *Distributed and Parallel Databases 3*, 2 (April 1995), 119–153.

[11] Kang, M. H., Park, J. S., and Froscher, J. N. Access control mechanisms for inter-organizational workflow. In *Proceedings of the Sixth ACM Symposium on Access control models and technologies* (2001), ACM Press, pp. 66–74.

[12] Koshutanski, H., and Massacci, F. An access control framework for business processes for Web services. In *Proceedings of the 2003 ACM workshop on XML security* (Fairfax, VA, October 2003), ACM Press.

[13] Koshutanski, H., and Massacci, F. A logical model for security of Web services. Tech. Rep. IIT TR-10/2003, First International Workshop on Formal Aspects of Security and Trust (FAST), Istituto di Informatica e Telematica, Pisa, Italy, September 2003. Editors: Theo Dimitrakos and Fabio Martinelli.

[14] Leone, N., Pfeifer, G., and et al. The DLV system. In *the 8th European Conference on Artificial Intelligence (JELIA)* (September 2002), vol. 2424 of *Lecture Notes in Computer Science*, Springer, pp. 537–540.

[15] Li, N., Grosof, B. N., and Feigenbaum, J. Delegation logic: A logic-based approach to distributed authorization. *ACM Transactions on Information and System Security (TISSEC) 6*, 1 (2003), 128–171.

[16] Li, N., Winsborough, W. H., and Mitchell, J. C. Distributed credential chain discovery in trust management. *Journal of Computer Security 11*, 1 (February 2003), 35–86.

[17] Niemelä, I., Simons, P., and Soininen, T. Stable model semantics of weight constraint rules. In *Proceedings of the Fifth International Conference on Logic Programming and Nonmonotonic Reasoning* (December 1999), Springer-Verlag.

[18] Weeks, S. Understanding trust management systems. In *IEEE SS&P-2001* (2001), IEEE Press.

[19] Yu, T., Winslett, M., and Seamons, K. E. Supporting structured credentials and sensitive policies through interoperable strategies for automated trust negotiation. *ACM Transactions on Information and System Security (TISSEC) 6*, 1 (2003), 1–42.

IDENTITY-BASED KEY INFRASTRUCTURES (IKI)

Yvo Desmedt,[1,2] and Mike Burmester,[2]

[1] *Department of Computer Science*
University College London, UK

[2] *Department of Computer Science*
Florida State University, USA

burmester@cs.fsu.edu

Abstract Kohnfelder realized in 1978 that public key schemes require a Public Key Infrastructure (PKI). X500/X509 were set up to standardize these ideas. PGP, proposed by Zimmermann is an alternative to the original PKI idea. Variants of the PGP based PKI were discussed independently by Reiter-Stubblebine and Burmester-Desmedt-Kabatianskii.

The goal of Shamir's 1984 idea of "identity-based" cryptography was to avoid a Public Key Infrastructure. Instead of having the users have their own public key, the identity of the user is the "public key," and a trusted center provides each party with a secret key. Several identity-based cryptosystems have been proposed, in particular recently.

We analyze Shamir's identity-based concept critically. We argue the need for at least a registration infrastructure, which we call a "basic Identity-based Key Infrastructure." Moreover, if secret keys of users can be stolen or lost, the infrastructure required to deal with this is as complex as the one of PKI. We make further comparisons between public key systems and identity-based ones.

Keywords: PKI, trust infrastructures, identity-based cryptosystems

INTRODUCTION

While in conventional cryptosystems the sender as well as the receiver's key must remain secret, in public key systems the privacy of one of the keys is not essential, and so can be made public. However, public keys must still be authenticated. Kohnfelder Kohnfelder, 1978 observed the need for some Public Key Infrastructure to authenticate the public key of a user. A lot of work has been done on Public Key Infrastructures since then. X500 and X509 standards were developed (see e.g. Menezes et al., 1996; Schneier, 1996).

Public-key certificates have two parts: data and a signature. The data contains information about the identity of an entity, the public-key, the validity period and other relevant details. The signature is a digital signature on the data by a certifying entity. For the X509 certificates this is a Certification Authority (CA). With X509, the infrastructure is hierarchical, i.e. a rooted tree. The *Root* is a certification authority, called Root Certification Authority (RCA). The public-key of the RCA is known *a priori* to all users, and this knowledge is used to induce confidence in the public-keys of CAs who authenticate the public key of the user. More than one tree can be used.

Alternatives to X509 have been proposed in PGP Zimmermann, 1995. Alternatives that make the infrastructure more robust to make it less vulnerable to attacks by outsiders, and maintain security when CAs are sloppy, were suggested Reiter and Stubblebine, 1997; Burmester et al., 1998 (see also Burmester and Desmedt,).

Several countries in South-East Asia are in the process of setting up national PKIs. A problem with PKI is the cost to run a proper PKI. To be of any use, the identity of the party must be properly verified. This cannot be done over the internet! The natural question is whether one can avoid using a public key and a PKI using alternatives.

From a web page of Microsoft Microsoft Security Bulletin MS01-017, 2003 we learn:

> In mid-March 2001, VeriSign, Inc., advised Microsoft that on January 29 and 30, 2001, it issued two VeriSign Class 3 code-signing digital certificates to an individual who fraudulently claimed to be a Microsoft employee. The common name assigned to both certificates is "Microsoft Corporation". The ability to sign executable content using keys that purport to belong to Microsoft would clearly be advantageous to an attacker who wished to convince users to allow the content to run.

So, it is no surprise that some question PKIs Ellison and Schneier, 2000.

In 1984 Shamir Shamir, 1985 suggested the concept of identity-based cryptosystem. In such system *the public key of the user is just the identity of the user.* To guarantee that this works, a trusted center will use a master (secret) key. Using this master key, the trusted center computes from the identity of the user u, the secret key of the user. It seems that this removes the need for PKI. The goal of this paper is to critically analyze this impression. Note that recently several identity-based cryptosystems have been proposed, e.g. by Boneh-Franklin Boneh and Franklin, 2001.

We will demonstrate in Section 2 that a basic Identity-based Key Infrastructure is at least required to verify the identity of the user before handing out the secret key. When keys can be stolen or lost, one needs to deal with revocation, which is discussed in Section 3. We conclude in Section 4 and also discuss robustness issues. We now start by reviewing the concept of identity-based cryptography.

1. BACKGROUND AND NOTATION

Shamir invented the notion of identity-based cryptography Shamir, 1985. The idea is that each individual could use his/her/its own identity as "public key." A Key Distribution Center would compute the secret key starting from the identity of the user, using an algorithm f and based on the Key Distribution Center's top secret master key, K. We denote the output of f as $f(\cdot,\cdot)$.

Note that using secure distributed computation (see e.g. Goldreich et al., 1987; Ben-Or et al., 1988; Chaum et al., 1988) the need to trust a single Key Distributing Center can be relaxed.

Let ID_u be the identity of the user u. Evidently, this must be unique. So, it may have to contain other information besides what we call the *natural ID*, being the first name(s) and the last name of the user.

To deal with the case $f(ID_u, K)$ does not[1] exist, Shamir envisioned using a short binary string j_u so that $f(ID_u||j_u, K)$ does exist, where $||$ indicates concatenation. j_u could be computed using deterministic exhaustive search. So, the secret key of the user is $SK_u = f(ID_u||j_u, K)$, which is provided to the user by the Key Distribution Center.

If $f(ID_u, K)$ always exists, i.e. for each each string ID_u, then it seems that there is no need for j_u. However, in Section 3 we will argue that in a real world situation this is false.

When using encryption and digital signatures, the secret key SK_u plays the same role as the secret key in public key systems. $ID_u||j_u$ plays the same role as the public key does in public key systems. So, to privately send a message M to user u, the sender will send the ciphertext $C = E_{ID_i||j_u}(M)$ and to decrypt, the receiver u will compute $D_{SK_u}(C)$. To digitally sign the message M the user u will compute the signature $sign = Sign_{SK_u}(M)$ and to verify its correctness the receiver will compute the Boolean verification function $V_{ID_u||j_u}(M, sign)$.

2. THE NEED FOR A BASIC IDENTITY -BASED KEY INFRASTRUCTURE

The impact of lost or stolen secret keys on the concept of identity-based cryptography is only discussed in details from Section 3 on. So, in this section only, we assume, hypothetically, that secret keys are never lost or stolen.

As well known by the information security community, it is essential in PKI, that the identity of the user be properly checked. The aforementioned example of VeriSign illustrates this. For the same reason is it required that the Key Distribution Center verifies the identity of the user requesting a secret key. Indeed, if a third party can fraudulently claim to be user u, then it can pretend being the user u, i.e. decrypt/sign as user u.

Since, the identity verification cannot be done remotely (so not over the internet), there is a need for what one could call a Local Registration Center. We now stipulate the steps that need to be taken at the stage of "registration." This will demonstrate the need for a true infrastructure.

1 The Local Registration Center needs to first of all verify the identity ID_u of the user in person, checking as much evidence as possible. The required evidence may be a birth certificate, a passport (and possibly old passports[2]), a driver's license, witnesses, etc.

  In the case the user is an organization, such as a company, and the user would like to obtain a secret key for the entity, one needs to verify that the people who represent this organization have the authority to do so and to obtain a secret key in the name of the organization. Local laws may differ from country to country on who can legally represent an organization.

  Note, if one does not trust the Local Registration Center, multiple could be used, in a similar way as pointed out in Burmester et al., 1998; Reiter and Stubblebine, 1997 (see also Burmester and Desmedt,).

2 One needs to verify that the proposed ID_u has not yet been assigned. Indeed, otherwise, (at least two) users may have the same secret key and the one can pretend to be the other. Note that since public keys are significantly longer and much more random in nature than identities, the probability two users have the same public key is almost zero, and so can be ignored. For identities, the situation is very different. The solution to this problem now depends on whether the format of ID_u is:

  variable length encoding. The user could append other relevant information, which may depend from country to country[3].

  fixed length encoding. (This is for example the case with the login name on many operating systems.) In this case if:

the user's name are sufficiently long the user could combine parts of his/her first name(s) with parts of the last name to obtain a compact unique ID. In the case of a corporation, there is evidently no first name, etc.

the user's names are not sufficiently long we basically fall in a combination of the variable length encoding case with the fixed length encoding. The user needs to append to his name and/or compact the string.

Evidently, one could use j_u to make the identity unique, but, j_u will need to be used for other purposes, as we discuss in Section 3.

The issue of checking the uniqueness of the ID causes several problems:

- The ID_u is no longer public information in the sense that any third party will *not* know a priori all the information appended or the compacted string. In the case of variable length encoding, one can evidently try to append known information. For example, one could append the affiliation of this person, or the network provider used, etc. However, this also implies that a new secret key will be required when the affiliation or network provider changes.

  Note that if a public key system is used, one also needs to uniquely identify the user to find the public key of that user. However, there is a major difference. If a user plays different roles in society, and is therefore known under different "affiliations," the different IDs in the PKI database[4] related to the same individual, could all point to the *same* public key. In identity-based cryptosystems, these different "affiliations" give rise to different secret keys. If the user relies on small handheld/handless devices, this may cause lack of memory problems.

- The need to check the uniqueness of ID_u, implies the need for an infrastructure. Indeed, before a secret key is issued, one must make certain it is globally unique.

  Note that if the identity contains a (work related) affiliation, the check for identity is much simpler to organize. First make sure that all affiliations have a unique representation. Then it is up to the Local Registration Center to make certain that within this organization the name is unique. However, this is no guarantee that the natural identity of the user can be used within this organization. Indeed, certain first names and last names are so popular it is not uncommon to find two people in the same organization with the same (first name, last name). Evidently, if one replaces the affiliation with such entities as network provider, etc. the same comment applies.

This check may imply a time delay. If it is not unique, the user is contacted and a new ID_u is suggested. Evidently an alternative solution is to provide several names at once. The time delay can for all practical purposes be eliminated if the user can in advance "reserve" a name, e.g. over the internet. This gives the user plenty of time to think about good strings to append or how to compact the identity, before making the final choice. *Note that if such an identity is reserved the user will still need to demonstrate in person to the Local Registration Center the validity of his/her/its identity before receiving a secret key.*

3 The user will need to obtain the secret key SK_u privately and in an authenticated fashion. This introduces a *key distribution problem*, in particular if the user wants to avoid that the Local Registration Center learns the secret key SK_u. The solution is to provide a temporary key that the Key Distribution Center will use to encrypt the secret key SK_u. For simplicity, we focus on the case the user does not like that the Local Registration Center knows SK_u. In that case the user could make a Temporary Public Key TPK_u, secret key pair, it will use to receive its secret key SK_u.

4 At this stage the Local Registration Center can forward to the Key Distribution Center a request for the user with unique identity ID_u to obtain a secret key SK_u. The request, signed by the Local Registration Center, will at least contain:

(a) the unique identity ID_u, and

(b) the temporary (public) key TPK_u that will be used by the Key Distribution Center to provide securely the key to the user.

5 The Key Distribution Center can now compute the key and send it signed. So the user receives:

$$(E'_{TPK_u}(SK_u), Sign_{SK_{KDC}}(E'_{TPK_u}(SK_u)))$$

where E' is a public key encryption algorithm.

6 The Key Distribution Center informs the hierarchy about the identity key: (ID_u, j_u).

3. THE IMPACT OF REVOCATION AND NON-UNIQUENESS OF THE NATURAL ID

The registration phase already introduces the need for an infrastructure. In certain circumstances this infrastructure can be kept relatively small.

From our discussions until now, it seems that a major difference between a Public Key Infrastructure and a Identity-based Key Infrastructure is that in the last case one only needs the infrastructure for the registration phase. However, in Public Key Infrastructures, when a third party (e.g. Bob) wants to learn the public key of Alice, he/she/it needs to consult the infrastructure or have obtained the public key from Alice[5]. Such an interaction does not seem necessary for Identity-based Key Infrastructures. There are two problems with above reasoning:

1 If ID_A is different from the natural identity of Alice, then this means that there are at least two people who have the same natural identity. So if Bob wants to use ID_A, e.g. to send an encrypted message, then Bob needs to find out what the correct ID_A is. If that is not natural, then Bob needs to consult an infrastructure and evidently, the reply given by this infrastructure must be authenticated, so signed.

   However, it would be incorrect to conclude that Bob only needs to consult the infrastructure when ID_A is not unique, because Bob may not know this. Consulting a WWW page about this, also requires that this WWW page to be authenticated.

2 If keys are stolen or lost, then continuing to use the old ID_A and j_A of Alice, clearly undermines the security. In the case of a public key system, the user could just provide in person (and his/her/its identity having been checked) the Public Key Infrastructure a new public key. However, providing the Identity-based Key Infrastructure a new ID_A may be undesirable, in particular, when ID_A is equal to the natural ID of the Alice. So, a solution to this issue, is that a new j_A is used.

   So, to deal with revocation, one needs a similar solution as in the case of Public Key Infrastructures.

So, to deal with non-uniqueness of natural identities and the use of j_u, a similar structure as used for Public Key Infrastructures is required, which we call Identity-based Key Infrastructure. In this context, one could call the Local Registration Centers: Certifying Authorities. Although their duties in an Identity-based Key Infrastructure are different, these are sufficiently equivalent to call them Certifying Authorities. The Identity-based Key Infrastructure also needs to be consulted by a third party before it will use (for the first time) an identity (ID_A, j_A). Moreover, to deal with lost and stolen secret keys, one needs a revocation mechanism. Similar mechanisms as in the case of Public Key Infrastructures can be used, this means the typical off-line or the more recent on-line approach Rivest, 1998 (see also McDaniel and Rubin, 2000).

Note that Boneh-Franklin Boneh and Franklin, 2001 already mentioned the revocation problem. However, they assume an underlying PKI. We argue that to deal with revocation in identity based cryptosystems, we need a PKI like structure.

4. CONCLUSION

We argued that to securely deploy identity-based cryptography, one needs a structure as complex as Public Key Infrastructures. This is primarily a consequence of the fact that the natural identity (first name, last name) may not be unique. Revocation just aggravates the problem. So, in many circumstances, there is a need for an Identity-based Key Infrastructure.

The problem with not properly verifying the identity of users, as in the aforementioned case of VeriSign, implies that CA's may be untrustworthy. Moreover, if on-line revocation is used, there is the risk of hackers breaking into the CA McDaniel and Rubin, 2000, making CA also untrustworthy. To deal with this problem, more robust PKIs have been suggested Zimmermann, 1995; Reiter and Stubblebine, 1997; Burmester et al., 1998; Burmester and Desmedt, . In all of these Bob (potentially) uses more than one CA[6] to obtain the public key. Adding robustness can also be done for Identity-based Key Infrastructures.

To conclude, we compare public key systems with identity-based ones. Except in the case one only wants low security or, one can guarantee that secret keys will not be lost or stolen and that the natural identities uniquely identify the user, there is a need for an Identity-based Key Infrastructure. Its role is very similar to the one of a Public Key Infrastructure. However, one needs to deal with a key distribution problem, i.e. how can the Key Distribution Center securely provide the secret key SK_A to Alice. This can be solved using a public key system with a temporary public key, roughly doubling the hardware/software needs. In the case identity-based public key is used in a handheld/handless device, its secret key may have to be uploaded securely from a PC that runs public key software.

Finally, we note that (ID_u, j_u) may be shorter than the pair: (identity information, public key). This may be the only real advantage of identity-based cryptography when used in secure environments that, as we argued, need an Identity-based Key Infrastructure. Evidently, a well known disadvantage is that the Key Distribution Center knows each user's secret key.

Acknowledgments

Part of this research was funded by NSF CCR-0209092 and was inspired by Cisco CIAG Research Wishlist topic: "The Internet Without PKI, What are the Alternatives"? The first author thanks Tanja Lange (Ruhr Universität Bochum, Germany) and Roberto Avanzi (University of Duisburg-Essen, Germany) for some discussions related to the topic of identity based cryptography.

Notes

1. For example, if f corresponds to computing the square root of ID_u modulo n and K is the prime factorization of n, then this may not always exist, since not all numbers modulo n are quadratic residues.
2. Certain countries return to their citizens old passports. To avoid that one attempts to reuse these, holes are punched through the passport or corners are cut off, etc.
3. Although Shamir Shamir, 1985 actually suggested to add such information as social security number, in some countries, such as the US, this is a bad idea, since a social security number is viewed as a global password! For example it allows an individual to get a loan.
4. If we were to use database terminology, this ID would be called the "key". However, in the case of a public key system, from a cryptographic viewpoint, this is not a key at all. Therefore, we avoided this confusing terminology.
5. We also view PGP as a Public Key Infrastructure.
6. In the case of PGP these are actually not "Authorities," but could be just friends, so one could object against the word CA. However, the VeriSign example questions whether such organizations are "Authorities," or just "self proclaimed authorities."

References

Ben-Or, M., Goldwasser, S., and Wigderson, A. (1988). Completeness theorems for non-cryptographic fault-tolerant distributed computation. In *Proceedings of the twentieth annual ACM Symp. Theory of Computing, STOC*, pages 1–10.

Boneh, D. and Franklin, M. (2001). Identity based encryption from the Weil pairing. In *Advances in cryptology – Crypto '2001*, volume 2139 of *Lect. Notes Comput. Sci.*, pages 213–229. Springer.

Burmester, M. and Desmedt, Y. Hierarchical public-key certification: The next target for hackers? Submitted October 2001 to Communications of the ACM, accepted February 21, 2003.

Burmester, M., Desmedt, Y., and Kabatianskii, G. (1998). Trust and security: A new look at the Byzantine generals problem. In Wright, R. N. and Neumann,

P. G., editors, *Network Threats, DIMACS, Series in Discrete Mathematics and Theoretical Computer Science, December 2–4, 1996, vol. 38*. AMS.

Chaum, D., Crépeau, C., and Damgård, I. (1988). Multiparty unconditionally secure protocols. In *Proceedings of the twentieth annual ACM Symp. Theory of Computing, STOC*, pages 11–19.

Ellison, C. and Schneier, B. (2000). Ten risks of PKI: What you're not being told about Public Key Infrastructure. *Computer Security Journal*, 16(1):1–7. See also `http://www.counterpane.com/pki-risks.html`.

Goldreich, O., Micali, S., and Wigderson, A. (1987). How to play any mental game. In *Proceedings of the Nineteenth annual ACM Symp. Theory of Computing, STOC*, pages 218–229.

Kohnfelder, L. M. (1978). Toward a practical public-key cryptosystem. BSC thesis, MIT Department of Electronical Engineering.

McDaniel, P. and Rubin, A. (2000). A response to "can we eliminate certificate revocations lists?". In Y, Frankel, editor, *Financial Cryptography, 4th International Conference, Proceedings (Lecture Notes in Computer Science 1962)*, pages 245–258. Springer-Verlag. Anguilla, British West Indies, February 20–24.

Menezes, A., van Oorschot, P., and Vanstone, S. (1996). *Applied Cryptography*. CRC, Boca Raton.

Microsoft Security Bulletin MS01-017 (March 22, 2001, updated: June 22, 2003). Microsoft security bulletin ms01-017, erroneous verisign-issued digital certificates pose spoofing hazard.
http://www.microsoft.com/technet/security/bulletin/MS01-017.mspx.

Reiter, M. K. and Stubblebine, S. G. (1997). Path independence for authentication in large scale systems. In *Proceedings of the 4th ACM Conference on Computer and Communications Security*, pages 57–66. Zurich.

Rivest, R. L. (1998). Can we eliminate certificate revocations lists? In Hirschfeld, R., editor, *Financial Cryptography, 2nd International Conference, Proceedings (Lecture Notes in Computer Science 1465)*, pages 178–183. Springer-Verlag. Anguilla, British West Indies, February 23–25.

Schneier, B. (1996). *Applied Cryptography*. J. Wiley, New York, second edition.

Shamir, A. (1985). Identity-based cryptosystems and signature schemes. In Blakley, G. R. and Chaum, D., editors, *Advances in Cryptology. Proc. of Crypto 84 (Lecture Notes in Computer Science 196)*, pages 47–53. Springer-Verlag. Santa Barbara, California, U.S.A., August 19–22.

Zimmermann, P. R. (1995). *The Official PGP User's Guide*. MIT Press, Cambridge, Massachussets.

MODINT: A COMPACT MODULAR ARITHMETIC JAVA CLASS LIBRARY FOR CELLULAR PHONES, AND ITS APPLICATION TO SECURE ELECTRONIC VOTING

Hiroaki Kikuchi and Junji Nakazato

[1] *Dept. of Information Media Technology*
Tokai University
kikn@tokai.ac.jp

[2] *Graduate School of Engineering*
Tokai University
nakazato@ep.u-tokai.ac.jp

Abstract ModInt is a compact Java class designed for the arbitrary-precision modular arithmetic operations used in secure cryptographical applications, including auctions and electronic voting. The library is small enough to port to a Java-enabled cellular phone. In this paper, we present a new electronic voting system using the ModInt library, in which voters have their ballots encrypted with the public key of a trusted agent and cast them to an untrusted server, which tallies them without decrypting the ballot and updates a linear-feedback shift register (LFSR) consisting of log n ciphertexts. The privacy of voters is therefore assured assuming that the security of the public-key crypto system is maintained, and the trusted party keeps their private key secure. This feature reduces the costs of management of the counting server, which is now free from the risks of being compromised.

Keywords: electronic voting, homomorphic encryption, Java applet, modular arithmetic

1. INTRODUCTION

The cellular phone is a widely used communication medium that allows us to communicate more easily and frequently than before. It has been developed to provide a number of attractive services to customers, including web browsing, and digital still camera and GPS (global positioning system) functions. The Java application platform is also one of these new services that expands the range of application of cellular phones. The platform can offer, for example, interactive computer games, instant messaging clients, or interfaces to servers that are more intelligent and user-friendly than simple web browsers.

Electronic voting is also a prime example of a function suitable for a Java-enabled cellular phone. With a mobile voting device, one may easily participate in decision-making processes whenever suitable. According to a Japanese governmental survey [4], more than 90% of Japanese teenagers have their own cellular phone, and they are expected to be potential voters who take part in elections only if they are allowed to cast a vote from the Internet. In the near future, the population of cellular phone users will exceed that of passport holders, and hence it is possible that the cellular phone could replace the passport as a nationwide user identification scheme. Additionally, Japanese regulation has been updated in 2001 making electronic voting legal throughout Japan[13]. In June 2002, the first Japanese electronic election was held in Niimi, Okayama, in the mayoral and local assembly elections[14]. In the elections, more than 19,000 eligible voters cast their ballots from the voting machine by inserting a voting card into a voting machine, and touching the names of their preferred candidates, which appeared on a video screen. Their votes were stored in the memories of the voting machines, and counted electronically after the voting ended. The electronic voting reduces the time and effort spent on tallying vote. The total time is just 25 minutes. According to the questioner, 97 % of voters said that the machine is better than the conventional paper-based election.

The biggest issue to be resolved about real-world Internet voting is the conflict between security and privacy. In order to prevent dishonest voters and potential attackers from manipulating the tally in malicious ways — such as double voting, altering or forging ballots, and unauthorized voting — the security technologies, using public-key infrastructure with public key certificates, provide secure authentication for eligible voters. However, the identification may be used to trace the voters, and the privacy of voters may be violated by the government or the administrators of the counting server.

To avoid these problems, a number of electronic voting protocols have been proposed to date. In 1997, Cramemer, Damgard and Schoenmakers presented an optimally efficient voting protocol using a homomorphic encryption, in which the size of the ballot is a single ciphertext[9] Fujioka et al. proposed an efficient protocol in [1], combining a blind signature and anonymous channels or mix servers, and performed the experimental trial over the Internet. In 2001, Furukawa and Sako proposed a protocol to prove the correctness of shuffled ballots, and implemented it with several mix servers.

In this paper, we present a new electronic voting protocol using a linear feedback shift register (LFSR), which makes a tally using the internal state of the LFSR. In the proposed protocol, a voter shifts the public register if they wish to vote "Yes", or does nothing if they wish to vote "No". However, using a homomorphic public-key encryption scheme, no one can determine whether the vote is "Yes" or "No" by watching the register except the trusted authority, who keeps the corresponding private key secure and reveals it to the public only when the time comes to open the tally. The communication cost for voters is of $O(\log n)$ where n is a number of voters. The idea of using an LFSR as a tally is not our own. In 2001, Katz et al. proposed cryptographic counters and constructed a concrete protocol using quadratic residue (QR) encryption in [7]. In contrast to this approach, our system adopts El Gamal encryption that allows the decryption process to be distributed across several (incompletely trusted) authorities who hold a share of the corresponding private key in an ordinary manner.

These cryptographic approaches are substantially effective in contemporary personal computers, but the restricted computational power of cellular phones is not sufficient to perform these cryptographical computations fully. Because of restricted storage, the `BigInteger` class library, which is a Java standard class for processing the arbitrary precision arithmetic used in cryptographical protocols, is not supported in any commercial Java-enabled cellular phone device. Moreover, the free memory capacity is 30kB at most in the NTT 504i, or 10kB in the 503i, which is too small to support the comprehensive methods in the `BigInteger` class.

In order to address the issue, we have developed a compact Java class library named `ModInt`, which is designed for the modular arithmetic used in cryptographic protocols. The `ModInt` class is

- specially designed for modular arithmetic, with all operations performed implicitly in modulo, and
- small enough to be ported to restricted environments such as cellular phones.

We are interested in how large overheads are generated in modular arithmetic in the cellular phone environment, and whether such arithmetic is suitable for performing electronic voting protocols.

Thus, in this paper, we demonstrate the performance of the ModInt class library and prove that the cryptographical protocol is feasible, even in an ordinary cellular phone environment. Based on the experimental results of the library, we clarify the necessary conditions for large-scale electronic voting.

The compact Java class library allows us to hold a variety of web-based electronic voting including large-scale election for governors, election for city mayor, and small-scale election for student president. In this sort of public election, an election committee manages a list of eligible voters and requires user authentication processes using appropriate identification protocols. The simplest is a password-based authentication. If we can assume the national identity numbering system where every citizen has assigned number and the X.509 public-key certificate, PKI can be used to identify the eligible voter. According to the "e-Japan Priority Policy Program"[12], a PKI in Asia region is going to be developed by the 2005. On August 5th 2002, Japanese government has started the Basic Resident Register Network, where the 11-digit resident's certificate codes are assigned to every citizen and a copy of resident's certificate is available at any office around Japan. Though there are some concerns about the protection of the privacy, the option of national-wide PKI is now realistic. Any of these authentication methods can be used in conjunction with the proposed voting system.

August 5th 2002. Basic Resident Register Network Individual info: Address, Names, Birthdays, Sex. 11-digit Regidentfs certificate code. A copy of residentfs certificate is available at any office around Japan. The concern about the protection of privacy by the government. Kokubunji, Hino, Kunitachi: potponement of its operation.

The other direction of the proposed electronic voting system is a questionnaire-based user evaluation. Web-based questionnaire have been used for many applications such as guest questionnaire, commercial products evaluation, or course evaluation. In any of questionnaire, privacy should be preserved though most of them don't take any account of the issue. The proposed Java-based system allows to any party to improve the privacy enhancement just replacing the casting and the tallying processes by the proposed system without developing from the scratch. In comparison with the public election mentioned above, the strong user authentication is not always necessary in this sort of questionnaire. Participation to evaluation is open to any users and checking the source IP address is good enough to prevent from malicious users from double casting.

The structure of this paper is as follows. In Section 2, we describe the protocol for oblivious vote counting with the LFSR. Following this, we demonstrate the system design and the performance of the `ModInt` Java class library in detail. Finally, we evaluate the implemented electronic voting system in terms of its complexity and communication requirements.

2. PRELIMINARY

2.1 Homomorphic Encryption Scheme

Let M be a set of plaintexts, $\{m_0, m_1\}$, where m_0 and m_1 represent boolean values corresponding to 'false' and 'true', respectively.

Let us suppose a *homomorphic encryption scheme* E that has the following properties.

- E is homomorphic over $GF(2)$, i.e., for elements a and b of $\{m_0, m_1\}$,

$$E[a \oplus b] = E[a] \times E[b] \tag{1}$$

  holds, where $a \oplus b$ is an exclusive-OR, defined as $m_0 \oplus m_1 = m_1 \oplus m_0 = m_1$ and $m_0 \oplus m_0 = m_1 \oplus m_1 = m_0$.

- E is semantically secure, that is, no one is able to distinguish ciphertexts of m_0 and m_1 with a probability significantly greater than a random guess.

- The key generation can be distributed among a certain number of players. The size of the public key should not depend on the number of shares.

- The decryption process can be distributed among t-out-of-n players who share the corresponding private key. The computational and communication (bandwidth and rounds) costs should be as small as possible.

El Gamal encryption satisfies all requirements under the Decision Diffie–Hellman (DDH) assumption, if we let $m_0 = 1$ and $m_1 = -1$ (mod p). Let p and q be large primes such that $p = 2q + 1$ and let $\mathcal{G}$ be the set of multiplicative groups of order q in Z_p^*. Let g be a primitive element of $\mathcal{G}$. If we have $p = 2q + 1$, which is commonly used for assignment, then message -1 does not belong to $\mathcal{G}$ and thereby testing quadratic residue of ciphertexts will reveal the correspondence of $E[m_0]$ and $E[m_1]$.

An El Gamal encryption of the message m with the public key $y = g^x$ is of the form $E_a[m] = (M, G) = (my^a, g^a)$, where a is a random number chosen from Z_q. To decrypt the ciphertext (M, G), we use the corresponding private key x to compute $M/G^x = mg^{xa-ax} = m$. By element-wise multiplication, we define $E[a] \times E[b] = (M_aM_b, G_aG_b)$, which yields a new ciphertext $E[a \oplus b]$, and it can be seen that the El Gamal encryption is homomorphic over $GF(2)$.

A distributed decryption is also feasible. A private key is jointly generated by the collaboration of t honest parties (key holders) out of n, and distributed among them using $(t-1)$-degree random polynomials $f(x)$ as $f(1), f(2), \ldots, f(n)$. To decrypt a provided ciphertext with the public key $y = g^{f(0)}$, the i-th party publishes $G^{f(i)}$ for $i = 1, \ldots, t$, and then computes $G^{f(1)\gamma_1} \cdots G^{f(t)\gamma_t} = G^{f(0)}$, where γ_i is the Lagrange coefficient for i. For verifiability, the players use Verifiable Secret Sharing (VSS) as the proof of possession of $f(i)$ such that $G^{f(i)}$.

Another instance of a homomorphic encryption scheme is QR encryption, used in [7].

2.2 Proof of Knowledge

We will use the proof of knowledge of the private input to the computer, which is based on the disjunctive and conjunctive proofs of knowledge in [10].

Conjunctive Proof of Knowledge By $PK\{(\alpha) : y_1 = g_1^\alpha \wedge y_2 = g_2^\alpha\}$, we denote a proof of knowledge of discrete logarithms of elements y_1 and y_2 with bases g_1 and g_2. Selecting random numbers r_1 and $r_2 \in Z_q$, a prover sends $t_1 = g_1^{r_1}$ and $t_2 = g_2^{r_2}$ to a verifier, who then sends back a random challenge $c \in \{0,1\}^k$. The prover shows $s = r - cx \pmod q$, which should satisfy both $g_1^s y_1^c = t_1$ and $g_2^s y_2^c = t_2$.

Disjunctive Proof of Knowledge We use $PK\{(\alpha, \beta) : y_1 = g^\alpha \vee y_2 = g^\beta\}$ to denote a proof of knowledge of one out of the two discrete logarithms of y_1 and y_2, to the base g. The prover can prove that they know a secret value under which either $y = y_1$ or $y = y_2$ must hold, without revealing which identity was used. Without loss of generality, we assume that the prover knows α for which $y = g^\alpha$ holds. The prover uniformly selects $r_1, s_2 \in Z_q$ and $c_2 \in \{0,1\}^k$ and sends $t_1 = g^{r_1}$ and $t_2 = g^{s_2} y_2^{c_2}$ to the verifier, who then provides a random challenge $c \in \{0,1\}^k$, where k is a security parameter. On receiving the challenge, the prover sends $s_1 = r_1 - c_1\alpha \pmod q$, s_2, c_1 and c_2, where $c = c_1 \oplus c_2$. The verifier can see if the prover is likely to have the knowledge by testing both $t_1 = g^{s_1} y_1^{c_1}$ and $t_2 = g^{s_2} y_2^{c_2}$ with provability $1 - 2^{-k}$. Note that the same test can be used when t_1 and t_2 are prepared for the other knowledge β.

2.3 Oblivious LFSR Protocol

An oblivious LFSR is an array of n ciphertexts that are updated according to a linear irreducible polynomial.

An LFSR consists of n flip flops, $A_1, \ldots, A_n$. A shift of LFSR is defined for $i = 1, \ldots, n$, by

$$A'_i = A_{i-1} \oplus a_{i-1} A_n \tag{2}$$

where a_i is of characteristic polynomial of t-degree

$$f(x) = 1 + a_1 x + a_2 x^2 + \cdots + a_n x^n$$

and $A_0 = 0$, $A'_1 = A_n$, and $a_j \in \{0, 1\}$. A (primitive irreducible) polynomial f has at most $2^n - 1$ states.

An oblivious LFSR is a protocol in which a counter $\mathcal{C}$ communicates with 2^n voters $\mathcal{V}$ once each, and sends the tally (register) to an authority $\mathcal{S}$, who decrypts it and publishes the result. The following are the steps required for oblivious LFSR.

Step 1: (initialization) Counter $\mathcal{C}$ initializes a register $A_1 = E[-1]$, $A_2 = E[1], \ldots, A_n = E[1]$.

Step 2: (voting) Let $b \in \{-1, 1\}$ denote a "Yes" and a "No" vote, respectively. Voter $\mathcal{V}$ sends the value to be updated to counter $\mathcal{C}$, $C_1, \ldots, C_n$, defined by

$$C_i = \begin{cases} E_{r_i}[1] \times A_{i-1} \times A_n^{a_{i-1}} & \text{if } b = -1, \\ E_{r_i}[1] A_i & \text{if } b = 1. \end{cases} \tag{3}$$

Note that when $b = 1$ ("No"), the voter does not change the state of the LFSR, but multiplying with the ciphertext of 1 makes the vote indistinguishable by a third party.

Step 3: (proof of availability) Voter $\mathcal{V}$ proves that the values to be updated are properly computed by Equation 3 in the following zero-knowledge proof,

$$PK\left\{(r_1, \ldots, r_n) : \begin{array}{c} C_1 = E_{r_1}[A_1] \wedge \cdots \wedge C_n = E_{r_n}[A_n] \\ \vee \\ C_1 = E_{r_1}[A'_1] \wedge \cdots \wedge C_n = E_{r_n}[A'_n] \end{array}\right\}$$

With the proof of the votes, counter $\mathcal{C}$ verifies that the new $C_1, \ldots,$ C_n sent from $\mathcal{V}$ are consistent with the internal states $A_1, \ldots, A_n$ and $f(x)$. If they are incorrect, $\mathcal{C}$ discards $C_1, \ldots, C_n$, otherwise, it replaces the register with the new $C_1, \ldots, C_n$.

Step 4: (opening) When the opening time comes, authority $\mathcal{S}$ decrypts the register as $B_1 = D[A_1], B_2 = D[A_2], \ldots, B_n = D[A_n]$ and publishes $B_1, \ldots, B_n$, which represent a k-th shifted pattern when k voters cast "Yes" votes.

3. IMPLEMENTATION ON CELLULAR PHONES

We implement the oblivious LFSR protocol with the NTT Docomo cellular phone series that supports the Java applications called i-appli [5].

3.1 System Design

Table 1 shows the system specification of the Java software development toolkit [6].

Table 1. Specification for the system

Language	JDK1.3,J2ME Wireless SDK for the DoJa release 2.2
Public-key encryption	ElGamal with 512 bit
Protocol	HTTP/CGI

In our implementation, a client (voter) is a simple Java applet and a server is a Java application equipped with an HTTP server. Both the Java applet and the application consist of seven Java classes, e.g., ModInt class for modular arithmetic integers, and ElGmal class.

3.2 ModInt class

ModInt is a compact Java class designed for arbitrary precision modular arithmetic operations. In order to reduce the size of the library, it omits the following functions supported in the BigInteger class.

- bit manipulation (AND, OR, NOT),
- a sign digit,
- a hexadecimal representation for input and output,
- prime number generation (which may be performed off-line by a more powerful computer).

The ModInt class provides the fundamental 15 methods in 6 kB of byte-code.

ModInt	BigInteger
`p = new ModInt("71a04e8b02");`	`p = new BigInteger("71a04e8b02", 16);`
`q = new ModInt("e0ae02e788");`	`q = new BigInteger("e0ae02e788", 16);`
`m = p.getInstance("dedf322");`	`m = new BigInteger("dedf322", 16);`
`y = p.getInstance("12af7cc");`	`y = new BigInteger("12af7cc", 16);`
`r = q.getInstance("98af3df");`	`r = new BigInteger("98af3df", 16);`
`c = m.multiply(y.power(r));`	`c = m.multiply(y.modPow(r, p)).mod(p);`

Figure 1. Examples of El Gamal encryption using the ModInt and BigInteger libraries

Table 2. Difference between BigInteger and ModInt

	`BigInteger`	`ModInt`
Size of bytecode	40kB	6kB
# of public methods	43	15
Integers	signed, variable length	unsigned, fix length (specified by p)
Input format	decimal, string, byte array	hex,int array
Internal object	big endian, int array $\times$ 1, sign, bit length	big endian, int array $\times$ 2

In the ModInt class, a value is an object implicitly bound to a modulo p and is computed in (mod n). In order to distinguish ModInt from BitInteger, let us consider an example of part of the ElGamal encryption algorithm, defined by

$$c = my^r \bmod p.$$

If this is executed in BigInteger, after instantiating four objects, m, y, r and p, we must write explicitly

```
c=m.multiply(y.modPow(r, p)).mod(p),
```

where two modular commutations are performed, incurring an overhead when c is extended to size $2p$, and then is reduced to size p.

In ModInt, we simply write

```
c=m.multiply(y.Pow(r))
```

where modulo p is implicitly computed. A more complete source is shown in Figure 1.

We illustrate the difference between `BigInteger` and `ModInt` in Table 2, and show some fundamental Java constructors and methods in Table 3 and 4, respectively.

Table 3. List of Constructors

Constructor	Description
`ModInt(String p)`	Set modulus p in hexadecimal format
`ModInt(int[] a, int[] p)`	Generates an instance a mod p

Table 4. List of instance methods

Return value	Method	Description
`ModInt`	`getInstance(String s)`	Generates an instance of `s` in hexadecimal format. The modulo p of `s` is identical to `this`.
`String`	`toString()`	Returns a string of `this` value represented in hexadecimal format.
`ModInt`	`subtract(ModInt b)`	Returns an instance of $this - b \pmod{p}$
`ModInt`	`add(ModInt b)`	Returns an instance of $this + b \pmod{p}$
`ModInt`	`multiply(ModInt b)`	Returns an instance of $this \times b \pmod{p}$
`ModInt`	`shiftLeft()`	Returns an instance of $this \ll 32 \pmod{p}$
`ModInt`	`power(ModInt e)`	Returns an instance of $this^e \pmod{p}$
`long`	`compareTo(ModInt b())`	Compares $this$ with b in number, and returns 0 if $this = b$

Table 5. Execution environment of the emulator

CPU	Pentium III 1.13G, 512 MB, 40GB
OS	Windows 2000 Server

4. EVALUATION

We evaluated our system on two platforms: an i-appli emulator and a commercially-available cellular phone (the NTT Docomo D503is).

4.1 Performance in the J2ME Emulator

Table 5 shows an environment running the i-appli emulator.

In Table 6 and 7, we show the performance of our system in the i-appli emulator for J2ME, where Key, Carry, Enc and Shift are the processing times for reading the public key, reading the register, computing re-encryption, shifting the LFSR, and sending votes, respectively. The $|p|$ represents the size of p in bits.

Table 6. Processing time for key size $|p|$ [s] ($n = 3$, Emulator)

$\|p\|$ [bit]	Key	Carry	Enc	Shift	Write	Total
128	0.6	0.3	0.7	0.00	0.03	6.4
256	0.5	0.4	2.7	0.01	0.05	8.6
512	0.7	0.6	18.9	0.10	0.10	27.4
1024	0.8	1.1	40.0	0.17	0.16	47.1

Table 7. Processing time for bit length of register n [s] ($|p| = 512$,Emulator)

n	Key	Carry	Enc	Shift	Write	Total
3	0.5	0.4	2.7	0.10	0.05	8.6
5	0.6	0.8	23.2	0.04	0.14	30.9
8	0.6	1.2	40.1	0.10	0.22	48.9

Table 8. Processing time for key sizes $|p|$ [s] ($n = 3$, D503is)

$\|p\|$ [bit]	Key	Vote	Total
128	2.8	25.4	41.7
256	5.4	92.7	145.8
512	3.2	426.7	443.6

We show the processing time varying with the size of p for $n = 3$ in Table 6, and n in $|p| = 512$ in Table 7.

From the tables, we notice that the performance bottleneck is within the cryptographical processes rather than the file input-output processes. At most, the encryption takes more than 80% of the total time.

4.2 Performance in the Docomo D503is

We demonstrated the implemented system on the NTT Docomo D503is (see Figure 2). The performance of the D503is is reported in Table 8 and 9 and in Figure 3 and 4. The value $|p| = 512$ is used for this system.

Based on the experimental performance, we observe that the time for processing depends on n and $|p|$. It should be noted that the D503is takes about 23 longer than the PC emulator does, as power consumption concerns limit the computational power of cellular phones in mobile environments.

4.3 Pre-computation to Reduce Processing Time

According to the experimental results, we can estimate an average processing time for any given parameter. For example, if $|p| = 512$ and

Figure 2. A screenshot of the implemented system

Table 9. Processing time for bit lengths of register n [s] ($|p| = 512$, D503is)

n	Key	Vote	Total
3	3.2	426.7	443.6
5	7.3	664.6	686.2
8	5.2	1055.4	1084.7

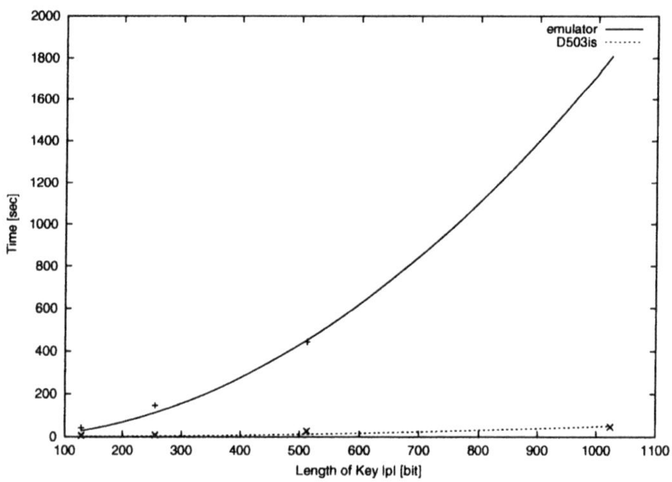

Figure 3. Processing time for key size $|p|$ [bit]

there are 0.1 million voters, then by calculating $n = \log_2 100,000 = 17$, we expect about 37 minutes per voter, which is not realistic.

Hence, we try to reduce the number of cryptographical processes. Recalling the updating formula

$$C_i = E[1] \times A_{i-1} \times A_n^{a_{i-1}},$$

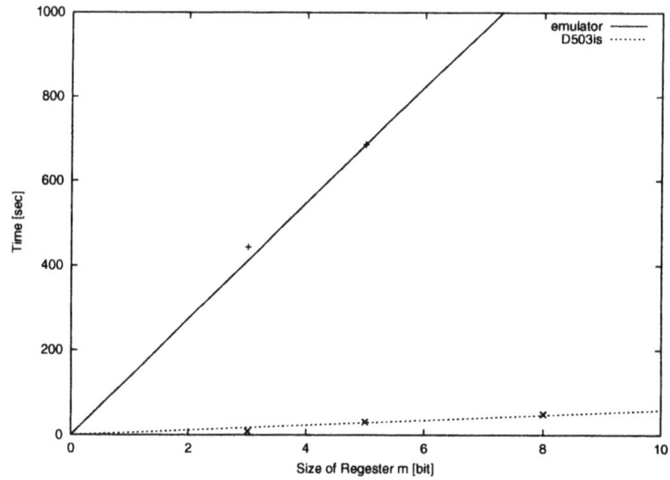

Figure 4. Processing time for length of register n

Table 10. The estimated reduction in voting time [s] ($|p| = 512$). Original: the performance in N503is, Enc: the estimated encryption time based on Table 7, and Reduced: the expected reduction by pre-computation

n	Original	Enc	Reduced
3	443.6	348.8	94.8
5	686.2	539.5	146.7
8	1084.7	852.8	231.9

where $E[1]$ is independent of both the register A_i and the vote b, and thus can be recomputed before casting. Unless the internal memory is compromised, we can store as many $E[1]$ values as possible to skip the encryption steps. Based on the experimental data in Table 7, we show the expected reduction in time in Table 10, which demonstrates that the highest reduction is about 22 %.

4.4 Scalability of the Proposed System

We estimate the scalability of the proposed system based on the experimental processing time. Let us assume that the maximum waiting time is 30 seconds. We may then estimate how many powerful computers would be required when elections are conducted at several scales, ranging from 32 voters for a tiny class president election to a 300 million-scale election for governors. Table 11 illustrates the estimation. It shows that computation time can be reduced by about 22%. If we hold a 1000-scale election, we need a Java processor in the cellular phone that is 10.1 times faster.

Table 11. Estimated improvement of CPU power requirement for elections at several scales

Election scale	# of users	# of registers	orig.		w. precomp.	
	m	n	Emu.	D503	Emu.	D503
Class president	32	5	1	23.5	1	5.17
Student body president	1000	10	1	45.7	1	10.1
City mayor	10000	14	2.6	60.7	1	13.4
State governor	3 million	22	4.2	98.4	1	21.6

Table 12. Comparison between [1] and the proposed system

System	System [1]	Proposed system
Voting	$O(1)$ 0.5 s	$O(n)$ 23.3 s
Mixing	$O(2^n)$ 2.52 s	N.A.
Opening	$O(1)$ 3.05 s	$O(n)$ 1 s

4.5 Comparison to the Conventional System [1]

In [1], an electronic voting system is reported. The system uses a blind signature with mix servers for preserving the privacy of voters, the Schnorr signature for the administrator, threshold El Gamal encryption for the public key cryptosystem, and hybrid mix networks combining the Diffie-Hellman and the DES algorithms for shuffling the votes.

The largest difference to our system is the existence of an anonymous channel, such as mix and shuffle. In our system, the anonymous channel is not required, but a greater computational cost is needed for updating registers.

We summarize the comparison in Table 12, where the performance with 10 users is evaluated on a Pentium II 400 MHz and the digital signature processes are omitted in [1]. The performance in our system is estimated from Table 4.

In conclusion, by removing the time-consuming process of mixing, which takes a time proportional to the number of voters, we reduce the processing time at the server. The cost is dependent on the number of registers, which is the log of the number of voters.

In comparison to the electronic voting in Niimi city, our system does not require any special devices such as the voting machine and the voting card. The cellular phone is a versatile tool and hence it is better choice as the voting device. Moreover, the black-box machine cannot convince

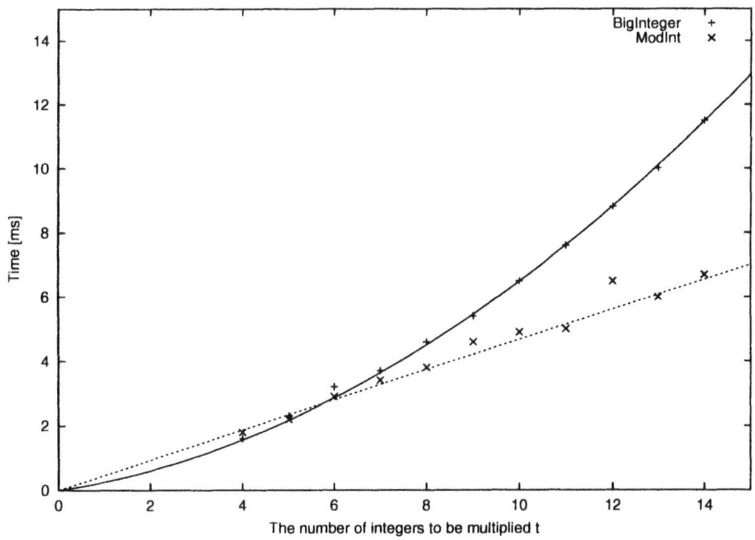

Figure 5. Processing time for **BigInteger** and **ModInt** [ms]

people that the tallying has been performed correctly and the privacy of voting is preserved. The every step in our system is publicly verifiable to any party who wish to verify the accountability.

4.6 Performance of the ModInt Library

The performance advantage of the ModInt library, compared to the BigInteger library, is due to improvements to the multiplication of a few integers at once. For instance, letting a, b and c be 1024-bit BigIntegers, we write

```
a.multiply(b).multiply(c).mod(p),
```

whose size is expanded up to 3072 bits once, before being reduced back to 1024 bits. The ModInt, in contrast, constantly maintains a size of 1024 bits. In general, the processing time with respect to the number of multiplications t is shown in Figure 5.

The result shows that the ModInt achieves better performance than BigInteger when t is greater than 6, although t is limited to less than 3 in the El Gamal encryption in our system.

5. CONCLUSION

In this paper, we have proposed an efficient cryptographic protocol for a secure electronic voting system and have shown that the proposed protocol is feasible for a Java-enabled cellular phone platform with limited computing power. From the experimental results, we determined that a cellular phone processor 22 times faster than those currently available is required to hold a 3 million voter election, and that pre-computation helps by saving 22% of the processing time. Our implemented ModInt Java class library is specially designed for modular arithmetic and provides better performance than BigInteger.

References

[1] A. Fujioka, M. Abe, M. Ohkubo and F. Hoshino, "Implementation and Experimentation of Practical Electronic Voting", Proc. of Symposium on Cryptography and Information Security, Vol 48, 2000 (in Japanese).

[2] M. Abe, "Universally Verifiable Mix-Net with Verification Work Independent of the Number of Mix-Servers," IEICE Trans. Fundamentals, Vol. E83-A, 2000.

[3] J. Furukawa and K. Sako, "An Efficient Scheme for Proving a Shuffle," Proc. of CRYPTO'01, LNCS 2139, pp. 368–387, 2001.

[4] Japanese Governmental Survey in Influence of Information Technologies at Home in 2002, (http://www5.cao.go.jp/seikatsu/2002/0405it-chousa/index.html).

[5] NTT Docomo Official Page, (http://www.nttdocomo.co.jp/ September 2002).

[6] i-mode Official Page, (http://www.nttdocomo.co.jp/p_s/imode/ September 2002).

[7] J. Katz, S. Myers and R. Ostrovsky, "Cryptographic Counters and Applications to Electronic Voting," in proc. of EUROCRYPT'02, pp. 78-92, 2001.

[8] R. Crammer, I. Damgard, and B. Schoenmakers, "Proofs of partial knowledge and simplified design of witness hiding protocols," Proc. of CRYPTO '94, pp. 174–187, 1994.

[9] R. Cramer, R. Gennaro and B. Schoenmakers, "A Secure and Optimally Efficient Multi-Authority Election Scheme," Proc. of *EUROCRYPT 1997*, 2001.

[10] R. Cramer, I. Damgård, and B. Schoenmakers, "Proofs of partial knowledge and simplified design of witness hiding protocols," Proc. of *CRYPTO '94*, pp.174–187, 1994.

[11] M. Jakobsson and A. Juels, "Mix and Match: Secure Function Evaluation via Ciphertexts," Proc. of *ASIACRYPTO 2000*, LNCS 1967, pp. 162–177, 2000.

[12] "E-Japan Priority Policy Program 2003", IT Strategic Headquarter, (http://www.kantei.go.jp/foreign/policy/it/0808summary/030808gaiyo_e.pdf), August 2003.

[13] "'The electromagnetic recording voting device" (the special law of electronic voting), February, 2002.

[14] "Regulation regarding Voting by Electromagnetic Recording Voting Device in the Election of Niimi City Assembly and Mayor", Niimi City, 2002.

DEPENDABLE SECURITY BY TWISTED SECRET SHARING

Semir Daskapan
Faculty of TBM, Delft University of Technology

Abstract: Large scale networked information systems are referred to as critical information infrastructures when they provide critical services to the critical physical infrastructures. Critical information infrastructures contain specific nodes that provide security services, like authentication servers; those are called security or trust centres. The goal of this research is to find an algorithm for enabling those centres to become sustainable by sharing their (superfluous) security resources and to resist Byzantine failures. The proposed secret sharing algorithm takes care of allocating in advance the secret content of the suffering centre to other healthy centres, so that only an arbitrary majority of them can reconstruct the content. This perfect t,n- threshold scheme is suitable in dynamic networks as it has an adaptive access structure. It is compared to existing schemes rather simple as it is purely based on permutations. It is efficient, i.e. favourable information rate, as all shares are much shorter than the secret itself. Secondarily, each secret share is even additionally protected (encrypted) against the holder as well against any outsider.

Key words: Trust, dependability, critical infrastructures, secret sharing, threshold cryptography

1. INTRODUCTION

An infrastructure consists of many dependent components, which enable together a service as a carrier. However, a dependable infrastructure contains usually one or more centralized nodes for certain control functions, like a centralized repository of metadata or an authentication server. Here the focus is on security distribution centers (SDCs). Any configuration that exhibits (multiple) single point of failures (SPF's) is avoided as much as possible. In order to resist failures, that cause discontinuity, those centralized nodes are set up in redundancy. The traditional approach of investing in redundancy by dedicated hardware has its limitation in scalability and costs.

Complementary to these traditional approaches, where each weak node is enriched with a few identical nodes, a resource sharing approach between all the critical nodes is already introduced in [1]. The objective of that is autonomous collaboration of all the nodes to back up each other when failures occur by offering their resources as temporary hosts. Destitute nodes will send their valuable secret assets to the allied nodes (pool members) with healthy and free redundant resources. However, whether this resource-sharing concept will also work in critical information infrastructures (i.e. distributed information systems) depends on the additional requirements of forward and backward confidentiality, but most on the timing of releasing the secret assets and their keys to the allied pool members. By applying a secret sharing algorithm it should be possible to distribute the secret assets without revealing the content prematurely. The reason is that no member will receive the complete secret asset, but only a fragment. Even if he manages to get all the fragments he will not have all the keys to decrypt the fragments, because of the uncorresponding fragments and keys. It was therefore baptised as *twisted secret sharing*. Only by collaborating with a majority of pool members the secret asset can be reconstructed.

In the second section other works are discussed. In the third section a guiding case is presented for the problem, by which a pattern is sought for the algorithm. This pattern is then introduced and discussed in the fourth section. Finally, in section 5 the paper is finalized by a conclusion.

2. OTHER WORKS

Survivability is traditionally enhanced by means of redundancy [2,3,4]. Redundancy means that in parallel similar systems are ready to replace the main system. From a security point of view, redundancy seems to decrease the vulnerability as it avoids (multiple) single points of failure (MSPFs). With many of such replicas, the system can withstand (number of replicas - 1) failures. However, with the increasing number of redundant systems, the number of targets and thus the possibility of confiscation increases. A typical form of that is the Byzantine failure [5] that occurs due to corrupted processors. As more systems provide the same functionality the danger is that the redundant systems suffer from asynchrony and unequal information, due to technical implementations.

Survivability is regarded as one of the attributes of dependability next to other aspects like reliability, availability and safety [6]. Critical information infrastructures (CIS) are designed according to high availability and reliability requirements. Ideally, this means that critical infrastructures should be available, even when the supporting hardware of the CIS is failing. Survivability of the core functionality of a system is often the main focus of research and the survivability of other dependency attributes are not considered. Although those services appear to decrease the vulnerability of the system, they are themselves not survivable and could therefore jeopardize the whole system when they fail. Therefore additional measures are taken for those overall services. Usually this is realized by applying in parallel redundant hardware, which will catch up the breakdown of the primary CIS. Besides that it is not very cost-effective, as redundancy is in economical terms an indispensable overhead, it is also insufficient due to the limited dedicated hardware.

One of the earliest works on survivable SDC's is the work of Gong [7], although specific for authentication servers. Here fault tolerance is achieved by collaboration of many SDC's in a group. Also the RAMPART system [8] provides secure communication in the presence of actively malicious processes and Byzantine failures for SDC's (authentication services) specific. Here, with the assumption of an asynchronous network up to a third of the members in a Rampart group may behave in Byzantine manner yet the group would still provide reliable multicast facilities. The system uses secure channels between two members of the protocol to provide authenticity. Protocols depend greatly on the *consensus* of processors to reach agreement on the course of action. Other works in this specific field are [9,10,11], which all deploy static redundancy with replicated services (srrs).

MAFTIA [12], SINTRA [13], COCA [14] and Ensemble [15] provide approaches for tolerating both accidental (Byzantine) faults and malicious attacks in large-scale distributed systems like on the Internet. This enables them to remain operational during attack, without requiring time-consuming and potentially error-prone human intervention. They assume asynchronous networks and use many updated techniques like, randomization by coin-tossing scheme [16] to overcome the impossible existence of a deterministic protocol [17]. The resilience is guaranteed for $t > n/3$ dependable members [18].

Those works also apply somehow (combinatorial) secret sharing [19,20] or threshold cryptography [21,22]. In distributed systems secrets play a special role, as they can bind the scattered components. A *secret* means some piece of information, be it a password, the private key of a public key cryptosystem, a solution to some mathematical problem or a set of credentials. With secret sharing, a single secret is divided into *w* parts and distributed across *n* nodes. To reconstruct the whole secret, one node must be able to collect or derive all the *w* pieces of the secret. The so-called t,n-threshold schemes have a specific access structure, i.e. the specification of participants which are authorized to reconstruct the secret. In that case at least $t<n$ are able together to reconstruct the secret, where $t<n$ is the threshold. Usually *t* is larger than *n/2* without gossip or larger than *2n/3* with gossip among the total nodes.

The twisted secret sharing algorithm that is proposed in this paper has a unique combination of characteristics, which have not been found in other works [19,23,24,25]. The *perfect* t,n- threshold scheme is suitable in dynamic networks as it has an *adaptive access structure*. It is compared to existing schemes rather *simple* as it is purely based on transpositions. It is *efficient* as all shares are much shorter than the secret itself and have thus a favorable information rate [26]. Each secret share is additionally *protected* (encrypted) against the holder as well against any outsider. Furthermore, whereas other works require numerical values this algorithm also applies easily for *alphanumerical values*. Due to proactive refreshing of keys and shares it is *unpredictable*.

3. TWISTED SECRET SHARING

In this section we will describe some scenarios to explain the problem and provide a corresponding solution to each scenario. This section describes as such the inductive phase to the general solution as is proposed in the next section 4.

3.1 The guiding case

Assume that in a pool of *p* leaders one leader suddenly collapses. Consider also the possibility that other arbitrary pool members (i.e. remaining minority) are also malicious. Each pool member has to decide on how to distribute his secret asset (token) and the belonging secret keys contrarily, such that his assets can be resurrected safely by the honest majority, despite the malicious minority in the pool.

The strategy of the survivability concept is to update replicas of the token frequently and to send them precautionary encrypted to other well functioning pool members without the decryption keys. After that the destitute SDC collapses one of those trust pool members receives the decryption keys and recovers the token so that he can continue the service.

The token has to be sent separately from the keys to uncorresponding locations before the collapse.

After the collapse the individual pool members sense the disability of the one particular SDC. They send subsequently their keys to exactly one predefined successor among them, so that he can recover the token. With the containing secrets he can prove to the clients to be the trusted leader. The clients will probably not notice the change of leadership because of this smooth masking. Those steps will be explained as follows.

Let T *be a set of* t *tokens, with* $T = \{t_1, t_2, \ldots, t_t)$ *and*
K *a set of* k *keys, with* $K = \{k_1, k_2, \ldots, k_k\}$ *and*
L *a set of* l *pool members, with* $L = \{l_1, l_2, \ldots, l_l\}$
Let TL *denote the allocation of* T *to* L*, with*
$TL = \{(t_1, l_1), \ldots, (t_t, l_x)\}$ *and*
KL *denote the allocation of* K *to* L*, with*
$KL = \{(k_1, l_1), \ldots, (k_k, l_y)\}$ *and*
$TKL = TL \cup KL$ *and* $\{t, k, l, x, y\} \in N$ *with* $\{x, y\} \leq l$

Step1. The destitute SDC sends frequently the encrypted tokens and the belonging keys before the collapse to many pool SDCs, such that the destination of the tokens does not correspond with the one of the keys.

$TKL = \{\ldots, (t_t, l_x;\ k_k, l_y)\}$, *with* $x \neq y$.

Step 2. The receiving many pool SDCs send subsequently the tokens and the keys after the collapse to exactly one predefined healthy pool SDC.

$TKL = \{\ldots, (t_t, l_x;\ k_k, l_y)_{ij}\}$, *with* $x = y = successor$

The result of this strategy is that each receiving pool member has a token, which he himself cannot open with the received key, so that no pool member can act malicious individually before the collapse. He has to find another pool member who has the key that belongs to his encrypted token. This third member should then also be corrupted to betray the destitute SDC. An additional requirement is therefore that up to a minority of the pool members should not be able to recover the token and is thus allowed to be corrupted.

Any majority is then able to recover the token and is thus assumed to be trusted. Why not to construct a strategy where exactly all the members are required to reconstruct the token. The reason for not requiring all the pool members (i.e. consensus) to cooperate in the recovery process is the increased power of the (corrupted) individual. If there is just one malicious member, he can block the recovery by not sending the token before and the key after the collapse. If that is the case the successor receives all the keys from the pool members, except from the malicious member with the corresponding key. The resurrection will not be completed and the clients of the collapsed SDC will become orphans.

3.2 An analogy: the sick old father

The question is how to achieve an allocation of pieces among the many pool members that meets the majority/minority and the uncorrsponding secret/key requirement in a most efficient way. With efficiency a minimum in the amount of interactions between the members and the minimum load per sent package is aimed. In order to explain this confusion by diffusion secret sharing algorithms an analogous example is worked out. From this an allocation matrix is derived that contains the keys and the destinations of the loads (secrets). The depicted pattern of the matrices of different scenarios will help to derive a general approach by induction.

Assume a sick old father who trusts nobody but himself and some of his sons. He has a treasure to leave behind and 3 sons (p1,p2,p3) waiting for his death and their inheritance. He has buried previously the treasure somewhere in the woods and drawn a map of the location. As he has not seen his sons for a long time, his problem is that he does not know which son(s) to trust. He cannot believe that all his descendants turned out to be bad and as such he is sure that the majority of them can be trusted. If he entrusts the map to one of his sons, he might risk that this particular son appears to be unreliable. This son might thus run away with the treasure before the father dies or even after he dies, so that the other (good?) sons inherit nothing. The same could happen when he entrusts his greedy wife, which he already put out of his testament.

Therefore he decides to tear the map in 3 fragments (v1,v2, v3) and to put each piece in a locked case (c1,c2,c3). To confuse his sons, copies of some fragments (v1',v2') are also put in the locked cases. Now we have more pieces of the fragments (c1(v2), c2(v3), c3(v1), c1(v1'), c3(v2')). Each son is respectively given one or more cases and exactly one uncorresponding distinguished key ((k1,k2,..)), which cannot open the cases the particular son has been given. The old man has thought this out well, but now he is wondering how many pieces to cut, which piece to put in each case and how to distribute the cases among all the sons so that exactly the minimum majority of the sons (i.e. the trusted ones = 2) would be able to reconstruct the map when they share cases and keys after his death?

To phrase it more formally: What is the optimal relation (with minimal cases, fragments and copies) between *v*, k and p if a minimum minority will be able to recover the map? Lets workout some cases.

The case with p = 3

With p = 3 sons the minimum majority is (min(q >3/2)=) 2. When applying the splitting method the secret T should be split in v pieces and distributed among the three, such that at least any two sons are required to reconstruct the secret. To phrase it otherwise, no maximum minority (i.e. 1 son) should be able to recover the secret T.

To solve v an inductive approach with increased efforts can be used starting with v =1. If v =1, then T is not split. The question is now how to distribute the one secret, such that no single son but at least two of them can recover it? If the father makes three copies (i.e. for each son one) of the secret and puts each of them in three different cases (with three different keys), then each son receives exactly one case and one key that does not correspond with the case. So each son has one decision variable to disturb: either one secret piece (per case) or one key to exchange. In this situation exactly at least two cooperating sons only can reconstruct the secret, so that for p = 3 the right v =1.

The case with p = 4

With p = 4 sons the minimum majority is (i.e. min(q > 4/2)=) 3. When applying the splitting method the secret T should be split in v pieces and distributed among the 4 sons such that at least any 3 sons are required to reconstruct the secret. To phrase it other wisely, no maximum minority (i.e. ≤ 2 sons) should be able to recover the secret T.

If v =1, then T is not split. The question is again how to distribute the one secret, such that at least three of them can recover it? This is not possible, as has been shown already in previous case that two sons are already able to recover a secret if not split.

If v =2, then T is split in two pieces. The question is now how to distribute the two pieces, such that at least three sons are needed to recover the main secret? Assume that the father shares a distinguished symmetric key with each son. The father has 2 decision variables for each son now: the two pieces. A brute force approach, in which all combinations are subsequently tried, results in tables 1 to 4.

All the corresponding options of symmetric keys $\{K_{0,1} .. K_{0,4}\}$ between father L_0 and p sons $\{L_1 .. L_2\}$ and pieces (T1,T2) are first depicted in table 1. A random search has been conducted and in table 2 the options are given to allocate those 2 pieces (in locked cases) to the p uncorresponding sons. In table 3 a configuration is given in which a minimal number of cases is distributed to achieve the requirements. It is obvious from table 3 that by one or by cooperation of two sons the main secret cannot be recovered. Reconstruction by any three sons is however possible. Thus for this case p = 4 and v = 2 meets the requirements. A solution with p = 4 and v>2 is also possible, but as then more pieces have to be distributed, it id considered less efficient and as such it is not preferred above v = 2. In table 4 only the distinct indexes are shown providing clearness. Testing whether the allocation meets the requirements is easy by hiding any minority of columns and then to check whether the rest will be able to reconstruct T.

Table 1

L_0	$\mathbf{K_{0,1}}$	$\mathbf{K_{0,2}}$	$\mathbf{K_{0,3}}$	$\mathbf{K_{0,4}}$
T1	$K_{0,1}$,T1	$K_{0,2}$,T1	$K_{0,3}$,T1	$K_{0,4}$,T1
T2	$K_{0,1}$,T2	$K_{0,2}$,T2	$K_{0,3}$,T2	$K_{0,4}$,T2

Table 2

L_0	$\mathbf{L_1; K_{0,1}}$	$\mathbf{L_2; K_{0,2}}$	$\mathbf{L_3; K_{0,3}}$	$\mathbf{L_4; K_{0,4}}$
T1	$K_{0,2}$,T1	$K_{0,1}$,T1	$K_{0,4}$,T1	$K_{0,3}$,T1
T2	$K_{0,4}$,T2	$K_{0,3}$,T2	$K_{0,2}$,T2	$K_{0,1}$,T2

Table 3

L_0	$\mathbf{L_1; K_{0,1}}$	$\mathbf{L_2; K_{0,2}}$	$\mathbf{L_3; K_{0,3}}$	$\mathbf{L_4; K_{0,4}}$
T1	$K_{0,2}$,T1		$K_{0,4}$,T1	
T2		$K_{0,3}$,T2		$K_{0,1}$,T2

Table 4

L_0	$\mathbf{L_1}$	$\mathbf{L_2}$	$\mathbf{L_3}$	$\mathbf{L_4}$
T1	2		4	
T2		3		1

The case with p = 5

For p = 5 solutions with v < 3 are neglected, because those do not meet the majority requirement as in the previous sections is discussed. However, applying the above described approach leads to an exponential search algorithm, which is computationally not achievable (NP). Considering that the keys and secrets frequently have to be refreshed and that each refreshment of the pieces requires a reallocation of them, it would make the application practically infeasible. Therefore this approach is facilitated by identifying and adopting a pattern of sequences in v, k and p, by which a polynomial solution domain is achieved.

4. THE PROPOSAL

In the previous section it was shown how to achieve the right distribution of secret shares by random searching, which is computationally infeasible with large p. In this section a random search is avoided by adopting a distribution pattern for the shares, which depicts the relation between p,k and

v. A sequence of key numbers is assumed from the upper left corner to the under right corner of the table as is depicted in table 5.

4.1 The pattern

Consider again p =3 and v =1 in tabel 5a, with CE_0 as the suffering father. The sequence starts with (L1,T1) = 2 or any other number which is not equal to 1 (due to the uncorrespondingness criterium). For frequent refreshment, each time the keys have to be (re)allocated the row rotates so that 231 becomes 123 and then 312 etc. Note that 123 will not be suitable due to the uncorrespondingness criterium. Each son (L) now needs to cooperate with another son to get his key, which results in this case in a majority of 2 (see for one example the shadowed column).

With 4 sons the majority is 3. This number excludes the option of not splitting the token (v =1), as is shown that an unsplit token can be reconstructed by a minority of two sons. So, for this reason the token is split in two pieces T1 and T2 and two rows will fill in the table. The horizontal sequence of the first row (T1) again starts with (L1,T1) = 2 and increases further. The horizontal sequence of the second share T2 (row) starts exactly with the next number of the vertical sequence of column L1 (L1,T2) = 3 and increases further by one. This double sequence (horizontal and vertical) can be extended for any n. In table 5 this linear approach is depicted for p = 3,5,7. Refreshment is achieved by rotating the table (both rows) horizontally.

Table 5 $v = p/2 - 1/2$

T/L	L_1	L_2	L_3
T1	2	3	1
T2			
T3			

a. p=3, v=1

L_1	L_2	L_3	L_4	L_5
2	3	4	5	1
3	4	5	1	2

b. p=5, v=2

L_1	L_2	L_3	L_4	L_5	L_6	L_7
2	3	4	5	6	7	1
3	4	5	6	7	1	2
4	5	6	7	1	2	3

c. p=7, v=3

This approach is computationally feasible, but is not optimal for even p numbers, as can be seen in the next tables 6 where for the cases p = 4,6,8 the same goal is achieved but with less pieces to distribute. For even numbers p a different pattern is adopted. The saving compared to the odd numbers is $(p*v)/2 = (2v*v)/2 = v^2$ so that for p = 4 and v = 2 not 8 pieces are distributed, like in table 4, but $2^2 = 4$ pieces. This counts not only for p = 4,6,8, but for all even numbers. The explanation for this saving is that for even numbers a roundup by 1 is done to achieve majority (p/2+1), whereas a majority is already achieved by the first rational number larger than p/2. The smuggle of +1 leads to redundancy in the number of distributed pieces. The odd numbers also smuggle, but up (+1/2) and down (-1/2), so that the net result is zero (honest).

Table 6 . $v = p/2$

T/L	L_1	L_2	L_3	L_4
T1	2		4	
T2		1		3
T3				
T4				

a. p= 4, v= 2

T/L	L_1	L_2	L_3	L_4	L_5	L_6
T1	2		4		6	
T2		5		1		3
T3	6		2		4	
T4						

b. p= 6, v= 3

T/L	L_1	L_2	L_3	L_4	L_5	L_6	L_7	L_8
T1	2		4		6		8	
T2		5		7		1		3
T3	6		8		2		4	
T4		1		3		5		7

c. p= 8, v= 4

4.2 The algorithm in pseudo code

The next list in pseudo code shows how to make such allocation tables. The pseudo code comprehends two main parts: for an even and for an odd number of pool members. They manage the key allocation according to the pattern of the previous tables 5 and 6. The even part resolves three contentious issues: 1) the hop from row to row of the sequence of keys, 2) the increase of sequential key number due to that hop and 3) continuously resetting the limited sequence. The odd part adds a fourth issue: alternation of blank spaces except for the first cell in a row.

```
ShiftO = 0                    ' Initialisation of the shifts
ShiftE = -1
Period = 600 sec              ' setting an example timer to refill the allocation tables
Do each Period
   Blank = false
   Count(p, mypool)           ' p = number of pool members (sons)
   If p = Odd then
      key = ShiftO            ' key = allocated key = content of cell in table
      v = n/2 –1/2                                ' v = number of pieces
      AllocateKeyPiece = ARRAY[1..v,1..p]         '
      For a = 1 to v do                           ' fill in rows from left to right
         key = key + 2                  ' jump two numbers when a new row starts
         For b = 1 to p do
            If b=! 1 then key = key + 1   ' setting the content of a cell
            If key > p then key = key-p   ' reset key sequence
            AllocateKeyPiece[a,b] = key
         End For
```

```
    End For
    ShiftO = ShiftO +1               'counter for rotating the rows each period

  Else
    key = ShiftE
    v = n/2
    AllocateKeyPiece = ARRAY [1..v,1..p]
    For a =1 to v do                      ' when a new row starts:
      key = key +3                        ' jump three numbers
      For b = 1 to p do
        If b=! 1 then key = key + 1       ' setting a key
        If key > p then key = key-p       ' reset key counter when all keys are set
        If Blank = false then             'alternating blank cells
          AllocateKeyPiece [a,b]  = key   'allocate key to cell= piece and location
          If b=! p then Blank = true      'block alternation when new row starts
        Else
          If b=! p then Blank = false
        Endif
      End For
    End For
    ShiftE = ShiftE +1

    End if
  End do
```

With the allocation table a leader can determine by which keys to encrypt the pieces and where to sent them. In order to commit the logistical operation efficiently the pieces with the same destination have to be clustered and sent together. The tables with indexes of key numbers are used to encrypt the shares first. This happens by sequentially reading the records from the table and replacing the indexes with the encrypted piece (eventually in a new table).

```
AllocateEncrPiece = ARRAY [1..v,1..p]
AllocateKeyPiece = ARRAY [1..v,1..p]
AllocateEncrPiece [ ] = 0
    For a =1 to v do
      For b =1 to p do
        If IsNotEmpty(AllocateKeyPiece [a,b]) then
          AllocateEncrPiece [a,b]) = E(k1,a; t1,b)
     Else
          AllocateEncrPiece [a,b]) = Empty
```

```
        Endif
      End For
    End For
```

The logistical preparation procedure will be completed by packaging each column as one message to one destination (pool member).

```
Package = Array[id_1 ..id_p]
Package[] =0
For a =1 to p do
  Package[id_a] = ""
  For b =1 to v do
    If IsNotEmpty(AllocateKeyPiece [a,b]) then
      Package[id_b] = Package[id_b] + ";"+AllocateEncrPiece[a,b])
    Endif
  End For
  ProcSend(Package[id_b]; id_b)
End For
```

The refreshment occurs usually by subsequently rotating the rows by one column as is shown in table 7 for n = 5. An exception occurs, when the keys in the first shift would correspond with the key owner. The rotation shifts then two columns at once.

Table 7. Refreshing by rotation

L_1	L_2	L_3	L_4	L_5
2	3	4	5	1
3	4	5	1	2

Shift=1

L_1	L_2	L_3	L_4	L_5
3	4	5	1	2
4	5	1	2	3

Shift=2

L_1	L_2	L_3	L_4	L_5
4	5	1	2	3
5	1	2	3	4

Shift=3

5. DISCUSSION

This paper proposes a scalable and efficient decentralized secret sharing algorithm (i.e. with no premature central facilitator) in order to avoid multiple single points of failures before and after a failure. It prevents premature exposition of the individual shares by the participants as they all host a decrypted share with a key of the others. The algorithm requires an arbitrary majority of participants to reconstruct the total secret and prescribes the exact keys for encryption and distribution of shares. The proposed algorithm has been adopted in a framework called Medusa that is meant to enhance autonomous survivability of security distribution centers in large-scale distributed information systems, like grid systems. The Medusa protocol set is subsequently added to an arbitrary authentication server and tested in a simulated environment by a discrete event network simulator NS+ with C++/TCL on Linux. The test case was a basic failure test in which randomly one of the joint authentication servers deliberately was stopped to assess how the secret sharing algorithm performed. It was obvious that the larger the allied pool was the more survivable the authentication service was, due to many shares. The extra performance overhead due to increasing numbers of participants and thus the increased number of shares was acceptable in comparison to the survivability increase. The complexity of the secret sharing algorithm is $O(n^2)$, due to the double loops of 1 to v and 1 to p. As such they were computable in polynomial time. However, there is a trade off between survivability and processing time. With a growing number of participants in the pool survivability increases at the expense of the increasing processing time for secret sharing. As such medium sized pools are aimed and by connecting them via one or more pool members this trade off can be avoided. The secret sharing algorithm is performed for the limited number of pool members in each pool, whereas survivability is determined by the range of all the interconnected pools.

A second test is intended where a distributed denial of service attack (ping-of-death) in which random adversaries send en masse ping messages to the authentication server.

REFERENCES

[1] S. Daskapan, W.G. Vree, A. Ali Eldin, "Trust metrics for survivable security systems", *Proc. of the IEEE International Conference on Systems, Man & Cybernetics*, Washington, 2003.

[2] A.E. Barbour, A.S. Wojcik, "A General Constructive Approach to Fault-Tolerant Design Using Redundancy", *IEEE Transactions on Computers*, pp. 15-29,1989.

[3] Matti A. Hiltunen, Richard D. Schlichting, Carlos A. Ugarte, "Enhancing Survivability of Security Services Using Redundancy", *The International Conference on Dependable Systems and Networks*, Sweden, 2001.

[4] Bruno Dutertre, Valentin Crettaz, Victoria Stavridou, "Intrusion-Tolerant Enclaves", *IEEE Symposium on Security and Privacy*, California, 2002.

[5] L. Lamport, R. Shostak, M. Pease, "The Byzantine Generals Problem", *ACM Transactions on Programming Languages and Systems*, vol.4(3), pp.382-401, July 1982.
[6] Jean-Claude Laprie, "Dependable Computing and Fault Tolerance: Concepts and Terminology," *15th International Symposium on Fault-Tolerant Computing*, pp. 2-11, 1985.
[7] L. Gong, "Increasing Availability and Security of an Authentication Service", *IEEE Journal on Selected Areas in Communications*, Vol 11(5), pp.657-662, 1993.
[8] M. Reiter, "Secure Agreement Protocols: Reliable and Atomic Group Multicast in Rampart", *Proc. of 2nd ACM Conference on Computer and Communications Security*, pp. 68–80, 1994.
[9] Thomas Hardjono, Jennifer Seberry, "Replicating the Kuperee authentication server for increased security and reliability", *Australasian Conference on Information Security and Privacy*, pp.14-26, 1996.
[10] Matti A. Hiltunen, Richard D. Schlichting, Carlos A. Ugarte, "Building Survivable Services Using Redundancy and Adaptation", *IEEE Transactions on Computers*, Vol. 52(2), pp. 181-194, 2003.
[11] Vivek Pathak, Liviu Iftode, "Byzantine Fault Tolerant Authentication", Rutgers University, Department of Computer Science *Technical Report*, DCS-TR-492, June 2002.
[12] P. Veríssimo, NF Neves, M. Correia, "The middleware architecture of MAFTIA: A Blueprint", *Proc. of the IEEE Third Information Survivability Workshop*, Boston, pp. 24-26, 2000.
[13] C.Cachin, J. Poritz, "Secure Intrusion Tolerant Replication on the Internet", *Proc. of the 2002 International Conference on Dependable Systems and Networks*, Washington, pp. 167-176, 2002.
[14] Lidong Zhou, Fred B. Schneider, Robbert van Renesse, "COCA: A Secure Distributed On-line Certification Authority", *Technical Report TR2000-1828*, 2000.
[15] Ohad Rodeh, Kenneth P. Birman, Danny Dolev ,"The architecture and performance of security protocols in the ensemble group communication system: Using diamonds to guard the castle", *ACM Transactions on Information and System Security*, Vol. 4(3), pp.289 – 319, 2001.
[16] R. Canetti and T. Rabin, "Fast asynchronous Byzantine agreement with optimal resilience", *Proc. 25th Annual ACM Symposium on Theory of Computing*, pp. 42-51,1993.
[17] M. J. Fischer, N. A. Lynch, M. S. Paterson, "Impossibility of distributed consensus with one faulty process", *Journal of the ACM*, Vol. 32(2), pp.374-382, 1985.
[18] G. Bracha. "An asynchronous [(n -1)/3]-resilient consensus protocol", *Proc. of 3rd ACM Symposium on Principles of Distributed Computing*, pp. 154-162, 1984.
[19] A. Shamir, "How to Share a Secret", *Communications of the ACM*, Vol 22(11), pp.612-613, 1979.
[20] K. Kurosawa, K. Okada , H. Saido , D. Stinson, "New combinatorial bounds for authentication codes and key predistribution schemes", *Designs, Codes and Cryptography*, Vol 15(1), pp.87-100, 1998.

[21] Y. Desmedt, "Threshold cryptography", *European Transactions on Telecommunications*, Vol 5(4), pp. 449–457, 1994.
[22] J. Seberry, C. Charnes, J. Pieprzyk and R. Safavi-Naini, "Crypto Topics And Applications II", Algori*thms and Theory of Computation Handbook*, CRC Press, 1999.
[23] G. J. Simmons, "An Introduction to Shared Secret and/or Shared Control Schemes and their Application", *Contemporary Cryptology: The Science of Information Integrity*, G. J. Simmons (Ed.), IEEE Press, pp.441-497, 1991.
[24] G. R. Blakley, "Safeguarding cryptographic keys", *Proc. of American Federation of Information Processing Societies 1979 National Computer Conference*, Vol.48, pp. 313-317, 1979.
[25] C. Cachin, On-line secret sharing, in "Cryptography and Coding V", C. Boyd (ed.), *Lecture Notes in Computer Science*, Vol. 1025, pp. 190-198, 1996.
[26] E.F. Brickell, D.R. Stinson, "Some improved bounds on the information rate of perfect secret sharing schemes", *Journal of Cryptology*, Vol. 5(3), pp. 153-166, 1992.

A LANGUAGE DRIVEN INTRUSION DETECTION SYSTEM FOR EVENT AND ALERT CORRELATION

Eric Totel, Bernard Vivinis, Ludovic Mé
Eric.Totel@supelec.fr, Bernard.Vivinis@supelec.fr, Ludovic.Me@supelec.fr
Supelec, BP 81127, 35511 Cesson-Sévigné Cedex, France

Abstract It is well known that security prevention mechanisms are not sufficient to protect efficiently an information system. Intrusion detection systems are required. But these systems present many imperfections. In particular, they can either generate false positives (i.e., alarms that should not be produced) or miss attacks (false negatives). However, the main problem is the generation of false positives that can overwhelm the information system administrator. In this paper, we follow the notion of correlation proposed by others. The objective is to aim at correlating either events in the analyser or alerts in the manager. We first present the ADeLe language, which provides a way to define the correlation properties. Then we present which algorithms have been carried out in our IDS to handle ADeLe signatures. Finally, we show the stress tests that have been applied to the probe algorithms that we have implemented.

Keywords: Intrusion detection, event correlation, alert correlation, attack signature recognition, site security monitoring

1. INTRODUCTION

In common information systems, security is achieved by carrying out prevention mechanisms, such as authentication and access control, to enforce security policy. But this is not sufficient, because vulnerabilities remain. That is why it is actually also necessary to carry out intrusion detection mechanisms (IDS) in order to detect intrusions in the system.

Two intrusion detection techniques have been proposed: misuse detection, which requires specific representation of computer system abuse (such as attack signatures), and anomaly detection. In this paper, we focus on misuse detection.

According to the IETF Intrusion Detection Working Group (IDWG), an IDS is composed of three elements: one or more sensors whose role is to get data from the monitored system and to format data to build events, the analyzer that analyses events from the sensors to generate alerts if needed, and the manager that gets alerts from the analyzer for further treatment (presentation, correlation and reaction).

However, most of the intrusion detection tools produce too many false alarms (false positives, i.e., alarms that are risen when no attack occurs). These false alarms are due to a combination of causes, among which signatures with too low a level of semantic, e.g, mono-event signature. A first approach consists in correlating alerts in the manager (see [10] for example). Several Correlation algorithms have been proposed (see [9–2]).

In this paper, we propose to correlate events or alerts in order to lower the number of false positives generated by the IDS. The correlation can be applied at the lowest possible level in the system, i.e., at the analyser level. Our goal is to correlate events from several sources or alerts generated by several analysers, in order to get a better level of semantic for the events or alerts. This provides a way of producing more precise alerts for the system administrator.

In our approach, a correlation language has been designed. This language [8] intends to provide operators that describes the correlation that can be applied either between events or alerts.

This paper aims to briefly describe the language we have defined (Section 2), and then to focus on the implementation of an IDS configured through this language (Sections 3). Finally, Section 4 focuses on testing the IDS.

2. THE ADeLe LANGUAGE

2.1 Introduction

The ADeLe language [8] aims at describing all functional aspects relative to an attack, either from an attacker or from a system administrator point of view. These different aspects, described using a unique language, provide a way to build an attack database containing all necessary information for understanding how to carry out, detect, and react to an attack. This means that the description of an attack contains:

- the description of the exploitation of one or more vulnerabilities in order to compromise the system security policy (i.e., the attack scenario);

- the description of the attack from the target viewpoint, that provides a way of detecting the attack signature when it occurs;

```
<EVENTS>
   CP : SNARE
   {
           event.name == "execve()";
           path       == "/bin/cp";
           arguments  MATCHES "cp /bin/sh *";
   } CP0;
   OPEN: SNARE
   {
           event.name    ==      "open(O_WRONLY|O_CREAT)"
           process.name ==       "cp";
           attributes    ==      "rwxr-xr-x";
   } OPEN0;
   ...
</EVENTS>
```

Figure 1. Event filter examples

- the description of the appropriate response to be given to a detected attack.

In the following sections, we focus on the second aspect of the language, that is dedicated to the IDS detection process control. The signature is organised into three parts: the EVENT part that contains the filter definitions, the ENCHAIN part defines ordering relations on events, and the CONTEXT part describes the correlation contexts between the events.

2.2 Filtered Events

The ADeLe language correlation process is based on the use of filters. These filters describe what events or alerts must be used as elementary elements in the correlation process.

As shown on Figure 1, an event filter is defined using its information source name (here *Snare* audit[1]) and a name (here *CP* or *OPEN* are naming the types of the filter). The event filter describes what values are taken by the event attributes. The IDS will use this information to filter the audit data items, and elect the data items that fit this definition. Moreover, we can name a given instance of a filtered element. This name

[1]Snare is an audit tool that works on Linux systems and provides OS audit data, http://www.intersectalliance/projects/Snare

will be handled as a variable by the correlation operators we describe below (e.g., *CP0* or *OPEN0*).

2.3 Correlation Operators

Once the events or alerts taken as inputs are filtered, ADeLe defines what are the correlations that must be applied between them. This leads to define ordering and occurences information on the filtered inputs.

This is defined by means of five operators: *Sequence*, *OneAmong*, *NonOrdered*, *Without* and *Repeat*. These operators are used in order to define the signature of the attack. In the following subsections, we explain what is the meaning of each operator.

2.3.1 Sequence operator. The sequence operator enables a description of a sequence of events. As shown on example 1 below, the sequence expression defines both temporal constraints on the element order, and the elements that are required to recognise the signature of the attack. In this example, we must detect three elements sequentially: A, B, then C in this given order. The formalism used to express the semantic of the operator is the chronicle formalism [4, 5]. This semantic shows us that three events must be detected, and gives the temporal constraints associated to their recognition.

EXAMPLE 1

```
Syntax:
<ENCHAIN>
   Sequence(A, B, C)
</ENCHAIN>
```

Semantic: $event(A) \wedge event(B) \wedge event(C) \wedge t(A) < t(B) \wedge t(B) < t(C)$

2.3.2 NonOrdered operator. The NonOrdered operator enables the description of a signature in which all elements must be recognised without any order constraint. The example below shows that all the three events A, B and C must be detected without time constraints between them.

EXAMPLE 2

```
Syntax:
<ENCHAIN>
   NonOrdered(A, B, C)
</ENCHAIN>
```

Semantic: $event(A) \wedge event(B) \wedge event(C)$

2.3.3 OneAmong operator. The OneAmong operator enables the description of the recognition of one expression among many. This actually defines a disjonction between the elements during a given detection. For a given attack, only the detection of one of the proposed elements is sufficient to ensure the detection of the signature. In the example below we show that A or B or C must be detected in order for the signature to be recognised.

EXAMPLE 3

```
Syntax:
<ENCHAIN>
   OneAmong(A, B, C)
</ENCHAIN>
```

Semantic: $event(A) \vee event(B) \vee event(C)$

2.3.4 Without operator. The Without operator enables the description of how a positive signature detection can be removed if a negative signature is detected. The positive signature is the signature part that must be recognised to report an alert, and the negative signature is the signature part that produces the stop of the recognition. In the example below, the sequence A and B must not be reported if C is detected during the sequence detection. This operator provides a way to cancel a partial detection if a given event or expression occurs. The detection of the negative signature (here C) cancels the detection of the positive signature(here *Sequence(A,B)*).

EXAMPLE 4

```
Syntax:
<ENCHAIN>
   Without(Sequence(A, B), C)
</ENCHAIN>
```

Semantic: $event(A) \wedge event(B) \wedge noevent(C) \wedge t(A) < t(B)$

2.3.5 Repeat operator. The *Repeat* operator enables the description of a repetition of an event or an expression. In the example below, we accept any set of 10 consecutive events A.

EXAMPLE 5

```
Syntax:
<ENCHAIN>
   Repeat(A, 10)
</ENCHAIN>
```

Semantic:

$$\prod_{i\in[1,10]} event(A_i) \wedge \prod_{i\in[1,9]} t(A_i) < t(A_{i+1})$$

2.3.6 Context correlation. The operators describe signatures where the order of appearance of the detected elements is strictly defined. This is the first part of the correlation information required. The second part of the correlation deals with the values of attributes of the filtered events. It is thus necessary to define additional correlation between the elements detected. Eventually, the correlation is both temporal (temporal relations between the filtered events) and at the level of inter event values (relation between the attributes of the filtered events).

Time Constraints. Defining timing constraints between the filtered events requires three operators: *TimeOut*, *MinDelay*, *MaxDelay*. These operators provide all required properties for defining temporal constraints in the attack detection. The *TimeOut* operator defines how much time a pending detection is valid until it is canceled. The *MinDelay* operator defines the minimum delay that must occur between two events. The *MaxDelay* defines how much time must exist between two events. Those operators are illustrated in Example 6.

EXAMPLE 6

```
<CONTEXT>
   TimeOut(100);
   MaxDelay(A, B, 20);
   MinDelay(A, B, 10);
</CONTEXT>
```

Semantic:

$$|(t(A) - t(B)| < 20s \wedge |(t(A) - t(B)| > 10s \wedge scenarioDetection < 100s$$

Constraints on inter event attributes. In order to correlate successfully filtered events, the language enables the description of constraints on the event attributes. These constraints require that some attribute values must be linked with other event attribute values. In the example 7 below, the snare audit event attributes *user.uid* must be identical for the events A, B and C.

EXAMPLE 7

```
<CONTEXT>
   A.user.uid == B.user.uid;
   B.user.uid == C.user.uid;
</CONTEXT>
```

2.4 Conclusion on ADeLe

Only the detection part of the ADeLe language has been described in the previous sections. This part of the language is used in the context of attack detection and in particular in the IDS presented in the next section. It must be noted that the ADeLe semantic is very close to the chronicles one, but that it is less general and much more adapted to the expression of signatures in the field of intrusion detection.

3. AN INTRUSION DETECTION SYSTEM

3.1 IDS Architecture

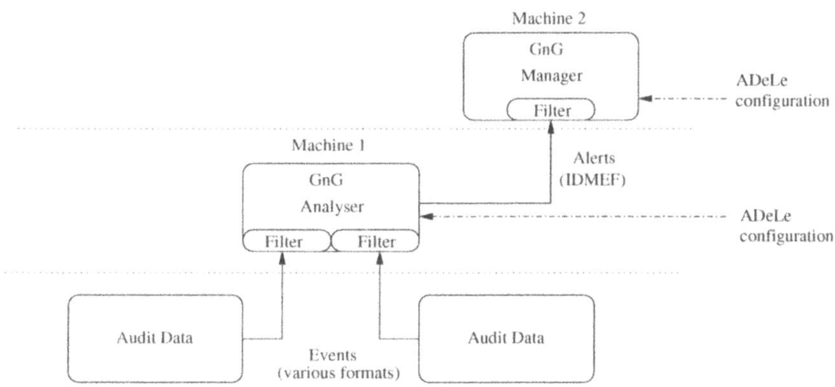

Figure 2. *GnG* architecture

The IDS that has been carried out is called *GnG* (for Gassata New Generation). It is designed to take both the role of an analyser or a manager inside the system, because it can take as input any audit data events as well as alerts in IDMEF [3] format.

The correlation can be applied at both levels (see Figure 2). Correlation of events puts in relation multiple source audit data items and elects data on which correlation can be applied. Correlation of alerts takes as input IDMEF format from one or several analysers and elects the alerts that can contribute to the recognition of an attack signature. In both cases, the data in IDS entry are filtered to retain all events or alerts that can contribute to the evolution of signature recognition.

The ADeLe language is used at both level to define how the events or alerts can be correlated.

3.2 Events and Alerts correlation using finite state automata

In order to recognise the signatures of the ENCHAIN section, the IDS uses finite state automata. These automata are built automatically using the signature definitions. Each operator inside the signature is declined into an automaton to ensure the recognition. The goal of this automaton is to implement the multi-pattern matching detection algorithm.

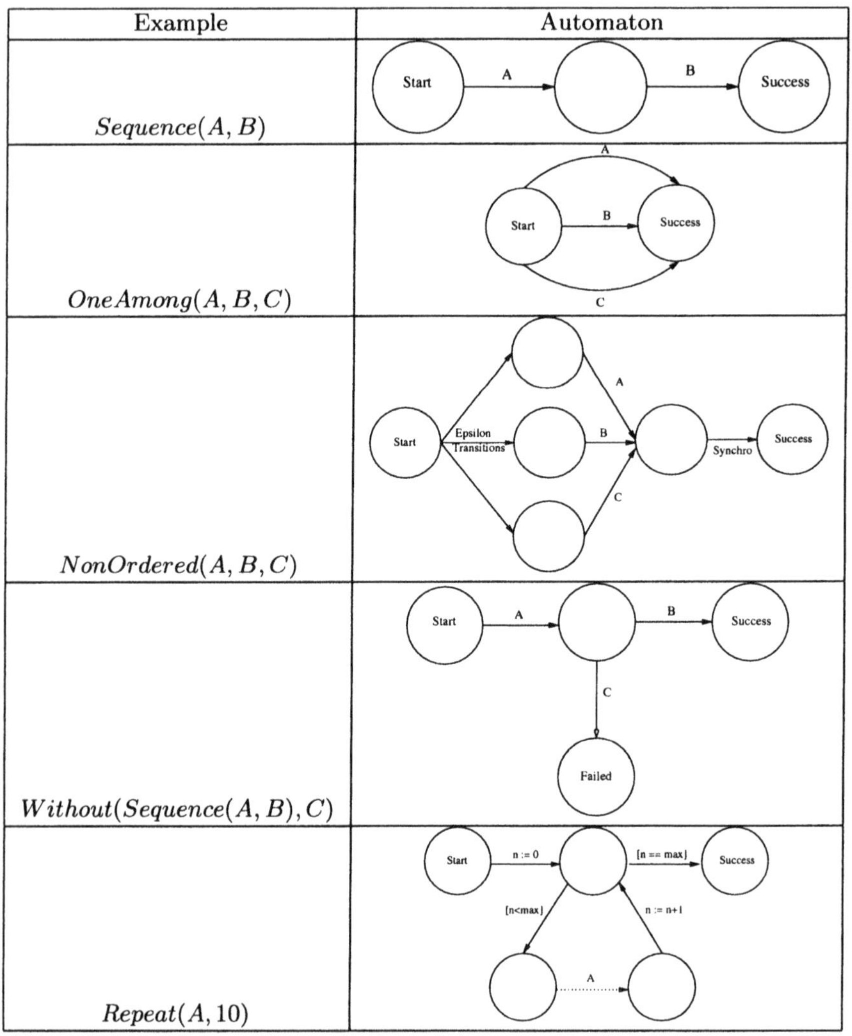

Figure 3. Operators translation to finite state automaton

The table above (Figure 3) shows how each operator is translated to an equivalent automaton. This representation gives the semantic of these operators, but using another description model than in section 2.3.

However, they do not show only a semantic of the operators, but how the intrusion detection is handled in *GnG* internally.

3.3 Automaton merge

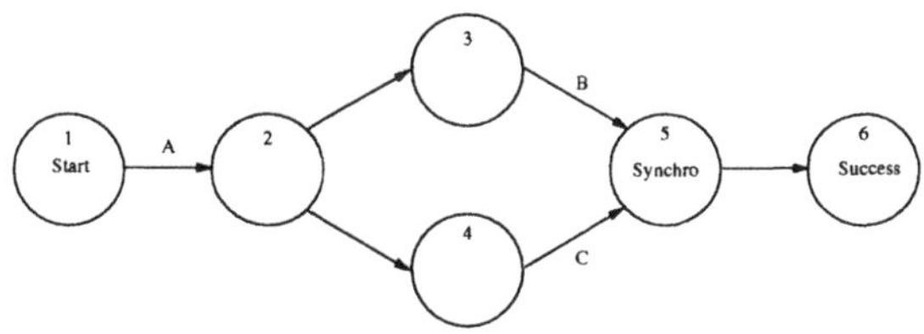

Figure 4. $Sequence(A, NonOrdered(B, C))$ automaton

The single operator automata that have been described in the previous section are merged to obtain the final signature automata. These final automata are used to perform detection of events or alerts. The algorithms required to merge the various single automata are not described in detail in this paper, but can be found in [7]. To illustrate the merge action, we show in Figure 4 the automaton equivalent to the signature *Sequence(A, NonOrdered(B, C))*.

3.4 Plan recognition

3.4.1 Plan definition. A plan is an active entity that stores partially recognised attacks. It stores both the current states of automata and a history of all elements that have been accepted in a given partial recognition.

A plan in our model is a pair of two n-uplets. The first n-uplet is a list of automaton states that are currently active for this plan. The second n-uplet is a list of history items that have been accepted.

3.4.2 Plan evolution. Our goal is to allow, if needed, the detection of all occurences of a given signature, e.g., for the *Sequence(A, NonOrdered(B, C))* and the audit events a_1, a_2, b, c, we intend to detect two attacks: a_1, b, c and a_2, b, c. We use as an example the scenario illustrated in Figure 4 with the state numbers as defined in this figure.

Event Accepted	Parent Plan	Plans Produced
		$p_0 = ((start), ())$
a_1	p_0	$p_1 = ((2), (a_1))$
ϵ-transition	p_1	$p_1 = ((3,4), (a_1))$
a_2	p_0	$p_2 = ((2), (a_2))$
ϵ-transition	p_2	$p_2 = ((3,4), (a_2))$
b	p_1	$p_3 = ((5,4), (a_1, b))$
	p_2	$p_4 = ((5,4), (a_2, b))$
c	p_1	$p_7 = ((3,5), (a_1, c))$
	p_2	$p_8 = ((3,5), (a_2, c))$
	p_3	$p_5 = ((5,5), (a_1, b, c))$
	p_4	$p_6 = ((5,5), (a_2, b, c))$
synchro	p_5	$p_5 = ((6), (a_1, b, c))$ (detection)
	p_6	$p_6 = ((6), (a_2, b, c))$ (detection)

Figure 5. Plan evolutions

The Figure 5 shows what plans are produced when each event is received. In the example given in the table, we see several properties of the plan handling algorithm:

- at the initialisation of the algorithm, a plan in the *start* state of the automaton with an empty history is initialised;
- a plan evolves when it faces an audit event that can contribute to its evolution;
- when a plan evolves via an event transition, the current plan is duplicated *and* produces a new plan;
- when a plan faces what we could call *fork* epsilon transitions (i.e., state duplication), it handles several states of automaton in parallel;
- when a synchronisation between two running states is required, it is processed when all states of the plan are in the same synchronisation state.

This way of handling plans is very close to the way chronicles are managed in [4]. It also has all the drawbacks of the chronicles. For example, no plan is deleted unless we have special operators that would provide a mean to break a recognition. It is thus necessary to define how to break partial plans and, thus, how to control the memory allocation. This is the purpose of the next section.

3.5 Partial Plan Deletion

We have designed several mechanisms in order to control the memory allocation, and thereby cancel partially recognised plans. The main problem of such an approach is ensuring that no important alert is missed. In order to ensure that the user of the ADeLe language knows exactly what he is doing, all plan deletions are explicitly controlled by the user. No implicit plan deletion is defined in the language.

3.5.1 Plan recognition break. Two operators exist whose goal is to reduce the number of partial pending plans.

The *Without* operator provides an explicit break rule: if the negative plan (cf Section 2.3.4) is recognised, positive plans that are pending are deleted from the set of pending plans.

The *First* operator ensures that a single event selected to fire a transition is the *first* of its type in the flow of audit events or alerts. Thus, if the previous example (cf. Figure 5) has been *Sequence(First(A), NonOrdered(B, C))*, the only detected signature would have been a_1, b, c. From the point of view of the finite state automaton, no replication of a plan is produced when the A transition is fired. And no further a event can be taken into account unless the whole signature has been detected (Moreover, the recognition is re-initialised when the detection succeeds, and new a events can again, in this case, be taken into account).

3.5.2 Temporal plan break. Moreover, the temporal properties that are defined in the temporal context provide means of breaking plan recognitions. If temporal deadlines (cf. Section 2.3.6) are reached, the corresponding plans are deleted. This is the only way to break recognitions that do not contain explicit break operators. However, breaking recognitions using temporal deadlines is dangerous, because an attack can be intentionnaly performed during a long time period. Thus, it is possible for an attacker to avoid an IDS detection. However, in some situations (e.g., some repetitive worm attacks) these deadlines have a meaning, and provide a good temporal correlation context.

4. IDS TEST

To Perform tests on an IDS, many points must be taken into account: false positives and false negatives generation, adequacy of the algorithms with real situation, etc. The tests that are proposed here only aim at stressing the detection algorithm.

4.1 IDS behaviour test

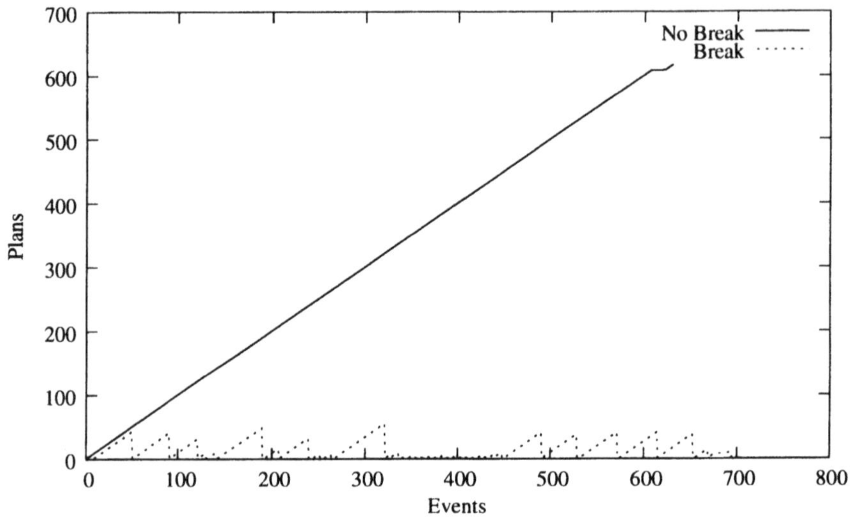

Figure 6. Plan number evolution

4.1.1 Plan Number Evolution. In this section, we illustrate the behaviour of the IDS in terms of plan number evolution. The Figure 6 describes the plan number evolution for a signature that describes the detection of a *Sequence* of two events and the same signature *Without* a recognition of one event. The signatures that we want to detect are mixed with a set of five hours of audit data items (these data were collected on a user machine in the laboratory, and reflect a common computer use). We can see that the evolution of the number of plans is approximately linear for the first signature. In the case of the second signature, the number of plans is limited to an upper bound. Thus, the "without" type signature provides a way of controlling the plan number evolution.

4.1.2 Plan Deletion Illustration. The Figure 7 shows the evolution of the number of plans when a *TimeOut* of one second is defined on the *Sequence* recognition. We see that every second, all plans that time out are deleted. This implies that the number of plans is contained under a maximum limit.

4.2 Advanced Tests

The tests that were defined include four attacks recognition in a flow of real audit data (five hours of audit data). Their role is to check if false

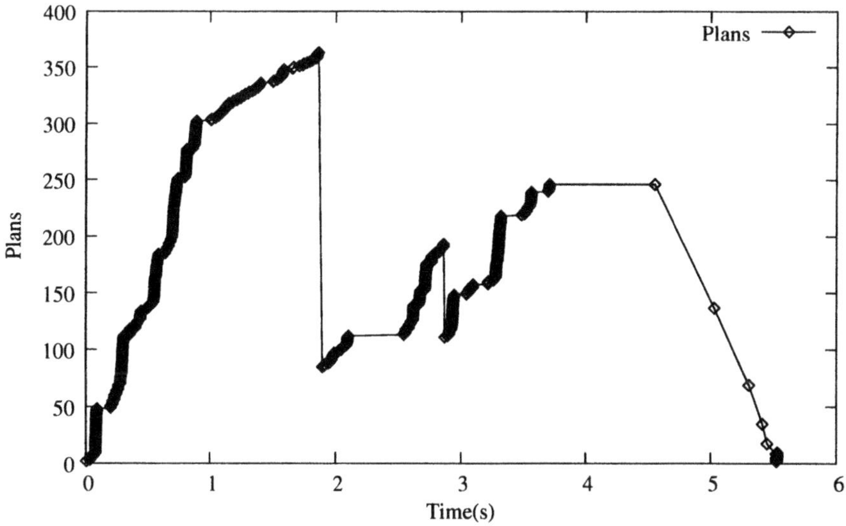

Figure 7. Time break

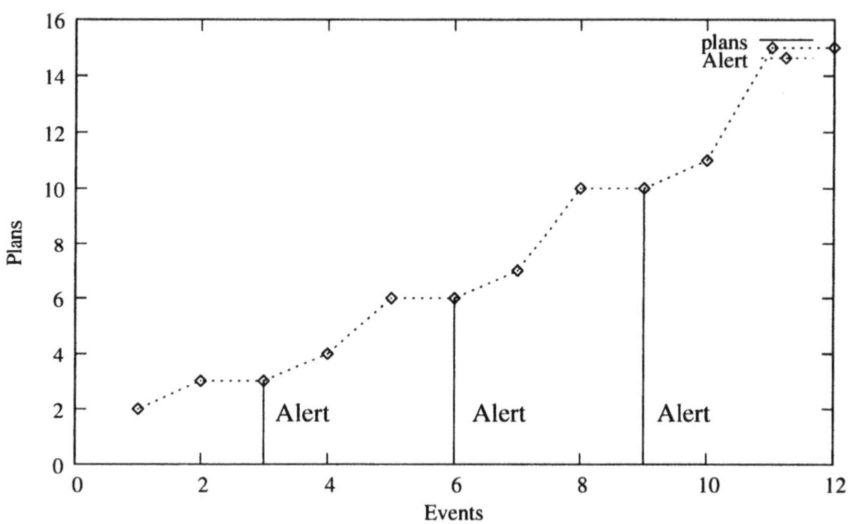

Figure 8. Attack detection

alarms are generated (either false positives or false negatives). In order to perform these tests, we used a program that mixes real data flow with the signatures of the four attacks described. The Figure 8 shows the behaviour of the IDS when the fourth attack (RPCINFO attack) we have defined is mixed in a real flow of audit data items. We see that three alerts are generated instead of one. This was the worst case in the tests we have defined.

The results can be summarized as follows: the more precise a signature is, the less false positives we obtain. However no false negatives were encountered during this test: the attack signatures described in ADeLe were successfully detected.

EXAMPLE 8 *Attack: RPC Info*

```
<EVENTS>
PORTMAP : SNARE
{
  process.name == "portmap";
} PORTMAP0;

RPCMOUNT : SNARE
{
  process.name == "rpc.mountd";
} RPCMOUNT0;

FINGER: SNARE
{
  process.name == "tcpd";
  arguments == "in.fingerd";
} FINGER0;
</EVENTS>

<ENCHAIN>
  Sequence(PORTMAP0, RPCMOUNT0, FINGER0)
</ENCHAIN>
```

5. RELATED WORK

The correlation approaches can be divided into three categories: the explicit, the semi-explicit and the implicit correlations.

The explicit correlation considers that an intrusion in a system can be described using several steps (the signature of events). Several works are using this type of correlation process: for example, Cuppens and Ortalo published an attack description language called Lambda [2]. It intends to describe all aspects of an attack against a system. Logweaver is a language driven tool that can analyse online event flows and correlate information [6]. Suthek is a logical language that is used to describe sequences of events in system audit traces [11, 12]. The chronicles have been used to correlate alerts [9]. They provide a way to describe sequences of events that can be detected in a system. This paper shows how correlation can be applied using the chronicles as detection and correlation language.

The semi-explicit correlation type generalises the explicit correlation by introducing the notion of attack pre-conditions and post-conditions. Lambda [2] proposes such notions.

The implicit correlation, unlike the explicit and semi-explicit approaches, tries to recognise and build attack scenarii. This can be, for example, found in [13].

6. CONCLUSION

This paper relates our efforts to propose a fully functional intrusion detection system based on event and alert correlations. Among the several approaches available in the literature, we chose to implement a language driven signature based correlation. The correlation information is explicit and describes what are the different constraints between the events, either from a value or temporal point of view.

We have focused both on the ADeLe language functionnalities and on their implementation through the description of the IDS algorithms. Unlike other papers this implementation shows how an efficient recognition algorithm may be carried out in our field. In particular, the algorithm stress test proves that the algorithm can be adapted to a real life use (e.g., it can handle real-time correlation for active OS events).

In the context of the DICO project[2], we are currently performing some IDS tests whose goal is to compare several similar IDS tools (Suthek, Logweaver, cf. previous section). These tests would eventually benefit to our work by proving that our language and implementation algorithms are consistent with the other existing works in this domain.

References

[1] F. Cuppens. Managing alerts in multi-intrusion detection environment. In *Proceedings of the 17th Annual Computer Security Applications Conference (ACSAC'01)*, 2001.

[2] F. Cuppens and R. Ortalo. Lambda: A language to model a database for detection of attacks. In H. Debar, L. Mé, and S. F. Wu, editors, *Proceedings of the*

[2]The DICO project is an RNTL project sponsored by the french ministry of research

Third International Workshop on the Recent Advances in Intrusion Detection (RAID'2000), number 1907 in LNCS, pages 197–216, October 2000.

[3] H. Debar, M. Huang, and D. Donahoo. Intrusion detection exchange format data model. Technical report, IETF, 2002.

[4] C. Dousson. *Suivi d'evolutions et reconnaissance de chroniques.* PhD thesis, 1994.

[5] C. Dousson. Alarm driven supervision for telecommunication networks. In *Annales des Telecommunications*, pages 501–508, 1996.

[6] J. Goubault-Larrec. An introduction to logweaver. Technical report, LSV, 2001.

[7] J. Goubault-Larrec. DICO Project SP3.3: Detection algorithms and signature languages. Technical report, DICO Project, July 2003.

[8] C. Michel and L. Mé. ADeLe: an attack description language for knowledge-based intrusion detection. In *Proceedings of the 16th International Conference on Information Security (IFIP/SEC 2001)*, pages 353–365, June 2001.

[9] B. Morin and H. Debar. Correlation of intrusion symptoms: an application of chronicles. In *6th International Conference on Recent Advances in Intrusion Detection (RAID'03)*, September 2003.

[10] B. Morin, L. Mé, and H. Debar. M2D2 : A formal data model for IDS alert correlation. In *5th International Conference on Recent Advances in Intrusion Detection (RAID'02)*, October 2002.

[11] J.-P. Pouzol and M. Ducassé. From declarative signatures to misuse ids. In W. Lee, L. Mé, and A. Wespi, editors, *Proceedings of the Fourth International Workshop on the Recent Advances in Intrusion Detection (RAID'2001)*, LNCS, October 2001.

[12] J.-P. Pouzol and M. Ducassé. Formal specification of intrusion signatures and detection rules. In *Proceedings of 15th Computer Security Foundations Workshop (CSFW'02)*, IEEE Computer Society Press, pages 64–76, 2002.

[13] A. Valdes and K. Skinner. Probabilistic alert correlation. In W. Lee, L. Mé, and A. Wespi, editors, *Proceedings of the 4th International Symposium on the Recent Advances in Intrusion Detection (RAID'2001)*, LNCS, October 2001.

INSTALL-TIME VACCINATION OF WINDOWS EXECUTABLES TO DEFEND AGAINST STACK SMASHING ATTACKS

Danny Nebenzahl,[1] and Avishai Wool,[2]

[1] *Dept. of Computer Science, Tel Aviv University, Ramat Aviv 69978, Israel.*
nenenzah@post.tau.ac.il

[2] *School of Electrical Engineering, Tel Aviv University, Ramat Aviv 69978, Israel.*
yash@acm.org

Abstract Stack smashing is still one of the most popular techniques for computer system attack. In this paper we present an anti-stack-smashing defense technique for Microsoft Windows systems. Our approach works at install-time, and does not rely on having access to the source-code: The user decides when and which executables to vaccinate. Our technique consists of instrumenting a given executable with a mechanism to detect stack smashing attacks. We developed a prototype implementing our technique and verified that it successfully defends against actual exploit code. We then extended our prototype to vaccinate DLLs, multi-threaded applications, and DLLs used by multi-threaded applications, which present significant additional complications. We present promising performance results measured on SPEC2000 benchmarks: Vaccinated executables were no more than 8% slower than their un-vaccinated originals.

Keywords: Computer security, Malware, Buffer-Overflow, Microsoft Windows Operating System

1. INTRODUCTION

1.1 Background

Stack smashing attacks, which exploit buffer overflow vulnerabilities to take control over attacked hosts, are the most widely exploited type of vulnerability. About half of the CERT advisories in the past few years have been devoted to vulnerabilities of this type [CER02]. Stack smashing is an old technique, dating back to the late 1980's. E.g, the Internet worm [Spa88] used stack smashing. However, this technique is still in current use by hackers: For instance, it is the underlying method of attack used by the MSBlast virus [CER03a],[CER03b]. In general, stack smashing works against a program that has a buffer overflow bug: A malicious attacker inputs a string that is too long for the buffer, thereby overwriting the program's stack. Since the program keeps return addresses on the stack, the overwriting string can modify a return address — and when the function returns, the attacker's injected code gets control. A detailed description of the stack smashing attack mechanism can be found in [BST00],[CPM+98].

1.2 Classification of Anti-Stack-Smashing Techniques

Various techniques have been developed to defend against stack smashing attacks. One way to classify these techniques is by the method they use to handle the vulnerability [WK03]:

- Techniques ensuring that software vulnerabilities exploitable by stack smashing attacks do not exist: i.e., they attempt to eradicate buffer overflows.
- Techniques that prevent an attacker from gaining control over the attacked host: i.e., they assume that buffer overflows will continue to occur, and attempt to ensure that the attack code will not be executed successfully.

Our tool is of the latter type. It does not detect buffer overflows, but defends against their exploitation.

A second classification of anti-stack-smashing techniques is based on the stage in the software life cycle in which the counter-measures are deployed [WK03]:

- Techniques that are deployed by the software *developer* at the software coding stage. These techniques include static code analysis and modified compilers.
- Techniques that are deployed by the software *user*, before, or while, using the vulnerable software. These techniques include wrappers, emulators and binary code manipulations.

Our tool is a user tool: It does not require access to the source code.

Note that anti-stack-smashing developer tools (static checkers, compilers) have the advantage of working at a high level of abstraction, e.g., with access to the C source code. In contrast, user tools have little or no information about the language or techniques used to create the program—all they have to work with is the binary executable. However, we argue that user tools are extremely valuable: Typically, the user has no control over the bugs in the software. Thus having the ability to *vac-*

cinate software, at the user site, at the user's discretion, is an important goal.

The vast majority of anti-stack-smashing tools are Unix-based. This is because source code is readily available for the operating system, the compilers, and application software. In contrast, our tool targets the proprietary Microsoft Windows operating systems.

The work closest to ours was recently suggested, independently, by Prasad and Chiueh [PC03]. They too add anti-stack-smashing instrumentation into Windows binary files. However, they only handle very simple Win32 executables. Our work shows that advanced features like multi-threading, DLLs (Dynamic Link Libraries), and their combination, significantly complicate the problem. Dealing with such features requires new and different methods. A more detailed comparison of our work and that of [PC03] can be found in Section 8.

1.3 Contributions

The contribution of our work is twofold. Our first contribution is that we created an anti-stack-smashing tool that works at *install time*, or whenever the software user wishes. Thus, our technique does not require access to the source code of the application and assumes nothing about the application, beyond it being written in a high level compiled language. The main idea is to equip existing binary files with additional machine code that can detect a stack smashing attack.

The second contribution is that our tool is a "system-wide" solution. Our tool handles simple applications, multi-threaded applications, and DLLs. Thus the user can instrument a vulnerable binary while keeping its interoperability: A DLL can be instrumented while the applications using it are not instrumented, and an application can be instrumented with or without instrumenting DLLs it uses.

We have built a working prototype implementing our approach, that instruments Win32 executables running on an x86 Intel Pentium platform. We vaccinated several Windows executables with known buffer overflows, and successfully defended them against real exploits. Our approach enjoys minimal overhead: In standard benchmarks we have not observed more than an 8% slowdown in the vaccinated program.

Organization: In Section 2 we describe our solution architecture. Section 3 describes the implementation of our technique to vaccinate simple Windows applications. Section 4 describes the techniques used to vaccinate Windows DLLs. Section 5 describes the techniques used to overcome the challenges imposed by multi-threaded applications. Section 6 describes the techniques used to vaccinate DLL's used by multithreaded applications. In Section 7 we evaluate our solution. In Section 8 we describe alternative approaches to stack-smashing protection, and we conclude in Section 9.

2. SOLUTION ARCHITECTURE

2.1 Design Choices

Our first choice was to use a separate stack (as done by LibVerify [BST00] on Linux), rather than insert so-called "canaries" into the stack, as done by StackGuard [CPM+98]. We believe that a separate stack offers better protection than canaries—e.g., [WK03] shows how an attacker can overcome a simple canary-based mechanism. Furthermore, the separate stack approach can be modified to support any mechanism that requires additional memory to be allocated and used in run-time (such as encrypted per-thread canaries).

We chose to insert all the instrumentation code into the executable file itself, rather than to rely on load time instrumentation code (cf. [HB99]). In a proprietary operating system like Microsoft Windows, modifying the binary is more robust and less OS-version dependent. Nevertheless, load-time instrumentation is a viable option.

Furthermore, we chose not to use the Detours library [HB99], and to implement our own instrumentation mechanism. This gave us the ability to instrument functions that Detours skips (such as short functions, and functions with jumps into entry or exit code).

We implemented our approach using static instrumentation: Our vaccination tool instruments the executable file when it is not used, not its image in memory during run-time. It should be noted that while this choice limits our solution to "normal" (i.e., non self-modifying) programs, it results in better performance than dynamic instrumentation.

Finally, we do not assume the presence of any debug symbols, map files, or any information beyond what is available in the raw Windows binary file — simply because such information is typically only available at compile time and rarely available to users at install time.

2.2 The Basic Method

Our anti-stack-smashing mechanism is based on instrumenting existing software. The instrumentation code is added at the function level—each function is instrumented with additional entry and exit code. The added entry code records the return address from the stack by pushing it onto a private stack. The added exit code tests whether the return address found on the stack just before returning from the function is identical to the one at the top of the private stack, the one that has been recorded upon the function entry.

If our instrumentation detects that the stack has been smashed, i.e., the return address has been overwritten, we halt the program by using a deliberate illegal memory access. Halting the program is not the only possible option. Since we have the original return address in the private stack, we could return to the correct caller of the function. However, we believe this to be an inferior choice because continuing to run after detecting that the stack is corrupt will result in unpredicted behavior. Implementing a messaging mechanism to inform the user about the attack, or to log data about the attack, can also be dangerous, as noted in [Ric02].

2.3 Disassembly

Since the disassembly process is not in the core of our research, we chose to use a commercial disassembly package, the IDA Pro [IDA03]. We chose IDA Pro since it has been recognized as a very accurate disassembler that is capable of distinguishing between code and data [CE01], [LD03]. The input to IDA Pro is the binary to be disassembled and the output is a listing file of the disassembled program.

2.4 Function Discovery

The next step in our process is the discovery of function boundaries. To do this we wrote a parser for the IDA Pro listing file. We identify a function entry when we find an address that is called from some other address. Thus, the listing file is scanned to detect calls, and the called addresses are marked as function entry addresses. Each function is then scanned, building a tree emulating all possible branches in the function, until all RET commands (exit addresses) of the function are detected. Note that a function can have more than one entry point, and more than one exit point. Note also that our function discovery will miss "non-standard" functions — e.g., functions that are not called by the CALL instruction, or that do not return by the RET instruction. We believe that this is not a significant issue since compilers generate standard call sequences.

2.5 Function Analysis and Classification

We instrument a function by overwriting the entry-code of the function with a jump instruction to an area that is not used in the binary. The jump-to area includes: (i) Our instrumentation code; (ii) The original entry code instructions that were overwritten by our jump instruction; and (iii) A jump back to the rest of the original function. A similar solution is used to instrument the function's exit-code.

However, in a CISC architecture, implementing this solution is not so simple. In a CISC architecture, different instructions may have different lengths. Replacing the original entry (or exit) code with a jump may replace several instructions of the original code with the one instruction of the added jump. Furthermore, the original program may include jumps to one of the instructions replaced by the added jump. Therefore, simply replacing the original code with a jump may result in an erroneous program that includes jumps to illegal instructions. In order to avoid this problem, we classify functions into three categories:

- Simple functions. Simple functions have one entry point, one exit point and do not include jumps into the first or last instructions of the function. These functions are handled as mentioned above.

- Complicated functions. These functions may have more than one entry or exit point, and may have jumps into entry code or jumps into exit code. To instrument complicated functions, we copy them in full to an unused area. Their entry code areas are replaced with jumps to the new, instrumented copy of the whole function. If the function has more than one entry point, multiple instances of the function will be created, one for each entry point. Each copy is instrumented to handle a call sequence that enters through *its* entry point.

- Un-handled functions. Functions that include indirect jumps (jumps whose destination address is determined by a value in a register or by data). The difficulty with such jumps is that, a-priori, the jump destination is not known, and thus determining the function's boundaries with certainty is impossible. Compilers use indirect jumps to efficiently implement C `switch` statements as jump tables. Although discovering jump tables in binary files is feasible, and done quite well by IDA-Pro [CE01], in our prototype we decided to not instrument such functions. In a real world product one should of course implement jump table discovery methods, such as the ones mentioned in [CE01]. In all of the programs we have instrumented, indirect jumps were used in less than 4% of the functions.

3. INSTRUMENTING A SIMPLE WIN32 APPLICATION

In our terminology, a *simple Win32 application* is a Windows application that is not multi-threaded. Examples of such applications are

Notepad, RegEdt32, Calc etc. These applications are loaded into fixed virtual memory addresses, and the memory allocated to them is used solely by them. As mentioned above, our instrumentation code manages a private stack. In simple Win32 applications, we allocate the memory used for this stack, statically, at the end of the original binary. Thus, the address of the private stack in known at instrumentation time, and the instrumentation code can directly access the private stack. The initialization code added at the beginning of the program initializes the private stack, and the instrumentation code of each function manages the private stack.

In our prototype implementation the jump instruction added to each function's entry or exit code takes 5 bytes, and the added instrumentation code takes 29 bytes for each entry point and 41 bytes for each exit point of the function. We use a private return address stack of 768 bytes, which allows a function nesting of depth 192.

To demonstrate the effectiveness of our instrumentation, we instrumented the RegEdt32 application. This application has a known buffer overflow [Bug03] for which exploit code is available. We first verified that the exploit indeed successfully attacks the application. Then we instrumented the application and checked that the instrumented application still worked correctly. Finally, we verified that the exploit caused the instrumented application to halt, as described in Section 2. We also instrumented other simple Win32 applications (such as Notepad.exe, WinHlp32.exe, Calc.exe), against which we did not have exploit code. We verified that all these applications worked correctly after vaccination, to demonstrate that our instrumentation does no harm.

4. INSTRUMENTING A DLL

DLLs are PE files that contain function libraries that can be used by multiple applications simultaneously. The use of DLLs is extremely common in the Microsoft Windows family of operating systems: e.g., there are over 1000 DLLs in a typical Win2000 `C:\WINNT\SYSTEM32` directory. Since DLLs may be used in different combinations, the DLL writer cannot predict the memory address that the DLL will be loaded to. When creating a DLL binary, the compiler creates a *relocation* table. The relocation table enables the OS to load the DLL to any available memory region.

Instrumenting a DLL imposes two main problems: Allocating memory for the private stack, and handling DLL relocation. Since a DLL can be used by more than one process, in order to prevent race conditions, memory for our private stack should be allocated per process. Thus, the address of our private stack needs to be determined at run time, as new processes access the DLL. Handling relocation means that either our instrumentation code must not use direct addressing, or we need to update the relocation table so our instrumentation code will relocate correctly.

We suggest a method to handle both problems simultaneously, based on an operating system paging policy named Copy-on-Write. When a process writes to a page that is managed according to the Copy-on-Write policy, the OS creates a copy of the page that is private to that process.

To use this feature we made the instrumented DLL into a self-modifying and self-relocating DLL. Upon loading or being attached to a process, the initialization code that we add at the entry point of the DLL determines where it is in memory, and updates the instrumented code so all references to the private stack will be to the correct addresses (thus handling relocation). This action, of rewriting part of the DLL in memory, causes the operating system to duplicate the rewritten pages. The rewritten pages contain the private stack and our instrumentation code. Thus, by making the instrumentation self-relocating, and hence self-modifying, the operating system gives us for free a separate memory block to store the private stack for each process.

However, our implementation did not completely escape the need to update the relocation table. Since we handle complicated functions by duplicating them, we must update the relocation table so the instructions in the new copied functions (e.g., an access to a global variable) will relocate correctly.

We implemented and tested a prototype utilizing this method. We first wrote a vulnerable DLL and our own exploit code, and checked that the exploit smashes the stack. Then we instrumented the DLL, and checked that it works correctly in a multi-process environment by deliberately causing race conditions between multiple processes that use the DLL. We tested our defense mechanism by causing a buffer overflow in the DLL. Our instrumentation code did indeed catch the stack-smashing and halt. We have not yet tested our defense against a real exploit of a real DLL.

5. INSTRUMENTING A MULTI-THREADED APPLICATION

In a multi-threaded environment, the process's memory is shared between threads. In contrast to multiprocess operation, in a multi-threaded application all the threads are allowed to change memory regions owned by the process, and it is the application's responsibility

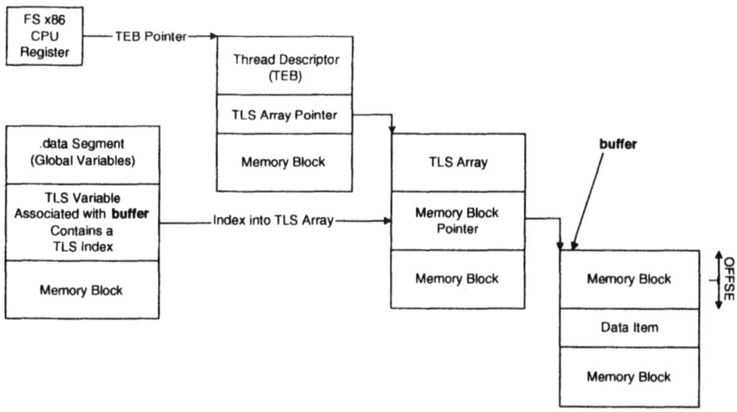

Figure 1. Accessing the TLS variable *void *buffer* at offset *OFFSET*

to ensure its correct behavior — with little or no help from the OS. Thus, the Copy-on-Write trick we relied on for solving the per-process memory allocation problem (recall Section 4) is not applicable for the multi-threaded scenario: Memory that is shared by multiple threads of the same application is not considered to be "shared" as far as the operating system is concerned.

Instead, we use another feature of the Windows operating system. Win32 has a memory allocation mechanism that is capable of allocating memory per thread, called Thread Local Storage (TLS). This mechanism facilitates defining memory structures (variables) that are allocated per thread. Allocation of such memory can be defined in the PE file by adding a special section (the .tls section) that describes the per-thread allocation and initialization functions. Accessing a TLS variable is a much more complicated task and thus has a performance penalty greater than accessing a standard variable, since at compile time the address of the variable is not known. Instead, when a programmer defines a TLS variable (using a special C language extension), another hidden variable is created. This added variable is set by the operating system to a value called the *tls index.* When accessing a TLS variable, the hidden variable is accessed in order to retrieve the tls index. The thread descriptor (A variable maintained by the operating system storing information about each thread) is accessed in order to retrieve the address of a table of pointers. This table is accessed using the tls index to retrieve a pointer to the actual TLS variable. This run time access sequence is shown in Figure 1.

We solve the need for per-thread private stack memory allocation by defining the private stack as a TLS variable. The allocation is done by adding a .tls section to the executable file. The .tls section describes the private stack and it's initialization function. This change causes our entry code to grow to 39 bytes (instead of 29) and our exit code to grow to 47 bytes (instead of 41).

We implemented and tested a prototype utilizing this instrumentation method. We wrote a vulnerable multi-threaded program and verified that overflowing a buffer causes stack smashing. Then we instrumented the program and verified that it still works correctly—including under race conditions between threads. Then we attacked the program by causing a buffer overflow, and confirmed that the instrumented code detected the stack smashing and halted. We also tested the correct operation of various instrumented applications such as Windows Explorer, Microsoft Internet Explorer and more.

6. INSTRUMENTING DLLS USED BY MULTITHREADED APPLICATIONS

Another implication of multi-threaded applications is the handling of DLLs used by multi-threaded applications. The technique used to vaccinate multi-threaded applications cannot be used to vaccinate DLLs used by multi-threaded applications. The reason is that TLS variables added to a PE file cannot be used in DLLs, because a DLL may be dynamically linked to a process that is running. Thus, at the moment of attachment, TLS variables for all existing threads must be allocated and initialized simultaneously. This is not possible for any initialization function, nor is it supported by Win32. Since neither the software end user, nor the software developer knows whether a DLL will be used by a multi-threaded application, a general purpose instrumentation method must address DLLs that are used by multi-threaded applications. In other words, the simpler mechanism we used in section 4 is not sufficient.

We solve the need for private stack memory allocation per-stack by using a Win32 mechanism for allocating dynamically TLS variables. This mechanism provides the programmer with the following API:

- TlsAlloc - a function for allocating a TLS index. After allocating this index, each thread will be allocated an uninitialized pointer.

- TlsSetValue - a function that allows a thread to set the value of the pointer allocated for a TLS variable. The programmer allocates a memory block for the TLS variable, and sets the pointer to it using this function.

- TlsGetValue - a function that allows the programmer to retrieve the pointer to the TLS data.

Benchmark	Original Run Time (sec)	Instrumented Run Time (sec)	Performance Penalty (%)
bzip2	67.6	70.5	4.33
gzip	15.6	16.3	4.36
mcf	435.62	446.5	7.09

Table 1. Run time performance measurements of standard and instrumented SPEC2000 applications.

Benchmark	Orig.Size (KB)	Instrumented Size(KB)	Increase (%)	IDA-Pro Listing(KB)	Instrumentation Time(sec)
bzip2	169	208	20.6	11,784	5.6
gzip	113	154	36.9	39,527	12.8
mcf	77	99	28.6	3795	1.2

Table 2. PE file size performance measurements of SPEC2000 and instrumented SPEC2000 applications.

The use of this API is as follows: At the initialization of a DLL one allocates a TLS index, and saves that index in a variable local to the DLL. When a process or a thread attaches the DLL, the DLL allocates the memory needed for the actual TLS variable, and uses the TlsSetValue to store a pointer to the memory allocated. When the DLL is to access the TLS variable, it first calls the TlsGetValue API function to retrieve a pointer to the thread's copy of the variable, and accesses that variable.

We use this TLS allocation and management scheme in order to allocate and access the private stack. In order to use this scheme we needed to instrument the DLL in a way that the memory allocation functions would be called when the DLL is loaded and when processes attach the DLL. We found a very simple way to implement this functionality. We wrote a DLL that implements only the TLS and memory allocation functions, and the private stack management functions. We add the stack management functions of this DLL to the Import Directory of the instrumented DLL — a Data Directory describing the DLLs used by an application or a DLL. This causes the operating system to call our DLL's initialization and memory allocation functions when the instrumented DLL is loaded or when a process or a thread attaches to it.

We enhanced our prototype to use our private stack allocation and management DLL. We developed a vulnerable DLL and a multithreaded application that uses that DLL. We tested that threads may exploit the DLL's vulnerability. Then we instrumented the DLL, and tested the correct operation of the multi-threaded application under controlled race conditions. Finally we caused the multi-threaded application to exploit the vulnerability of the DLL. Our stack-smashing detection code detected the exploit.

7. EVALUATION

7.1 Performance

Instrumenting an application, especially by adding code to all program functions, results in some performance degradation. This performance degradation is caused by several factors: (i) The added code that needs to be run. (ii) The larger address space that may change the paging performance for the program. (iii) The change of memory locations in the program that results in a change of the cache performance.

Since the performance penalty is a result of multiple program and system parameters, we decided to evaluate the performance of whole instrumented programs rather than measure micro-benchmarks. To do this, we instrumented several SPECINT applications from the SPEC2000 suite [SPE00] using the instrumentation method for handling multi-threaded applications (even though all the applications were simple Win32 applications). This evaluation results in a worst case measurement:

- The performance penalty of the multi-threaded instrumentation code is the highest due to the TLS variable accessing sequence.
- The SPEC2000 applications are in general number-crunching applications, designed to benchmark compilers and CPUs. These application are much more demanding than the usual applications used by end-users of the Win32 environment.

We ran the original and instrumented programs on a 2GHz Pentium 4 with 256MB RAM and measured their average run time. We verified that the instrumented code produced the same output as the standard un-instrumented code. The measured performance penalty was less than 8% for the SPEC2000 applications we measured (see Tables 1 and 2 for details), and the increase in program size was 20-37%. We consider these results very positive: an 8% slowdown most likely will not be noticeable in an interactive program. Table 2 also shows the size of the intermediate IDA-Pro listing file, which can grow quite large, and the instrumentation time, which is roughly proportional to the size of listing file. Note that our focus was on demonstrating a working prototype, so we did not make any effort to reduce the instrumentation time. We believe that the instrumenting tool can be greatly optimized through the use of better data structures.

7.2 Limitations of the approach

We can identify two limitations caused by the fact that we instrument a binary executable file, rather than its loaded image.

1 Known Un-handled Functions: As noted in Section 2.5, we do not instrument functions that use indirect jump instructions. Thus, our tool does not defend against attacks that exploit vulnerabilities in these functions.

2 Unknown Un-handled Functions: The function discovery process may miss functions, for example functions called by indirect calls such as function call tables. Vulnerabilities in such functions are not defended by our defense mechanism.

The first limitation can be addressed by introducing techniques of jump table discovery, which will reduce the number of un-handled functions. The second limitation can be handled at run time, for example by checking that the destination of an indirect call is indeed instrumented. These improvements are left for future work.

Finally, we need to test our tool on more Win32 applications and DLLs, and validate its success in defending against more real exploits.

8. ALTERNATIVE APPROACHES AND TOOLS

Developer tools. Probably the most influential anti-stack-smashing tool is StackGuard. StackGuard [CPM+98] is a compiler enhancement, that equips the generated binary code with facilities that can detect a stack smashing attack. StackGuard works by having each function's entry code place a per-run constant, so called a *canary*, on the stack. The function's exit code verifies the canary's existence. The assumption is that a buffer overflow which overwrites the return address would also overwrite the canary. StackGuard has been commercialized by Immunix [Imm03] and has been used to produce a full hardened Unix system. A similar compiler option is now supplied as a standard feature in Microsoft Visual C++ .NET compilers [Mic01],[HL02].

StackShield [Sta00] is another GNU compiler enhancement. In StackShield, the attack detection is based on tracking changes of the actual return address on the stack. Each function's return address is recorded in a private stack upon function entry, and the function's exit code verifies that the return address has not changed.

Another tool that inserts instrumentation at compile time is PointGuard [CBJW03] (which encrypts pointers to code). Static source code analysis techniques have also been developed to detect software vulnerabilities that may be exploited by stack smashing attacks [GO98], [DRS01], [EL02], [WFBA00].

User tools. A user tool, implemented as a Linux patch, prevents stack smashing exploits by setting the stack memory pages to be non-executable [Sol].

Libsafe [BST00] is a run time attack detection mechanism that can discover stack smashing attacks against standard library functions. It is implemented as a dynamically loaded library that intercepts calls to known vulnerable library functions, and redirects them to a safe implementation of these functions.

Libverify [BST00] is a Unix-based attack detection technique similar to StackShield, but in which the attack detection code is embodied into a copy of the executable image, which is created on the heap at load time. However, the authors only handled the simplest case (single threaded programs, no DLLs).

Recently, a technique based on randomized instruction set emulation, employed at run time, has shown success in detecting various code injection attacks, including stack smashing attacks [KKP03],[BAF+03]. This technique has a high performance penalty because of the emulation overhead.

Comparison with the work of Prasad and Chiueh. The work closest to ours was recently suggested, independently, by [PC03]. They too add anti-stack-smashing instrumentation into Windows binary files. However, their work leaves several key issues unresolved. Most notably, they only handle simple Win32 executables (single thread programs, no DLLs). Furthermore, they demonstrated that their approach successfully vaccinates a test program with a deliberate buffer overflow, rather than a live exploit. Finally, they use the Detours library [HB99], which slightly limits the class of functions they are able to instrument, and requires load-time activation of the instrumentation. Beyond the work of [PC03], our work has the following features:

- We are able to instrument DLLs.
- We handle multi-threaded applications.
- We handle DLLs called by multi-threaded applications.

- We can instrument a wider set of functions, including functions with jumps into the middle of entry/exit code, and very short functions. This is because we wrote our own instrumentation code rather than using Detours.
- We demonstrated that our approach protects against a real buffer overflow vulnerability found in the wild.

9. CONCLUSIONS

We presented an install-time vaccination technique as a countermeasure against stack-smashing-attacks on the proprietary Microsoft Windows OS. Our technique enables software users to be protected without access to the source-code. We developed a prototype that successfully instruments simple Win32 applications, DLLs, multi-threaded applications, and DLLs used by multi-threaded applications thus providing a system-wide solution. We have shown that the prototype can defend against real exploit code.

References

[BAF+03] E. G. Barrantes, D. H. Ackley, S. Forrest, T. S. Palmer, D. Stefanovic, and D. D. Zovi. Randomized instruction set emulation to disrupt binary code injection attacks. In *Proc. 10th ACM Conf. Computer and Communications Security (CCS)*, Washington, DC, 2003.

[BST00] Arash Baratloo, Navjot Singh, and Timothy Tsai. Transparent run-time defense against stack smashing attacks. In *Proc. USENIX Annual Technical Conference*, 2000.

[Bug03] Microsoft Windows RegEdit.exe registry key value buffer overflow vulnerability. Bugtraq id 7411, 16 April 2003. `http://www.securityfocus.com/bid/7411`.

[CBJW03] Crispin Cowan, Steve Beattie, John Johansen, and Perry Wagle. PointGuard: Protecting pointers from buffer overflow vulnerabilities. In *Proc. 12th USENIX Security Symposium*. USENIX, 2003.

[CE01] Cristina Cifuentes and Mike Van Emmerik. Recovery of jump table case statements from binary code. *Science of Computer Programming*, 40(2–3):171–188, 2001.

[CER02] CERT/cc statistics 1988-2001, 2002. http://www.cert.org/stats/.

[CER03a] CERT advisory CA-2003-16: Buffer overflow in Microsoft RPC, 17 July 2003. `http://www.cert.org/advisories/CA-2003-16.html`.

[CER03b] CERT advisory CA-2003-20: W32/Blaster worm, 11 August 2003. `http://www.cert.org/advisories/CA-2003-20.html`.

[CPM+98] Crispan Cowan, Calton Pu, Dave Maier, Jonathan Walpole, Peat Bakke, Steve Beattie, Aaron Grier, Perry Wagle, Qian Zhang, and Heather Hinton. StackGuard: Automatic adaptive detection and prevention of buffer-overflow attacks. In *Proc. 7th USENIX Security Symposium*, pages 63–78, San Antonio, Texas, January 1998.

[DRS01] Nurit Dor, Michael Rodeh, and Mooly Sagiv. Cleanness checking of string manipulations in C programs via integer analysis. In *Proc. 8th International Static Analysis Symposium (SAS), LNCS 2126*, Paris, France, 2001. Springer-Verlag.

[EL02] David Evans and David Larochelle. Improving security using extensible lightweight static analysis. *IEEE Software*, 19(1):42–51, 2002.

[GO98] A. K. Ghosh and T. O'Connor. Analyzing programs for vulnerability to buffer overrun attacks. In *Proc. 21st NIST-NCSC National Information Systems Security Conference*, pages 274–382, 1998.

[HB99] Galen Hunt and Doug Brubacher. Detours: Binary interception of Win32 functions. In *Proc. 3rd USENIX NT Symposium*, pages 135–144, 1999.

[HL02] Michael Howard and David LeBlanc. *Writing Secure Code.* Microsoft Press, 2nd edition, 2002.

[IDA03] The IDA Pro disassembler and debugger, v4.51, 2003. `http://www.datarescue.com/idabase/`.

[Imm03] Immunix secured solutions, 2003. `http://www.immunix.com`.

[KKP03] Gaurav S. Kc, Angelos D. Keromytis, and Vassilis Prevelakis. Countering code-injection attacks with instruction-set randomization. In *Proc. 10th ACM Conf. Computer and Communications Security (CCS)*, Washington, DC, 2003.

[LD03] Cullen Linn and Saumya Debray. Obfuscation of executable code to improve resistance to static disassembly. In *Proc. 10th ACM Conf. Computer and Communications Security (CCS)*, Washington, DC, 2003.

[Mic01] Microsoft Visual C++ compiler options: /gs (control stack checking calls). Online documentation, 2001. `http://msdn.microsoft.com/library/default.asp?url=/library/en-us/vccore/html/_core_.2f.gs.asp`.

[PC03] Manish Prasad and Tzi-cker Chiueh. A binary rewriting defense against stack based overflow attacks. In *Proceedings of the USENIX 2003 Annual Technical Conference*, 2003.

[Ric02] Gerardo Richarte. Four different tricks to bypass StackShield and StackGuard protection. Core Security Technologies, 2002. `http://downloads.securityfocus.com/library/StackGuard.pdf`.

[Sol] Solar Designer. Nonexecutable user stack. `http://www.false.com/security/linux-stack/`.

[Spa88] Eugene H. Spafford. The Internet worm program: An analysis. Technical Report CSD-TR-823, Purdue University, West Lafayette, IN 47907-2004, 1988.

[SPE00] SPEC CPU2000 V1.2. Standard Performance Evaluation Corporation, 2000. `http://www.specbench.org/osg/cpu2000/`.

[Sta00] Stackshield, 2000. `http://www.angelfire.com/sk/stackshield`.

[WFBA00] David Wagner, Jeffrey S. Foster, Eric A. Brewer, and Alexander Aiken. A first step towards automated detection of buffer overrun vulnerabilities. In *Proc. Network and Distributed System Security Symposium (NDSS)*, pages 3–17, San Diego, CA, February 2000.

[WK03] John Wilander and Mariam Kamkar. A comparison of publicly available tools for dynamic buffer overflow prevention. In *Proceedings of the 10th Network and Distributed System Security Symposium (NDSS)*, pages 149–162, San Diego, California, February 2003.

EIGENCONNECTIONS TO INTRUSION DETECTION

Yacine Bouzida and Sylvain Gombault
Departement RSM GET/ ENST Bretagne
2, rue de la Châtaigneraie, CS 17607
35576 Cesson Sevigne CEDEX, FRANCE
{Yacine.Bouzida,Sylvain.Gombault}@enst-bretagne.fr

Abstract Most current intrusion detection systems are signature based ones or machine learning based methods. Despite the number of machine learning algorithms applied to KDD 99 cup, none of them have introduced a pre-model to reduce the huge information quantity present in the different KDD 99 datasets. We introduce a method that applies to the different datasets before performing any of the different machine learning algorithms applied to KDD 99 intrusion detection cup. This method enables us to significantly reduce the information quantity in the different datasets without loss of information. Our method is based on Principal Component Analysis (PCA). It works by projecting data elements onto a feature space, which is actually a vector space R^d, that spans the significant variations among known data elements. We present two well known algorithms we deal with, decision trees and nearest neighbor, and we show the contribution of our approach to alleviate the decision process. We rely on some experiments we perform over network records from the KDD 99 dataset, first by a direct application of these two algorithms on the rough data, second after projection of the different datasets on the new feature space.

Keywords: Intrusion Detection, Principal Component Analysis, KDD 99, Decision Trees, Nearest Neighbor

1. Introduction

A modern computer network should acquire many mechanisms to ensure the security policy of data and equipment inside the network. Intrusion detection systems (IDSs) are an integral package in any well configured and managed computer system or network. IDSs may be some software or hardware systems that monitor the different events occuring in the actual network and analyze them for signs of security threats.

There are two major approaches in intrusion detection: anomaly detection and misuse detection. Misuse detection consists of first recording and representing the specific patterns of intrusions that exploit known system vulnerabilities or violate system security policies, then monitoring current applications or network traffic activities for such patterns, and reporting the matches. There are several developed models in misuse intrusion detection [Ilgun, 1993; Kumar and Spafford, 1994]. They differ in representation as well as the matching algorithms employed to detect such threat patterns. Anomaly detection, on the other hand, consists of building models from normal data and then detect variations from the normal model in the observed data. Anomaly detection was originally introduced by Anderson [Anderson, 1980] and Denning [Denning, 1987]. The main advantage with anomaly intrusion algorithms is that they can detect new forms of attacks, because these new intrusions will probably deviate from the normal behavior [Denning, 1987].

There are many IDSs developed during the past three decades. However, most of the commercial and freeware IDS tools are signature based [Roesch, 1999]. Such tools can only detect known attacks previously described by their corresponding signatures. The signature database should be maintained and updated periodically and manually for new attacks. For this reason, many data mining and machine learning algorithms are developed to discover new attacks that are not described in the training labeled data.

Literature survey on intrusion detection indicates that most researchers applied an algorithm directly [Pfahringer, 2000; Agrawal and Joshi, 2000; Levin, 2000] on the rough data obtained from network traffic or other local or remote applications. The majority of the machine learning algorithms applied to anomaly intrusion detection suffers from the high consuming time [Pfahringer, 2000] when applied directly on rough data. The KDD 99 cup intrusion detection datasets [KDD Cup, 1999] are an example where many machine learning algorithms, mostly inductive learning based, were applied directly on the data which is a binary TCPdump data processed into connection records. Each connection record corresponds to a normal connection or to a specified attack as described in section 2.

Much of the previous work on anomaly intrusion detection in general and on the KDD 99 cup datasets in particular ignored the issue of just what measures of the user, application and/or network traffic behavior stimulus are important for intrusion detection. This suggested to us that an information theory approach coding and decoding user/application or connection record behaviors may give new information content of user/attack behaviors, emphasizing the significant local or global *"features"*. These features may or may not be directly related to the actual used metrics or attributes such as CPU consumed time, number of web pages visited during a session in the case of user behaviors and such as the used protocol, service in the case of network connection records. In the remaining of this paper, we will be just interested in network connection records (for more details on profiles' behaviors, see [Bouzida and Gombault, 2003]).

In the language of information theory, we want to extract the relevant information in a network connection record, encode it efficiently, and compare one network connection record encoding with a database of network connection records encoded similarly. A simple approach to extract the information contained in a network connection record is to capture the variation in a collection of connection records, independently of any judgement of feature, and use this information to encode and compare network connection records.

In mathematical terms, we wish to find the principal components of the distribution of the connection records, or the eigenvectors of the covariance matrix of the set of the connection records [Jolliffe, 2002]. These eigenvectors can be thought of as a set of features which together characterize the variation between records connections. Each connection record location contributes more or less to each eigenvector which we call *"eigenconnection"*. Each connection record can be presented exactly in terms of linear combination of the eigenconnections. Each connection can also be approximated using only the best -*"eigenconnections"*- those that have the largest eigenvalues, and which therefore account for the most variance within the set of connection records. The best N eigenconnections span an N dimensional subconnection -*"connection space"*- of all possible connection records.

This new space is generated by an information theory method called Principal Component Analysis (PCA) [Jolliffe, 2002]. This method has proven to be an exceedingly popular technique for dimensionality reduction and is discussed at length in most texts on multivariate analysis. Its many application areas include data compression [Kirby and Sirovich, 1990], image analysis, visualization, pattern recognition [Turk and Pentland, 1991] and time series prediction.

The most common definition of PCA, due to Hotelling (1933) [Hotelling, 1933], is that, for a set of observed vectors $\{v_i\}$, $i \in \{1, \ldots, N\}$, the q principal axes $\{w_j\}$, $j \in \{1, \ldots, q\}$ are those orthonormal axes onto which the retained variance under projection is maximal. It can be shown that the vectors w_j are given by the q dominent eigenvectors (i.e. those with largest associated eigenvalues) of the covariance matrix $C = \sum_i \frac{(v_i - \overline{v})(v_i - \overline{v})^T}{N}$ such that $Cw_j = \lambda_i w_j$, where $\overline{v}$ is the simple mean. The vector $u_i = W^T(v_i - \overline{v})$, where $W = (w_1, w_2, \ldots, w_q)$, is thus a *q-dimensional* reduced representation of the observed vector v_i.

We investigate, in this paper, an eigenconnection approach based on principal component analysis for anomaly intrusion detection applied to the different KDD 99 intrusion detection cup datasets.

This paper is organized as follows: Section 2 describes the different KDD 99 intrusion detection cup datasets. Sections 3 and 5 introduce the application of two algorithms; the nearest neighbor and decision trees, where in section 3 these algorithms are briefly presented and in section 5, we present and discuss the different results obtained by using these two algorithms on rough data or after reduction of the feature space using PCA. Section 4 provides the eigen-connection approach for dimensionnality reduction of data. Finally, section 6 concludes the paper.

2. Description of KDD 99 intrusion detection datasets

The main task for the KDD 99 classifier learning contest [KDD Task, 1999] was to provide a predective model able to distinguish between legitimate (normal) and illegitimate (called intrusion or attacks) connections in a computer network. The training dataset contained about 5,000,000 connection records, and the training 10% dataset consisted of 494,021 records among which there were 97,278 normal connections (i.e. 19.69%). Each connection record consists of 41 different attributes that describe the different features of the corresponding connection, and the value of the connection is labeled either as an attack with one specific attack type, or as normal. The 39 different attack types present in the 10% datasets are given in table 1.

Each attack type falls exactly into one of the following four categories:

1 Probing: surveillance and other probing, e.g., port scanning;

2 DOS: denial-of-service, e.g. syn flooding;

3 U2R: unauthorized access to local superuser (root) privileges, e.g., various *"buffer overflow"* attacks;

4 R2L: unauthorized access from a remote machine, e.g. password guessing.

The task was to predict the value of each connection (normal or one of the above attack categories) for each of the connection record of the test dataset containing 311,029 connections.

It is important to note that

- the test data is not from the same probability distribution as the training data and
- the test data includes some specific attack types not in the training data. There are 22 different attacks types out of 39 present in the training dataset. The remaining attacks are present in the test dataset with different rates towards their corresponding categories. There are 4 new U2R attack types in the test dataset that are not present in the training dataset. These new attacks correspond to 92.90% (189/228) of the U2R class in

the test dataset. On the other hand, there are 7 new R2L attack types corresponding to 63% (10196/16189) of the R2L class in the test dataset. In addition, there are only 104 (out of 1126) connection records present in the training dataset corresponding to the known R2L attacks present simultaneously in the two datasets. However, there are 4 new DOS attack types in the test dataset corresponding to 2.85%(6555/229853) of the DOS class in the test dataset and 2 new Probing attacks corresponding to 42.94% (1789/4166) of the Probing class in the test dataset.

Table 1. The different attack types.

Probing	DOS	U2R	R2L
ipsweep, mscan, nmap, portsweep, saint, satan.	apache2, back, land, mailbomb, neptune, pod, processtable, smurf, teardrop, udpstorm.	buffer_overflow, httptunnel, loadmodule, perl, ps, rootkit, sqlattack, xterm.	ftp_write, guess_passwd, imap, multihop, named, phf, sendmail, snmpgetattack, snmpguess, spy, warezclient, warezmaster, worm, xlock, xsnoop.

We ran our experiments using two different machine learning algorithms; the nearest neighbor and decision trees, on the 10 % KDD 99 intrusion detection cup [KDD Cup, 1999] generated by the MIT Lincoln Laboratory. Lincoln Labs set up an environment to acquire nine weeks of raw TCPdump data for a local-area network (LAN) simulating a typical U.S. Air Force LAN. They operated the LAN as if it were a true Air Force environment, but peppered it with the 39 different attacks types. The TCPdump data collected from the network traffic was transformed into connection records using some data mining techniques [Lee et al., 1999].

3. Nearest neighbor and decision trees

3.1 Nearest Neighbor NN

One of the easiest method in machine learning field is the nearest neighbor method or *NN*. It consists of classifying new observations into their appropriate categories by a simple comparison with the known well classified observations. Recall that the only knowledge we have is a set of $x_{i,\ i=1,..,M}$ points correctly classified into categories. It is reasonable to assume that observations which are close together -for some appropriate metric- will have the same classification. Thus, when classifying an unknown sample x, it seems appropriate to weight the evidence of the nearby's heavily. One simple non-parametric decision procedure of this form is the nearest neighbor rule or *NN-rule*. This rule classifies x in the category of its nearest neighbor. More precisely, we call x'

a nearest neighbor to x if $min\ d(x, x_i) = x', where\ i = 1,.., M$ and d is the distance between the two considered points such as the Euclidean distance.

After its first introduction by Fix and Hodges [Fix and Hodges, 1951], the *NN* classifier has been used and improved by many researchers [Bay, 1998; Dasarathy, 1991] and employed on many data sets from UCI repository [Hettich and Bay, 1999]. A common extension is to choose the most common class in the *kNN*. The *kNN* is performed on KDD 99 intrusion detection datasets by Eskin et. al [Eskin et al., 2003]. It was applied for another purpose where the dataset is filtered and the percentage of attacks is reduced to 1.5% in order to perform unsupervised anomaly detection. In the following, we are interested in applying the NN classifier on the different datasets with its simplest form. That is compute all possible distance pairs between all the training data set and the test dataset records.

Since our datasets consist of continuous and discrete attributes values, we have converted the discrete attibutes values to continuous values following the following idea. Consider we have Σ_i possible values for a discrete attribute i. For each discrete attribute correpond $\mid \Sigma_i \mid$ coordinates. There is one coordinate for every possible value of the attribute. Then, the coordinate corresponding to the attribute value has a value of 1 and all other remaining coordinates corresponding to the considered attribute have a value of 0. As an example, if we consider the protocol type attribute which can take one of the following discrete attributes tcp, udp or icmp. Then, there will be three coordinates for this attribute. If the connection record has a tcp (resp. udp or icmp) as a protocol type then the corresponding coordinates will be $(\ 1\ 0\ 0\ )$ (resp. $(\ 0\ 1\ 0\ )$ or $(\ 0\ 0\ 1\ )$). With this transformation, each connection record in the different KDD 99 datasets will be represented by 125 (3 different values for the *protocol_type*, 11 different values for the *flag* attribute, 67 possible values for the *service attribute* and 0 or 1 for the other remaining 6 discrete attributes) coordinates instead of 41 according to the above discrete attributes values transformation.

3.2 Decision trees

Decision tree induction has been studied in details in both areas of pattern recognition and machine learning. In the vast area concerning decision trees, also known as classification trees or hierarchical classifiers, at least two seminal works are to be mentioned, those by Quinlan [Quinlan, 1986] and those by Breiman et al. [Breiman et al., 1984]. The former synthesizes the experience gained by people working in the area of machine learning and describes a computer program called ID3, which has evolved in a new system, named C4.5 [Quinlan, 1993]. The latter originated in the field of statistical pattern recognition and describes a system, named CART (Classification And Regression Trees), which has mainly been applied to medical diagnosis. A decision tree is a tree that has three main components: nodes, arcs, and leaves. Each node is labeled with a feature attribute which is most informative among the attributes not yet considered in the path from the root, each arc out of a node is labeled with a feature value for the node's feature and each leaf is labeled with a category or class.

Most of the decision trees algorithms use a top down strategy; i.e from the root to the leaves. Two main processes are necessary to use the decision trees:

- Building process: it consists of building the tree by using the labeled training dataset. An attribute is selected for each node based on how it is more informative than others. Leaves are also assigned to their corresponding class during this process.
- Classification process: A decision tree is important not because it summarizes what we know, i.e. the training set, but because we hope it will classify correctly new cases. Thus, when building classification models, one should have both training data to build the model and test data to verify how well it actually works. New instances are classified by traversing the tree from up to down based on their attribute values and the node values until one leaf is reached that corresponds to the class of the new instance.

We use the C4.5 algorithm [Quinlan, 1993] to construct the decision trees where Shanon Entropy is used to measure how informative is a node. The selection of the best attribute node is based on the gain ratio $GainRatio(S, A)$ where S is a set of records and A a non categorical attribute. This gain defines the expected reduction in entropy due to sorting on A. It is calculated as the following [Mitchell, 1997]:

$$Gain(S,A) = Entropy(S) - \sum_{v \in Values(A)} \frac{|S_v|}{|S|} Entropy(S_v) \qquad (1)$$

In general, if we are given a probability distribution $P = (p_1, p_2, .., p_n)$ then the information conveyed by this distribution, which is called the Entropy of P is :

$$Entropy(P) = -\sum_{i=1}^{n} p_i log_2 p_i \qquad (2)$$

If we consider only (1) then an attribute with many values will be automatically selected. One solution is to use $GainRatio$ instead [Quinlan, 1986]

$$GainRatio(S,A) = \frac{Gain(S,A)}{SplitInformation(S,A)} \qquad (3)$$

where

$$SplitInformation(S, A) = -\sum_{i=1}^{c} \frac{| S_i |}{| S |} log_2 \frac{| S_i |}{| S |} \qquad (4)$$

where S_i is a subset of S for which A has a value v_i.

4. Eigenconnection approach

Principal component analysis (PCA) is a mathematical procedure that transforms a number of (possibly) correlated variables into a (smaller) number of uncorrelated variables called principal components. The objective of principal component analysis is to reduce the dimensionality (number of variables) of the dataset but retain most of the original variability in the data. The first principal component accounts for as much of the variability in the data as possible, and each succeeding component accounts for as much of the remaining variability as possible. In this section, we investigate the eigenconnection approach based on the principal component analysis. In our case, each connection record corresponds to one vector of n variables corresponding to the different attributes in the different datasets. The procedure is the following:

The set of n different measures are collected in a vector called connection record vector representing the corresponding connection. So if Γ is a connection vector then we can write

$$\Gamma = \begin{pmatrix} m_1 \\ m_2 \\ \vdots \\ m_n \end{pmatrix} \qquad (5)$$

where m_i, $i = 1, \ldots, n$ correspond to the different measures. In most cases, the connection vectors are very similar and they can be described by some *basic connection vectors*.

This approach involves the following initialization procedure:

1 acquire an initial set of connection records (this set is called the training set). In this paper, we use the kdd 99 10% training dataset containing $M = 494,021$ connection records;

2 calculate the eigenconnections from the training set, keeping only n' ($n' \ll n$) eigenconnections that correspond to the highest eigenvalues. These n' connections define the connection space.

3 calculate the corresponding distribution in $n' - dimensional-$ weight space for each known connection record, by projecting their connection vectors onto the *connection space*;

4 (optional) perform a machine learning algorithm (building process) on the new datasets in the new connection space; for the decision tree algorithm, it is necessary to build the tree which will be used in the detection process. However, there is no need to perform the *NN* algorithm at this stage. This is the reason why this step is optional depending on the machine learning algorithm being used.

Having projected the training data onto the new feature space, the following steps are then used to classify and detect intrusions from the new connection records in the test data:

1 calculate a set of weights based on the input connection record and the n' eigenconnections by projecting the input connection record vector onto each eigenconnection,

2 use one of the different machine learning algorithms (classification process) to detect intrusions from the new connection records represented in the new feature space.

4.1 Calculating the eigenconnections

Let the training set of connection vectors be $\Gamma_1, \Gamma_2, \ldots, \Gamma_M$. The average connection Ψ of this set is defined by:

$$\Psi = \frac{1}{M} \sum_{i=1}^{M} \Gamma_i \tag{6}$$

Each connection record vector Γ_i differs from the average Ψ by:

$$\Phi_i = \Gamma_i - \Psi \tag{7}$$

The eigenconnections are the eigenvectors of the covariance matrix C where

$$C_{(n \times n)} = \frac{1}{M} \sum_{i=1}^{M} \Phi_i \Phi_i^T = AA^T \tag{8}$$

$$A_{(n \times M)} = \frac{1}{\sqrt{M}} [\Phi_1 \Phi_2 \ldots \Phi_M] \tag{9}$$

Let U_k be the kth eigenvector of C, λ_k the associated eigenvalue and $\mathbf{U}_{(\mathbf{n} \times \mathbf{n}')} = [\mathrm{U}_1 \mathrm{U}_2 \ldots \mathrm{U}_{\mathrm{n}'}]$ the matrix of these eigenvectors (eigenconnections). Then

$$CU_k = \lambda_k U_k \tag{10}$$

such that

$$U_k^T U_l = \begin{cases} 1 & \text{if } k = l \\ 0 & \text{if } k \neq l \end{cases} \quad (11)$$

The feature vector corresponding to the connection record Γ_i is :

$$\Omega_i = U^T \times \Phi_i = \begin{pmatrix} \omega_1 \\ \omega_2 \\ \vdots \\ \omega_{n'} \end{pmatrix} \quad (12)$$

If the length of the connection record vector is n (number of considered attributes), the matrix C is $n \times n$. The principal component analysis of the contextual covariance matrix C is obtained by calculating its eigenvalues and eigenvectors, and ordering the eigenvalues (and the corresponding eigenvectors) in decreasing order. The sub-space generated by the eigenvectors corresponding to the highest eigenvalues has the highest inertia. By construction, all the directions of the eigenvectors are orthogonal. The principal components associated with the smallest eigenvalues often correspond to not interesting information [Jolliffe, 2002]. Therefore, they are usually removed. Other strategies can be adopted to select the components [Jolliffe, 2002].

The quantity given by

$$\sum_{i=1}^{n'} \lambda_i \quad (13)$$

is called inertia explained by the subspace generated by the first n' ($n' \ll n$) eigenvectors of C.

In practice, the number of the principal factorial components chosen depends on the precision we wish to reach. In general, we can limit to 2, 3 or 4 considered principal factors (axes). The inertia ratio explained by these axes is

$$\tau = \frac{\sum_{i=1}^{n'} \lambda_i}{\sum_{i=1}^{n} \lambda_i} \quad (14)$$

This ratio defines the information rate kept, from the whole rough input data, by the corresponding n' eigenvalues.

5. Experimental methodology and results

In this section, we will present the different results and experiments obtained when directly applying the two methods discussed in section 3 on the different KDD 99 cup datasets or with a combination with Principal Component Analysis; first by projecting all data on the new space generated by the few number

PCA's principal axes then applying the nearest neighbor or decision trees algorithm on the datasets but after their projection on the new reduced PCA's space.

The accuracy of each experiment is based on the percentage of successful prediction (PSP) on the test dataset.

$$PSP = \frac{number\ of\ successful\ instance\ classification}{number\ of\ instances\ in\ the\ test\ set} \tag{15}$$

5.1 Nearest neighbor with/without PCA

The first experiment, we perform, consists in evaluating the nearest neighbor algorithm on the KDD 99 database. The main problem encountered when computing the nearest neighbor is that it is computationally expensive to compute the nearest neighbor of each point. The complexity of this computation is $O(nm)$ where n is the number of connection records in the training dataset and m in the test dataset. Each distance computation between two connection records depends on the number of space coordinates where they are represented. Of course, this algorithm may be approximated with a cluster based estimation algorithm [Han and Kamber, 2001]. However, the distance between two connection records remains dependent on the coordinates number of the feature space. For this reason, we have projected the different datasets connections on the new feature space generated by the principal component axes. There are 125 coordinates of each connection in KDD 99 after transformation for discrete attribute values as explained in section 3 (41 attributes added to the representation of each discrete value that has at least two coordinates).

Table 2 presents the confusion matrix when applying directly the nearest neighbor on the feature space generated by these 125 coordinates.

Table 2. Confusion matrix obtained with the nearest neighbor algorithm on 125 coordinates.

Predicted as Actual	Normal	Probing	DOS	U2R	R2L
Normal(60593)	**99.50%**	0.26%	0.24%	0.00%	0.00%
Probing(4166)	17.21%	**72.01%**	10.28%	0.00%	0.50%
DOS(229853)	2.87%	0.12%	**97.01%**	0.00%	0.00%
U2R(228)	39.96%	18.80%	32.01%	**6.60%**	2.63%
R2L(16189)	96.12%	2.65%	0.00%	0.02%	**1.21%**
PSP=92.05%					

Using the same algorithm (the nearest neighbor), we have experimented the test dataset on a new feature space generated by at most seven PCA's axes. We have performed the different experiments by considering 2, 3, ..., or 7 axes. The results are not much different from each other when we consider from 2 to 7 axes. Table 3 shows the confusion matrix when we consider four axes (i.e. each connection record in the different datasets is represented by only four coordinates).

Table 3. Confusion matrix obtained with the nearest neighbor on 4 coordinates after performing PCA .

Predicted as Actual	Normal	Probing	DOS	U2R	R2L
Normal(60593)	**99.50%**	0.27%	0.23%	0.00%	0.00%
Probing(4166)	13.87%	**74.40%**	11.37%	0.00%	0.36%
DOS(229853	2.68%	0.18%	**97.14%**	0.00%	0.00%
U2R(228)	35.96%	14.47%	39.03%	**7.91%**	2.63%
R2L(16189)	97.49%	1.71%	0.00%	0.00%	**0.80%**
PSP=92.22%					

The confusion matrix in table 3 shows that the results after PCA application are slightly better. In addition, the computation time is reduced by a factor of approximately thirty ($\sim 125/4$) when considering 4 principal components. Hence, it is better to reduce the space on which the connection records are represented before applying any machine learning algorithm. This first experimentation is used to show that a combination between PCA and the nearest neighbor performs well even if a few axes are considered (at most seven) to represent the records. According to equation (14), the inertia ratio is close to 1 (0.999) when considering only 4 axes. This is the reason why a representation with only four axes provides a good prediction rate.

In the two experiments, the two last classes R2L and U2R are not well detected. The maximum PSP for U2R class is 7.91% and 1.21% for R2L.

5.2 Decision trees with/without PCA

This section presents experimental results using decision trees with the C4.5 algorithm. This latter is applied, in the first experiment, directly on the different datasets using the whole 41 attributes and then compared to its application on the datasets but after their projection onto the new space generated by the few principal component axes number.

During our experiments, we have considered, as in [BenAmor et al., 2004], two cases. The first consists in grouping the whole 39 attacks types into four attack categories before training. In the second case, they are gathered after classification.

Decision trees without PCA. In this section, we present the different results obtained when applying directly the C4.5 algorithm on rough data. Table 4 evaluates the application of the C4.5 algorithm on the dataset by gathering the whole attacks into four categories before the training step and its application on the rough dataset after classification.

Table 4. Confusion matrix relative to five classes using the C4.5 algorithm. The values between parentheses correspond to gathering the whole attacks results into five categories after classification.

Predicted as Actual	%Normal	%Probing	%DOS	%U2R	%R2L
Normal(60593)	**99.49(99.42)**	0.36(0.39)	0.12(0.15)	0.00(0.00)	0.02(0.03)
Probing (4166)	21.32(15.75)	**74.70(78.80)**	3.98 (5.45)	0.00(0.00)	0.00(0.00)
DOS (229853)	2.68(2.58)	0.00(0.46)	**97.31(96.96)**	0.00(0.00)	0.00(0.00)
U2R (228)	90.79(56.58)	1.75(28.51)	0.44(0.88)	**4.39(5.26)**	2.63(8.77)
R2L (16189)	92.03(94.63)	2.10(0.07)	0.01(0.00)	0.02 (0.03)	**5.84(5.27)**
PSP=92.60%,(PSP=92.35%)					

The different results obtained in this first experiment show that gathering the attacks before training or after classification does not influence the percentage of successful prediction. The two classes U2R and R2L are classified with a percentage of successful prediction of at most 5.84%. This is due to the low number of samples of these two classes in the training set; 0.01% examples of U2R in the training set (resp. 0.23% of R2L) versus 0.07% of U2R (resp. 5.20% of R2L) in the test dataset and to the new forms of these two attacks classes that appear in the test dataset which are not present in the training set.

Decision trees with PCA. We now apply the C4.5 on the new feature space generated by the principal axes. All the training dataset and the test dataset connection records are projected onto the new feature space. This new feature space is generated by at most 7 axes in our different experiments to validate the results of combining PCA with the C4.5 decision trees algorithm.

According to tables 5 and 8 [1], there is a slight difference between the use of decision trees on rough data and their combination with PCA on the new feature space.

However, it is important to mention that the number of nodes in the decision tree generated when we apply C4.5 on rough data is greater than that of nodes in the decision tree when applied with the different datasets but in the new

[1] PCAi corresponds to the projection of the data onto the first i principal axes corresponding to the first i highest eigenvalues.

Table 5. Confusion matrix relative to five classes using the C4.5 algorithm after dataset projection onto two principal component axes. The values between parentheses correspond to gathering the whole attacks results into five categories after classification.

Predicted as Actual	%Normal	%Probing	%DOS	%U2R	%R2L
Normal(60593)	**99.00(98.99)**	0.85(0.84)	0.12(0.12)	0.00(0.00)	0.03(0.04)
Probing(4166)	29.60(30.20)	**66.80(66.30)**	3.50(3.50)	0.10(0.00)	0.00(0.00)
DOS(229853)	2.42(2.42)	0.33 (0.33)	**97.25(97.25)**	0.00 (0.00)	0.00 (0.00)
U2R(228)	92.98(91.23)	0.00(0.00)	0.44(0.00)	**6.58(8.33)**	0.00(0.44)
R2L(16189)	99.94(97.69)	0.00(0.00)	0.06(0.01)	0.00(0.00)	**0.01(2.30)**
PSP=92.05%(PSP=92.16%)					

feature space generated by the PCA. In addition, the training time consumed to construct the decision tree with the new data in the feature space, generated by at most *seven* principal components, is more interesting as presented in table 6.

Table 6. Time and tree size with/without PCA.

	Decision trees without PCA	Decision trees with PCA
Number of nodes	$\sim$ 1500 before pruning $\sim$ 700 after pruning	$\sim$ 330 before pruning $\sim$ 211 after pruning
Training time	$\sim 3mn40sec$	$\sim 50sec$

Furthermore, the problem of the prediction ratio with the last two classes persists always as mentioned in the previous subsection. The highest predicted successful ratio obtained with the R2L class does not exceed 2.30%. This is because the principal components associated with the smallest eigenvalues often correspond to not interesting information [Jolliffe, 2002] that corresponds in reality in our case to the classes that are not present with high rates in the training set and we are taking into account only the highest eigenvalues. This is the reason why the R2L class prediction ratio is very small. To circumvent this problem, we have considered other axes corresponding to lower principal axes. In this case, we have obtained 5.86% as the highest prediction ratio for the R2L class.

Table 7 presents the best prediction ratios when considering other components axes not presented in tables 5 and 8.

According to table 7, the results obtained by combining decision trees with PCA are slightly better than those in table 4 when applying directly decision trees on rough data. We may improve the prediction ratio, when combining

Table 7. The best prediction ratios obtained for each class when considering different principal components.

Attack Category	Normal (60593)	Probing (4166)	DOS (229853)	U2R (228)	R2L (16189)
Detection Ratio	99.52%	78.84%	98.26%	12.72%	5.86%
PSP=92.63%					

decision trees with PCA, of the last two classes by duplicating their different samples in the training set. By this reasonning, they will not be considered as less interesting information.

6. Conclusion

We have presented in this paper a new idea on how to reduce the different representation spaces before applying some machine learning algorithms on the different KDD 99 intrusion detection datasets. This new representation permits to improve the learning time and space representation of the different datasets with a similar successful prediction in the whole experiments.

The main drawback which persists in combining decision trees or the nearest neighbor with PCA is the poor prediction ratio rate of the R2L class which is in most of the time classified as normal. This is due to its low presence in the training dataset (0.23%). We may improve this ratio by boosting the number of samples of this class in the training dataset before applying the PCA algorithm in order to transform it into an interesting information class represented by a principal component corresponding to a higher eigenvalue.

However, the new attacks types that are present in the two last classes, R2L and U2R, in the test data set could not be detected by machine learning algorithms [Sabhnani and Serpen, 2004]. This suggests to perform other unsupervised machine learning or data mining algorithms to deal with these new attacks that should be detected as new attacks. Hence, we may add a new class that we call new attacks class to which new attacks, which are not looking similar to the known attacks in the training dataset, will be classified. This will be discussed in a forthcoming paper.

Table 8. Confusion matrix relative to five classes using the C4.5 algorithm after dataset projection onto 3,4,5,6 or 7 principal component axes. The values between parentheses correspond to gathering the whole attacks results into five categories after classification.

Predicted as Actual	%Normal	%Probing	%DOS	%U2R	%R2L
Normal	**(60593)**				
PCA3	**99.47(99.52)**	0.35(0.32)	0.13(0.12)	0.00(0.00)	0.05(0.04)
PCA4	**99.43(99.51)**	0.36(0.34)	0.13(0.12)	0.01(0.01)	0.07(0.02)
PCA5	**99.23(99.19)**	0.39(0.37)	0.22(0.27)	0.00(0.16)	0.15(0.02)
PCA6	**99.46(99.47)**	0.36(0.34)	0.16(0.15)	0.00(0.02)	0.02(0.02)
PCA7	**99.39(99.42)**	0.35(0.14)	0.24(0.02)	0.00(0.06)	0.02(0.36)
Probing		**(4166)**			
PCA3	28.88(30.63)	**67.67(62.12)**	3.36(7.25)	0.10(0.00)	0.00(0.00)
PCA4	24.20(30.56)	**69.40(63.08)**	6.39(6.34)	0.00(0.00)	0.02(0.02)
PCA5	13.85(16.30)	**76.26(73.09)**	8.93(10.30)	0.00(0.29)	0.96(0.02)
PCA6	18.31(22.56)	**76.48(69.40)**	5.21(8.02)	0.00(0.02)	0.00(0.00)
PCA7	14.98(22.71)	**75.76(69.28)**	9.27(8.02)	0.00(0.00)	0.00(0.00)
DOS			**(229853)**		
PCA3	2.60(2.75)	0.20(0.00)	**97.18(97.25)**	0.00(0.00)	0.02(0.00)
PCA4	2.53(2.76)	0.27(0.05)	**97.17(97.17)**	0.00(0.00)	0.02(0.02)
PCA5	2.70(2.69)	0.02(0.06)	**97.26(97.16)**	0.00(0.00)	0.02(0.09)
PCA6	2.76(2.77)	0.02(0.03)	**97.20(97.06)**	0.00(0.00)	0.02(0.14)
PCA7	2.71(2.75)	0.04(0.03)	**97.22(97.06)**	0.00(0.00)	0.02(0.16)
U2R				**(228)**	
PCA3	85.09(65.35)	1.75(9.65)	0.44(14.91)	**12.72(8.33)**	0.00(1.75)
PCA4	68.86(77.63)	9.65(5.70)	9.65(9.65)	**5.26(6.58)**	6.58(0.44)
PCA5	35.53(62.28)	47.37(19.74)	11.40(13.60)	**4.39(3.95)**	1.32(0.44)
PCA6	38.16(41.23)	47.37(3.51)	7.46(48.25)	**5.70(4.39)**	1.32(2.63)
PCA7	38.16(41.23)	47.37(3.51)	7.46(48.25)	**5.70(4.39)**	1.32(2.63)
R2L					**(16189)**
PCA3	99.60(99.60)	0.02(0.02)	0.02(0.02)	0.01(0.01)	**0.35(0.35)**
PCA4	99.83(99.81)	0.03(0.01)	0.03(0.03)	0.01(0.02)	**0.11(0.12)**
PCA5	99.81(99.38)	0.02(0.14)	0.04(0.04)	0.03(0.04)	**0.10(0.41)**
PCA6	99.82(99.43)	0.02(0.14)	0.03(0.02)	0.04(0.05)	**0.09(0.36)**
PCA7	99.81(99.42)	0.02(0.14)	0.02(0.02)	0.06(0.06)	**0.09(0.36)**
PSP3=92.12%(PSP3=92.11%),PSP4=92.12%(PSP4=92.05%) PSP5=92.23%(PSP5=92.13%),PSP6=92.24%(PSP6=92.06%) PSP7=92.23%(PSP7=92.05%)					

Acknowledgements

The authors thank Nora and Frédéric Cuppens for the different discussions and their helpful comments on early versions of this paper.

References

Agrawal, R. and Joshi, M. V. (2000). PNrule: A New Framework for Learning Classifier Models in Data Mining A Case-Study in Network Intrusion detection. Technical Report RC-21719, IBM Research Division.

Anderson, J. P. (1980). Computer Security Threat Monitoring and Surveillance. Technical report, James. P. Anderson Co., Fort Washington, Pennsylvania.

Bay, S. D. (1998). Combining Nearest Neighbor Classifiers Through Multiple Feature Subsets. In *Proceedings of the 15th International Conf. on Machine Learning*, pages 37–45, San Francisco, CA. Morgan Kaufmann.

BenAmor, N., Benferhat, S., and ElOuedi, Z. (2004). Naive Bayes vs Decision Trees in Intrusion Detection Systems. In *The 19th ACM Symposium On Applied Computing - SAC 2004*, Nicosia, Cyprus.

Bouzida, Y. and Gombault, S. (2003). Intrusion Detection Using Principal Component Analysis. In *Proceedings of the 7th World Multiconference on Systemics, Cybernetics and Informatics*, Orlando, Florida.

Breiman, L., Friedman, J. H., Olshen, R. A., and Stone, C. J. (1984). Classification and Regression Trees.

Dasarathy, B. V. (1991). A Computational Demand Optimization Aide for Nearest-Neighbor-Based Decision systems. In *IEEE International Conference on Systems, Man and Cybernetics*, pages 1777–1782. Morgan Kaufmann.

Denning, D. (1987). An Intrusion Detection Model. *IEEE Transactions on Software Engineering*, 13(2):222–232.

Eskin, E., Arnold, A., Prerau, M., Portnoy, L., and Stolfo, S. (2003). A Geometric framework for unsupervised anomaly detection: Detecting intrusions in unlabeled data. *Applications of Data Mining in Computer Security.*

Fix, E. and Hodges, J. L. (1951). Discriminatory analysis: Nonparametric discrimination: Consistency properties. Technical Report 21-49-004, USAF School of Aviation Medecine, Randolf Field, Texas.

Han, J. and Kamber, M. (2001). *Data Mining: Concepts and Techniques.* Morgan Kaufmann Publishers.

Hettich, S. and Bay, S. D. (1999). The UCI KDD Archive. Available at: http://kdd.ics.uci.edu/.

Hotelling, H. (1933). Analysis of a complex statistical variables into principal components. *Journal of Educational Psychology*, 24:417–441.

Ilgun, K. (1993). Ustat, a real time intrusion detection system for UNIX. In *IEEE Symposium on Security and Privacy*, pages 16–28, Oakland, CA.

Jolliffe, I. T. (2002). *Principal Component Analysis.* Springer Verlag, New York, NY, third edition.

KDD Cup (1999). KDD Cup 99 Intrusion Detection Datasets. Available at: http://kdd.ics.uci.edu/databases/kddcup99/kddcup99.html.

KDD Task (1999). KDD 99 Task. Available at: http://kdd.ics.uci.edu/databases/kddcup99/task.html.

Kirby, M. and Sirovich, L. (1990). Application of the KarhunenLoeve Procedure for the Characterization of Human Faces. *IEEE Transactions On Pattern Analysis and Machine Intelligence*, 12(1):103–107.

Kumar, S. and Spafford, E. (1994). A pattern matching model for misuse intrusion detection. In *Proceedings of the 17th National Computer security Conference*, pages 11–21.

Lee, W., Stolfo, S. J., and Mok, K. (1999). Mining in a data flow environment: Experience in intrusion detection. In *Proceeding of the 1999 Conference on Knowledge Discovery and Data Mining KDD-99*.

Levin, I. (2000). KDD-99 Classifier Learning Contest LLSoft's Results Overview. SIGKDD Explorations. *ACM SIGKDD*, 1:67–71.

Mitchell, T. M. (1997). *Machine Learning*. McGraw Hill.

Pfahringer, B. (2000). Winning the KDD Classification Cup: Bagged Boosting. SIGKDD Explorations. *ACM SIGKDD*, 1:65–66.

Quinlan, J. R. (1986). Induction of decision trees. *Machine Learning*, 1:1–106.

Quinlan, J. R. (1993). *C4.5: Programs for machine learning*. Morgan Kaufmann Publishers.

Roesch, M. (1999). Snort - Lightweight Intrusion Detection for Networks. In *13th Systems Administration Conference - LISA 99*.

Sabhnani, M. and Serpen, G. (2004). On Failure of Machine Learning Algorithms for detecting Misuse in KDD intrusion Detection Data Set. *Intelligent Analysis*. To Appear.

Turk, M. and Pentland, A. (1991). Eigenfaces for Recognition. *Cognitive Neuroscience*, 13(1):71–96.

VISUALISING INTRUSIONS: WATCHING THE WEBSERVER

Stefan Axelsson
Dept. of Computer Science, Chalmers University of Technology, Sweden
sax@cs.chalmers.se

Abstract Despite several years of intensive study, intrusion detection systems still suffer from a key deficiency: A high rate of false alarms. To counteract this, this paper proposes to visualise the state of the computer system such that the operator can determine whether a violation has taken place. To this end a very simple anomaly detection inspired log reduction scheme is combined with graph visualisation, and applied to the log of a webserver with the intent of detecting patterns of benign and malicious (or suspicious) accesses.

The combination proved to be effective. The visualisation of the output of the anomaly detection system counteracted its high rate of false alarms, while the anomaly based log reduction helped reduce the log data to manageable proportions. The visualisation was more successful in helping identifying benign accesses than malicious accesses. All the types of malicious accesses present in the log data were found.

1. INTRODUCTION

A significant problem with intrusion detection systems is the high number of false alarms [2, 8]. To address this *information visualisation* [4, 10] was applied to the problem of intrusion detection. The main problem with applying information visualisation to intrusion detection is the large amount of data that the user is faced with. To address this an anomaly detection inspired method was used to reduce the log to manageable proportions before utilising graph visualisation to understand the actual data. The hypothesis was that this would enable the user to benefit from the strengths of both visualisation—quickly making sense of medium size data sets, and anomaly detection—summarily discarding large amounts of uninteresting data, all the while avoiding the problems of visualisation having a limit to the amount of data that can reasonably be handled, and anomaly detection having a high false alarm rate for decent detection rates.

An overview of this paper is as follows: First the experimental system is described in section 2. Then the anomaly based log reduction system is introduced in section 3 followed by a description of the visualisation and the report of the results from the visualisation and anomaly based log reduction in sections 4 and 5. The paper concludes with discussion, conclusion and future and related work in the remaining sections.

2. THE EXPERIMENTAL SYSTEM

For the experiment, a webserver access log was studied. It should be stressed that the primary interest is in experimenting with the effectiveness of the combination of visualisation and anomaly based log reduction, *not* in producing a realistic tool for usage in the field. Unfortunately there is a dearth of publicly available corpora useful for intrusion detection research. The most popular such corpora is the Lincoln Labs DARPA evaluation data. As it is export controlled it is unavailable to us.

The webserver under study serves a university computer science department. The server was running Apache version 1.3.26, and set to log according to the *common* log format. The log consists of a line based text file with each line representing a single HTTP access request. The *request* field i.e. the actual HTTP request sent to the server, is important as it is the central point of many attacks against a web server. The request field consists of the request method (‘GET’, ‘HEAD’, ‘CONNECT’, etc), followed by the *path* to the resource the client is requesting, and the method of access (e.g. ‘HTTP 1.1’). The *path* in turn can be divided into components separated by certain reserved characters.

We studied the log for the month of November 2002, since it was believed that it would contain security relevant incidents, and we had access to later logs with which to compare the results. The access log contained ca. 1.2 million records. Selecting the actual request fields and removing duplicates ca. 220000 unique requests were identified. Because of their importance it is these unique requests that will be studied in the rest of the paper.

3. THE LOG REDUCTION SCHEME

Intrusion detection strategies can be roughly divided into two major categories, *signature based* and *anomaly based.* Anomaly based schemes rely on detecting *unusual* behaviour. Anomaly detection possesses the capability to detect *new attacks*, since it detects unusual patterns of activity, not illegal patterns per se. This capability comes at the cost of a much higher false alarm rate, to the point where the false alarms could drown all true alarms.

The log reduction scheme is based on descriptive statistics; in this case the frequencies with which events occur. This is in the same vein as seminal intrusion detection systems such as NIDES [1], though the approach here is simpler still. In order to classify the requests according to how unusual they are they are first cut up into components letting the reserved characters " ?:&=+$," separate the fields. For example a request such as 'GET /pub/index.html HTTP 1.1', is separated into the components 'GET', 'pub','index.html', 'HTTP' and '1.1'. The absolute frequencies of the fields as they appear in different unique request strings are counted.

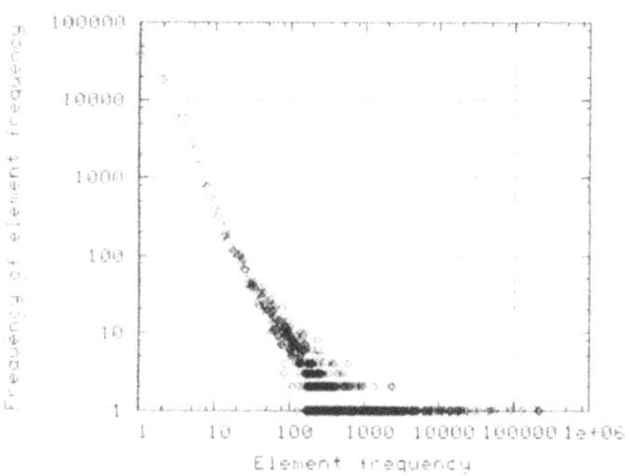

Figure 1. Frequencies of component frequencies

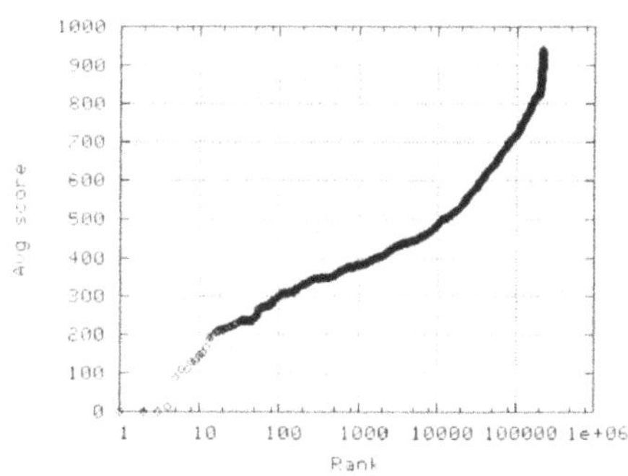

Figure 2. Requests sorted by lowest score

The request as a whole is scored by calculating the average of the absolute frequencies of the path components and hence requests consisting of unusual components have a low score, signifying that they are viewed as *anomalous*.. However, studying the *frequencies* of the *component frequencies* we see that a few high scoring elements (such as 'GET') could skew (i.e. drive up) the average. Therefore a *cutoff* is applied. Figure 1 lists the frequencies of the frequencies of the components. Studying the figure we see that a cutoff of 1000 seems reasonable since most of the activity appears to have died off by then. There are very few components with frequencies above 1000 and since they represent elements that are very common, they would tend to drown the lower frequency components we are interested in. Figure 2 plots the scores of the requests as a function of the ordering. The lowest scoring 5200 accesses are selected since that gives us a manageable amount of data to visualise.

4. VISUALISING THE LOWEST SCORING REQUESTS

The idea is to visualise the structure (and clusterings) of the various requests, the hypothesis being that differences in structure will enable the user to (relatively) quickly identify patterns of benign and malicious access. To accomplish this, the requests are cut into components as described in the previous section. The resulting components are visualised as a general graph where adjacent components in the request string are linked via directed edges in the graph. Using the same example as before: The request 'GET /pub/index.html HTTP 1.1', is cut up into nodes ('GET' etc.) with directed edges connecting 'GET' with 'pub', 'pub' with 'index.html' etc.

To visualise the resulting graph the graph visualisation tool *Tulip* was chosen.[1] Tulip has extensive features for interactive viewing and manipulation of general graphs. These aspects are unfortunately difficult to capture in writing (even with illustrations).[2]

To perform the actual detection the 5200 lowest scoring accesses is visualised in figure 3 [3] as a three dimensional general graph. The circular structure at the top of the graph that can be seen to reach almost all of the rest of the graph is the 'GET'-node. Note that the edges are not drawn as solid lines, since this would completely occlude the view.

At first figure 3 may look daunting, but closer scrutiny reveals several large features. Close to the center of the picture for example, we see a large double ring structure. Contrasting it with all other features, it looks rather unique, there is no other structure that looks similar (at least on this level), so we decide to investigate further.

Isolating the feature in question leads to figure 4. Following the links (which is somewhat difficult to do in the static display here) we learn of a loose structure that starts with either 'cgi-bin' or 'cgi-local' and progresses via 'recipient', 'subject' and then the unlikely looking random text strings. It turns out that these text strings are in fact recipient email addresses for *aol.com* and *hotmail.com* email users. The message to be mailed in many cases (but not all) purports to be from John Doe, and is simply "Is anybody out there?" So this particular access pattern seems

---

[1] Tulip is freely available under the GPL from 'http://www.tulip-software.org'.

[2] A more detailed version of this paper will be available on the author's website at the time of publication 'http://www.cs.chalmers.se/~sax'.

[3] The PDF-rendition of this graph may be clearer than a printed image.

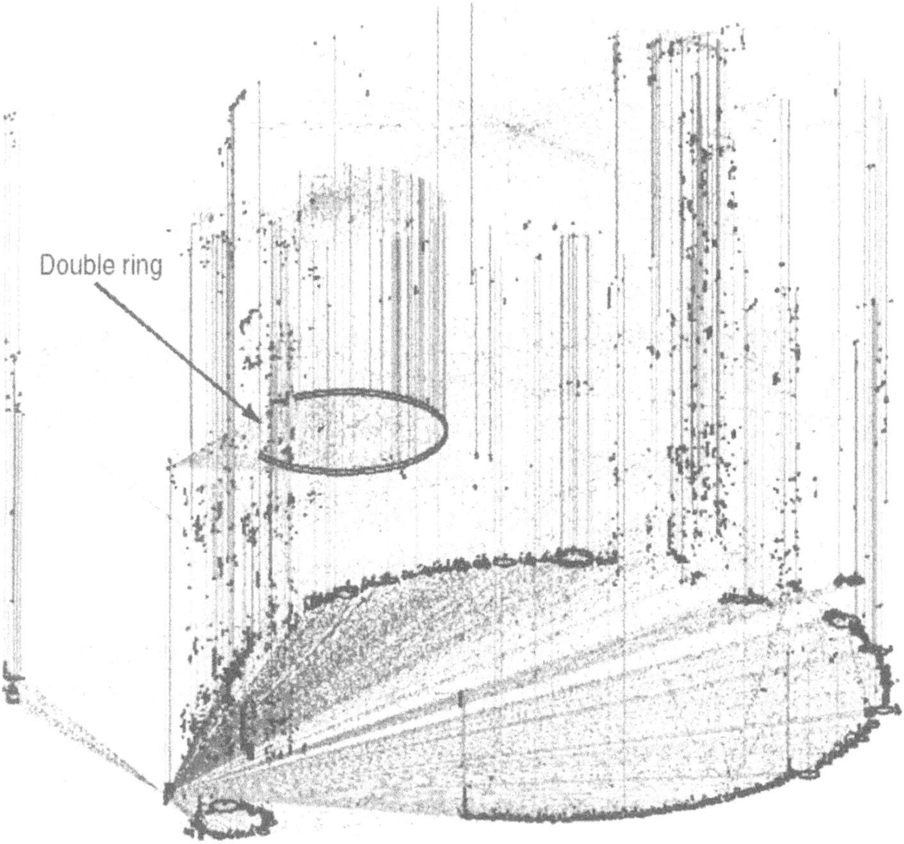

Figure 3. Graph of the lowest scoring requests

to be a *spam attack*, trying to use a misconfigured HTML to mail gateway that is commonly available. It was not active on the server however. The unlikely looking recipient names are probably automatically generated and the messages sent in the hope of eliciting a response and in doing so finding legitimate email addresses. Note that a tree visualisation would have been less powerful here, since there are two major (early in the request) parents of this particular pattern: 'cgi-bin' and 'cgi-local'. When visualised as a tree these branches would not have shared the latter features, even when they would have been the same.

We are usually not so fortunate as to identify attacks as we first lay our eyes on the graph. Instead we have to repeat the above detailed analysis. It so happens that all the other major features that are identifiable in figure 3 are uninteresting. However, they are all also much more regular

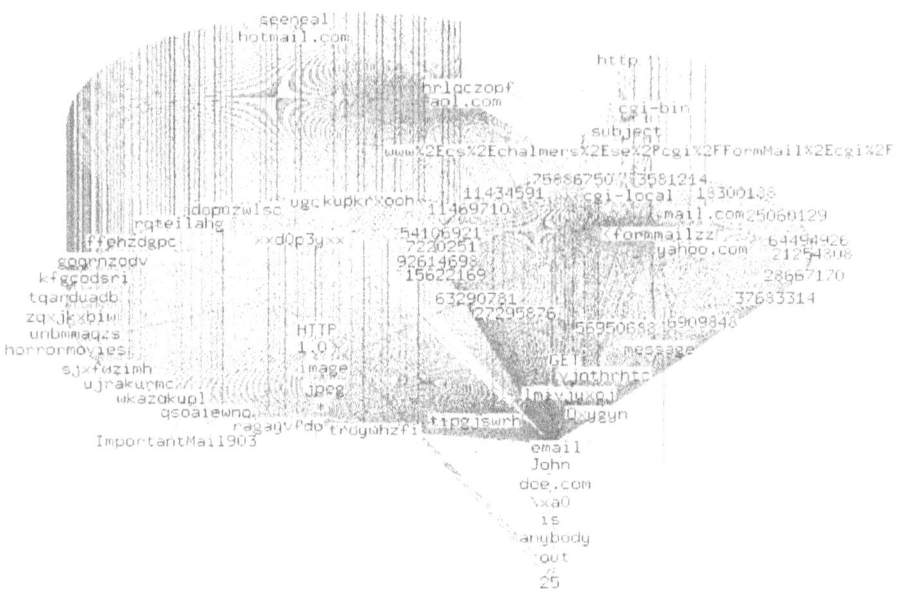

Figure 4. Zoom on feature (Spam attack in this case)

than the pattern we have just seen. This makes it possible for the user to eliminate many edges by eliminating fewer key parent nodes, slashing away what amounts to whole (improper) subtrees, and this fairly large graph can be withered down with a relatively modest amount of effort.

An example of this process is depicted in figure 5.[4] The accesses to files belonging to the user *carlsson* form a regular subtree that is easily identified as having nothing of a security relevant nature and hence easily discarded. This subgraph contains the data from 145 access requests, letting us discard some 3% of the reduced log in one fell swoop. Looking at the distribution of accesses in similar subgraphs, the twenty largest benign subgraphs (ranging from 213 to 59 access requests respectively) contain 42% of the access requests under scrutiny, and the first 180 benign subgraphs contain some 80% of all access requests in the reduced log. Hence most benign requests can be easily discarded. Unfortunately lack of space precludes more pictures of the details of this process than figure 5.

[4]For the purpose of illustration the corresponding access requests here have been isolated to make the graph clearer.

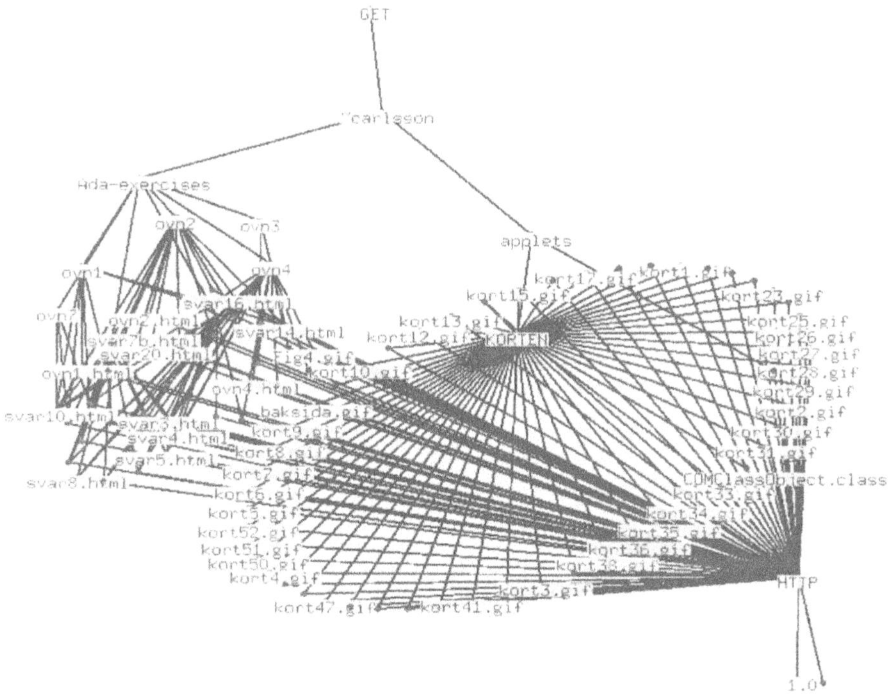

Figure 5. Zoom on feature (benign accesses forming a subgraph)

After about one to two hours the distilled requests that we cannot eliminate as being benign are arrived at, (since user experiments are yet to be performed on this method, a more precise estimate cannot be given with any certainty). These requests will be discussed in more detail in section 5.

5. DETAILED ANALYSIS OF THE FEATURES FOUND

The remaining accesses were classified into two categories, *suspect* and *intrusive*. The reason for using a *suspect* class is that since this is data from the field and the intentions of the entity submitting the request is not known, it is sometimes difficult to decide whether a request is the result of an intrusive process, the result of a flaw in the software that submitted it or a mistake by its user.

The *intrusive* class was further subdivided into seven different subclasses that correspond to metaclasses of the attacks that were observed:

Formmail attacks As already mentioned the server was subjected to a spam attack, in where spammers tried to exploit a commonly available web mail form to send unsolicited mail via the server.

Unicode attacks These are attacks against the Microsoft IIS web server, where the attacker tries to gain access to shells and scripts by providing a path argument that steps backward (up) in the file tree and then down into a system directory by tricking IIS into allowing e.g. '\..\' by escaping (as per the HTTP RFC) the offending backslash character in various ways. Many variations on the same basic scheme are present in the log data. Their visual structure is quite distinct from benign accesses.

Proxy attacks The attacker tried to access other web servers or IRC servers via the web server, hoping that it would be misconfigured to proxy such requests.

Pathaccess attacks These are more direct attempts to access command interpreters, cgi scripts or sensitive system files such as password files. A major class here is trying to access back doors known to be left behind by other successful system penetrations (mainly by worms). Configuration files of web applications, web store software for example were also targeted.

Cgi-bin attacks Attacks against cgi scripts that are commonly available and may contain security flaws.

Buffer overrun Only a few types of buffer overruns were found in the log data. All of these are known to be indicative of worms targeting the Microsoft IIS web server.

Misc This class contains seven accesses that probably are malicious in nature, but which do not fit in any of the previous classes. They are various probes using the *OPTIONS* request method, and a mixture of *GET* and *POST* request method calls that target the root of the file system.

The *intrusive* class (minus the *formmail* attack in figure 4) is depicted in figure 6. We directly see a few security relevant features (i.e. features that stand apart from benign accesses). One such feature is the 'system32'/'command.exe' tail in the lower middle of figure 6. This tail is common to many different branches from one of the roots of the graph and turns out to be a strong indicator of the *unicode* attack discussed above. Another peculiar cluster is at the top, second from the left, and it turns out to be instances of the meta class *proxy attacks*.

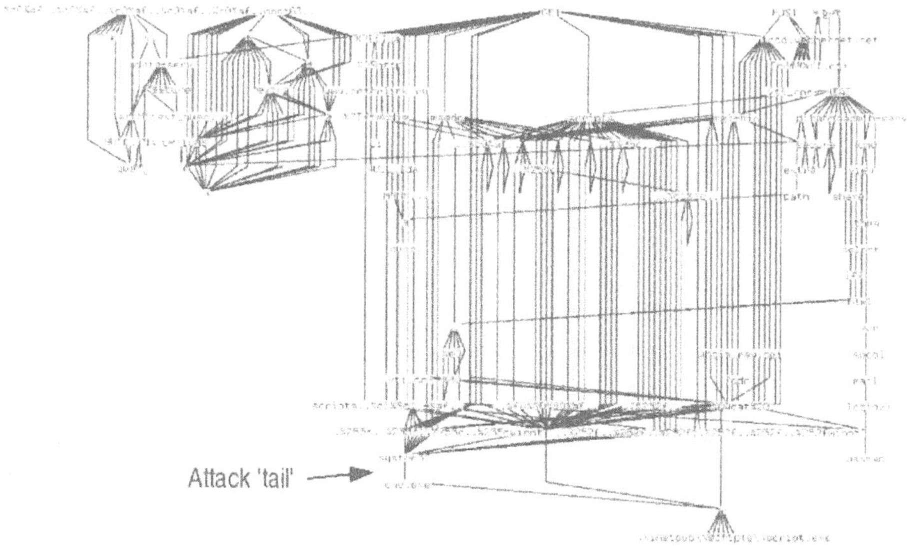

Figure 6. The remaining accesses deemed to be intrusion attempts.

As we can see from the graph there are many variations on the basic IIS encoding attack being tried against the server, although the graph of course does not list the actual attacks themselves. It would be interesting to study to what extent the graph predicts (all/most) possible attacks (i.e. by following one branch from the root to a leaf, do we get an executable attack that is not present in the log file proper). The overflow attacks are simple buffer overruns against fixed length buffers without adequate overflow protection, in our case these take the form of very long components (formed of a string beginning with repetitions of the character 'N' or 'A') followed by escaped shell code. Searching security sources we find that the first of these is characteristic of the Code Red worm.

Empirically in the data set studied here, the uninteresting access patterns are almost without fail very treelike in their appearance, with no common tails to speak of, while the opposite is true for the attacks. The reason seems to be that attacks that show some diversity nevertheless share common features that may come at any position in the request, while this is not true for normal accesses. In the case of unicode attacks we see it is the 'payload' i.e. the commands that the attacker wishes to execute that show many similarities, greater similarities in fact than the encoded path that leads to it. In the case of the spam attacks, it is the stereotype message delivered that is the key to the late similarities. It is conjectured that this would probably be difficult to avoid at least for the first kind of attack. There are only so many different commands to execute with the desired effect, and only so many file system paths to get to them, so there is bound to be a bottleneck (where the paths converge to a smaller set, and the command set starts), giving an hour glass shape. On the other hand the access to static legitimate web pages is highly tree-like in nature, and hence does not elicit the same hour glass shape. Cgi-bin arguments show more variation, but they do not in our experience contain the same clearly identifiable tails as the attack patterns. In the case of the scripts for collecting student opinion of courses taken for example, the text of the messages, while sharing many words between different opinions, still display much more variation, giving an almost random appearance, not the typical hour glass shape of the suspect patterns. However, not all attack patterns show this hour glass shape, the attacks with very little variation in the attack types does not provide enough data for a pattern to emerge. It should be noted that being subject to a large number of different variations of the same kind of attack *increases* the possibility of the attack being detected using the scheme presented here, since there is more structure apparent. The opposite is typically true of signature based intrusion detection systems that have been programmed to detect one type of attack (or at least a smaller range of attacks).

However, despite the structure apparent in the attacks, comparing e.g. figure 5 to figure 6 indicates that the visualisation method is better suited to *eliminate* benign requests than to *detect* malicious requests. Which is just as well given that there are many more benign requests than malicious ones; and in fact the *false alarms* suppression capability of an intrusion detection system determines its effectiveness [2].

6. EFFECTIVENESS OF THE LOG REDUCTION SCHEME

In order to evaluate the effectiveness of the log reduction scheme we analysed the entire 220000 unique web requests by hand. This proved to be an extremely tedious task, the author spent a week classifying and investigating the data.

The access requests were classified into the three categories: *normal*, *intrusive* and *suspect* as before. Furthermore the *intrusive* class was divided into the seven subclasses discussed earlier. This further subdivision benefits the investigation of the effectiveness of the log reduction mechanism since it avoids reporting results that might indicate an overall satisfactory detection rate, but on closer study turn out to be lacking in detection capability in any of the seven areas[5]. A summary of how many of the different classes of attacks the log reduction mechanism detected can be found in table 1.

Instances of the class *suspect* are not considered false alarms when they occur in the output of the log reduction tool, but on the other hand will not be considered missed attacks when they do not.

Attack	*Total*	*Alarms*	*Detection rate (%)*
Formmail	285	282	99
Unicode	79	79	100
Proxy	9	5	56
Pathaccess	71	16	23
Cgi-bin	219	17	8
Buffer overrun	3	2	66
Miscellaneous	7	2	29
All attacks	673	403	60

Table 1. Detection rates of the log reduction mechanism.

As we can see in table 1, the log reduction mechanism faired well, it managed to preserve evidence of all seven classes of attacks in the reduced log. In light of these results the author would not hesitate to claim the log reduction mechanism a success from the point of view of detection rate, even though it did not do spectacularly well in all classes, especially in the classes that were most similar to normal traffic, which is only to be expected.

The discussion of false alarm rates is complicated by the fact that anomaly detection is not actually done, but log reduction. The difference is that an anomaly detection system reports on the *absolute* abnormality of a request, while the log reduction system reports on the *relative* abnormality of a request. The corresponding anomaly detection system would produce a variable number of alarms indicating the level of intrusive activity, but the log reduction scheme will always report the same number of 'alarms' regardless, as a suitable number of 'alarms' are selected to perform visualisation on. Thus the notion of *false alarm rate* is not well defined in this context. Two papers discuss the issue of alarms, and false alarms in intrusion detection systems in more detail [2, 8].

Table 2 below lists the absolute values of the number of *attacks*, *normal traffic* and *suspect traffic* that are evident in the whole data set and

[5]This might well have been the case since the distribution of different accesses in each of the seven subclasses turns out to be highly skewed and so doing well in one class would skew the overall result.

in the reduced log. As might have been expected the number of benign accesses in the reduced log is quite high. Circa 86% of the reduced log contains benign traffic. (Note that this number would not correspond to the *false alarm rate*, but rather the *Bayesian false alarm rate* as defined in [2]. The true false alarm rate would be even higher.) This is acceptable since the number of accesses are kept within reasonable limits for the chosen visualisation method and the log reduction mechanism has a sufficiently high detection rate.

Access type	*All accesses*	*Alarms*
Attacks	673	403
Normal traffic	215504	4499
Suspect	115	28
Total	220000	5200

Table 2. Summary of the true and false alarms of the log reduction mechanism.

Note that it is pointless to compare the results from the system presented here with that of popular signature based systems such as *Snort*, since these rely on previous external knowledge of intrusions. There is unfortunately a dearth of suitable publicly available anomaly detection systems with which to compare the results, though that would be more useful.

7. DISCUSSION

Comparing the presented method with spending the same amount of time going through a false alarm list from an intrusion detection system, the user has been spent time actually learning about the type of traffic the webserver sees, knowledge that can be used to make the site run better/smoother. This is not generally the case when watching the output of an intrusion detection system.

This generalized knowledge—patterns that are distinct from benign traffic have been found, no benign traffic contains the kind of encoding that the encoding attacks does—can of course be used to program a signature based intrusion detection system. The author conjectures it would work well since it is known what elements of these intrusions that are unique to attacks and do not occur in normal traffic to the specific site. When performing intrusion detection the best results are achieved when having a model of both the normal and the intrusive traffic [8]. At the very least this knowledge can be used when tuning a signature based intrusion detection system, something which is always necessary.

The anomaly based log reduction system faired well, and it is believed that it would furthermore be difficult to try to circumvent it by injecting similar fields in other fake access requests to try and drive up the frequencies of the interesting fields. This is because such accesses would by their very nature not be successful either as attacks, or as access requests. Taking the result code (i.e. 404) into account when performing the log reduction would eliminate these fake accesses. This would perhaps come at the cost of a decreased detection of *attempted* intrusions. This could be addressed by looking at access requests that were either successful, or unsuccessful, or similar access requests, some of which were successful and some of which were not. As any 'chaff' access requests by their very nature must be unsuccessful there is a difference between them and other access requests (either benign or malicious) that can be put to work for detection.

On a different tack, the cost of going through the unique requests can be amortised. Studying the logs for the months following November, i.e. December through February, they contain on the order of the same number of unique requests in themselves, viewed in isolation. However, amortised there are 200000 previously unseen requests in Nov., 87000 in Dec., 66000 in Jan., and only 40000 in Feb.

A disadvantage with this approach compared to that of automated intrusion detection systems is that the detection is not necessarily real time or near real time. Especially if is decided to visualise the log in batches of one month at a time. Of course, nothing prevents the operator from performing the visualisation more often.

8. FUTURE WORK

The discussion has been limited to the *types* of different attacks seen in the web log (the same request string could emanate from many different sources). No attempt has been made to correlate the actual attacks with each other, or cluster the same attacks originating from different sources to try and identify the entity behind the attacks. Methods to do so (even visual methods [3]) already exist.

It would also be interesting to try the (few) other data sets such as the KDD'99 data set and to see if the structural differences found here carries over to other types of data.

It would be interesting to devise methods of evasion as noted above, and implement the suggested improvements to the log reduction to thwart them. A lack of time and space has prevented us from doing so here.

The intention is also to perform user experiments. These are more difficult than one might at first think, since training on the specific tool often is very effective for the outcome, and the task to be performed is complex and demands some skill. This makes the experiment prohibitively costly.

9. RELATED WORK

The idea of bringing visualisation to intrusion detection was first published in 1998, and a body of work has started to collect.

A small subfield (see e.g. [9, 5]) of anomaly detection and visualisation has arisen through the application of Kohonen maps to intrusion detection which itself is a visual representation of an underlying neural network model. These systems all build some neural network model of network traffic or host data and then present the resulting two dimensional scatter plot to the user. The interpretation of the map is known to be difficult [6].

These approaches all differ from the one presented here in that none perform log reduction similar to the scheme presented in this paper, but rather anomaly based intrusion detection, and that neither paper argue the effect the visual presentation has on the understanding of the security result by the user of the system. Furthermore they all use the raw data as input, not divided into categories (i.e. *accesses* in this case), and in doing so detect instances of attacks, but not necessarily categories of attacks as done here.

A quick survey of the available commercial intrusion detection systems was also made. Only three systems uses any degree of visualisation in our sense of the word. Two are uninteresting from our perspective but the remaining, *CA Network Forensics*: 'http://www.silentrunner.com', uses N-gram clustering followed by a three dimensional visual display of the clusters. On the surface the visual representation of the data in the clusters is similar to the one presented here (i.e. a general 3D network) but while the graphs may look similar they express very different relations. There is no discussion as to the interpretation of these graphs and the underlying structure of the data is not allowed to influence the visualisation.

The main difference between these works and the one presented here is that the one presented here combine anomaly based log reduction techniques with visualisation techniques, while they rely on clustering techniques combined with a visual display of the clusters, i.e. they visualise the result of the anomaly detection method, while the *data* that is singled out by the log reduction method is visualised here.

Theo et. al. [11] visualise communication between pairs of routers on the Internet using the BGP (Border Gateway Protocol) routing protocol. Their choice of problem and visualisation techniques are different from the one presented here, and they do not delve as deeply into the analysis of the security problems they detect (they are not as clearly *security problems*), but they do visualise a greater amount of data more compactly than done here and still manage to detect anomalies in the BGP traffic.

There has not been much research into anomaly detection of web accesses besides that by Kruegel et.al. [7]. They develop (as is done here) ad hoc statistical methods for detecting anomalous request strings. Their model is much more complex than the one presented here, taking many more parameters into account while only one (the element frequency) is taken into account here. As a result—as far as false alarm rates can be compared between a detector and a log reducer—they are rewarded with a false alarm rate about a factor of forty lower than the one reported here (and possibly a detector that's more resistant to evasion attempts). Even so the authors report a problem with handling even this level of false alarms, while the visualisation method presented here enables the user to quickly discard the uninteresting entries.

10. CONCLUSIONS

In summary, the hypothesis that the combination of anomaly based log reduction and visualisation would provide us with the benefits of both approaches while counteracting the drawbacks was supported. Furthermore the anomaly based log reduction system could indeed be very simple and still successfully serve as a front end to the visualisation system. The hypothesis that visualising the structure of the requests strings themselves cut into components would enable the operator to discard benign accesses with relative ease was supported. There was less evidence for the corresponding hypothesis: That one could just as easily identify malicious patterns. A few meta classes of attacks did exhibit features that set them apart from the benign traffic, but the others did not to a significant degree.

The presented method is relatively time efficient, and the operator learns about the usage of the website. Notably unusual but benign (often dynamic) traffic that is more varied and hence more prone to misclassification is studied in more detail.

The work invested in parring down the graph can be amortized over subsequent investigations, where the webserver logs for the following months contain less and less new traffic, and hence can be visualised more quickly, especially if one remembers what accesses were seen previously and why it was decided to discard them as uninteresting.

References

[1] D Anderson, T Frivold, and A Valdes. Next-generation intrusion-detection expert system (NIDES). Technical Report SRI-CSL-95-07, Computer Science Laboratory, SRI International, Menlo Park, CA 94025-3493, USA, May 1995.

[2] Stefan Axelsson. The base-rate fallacy and the difficulty of intrusion detection. *ACM Transactions on Information and System Security (TISSEC)*, 3(3):186–205, 2000.

[3] Stefan Axelsson. Visualization for intrusion detection: Hooking the worm. In *In the proceedings of the 8th European Symposium on Research in Computer Security (ESORICS 2003)*, volume 2808 of *LNCS*, Gjøvik, Norway, 13–15 October 2003. Springer Verlag.

[4] Stuart K. Card, Jock D. MacKinlay, and Ben Shneiderman. *Readings in Information Visualization—Using Vision to Think.* Series in Interactive Technologies. Morgan Kaufmann, Morgan Kaufmann Publishers, 340 Pine Street, Sixth Floor, San Fransisco, CA 94104-3205, USA, first edition, 1999. ISBN 1-55860-533-9.

[5] Luc Girardin and Dominique Brodbeck. A visual approach for monitoring logs. In *In the Proceedings of the 12th Systems Administration Conference (LISA '98)*, pages 299–308, Boston, Massachusetts, USA, 6–11 December 1998. The USENIX Association.

[6] Teuvo Kohonen. *Self-Organizing Maps*, volume 30 of *Springer Series in Information Sciences.* Springer Verlag, Third edition, 2001. ISBN 3–540–67921–9, ISSN 0720–678X.

[7] C. Kruegel and G. Vigna. Anomaly detection of web-based attacks. In *Proceedings of the 10th ACM Conference on Computer and Communication Security (CCS '03)*, pages 251–261, Washington DC, USA, October 2003. ACM Press.

[8] Wenke Lee and Dong Xiang. Information-theoretic measures for anomaly detection. In *IEEE Symposium on Security and Privacy*, Oakland, California, USA, 14–16 May 2001. IEEE.

[9] P. Lichodzijewski, A.N. Zincir-Heywood, and Heywood M.I. Host-based intrusion detection using self-organizing maps. In *In the proceedings of the IEEE International Joint Conference on Neural Networks.* IEEE, May 2002.

[10] Robert Spence. *Information Visualization.* ACM Press Books, Pearson education ltd., Edinburgh Gate, Harlow, Essex CM20 2JE, England, first edition, 2001. ISBN 0-201-59626-1.

[11] Soon Tee Teoh, Kwan-Liu Ma, S. Felix Wu, and Xiaoliang Zhao. Case Study: Interactive Visualization for Internet Security. In *Proceedings of IEEE Visualization 2002*, The Boston Park Plaza hotel, Boston, Massachusetts, USA, 27 October to 1 November 2002. IEEE Computer society.

A LONG-TERM TRIAL OF KEYSTROKE PROFILING USING DIGRAPH, TRIGRAPH AND KEYWORD LATENCIES

Paul S. Dowland and Steven M. Furnell
Network Research Group, School of Computing, Communications and Electronics, University of Plymouth, Drake Circus, Plymouth, PL4 8AA, United Kingdom, info@network-research-group.org

Abstract: A number of previous studies have investigated the use of keystroke analysis as a means of authenticating users' identities at the point of initial login. By contrast, relatively little research has focused upon the potential of applying the technique for identity verification during the logged-in session. Previous work by the authors has determined that keystroke analysis is a viable metric for continuous monitoring, provided that sufficient data is captured to create a reliable profile. This paper presents a series of results from a three-month trial in which profiles were created using digraph, trigraph and keyword-based keystroke latencies. The profiles were based upon a total of over 5 million keystroke samples, collected from 35 participants. The results demonstrate that the techniques offer significant promise as a means of non-intrusive identity verification during keyboard-related activities, with an optimum false acceptance rate of 4.9% being observed at a rate of 0% false rejection.

Key words: Authentication, Misuse Detection

1. INTRODUCTION

Over the last twenty years, the concept of keystroke analysis has been the focus of considerable research as a means of user authentication. The potential for profiling of keypresses was first identified by Gaines et al (1980). Since then, a number of research projects have been conducted to evaluate different methods of data gathering (using a range of operating systems and considering a variety of metrics) and post-processing techniques (ranging from purely statistical to AI/neural network approaches).

To date, however, virtually all published studies have focussed upon looking at the application of static strings, such as username and password pairs using the inter-keystroke digraph latency timing method. From the earliest studies in 1980 (Card et al & Gaines et al), the focus has been on the analysis of digraph latencies. Later studies, such as those by Joyce & Gupta (1990) and Mahar et al (1995) further enhanced the work, identifying additional statistical analysis methods that provided more reliable results.

In Legget et al. (1991), the concept of dynamic keystroke analysis was first proposed, with the introduction of a reference profile that could be used to monitor a live user session. Brown and Rogers (1993) also explored the idea of dynamic analysis, presenting preliminary results.

The authors' previous research (Dowland et al., 2002) described an experiment evaluating keystroke analysis based on inter-keystroke digraph latencies under Windows. This earlier trial concentrated upon the capture and subsequent analysis of digraph latencies using inter-keystroke timings. The trial results demonstrated the viability of this method, but suggested that, to be a reliable authentication measure, user profiles would need to be based upon much larger sample sizes. The previous trial was also based on a limited number of users in order to quickly evaluate the viability of the technique.

This paper presents the results of a long-term trial that was aimed at evaluating a range of techniques using a larger number of participants. This trial captured and evaluated trigraph and keyword latencies, in addition to digraph timings, under the Windows operating system. The paper begins by introducing the technical aspects of the trial conducted over a three month period before considering the statistical approach taken with the data analysis stage. The results are presented and discussed, leading to some overall conclusions, and proposals for future work.

2. CAPTURING KEYSTROKE DATA IN WINDOWS

While keystroke analysis has been investigated (and hence implemented) in previous studies, a GUI environment (e.g. Microsoft Windows) introduces new challenges. In previous published studies, the user has been required to type and interact with a specific application (typing either pre-defined or free-form text). While this approach makes the development of the keystroke monitoring software simple, and maintains the consistency of the test environment, it is not representative of normal typing behaviour as the user becomes focussed upon the task of typing, rather than focussed upon a task that *involves* typing. If the aim is to produce static keystroke analysis for occasional authentication judgements (e.g. supplementing login authentication) then this approach will work well. However, to implement continuous supervision using dynamic keystroke analysis it is necessary to monitor the users' normal behaviour when interacting with their normal applications and operating system environment. Even providing a simulation of these environments may not be sufficient to obtain valid sample data upon which to base a profile.

In order to address this problem, software was developed that would transparently monitor and log all typing activity. The system was designed to allow keystroke data to be collected under the Microsoft Windows XP environment (although the technique is equally applicable in all Windows operating systems). In order to collect the required data, it was necessary to implement a mechanism for acquiring user typing patterns across all applications running within a users' active session. This is important as the experiment was designed to create a profile for each user based upon their typical typing patterns when using their computer (not constrained to a specific application or task). The implementation of the keylogger utilised several key features of the Windows operating system and the underlying chains of messages on which the operating system is built (these are briefly discussed in the following section). The authors have not investigated the applicability of these techniques under other operating systems but it is likely that the same system could be developed under other systems providing access is given to the keypress events at an appropriate level.

Figure 1 illustrates the software architecture used to capture and log keystroke activity under Windows. As keys are pressed, messages are generated by Windows for both key up and down events. These messages are captured through the use of a hook function that redirects messages to a nominated program. The messages are passed from the hook (implemented as a system-wide DLL written in C) to the keylogger (implemented in Visual Basic and deployed as a system tray application). The keylogger functioned completely transparently to the user, requiring no user action to start or stop the logging process. The application was automatically started when the operating system (O/S) booted (run from the Startup program group on the start menu) and shut down automatically when the O/S closed. Gathered data was automatically saved after every 500 digraphs pairs and when the application was closed. To reassure users, an option was included to suspend logging of keystrokes. This was included due to concerns expressed by some users about monitoring of specific inputs – e.g. the typing of on-line banking login details.

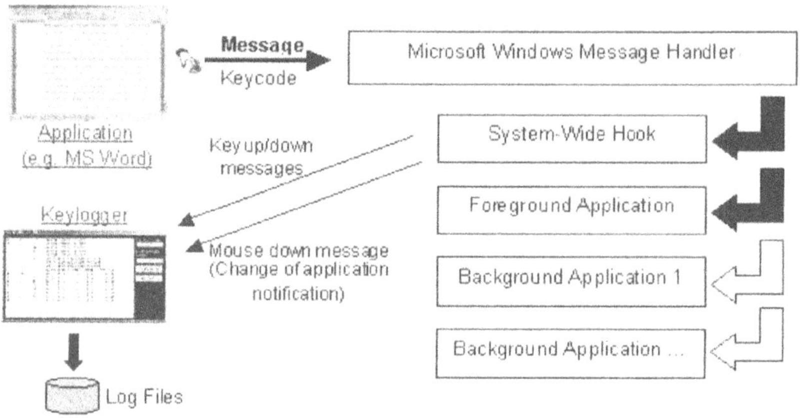

Figure 1. Implementation of keylogger

For each digraph pair logged, the application stored five items of information (Table 1) – these being written to an Access database after every 500 digraphs. This process was also repeated for each trigraph and keyword latency (i.e. trigraphs were stored as three consecutive characters and keywords as a string).

Table 1. Keylogger attributes logged per digraph

Item	Data types
AutoID	Auto-incrementing record number. This is used to maintain the order of the keystrokes typed as the timestamp is only accurate to 1 second.
Left character (C1)	ASCII code representing character
Right character (C2)	ASCII code representing character
Latency	Integer representing inter-keystroke latency in milliseconds
Application	String containing the window title from the foreground application.
Timestamp	A timestamp is added to every keystroke logged for later use.

While digraph and trigraph logging were based upon all keystrokes entered, keyword logging was based on a look up list. The top 200 commonly occurring words in the English language were monitored, and as each word was entered, its latency was recorded.

3. EXPERIMENTAL PROCEDURE

For this experiment a total of 35 users were profiled over a period of three months. Unfortunately several users disabled the keylogger when entering sensitive information and consequently forgot to re-enable it. Despite this, the key-logging trial collected considerable volumes of data with nearly six million samples collected across digraphs, trigraphs and keywords (Table 2).

Table 2. User profile results

User	Mean Digraph Latency (ms)	Typing Skill Classification	Digraphs	Trigraphs	Words
User 1	91	Best	34352	23352	1403
User 2	156	Average (skilled)	53306	36912	2599
User 3	99	Best	156718	107107	6154
User 4	251	Average (non-skilled)	27324	18688	1310
User 5	112	Good	50822	36713	1465
User 6	154	Average (skilled)	50167	34484	1885
User 7	106	Good	78579	54959	4349
User 8	130	Good	50102	35102	2932
User 9	97	Best	37618	24755	1741
User 10	145	Average (skilled)	70337	48942	4643
User 11	147	Average (skilled)	227660	145846	10617
User 12	102	Good	20216	14142	1032
User 13	157	Average (skilled)	65312	43015	1730
User 14	141	Average (skilled)	33639	23090	1784
User 15	139	Good	15951	11159	1068
User 16	150	Average (skilled)	42839	30299	2037
User 17	106	Good	105543	68068	3173
User 18	177	Average (skilled)	89730	59292	3121
User 19	117	Good	103876	71635	4617
User 20	121	Good	78597	53495	4479
User 21	141	Average (skilled)	80626	55881	2807
User 22	110	Good	117365	79534	6557
User 23	131	Good	118805	77013	5682
User 24	89	Best	201260	131954	8517
User 25	203	Average (skilled)	38944	26655	2266
User 26	192	Average (skilled)	48469	33907	2555
User 27	125	Good	33068	23115	1679
User 28	91	Best	70217	47033	2128
User 29	104	Good	88059	55707	3815
User 30	202	Average (skilled)	40741	28789	1007
User 31	86	Best	310823	211419	19726
User 32	93	Best	353867	237274	18056
User 33	144	Average (skilled)	276669	183455	6057
User 34	143	Average (skilled)	124409	87079	953
User 35	130	Good	140044	85413	6240
		Totals	**3,436,054**	**2,305,283**	**150,184**

Before considering the data from each user, the typing skill for each participant was evaluated based on the categorisations proposed by Card et al. (1980) where typists are broadly categorised into one of six categories. The results are presented in Table 2 together with the quantity of samples for each user (shown separately for digraph, trigraph and keywords). The results are weighted towards typists with above average skills due to the nature of the test subjects (i.e. all subjects were regular computer users who spent prolonged periods typing). This was considered acceptable as the likely use for a fully implemented system would be in environments with semi-skilled users (i.e. relatively few unskilled/poor typists).

4. STATISTICAL ANALYSIS

To eliminate extreme short/long digraph latencies that may adversely affect the distribution of digraph times, any digraph pair whose latency fell outside a nominal range was excluded from the log files. For the purpose of this experiment the range was restricted to times above 10ms and below 750ms. In an earlier trial the range was restricted to 40ms – 750ms, with these thresholds based on previous work conducted by Furnell (1995), and were designed to eliminate samples where two keys may have been accidentally struck together (thus, producing an infeasibly small latency) or, where the user may have made a pause in their typing and thus introduced an unnaturally large inter-keystroke latency. Unfortunately, the low pass filter was responsible for substantial quantities of data being removed from the user profiles and, as such, was reduced to 10ms for the purposes of this trial. If a digraph was removed due to the filtering, this also reset the trigraph and keyword logging so no further thresholds were needed for these two measures.

Following the initial filtering, the experimental data for each user was processed off-line to calculate the mean and standard deviation values for each unique digraph, trigraph or keyword. In the event that any profiled sample had a standard deviation greater than its mean value, the samples were sorted and the top/bottom 10% was then removed, followed by subsequent re-calculation of the mean and standard deviation values. The reason for this additional step was to remove samples where the latencies would have an adverse affect on the standard deviation (i.e. the distribution of samples was tightened).

Once all the user profiles were calculated, another application (the data comparator) was used to generate tables of results for each of the methods. The data comparator (Figure 2) was based on the original analyser developed in the previous trial (Dowland et al., 2002). A small number of additional features were introduced to the comparator to cater for the inclusion of trigraphs and keyword profiles.

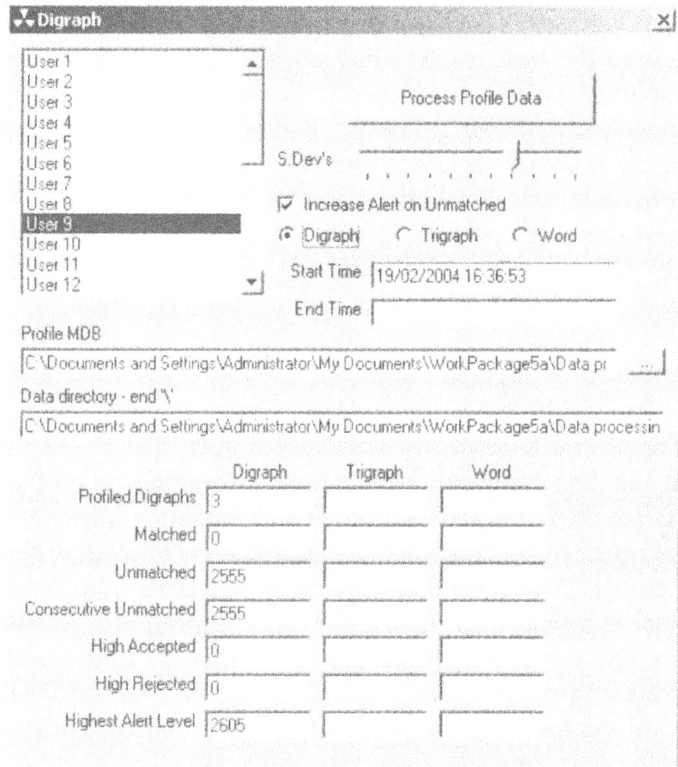

Figure 2. Data comparator (running)

In the previous trial, when a digraph was processed that did not exist in the reference profile, the alert level remained static (simply increasing the count of unmatched digraphs). This trial considered the role of unmatched samples as they are a potential indicator of impostor activity. I.e. if a user types a specific sample infrequently (to the extent that there is insufficient data on which to base a profile), it is reasonable to assume that these occurrences are un-representative of that user's normal typing behaviour. By default, in this trial, an unmatched sample increased the alert level by one, whilst a matched accepted/rejected sample varied the alert level by two accordingly. This behaviour can be adjusted by selecting the checkbox in the comparator – once unchecked; the alert level was not affected by unmatched samples.

Before starting the full profile comparisons a trial comparison was conducted based on a random selection of five users in order to determine the optimum settings for the deviation threshold. In the previous study the deviation settings were chosen from a range of 0.5, 1.0, 1.5 and 2.0 standard deviations with the best results obtained at 0.5. In order to determine an optimum setting, profile comparisons were made between 0.5 and 1.0 standard deviations (values below 0.5 had already been assessed in earlier trials). For the randomly selected users the best results were obtained at 0.7 with an increase in alert level above and below this threshold. As such, the later comparisons were performed with standard deviations settings of 0.6, 0.7 and 0.8. The permitted deviation was determined by the slider control that selects the number of standard deviations from the mean.

digraph mean ± (digraph standard deviation * permitted deviation)

Once the profile comparison was started each users' reference profile was loaded and then compared against the raw keylogger data files for all 35 users. This resulted in a table of 35 sets of statistics for each user. This process was repeated for trigraphs and keywords with three different profile deviation settings (0.6, 0.7 and 0.8 standard deviations from the mean). NB a setting of 0.5 standard deviations was introduced to the trigraph comparisons due to poor performance at 0.6 and 0.7 and unmatched alert increases were optionally applied to digraphs and trigraphs (hence doubling the number of comparisons for these metrics). With an average of nearly 100,000 samples per data file, each data comparison took approximately two hours with a total of 17 comparisons conducted – six for digraph, eight for trigraph and three for keywords (see Table 3).

Table 3. Profile comparison settings

Metric	Standard Deviations (S.D.)
Digraphs	0.6, 0.7, 0.8 S.D.
Trigraphs	0.5, 0.6, 0.7, 0.8 S.D. *0.5 added due to poo performance at 0.6 and 0.7*
Keyword	0.6, 0.7, 0.8 S.D.

Once the profile comparison was completed, the results were exported and a number of functions were used to derive 2-dimensional tables of data from the raw results from the comparator from which the FAR/FRR figures could be derived.

Following the basic analysis described in this section, a further modification was made to the comparator to determine how many keystrokes were needed before either the valid user was challenged or an impostor detected. The threshold for this challenge was based upon the best performance thresholds from the earlier trials and was initially set at an alert level of 70. The results from this trial using the digraph keylogger files at a threshold of 0.7 standard deviations is presented in Table 4. The results from this trial were somewhat variable, while some users had good results (e.g. user 7, 10 and 26), most user profiles had only moderately successful results. If we consider user 2, while 29/34 (85%) impostors were challenged in less than 100 digraphs, user 16 (when acting as an impostor against user 2's reference profile) was able to type over 40,000 digraphs before being challenged.

The results in Table 4 can also be considered in terms of the average number of keystrokes required before a challenge is issued. The results show that an average of 6,390 digraphs were accepted before an impostor was challenged compared with an average of 68,755 digraphs before the valid user was challenged. While these results seem to provide the appropriate differentiation between impostor and valid user, giving an impostor the opportunity to type over 6,000 digraphs presents a major security risk.

Table 4. Number of keypresses before a challenge

0.7 SD Reference Profiles

	User 1	User 2	User 3	User 4	User 5	User 6	User 7	User 8	User 9	User 10	User 11	User 12	User 13	User 14	User 15	User 16	User 17
User 1	56	21	72	22	22	22	50	24	24	38	31	23	23	22	21	21	66
User 2	29	53306	23	18	26	280	65	29	28	320	53306	26	53306	57	54	50	125
User 3	39	15	156718	19	26	22	20	30	20	14	28	19	17	21	21	20	127
User 4	22	68	15	300	15	37	16	23	19	55	22	27	53	17	22	32	22
User 5	25	31	34	19	73	31	31	30	35	22	22	26	29	32	27	29	145
User 6	18	54	26	22	70	141	43	19	23	60	347	25	156	22	23	32	44
User 7	43	101	372	27	145	108	78579	54	50	36	78579	55	247	42	38	39	78579
User 8	21	30	27	26	25	39	29	225	27	30	257	25	96	34	26	24	72
User 9	46	15	116	19	45	28	40	64	105	19	19	19	27	19	24	20	125
User 10	23	22	20	18	21	17	25	39	18	70337	83	21	26	32	21	17	41
User 11	20	84	25	17	19	88	15	63	16	144	227660	43	117	26	21	35	22
User 12	26	103	42	33	21	102	40	79	22	38	180	94	120	30	21	46	30
User 13	24	31	24	33	27	41	23	46	15	30	265	33	107	36	27	28	52
User 14	36	33	19	16	35	34	64	64	40	192	33639	20	49	78	32	31	217
User 15	17	44	12	20	23	29	23	15	17	17	15951	19	15	18	23	17	28
User 16	19	42839	17	21	18	37	17	46	26	1085	42	43	42839	39	40	60	22
User 17	33	21	85	24	51	35	89	83	53	17	56	23	49	45	23	21	105543
User 18	32	28	51	25	23	30	47	17	15	28	144	32	46	33	23	46	51
User 19	23	36	121	21	31	32	59	50	26	44	372	20	83	51	24	48	104
User 20	21	67	54	20	34	64	62	55	48	541	78597	21	102	81	36	28	66
User 21	19	2555	38	22	39	206	106	129	25	149	80626	48	2850	89	22	648	84
User 22	28	32	1160	18	29	90	29	28	24	30	343	23	27	30	19	36	23125
User 23	20	31	14	20	16	20	14	21	20	29	30	28	18	30	25	16	21
User 24	45	18	602	14	34	48	45	86	33	22	39	19	52	26	28	20	90
User 25	19	14	13	24	16	18	13	17	15	22	47	22	14	26	19	17	13
User 26	16	16923	15	38	37	162	20	56	22	58	18527	84	902	31	26	489	38
User 27	27	22	110	17	38	33	40	41	25	29	55	26	69	22	22	20	100
User 28	27	20	70217	18	28	17	51	17	29	19	18	22	16	22	22	21	18
User 29	27	32	58	20	27	24	33	33	17	33	35	20	24	24	20	24	44
User 30	21	42	38	23	27	44	34	53	20	50	43	31	50	37	20	49	34
User 31	43	26	310823	16	61	34	66	25	39	22	14	20	26	17	22	21	310823
User 32	23	13	52	17	36	35	53	32	25	14	273	21	23	23	25	13	112
User 33	39	25	39	23	26	34	32	81	20	25	27	23	26	28	20	33	76
User 34	28	78	27	19	81	95	52	26	43	64	923	22	96	30	33	32	68
User 35	21	25	20	22	16	19	41	18	24	40	87	27	20	31	21	17	73

	User 18	User 19	User 20	User 21	User 22	User 23	User 24	User 25	User 26	User 27	User 28	User 29	User 30	User 31	User 32	User 33	User 34	User 35
User 1	33	43	22	19	94	116	52	18	21	40	41	52	21	16	93	135	18	106
User 2	53306	79	321	64	27	127	23	49	16	27	24	54	120	27	19	53306	53306	524
User 3	14	28	15	14	50	98	79	15	19	23	79	29	18	49	325	22	23	25
User 4	178	18	19	29	18	63	15	80	86	25	16	16	120	15	18	34	53	20
User 5	17	20	29	21	141	50822	51	20	31	34	24	73	29	74	124	50822	30	34
User 6	3435	77	104	63	58	76	17	26	26	39	17	40	37	30	43	50167	209	145
User 7	90	78579	78579	28	78579	78579	2189	16	25	58	205	408	68	78579	78579	78579	280	78579
User 8	63	128	24	29	30	126	24	34	26	18	77	94	22	27	107	50102	93	489
User 9	20	27	34	15	296	11107	110	16	16	21	252	250	17	400	316	26	34	14
User 10	40	48	41	28	49	29	20	22	22	49	17	21	16	34	26	70337	44	101
User 11	113	11	63	65	61	725	26	51	48	19	12	21	60	21	17	227660	145	227660
User 12	20216	64	59	82	46	83	21	25	18	41	28	32	200	26	49	177	103	48
User 13	86	28	40	51	97	182	52	20	33	43	23	25	53	15	23	65312	40	172
User 14	51	91	150	81	139	1162	59	17	15	63	35	169	71	57	38	541	38	33639
User 15	15951	20	46	51	18	15951	20	14	34	32	23	57	62	14	14	15951	62	15951
User 16	42839	21	23	35	21	1111	15	144	61	42	15	19	642	15	15	42839	42839	31
User 17	59	51	106	64	57	209	38	19	20	33	29	55	22	100	214	105543	38	1954
User 18	166	44	33	46	44	43	21	34	23	22	24	67	33	45	36	191	236	51
User 19	45	103876	172	33	103876	1606	51	30	22	47	56	1229	30	101	91	157	60	103876
User 20	65	35	78597	73391	131	121	47	32	32	63	34	48	26	49	44	78597	132	78597
User 21	3444	199	211	80626	131	120	51	52	69	77	38	104	370	78	105	80626	80626	80626
User 22	30	315	70	32	117365	117365	51936	19	19	30	37	3608	19	70454	190	117365	91	117365
User 23	42	30	24	24	19	23	24	29	24	22	13	25	30	32	30	30	31	19
User 24	18	29	15	18	347	201260	201260	14	16	26	135	29	20	58	201260	153	24	201260
User 25	15	13	13	13	13	13	13	39	41	19	14	14	49	13	13	44	19	15
User 26	48469	65	158	417	56	146	21	189	48469	30	16	56	2127	17	17	48469	48469	271
User 27	53	19	41	37	33068	33068	111	13	36	69	20	98	37	2079	85	33068	40	33068
User 28	20	18	19	19	15	95	426	18	18	22	40	33	17	102	23	62	15	55
User 29	32	17	18	17	42	68	27	16	16	18	32	39	27	49	35	58	29	129
User 30	56	37	35	43	34	36	27	47	31	34	37	37	71	23	41	92	50	45
User 31	37	106	83	20	310823	310823	310823	17	19	38	170	3E+05	23	310823	310823	96	59	310823
User 32	39	59	40	13	67	404	32	18	15	32	52	267	31	32	353867	349	14	353867
User 33	22	22	26	26	87	122	99	17	32	23	25	76	26	74	84	276669	26	26
User 34	120	54	65	148	88	664	49	38	26	40	39	55	39	90	91	954	990	819
User 35	57	41	33	33	128	126	52	16	20	24	34	31	54	52	47	59	51	140044

For this trial the False Rejection Rate (FRR) was fixed at 0% (i.e. the valid user would not be rejected by the system). The False Acceptance Rates (FAR's) were then calculated for each user at the deviation thresholds specified in Table 3, and are shown in Table 5.

Table 5. Results from single-metric measures

	Digraph FAR						Trigraph FAR								Keyword FAR			
Standard Deviation	**0.6**		**0.7**		**0.8**		**0.5**		**0.6**		**0.7**		**0.8**		**0.5**	**0.6**	**0.7**	**0.8**
Unmatched Alert	**N**	**Y**	**N**	**Y**	**N**	**Y**	**N**	**Y**	**N**	**Y**	**N**	**Y**	**N**	**Y**	**N**	**N**	**N**	**N**
User 1	2.9	2.9	2.9	8.6	2.9	8.6	8.6	17.1	8.6	8.6	8.6	25.7	8.6	31.4	97.1	97.1	91.4	91.4
User 2	0.0	2.9	2.9	5.7	2.9	11.4	34.3	0.0	34.3	34.3	34.3	5.7	31.4	8.6	2.9	2.9	2.9	5.7
User 3	0.0	0.0	2.9	5.7	5.7	11.4	74.3	5.7	62.9	62.9	54.3	20.0	40.0	25.7	0.0	8.6	20.0	28.6
User 4	0.0	0.0	0.0	0.0	0.0	2.9	5.7	0.0	5.7	5.7	5.7	0.0	5.7	0.0	5.7	0.0	0.0	0.0
User 5	0.0	0.0	0.0	8.6	0.0	11.4	28.6	2.9	28.6	28.6	28.6	14.3	28.6	22.9	8.6	11.4	14.3	20.0
User 6	0.0	2.9	0.0	8.6	0.0	22.9	28.6	2.9	28.6	28.6	28.6	5.7	25.7	11.4	34.3	11.4	2.9	8.6
User 7	0.0	0.0	0.0	0.0	0.0	2.9	34.3	0.0	28.6	28.6	25.7	0.0	20.0	0.0	0.0	0.0	0.0	0.0
User 8	11.4	5.7	2.9	5.7	2.9	40.0	28.6	17.1	28.6	28.6	25.7	28.6	25.7	37.1	0.0	0.0	2.9	8.6
User 9	2.9	2.9	2.9	0.0	0.0	14.3	17.1	5.7	17.1	17.1	17.1	11.4	14.3	20.0	22.9	25.7	34.3	34.3
User 10	0.0	0.0	0.0	8.6	2.9	14.3	37.1	0.0	31.4	31.4	28.6	2.9	22.9	8.6	2.9	0.0	0.0	0.0
User 11	0.0	0.0	2.9	14.3	17.1	22.9	82.9	2.9	65.7	65.7	51.4	0.0	11.4	11.4	37.1	5.7	2.9	28.6
User 12	0.0	5.7	0.0	5.7	0.0	14.3	2.9	14.3	2.9	2.9	2.9	22.9	2.9	28.6	22.9	25.7	31.4	40.0
User 13	8.6	11.4	5.7	11.4	14.3	11.4	40.0	2.9	40.0	40.0	37.1	11.4	37.1	11.4	8.6	8.6	11.4	14.3
User 14	2.9	2.9	2.9	8.6	0.0	34.3	8.6	8.6	8.6	8.6	8.6	22.9	5.7	28.6	0.0	2.9	5.7	5.7
User 15	0.0	0.0	0.0	2.9	0.0	8.6	0.0	14.3	0.0	0.0	0.0	17.1	0.0	20.0	22.9	17.1	25.7	25.7
User 16	0.0	5.7	0.0	2.9	0.0	5.7	22.9	0.0	20.0	20.0	14.3	2.9	14.3	5.7	0.0	0.0	2.9	14.3
User 17	0.0	0.0	2.9	11.4	14.3	25.7	48.6	2.9	48.6	48.6	45.7	8.6	40.0	11.4	5.7	5.7	5.7	11.4
User 18	8.6	11.4	17.1	8.6	17.1	25.7	48.6	17.1	42.9	42.9	40.0	0.0	37.1	5.7	28.6	8.6	11.4	17.1
User 19	0.0	2.9	2.9	5.7	2.9	8.6	65.7	0.0	62.9	62.9	62.9	8.6	62.9	11.4	5.7	14.3	20.0	17.1
User 20	0.0	0.0	0.0	5.7	5.7	22.9	42.9	0.0	40.0	40.0	31.4	5.7	28.6	8.6	0.0	2.9	2.9	5.7
User 21	0.0	0.0	0.0	0.0	2.9	2.9	40.0	0.0	28.6	28.6	17.1	2.9	5.7	2.9	0.0	2.9	5.7	8.6
User 22	2.9	8.6	5.7	8.6	8.6	25.7	65.7	0.0	57.1	57.1	45.7	5.7	31.4	17.1	0.0	5.7	5.7	11.4
User 23	68.6	54.3	51.4	62.9	60.0	80.0	71.4	45.7	71.4	71.4	68.6	34.3	68.6	45.7	37.1	37.1	45.7	45.7
User 24	2.9	0.0	2.9	5.7	5.7	8.6	62.9	2.9	54.3	54.3	25.7	14.3	14.3	28.6	0.0	2.9	8.6	14.3
User 25	0.0	2.9	0.0	2.9	0.0	14.3	14.3	0.0	8.6	8.6	8.6	2.9	8.6	0.0	2.9	0.0	0.0	0.0
User 26	0.0	0.0	0.0	0.0	0.0	5.7	22.9	0.0	22.9	22.9	20.0	0.0	17.1	0.0	5.7	0.0	0.0	5.7
User 27	2.9	2.9	2.9	17.1	2.9	48.6	11.4	11.4	11.4	11.4	11.4	28.6	11.4	45.7	8.6	2.9	5.7	5.7
User 28	2.9	2.9	2.9	5.7	2.9	17.1	42.9	8.6	42.9	42.9	42.9	17.1	42.9	14.3	8.6	17.1	28.6	37.1
User 29	5.7	5.7	2.9	22.9	20.0	48.6	48.6	11.4	42.9	42.9	42.9	20.0	42.9	22.9	11.4	8.6	14.3	31.4
User 30	14.3	8.6	14.3	14.3	11.4	20.0	22.9	20.0	22.9	22.9	22.9	22.9	20.0	37.1	45.7	28.6	28.6	8.6
User 31	0.0	0.0	2.9	2.9	5.7	8.6	0.0	0.0	0.0	0.0	0.0	2.9	0.0	5.7	0.0	0.0	2.9	5.7
User 32	0.0	0.0	2.9	8.6	11.4	20.0	54.3	0.0	0.0	0.0	0.0	8.6	0.0	14.3	2.9	5.7	14.3	14.3
User 33	17.1	14.3	17.1	14.3	22.9	28.6	88.6	77.1	82.9	82.9	65.7	54.3	57.1	71.4	80.0	68.6	20.0	28.6
User 34	5.7	2.9	11.4	14.3	14.3	14.3	60.0	0.0	48.6	48.6	42.9	0.0	37.1	2.9	97.1	97.1	97.1	97.1
User 35	22.9	14.3	5.7	34.3	25.7	54.3	71.4	25.7	65.7	65.7	65.7	25.7	62.9	45.7	34.3	5.7	11.4	14.3
Average	**5.2**	**5.0**	**4.9**	**9.8**	**8.1**	**20.5**	**38.2**	**9.1**	**33.3**	**33.3**	**29.5**	**13.0**	**25.2**	**18.9**	**18.3**	**15.2**	**16.5**	**20.2**

When the results were calculated, the False Acceptance Rates per user were averaged across all users to provide an average FAR for each metric. The averaged results for the statistical approach are shown in Table 6. It should be noted that the keyword latencies did not use the unmatched alert increase due to the use of a word list/dictionary – i.e. many words would not be matched in the users' profile.

Table 6. Final statistical results (best results highlighted)

Metric	S.D.	Unmatched Alert	FAR
Digraphs	0.6	No	5.2%
	0.6	Yes	5.0%
	0.7	No	4.9%
	0.7	Yes	9.8%
	0.8	No	8.1%
	0.8	Yes	20.5%
Trigraphs	0.5	No	38.2%
	0.5	Yes	9.1%
	0.6	No	33.3%
	0.6	Yes	33.3%
	0.7	No	29.5%
	0.7	Yes	13.0%
	0.8	No	25.2%
	0.8	Yes	18.9%
Words	0.5	No	18.3%
	0.6	No	15.2%
	0.7	No	16.5%
	0.8	No	20.2%

5. DISCUSSION

While the results shown in Table 6 show some encouraging FAR levels there is still significant variation with the best results obtained at 0.7 standard deviations for digraphs, 0.5 standard deviations for trigraphs (with increased alert levels for unmatched digraphs) and 0.6 for keywords. However, when the full results are considered (as shown in Table 5), even at the optimum settings, certain users show high FAR levels (e.g. user 23's profile returned FAR levels of 51.4%, 45.7% and 37.1% respectively for digraph, trigraph and keywords at the average optimum settings). It can also be clearly observed that the results for trigraphs and keywords are significantly worse when compared with those for digraphs – this is most likely to be related to the number of underlying samples used for these techniques (i.e. the number of sampled digraphs were significantly higher than that for trigraphs and keywords, with a corresponding increase of samples per digraph). It is probable that over a longer period of time, the profiles could be refined for trigraphs and keywords to produce a more distinct user profile with a corresponding reduction in the FAR.

These results also demonstrate that the techniques can be very effective for some users while very ineffective for others. For example, when considering digraph FAR's at 0.6 standard deviations (where 0% FAR was actually experienced for 19 out of the 35 users – 54.3%) the average FAR (5.2%) has been heavily influenced by a single user (user 23) whose 68.6% FAR dramatically increases the average. In a full implementation, the authors propose that the use of keystroke analysis should only form a part of a comprehensive user monitoring system. As such, a users' typing would only be monitored if the method was shown to be a discriminating authentication technique for that user. The removal of user 23 from the results in Table 5 significantly affects the average FAR's presented in Table 6, reducing the best digraph results from 4.9% to 3.5% , trigraph results from 9.1% to 8.0% and keywords from 15.2% to 14.5%.

Further optimisation can be achieved by removing the worst 5 participants (15%) from the trial results. This provides a significant improvement in the results of the technique with average FAR's as low as 1.7% for digraphs, 4.4% for trigraphs and 12.8% for keywords (Table 7). While the keyword FAR in particular remains unacceptably high, a reference back to Table 5 reveals that there were still almost a third of users for whom 0% FAR was observed at the 0.5 standard deviation threshold. This suggests a clear potential for using the technique in a subset of cases – which could also increase if additional keyword typing samples were obtained to support the profiling.

Table 7. Optimised results

Metric	S.D.	Unmatched Alert	FAR
Digraphs	0.6	No	1.7%
	0.6	Yes	2.4%
	0.7	No	2.2%
	0.7	Yes	7.0%
	0.8	No	4.9%
	0.8	Yes	17.0%
Trigraphs	0.5	No	34.5%
	0.5	Yes	4.4%
	0.6	No	29.3%
	0.6	Yes	29.3%
	0.7	No	25.6%
	0.7	Yes	10.6%
	0.8	No	21.2%
	0.8	Yes	15.2%
Words	0.5	No	13.8%
	0.6	No	12.8%
	0.7	No	15.3%
	0.8	No	19.7%

The removal of a number of specific user accounts from the keystroke monitoring process is not an ideal solution to the problem of poor user authentication. Keystroke analysis is unlikely to be used as a sole-method of user authentication, instead, it is envisaged that the methods described in this paper would form a part of a larger authentication system and would be only one of a range of authentication metrics that each user could be monitored with. With a larger number of users (and hence a wider range of user typing abilities and corresponding authentication rates) there is likely to be a proportional increase in the number of users for whom keystroke analysis does not produce appropriate FAR/FRR rates – in these cases other, more appropriate, techniques would have to be used. Identifying the cause of poor user performance when using keystroke analysis is vital; on-going work within the authors' research group will conduct further analysis on the gathered data sets to try to determine the cause of the variation between users and identify common factors (e.g. users' typing abilities, differences between application usage etc.).

6. CONCLUSIONS

It is clear from the results presented in this paper that there is considerable potential for continuous user authentication based on keystroke analysis. The long-term sampling of digraph keystrokes has served to reinforce the validity of the technique, while the introduction of trigraph and keyword monitoring has provided additional metrics that can be used as alternative (or complimentary) techniques. In particular, the use of keyword monitoring has considerable potential when used to monitor for specific, high-risk typed words (e.g. delete, format etc.).

It is also clear that the simple statistical approach does not provide sufficient distinction for all users and a live implementation would have to consider which metric (if any) is most appropriate for each user. It is envisaged that keystroke analysis would become only one of a number of monitoring characteristics used by a more comprehensive system with other authentication and supervision techniques.

Future work will also consider how the individual keystroke metrics can be combined together. For example, by combining the confidence measures of multiple metrics (e.g. monitoring digraphs and trigraphs), coupled with monitoring specific keywords (e.g. the typing patterns for high-risk words – format, delete etc.), it may be possible to provide a higher level of confidence in the authentication of the user. The potential for this method will be considered in a later paper.

7. ACKNOWLEDGMENTS

The authors would like to thank the trial participants who assisted in the collection of the keystroke data. In particular, we would like to acknowledge the assistance of the staff and researchers of the Network Research Group, the Department of Psychology and the staff of TMA Global and John Nicholls Builders.

8. REFERENCES

Brown M. and Rogers S.J., 1993, "User identification via keystroke characteristics of typed names using neural networks", *International Journal of Man-Machine Studies*, vol. 39, pp999-1014.

Card S.K., Moran T.P. & Newell A., 1980. "Computer text-editing: An information-processing analysis of a routine cognitive skill", *Cognitive Psychology*, vol. 12, pp32-74.

Dowland P.S., Furnell S.M. & Papadaki M., 2002, "Keystroke Analysis as a Method of Advanced User Authentication and Response", *Proceedings of IFIP/SEC 2002 - 17th International Conference on Information Security*, Cairo, Egypt, 7-9 May, pp215-226.

Furnell, S.M., 1995, "Data Security in European Healthcare Information Systems", PhD Thesis, University of Plymouth, UK.

Gaines R., Lisowski W., Press S. and Shapiro N., 1980, "Authentication by Keystroke Timing: some preliminary results", *Rand Report R-256-NSF*, Rand Corporation.

Joyce R. and Gupta G. 1990. "Identity authentication based on keystroke latencies", *Communications of the ACM*, vol. 33, no. 2, pp168-176.

Legett J., Williams G., Usnick M. and Longnecker M., 1991, "Dynamic identity verification via keystroke characteristics", *International Journal of Man-machine Studies*, vol. 35, pp859-870.

Mahar D., Napier R., Wagner M., Laverty W., Henderson R.D. and Hiron M., 1995, "Optimizing digraph-latency based biometric typist verification systems: inter and intra typist differences in digraph latency distributions", *International Journal of Human-Computer Studies*, vol. 43, pp579-592.

TRUSTED COMPUTING, TRUSTED THIRD PARTIES, AND VERIFIED COMMUNICATIONS

Martín Abadi
University of California at Santa Cruz

Abstract Trusted Computing gives rise to a new supply of trusted third parties on which distributed systems can potentially rely. They are the secure system components (hardware and software) built into nodes with Trusted Computing capabilities. These trusted third parties may be used for supporting communications in distributed systems. In particular, a trusted third party can check and certify the data sent from a node A to a node B, so that B can have some confidence in the properties of the data despite A's possible incompetence or malice. We present and explore this application of Trusted Computing, both in general and in specific instantiations.

1. INTRODUCTION

Trusted third parties can be useful in a variety of tasks in distributed systems. For instance, certification authorities are helpful in associating public keys with the names of users and other principals; in multi-player games, servers can contribute to preventing some forms of cheating; and smart-cards with limited resources may rely on trusted, off-card servers for verifying downloaded bytecode class files. Unfortunately, resorting to trusted third parties is not always practical, as it typically results in deployment difficulties, communication overhead, and other costs. Moreover, well-founded trust is scarce in large-scale distributed systems, and so are reliable trusted third parties.

This paper considers new trusted third parties that may appear in general-purpose computing platforms as a result of several current efforts. Those efforts include substantial projects in industry, such as the work of the former Trusted Computing Platform Alliance (TCPA) and its successor the Trusted Computing Group (TCG), and Microsoft's Next Generation Secure Computing Base (NGSCB, formerly known as Palladium) [England et al., 2003]. They also include research projects such as XOM [Lie et al., 2000] and Terra [Garfinkel et al., 2003]. The trusted third parties are the secure system components (hardware and software) built into nodes with Trusted Computing capabilities.

These trusted third parties can contribute to both secrecy and integrity properties in distributed systems. In particular, when two nodes A and B communicate, the trusted third party embedded in A can check and certify the messages that A sends to B. This verification may have a variety of meanings—it can for example ensure the well-formedness of data fields, the absence of known viruses, the safety of mobile code, or the validity of certificate chains. The verification can offer security guarantees to B, often more efficiently than if B performed the check itself. Although the verification clearly depends on A's secure system components, it is protected against malfunctions in the rest of A, and can prevent their spread to B. The description and study of this scenario are the main contents of this paper.

The next section discusses efforts such as TCPA, the appearance of new trusted third parties, and (briefly) the applications that they may enable. Section 3 sets out our assumptions. Section 4 explains the use of a trusted third party for verified communications. Section 5 considers some examples, and section 6 summarizes benefits and drawbacks. Section 7 develops an example. Section 8 discusses extensions in which data is partly secret or generated by the trusted third party. Section 9 concludes. An extended version of this paper contains additional details and outlines more general mechanisms for verified communications, relying on machinery for remote invocation and on extensible runtimes.

2. NEW TRUSTED THIRD PARTIES?

Next we identify more precisely the new third parties described in the introduction, and consider whether they should be trusted. We also discuss the applications (some old, some new) that may rely on this trust.

2.1 The new third party

With systems such as NGSCB, a computing platform includes a protected execution environment, with protected memory, storage, and I/O. The platform is open in that it can run arbitrary programs like today's ordinary PCs, but those arbitrary programs should not compromise the security kernel or any subsystem under its protection. Moreover, the security kernel can authenticate the programs, and it in turn can be remotely authenticated.

Therefore, the security kernel may serve as a trusted third party for an interaction in a distributed system. Conveniently, this trusted third party is local to a node. In particular, the security kernel may assist a remote principal in interactions with the rest of the node, which may be arbitrarily corrupted. Moreover, the security kernel may communicate directly with a local human user, through

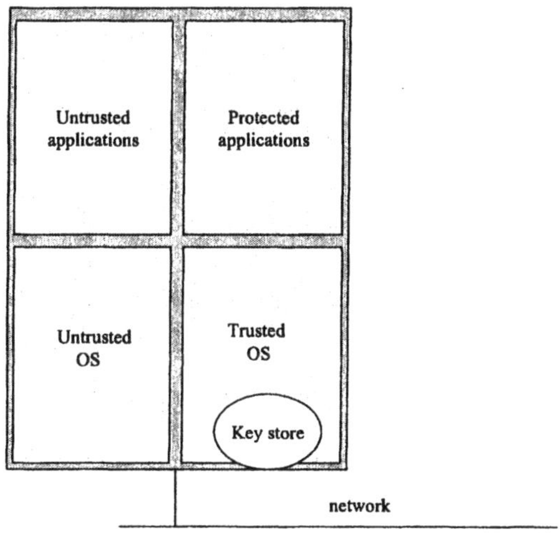

Figure 1. A typical picture of a system with NGSCB

secure I/O; it may therefore assist the user in its interactions with the rest of the node.

A subsystem protected by the security kernel may also play the role of trusted third party. Through standard delegation techniques (e.g., [Lampson et al., 1992]), the protected subsystem can act on behalf of the security kernel and its clients. The main advantage of relying on a protected subsystem is to retain, to the extent possible, the simplicity, manageability, and security of the kernel proper.

Figure 1 is a typical picture of a system with NGSCB. It shows a system with two sides: a left-hand side with arbitrary software (not necessarily trusted) and a right-hand side with secure system components, including an operating system and user-mode code.

2.2 Applications

This trusted third party can contribute to security in distributed systems, in several ways. The trusted third party can contribute to secrecy properties, for example holding secrets for a user, and presenting those secrets only to appropriate remote servers. The secrets would be kept from viruses that may come with arbitrary programs. The trusted third party can also contribute to integrity properties, for example checking incoming and outgoing data. In particular, as suggested in the introduction and explained in section 4, the trusted third party embedded in a node A can check and certify the messages that A sends to another node B. The trusted third party can protect B against A's incompetence or malice, for example against A's viruses. While the secrecy properties have received a fair amount of attention, we believe that the opportunities and problems related to integrity are also important. They are the focus of this paper. One may wonder also about availability properties—for example, asking whether the trusted third party can help protect against denial-of-service attacks. We address availability only indirectly (see section 6).

Trusted Computing is often narrowly associated with protecting movies and other proprietary content on commodity platforms, but it enables other significant applications. Several of those applications remain in the broad realm of Digital Rights Management (DRM). For instance, users may want to attach rights restrictions to their e-mail messages and documents; protected execution environments can help in enforcing those restrictions. Similarly, however, it has been argued that protected execution environments enable censorship and other worrisome applications [Anderson, 2003b]. Beyond DRM, NGSCB could be employed for secure document signing and transaction authorization [England et al., 2003], for instance. Notwithstanding such intriguing ideas, it appears that the thinking about applications remains active, and far from complete. One of the goals of this paper is to contribute to this thinking.

2.3 Limits on trust

TCPA, TCG, and NGSCB have been rather controversial. While they are associated with the phrases "Trusted Computing" or "Trustworthy Computing", they have also been called "Treacherous Computing" [Stallman, 2002]. Relying on them in the manner described in this paper will perhaps be considered naive. Even putting aside any consideration of treachery, trust should not be absolute, but relative to a set of properties or actions, and it is dangerous to confuse trusted and trustworthy.

Following Anderson [Anderson, 2003a], we mostly use an acronym rather than "Trusted Computing" or a similar name. We pick SCB, which may stand for "Secure Computing Base" (or "Sneaky Computing Base") because the descriptions in this paper focus on NGSCB, as we explain in section 3. By an SCB we loosely mean a collection of system components, hardware and software, including a security coprocessor with cryptographic keys and capabilities, a microkernel or other operating system, and possibly some protected subsystems running on top of these. Section 3 lists our assumptions more specifically.

The trust that one places on an SCB may be partly based on the properties of its hardware. If this hardware is easy to subvert, then assurances by the SCB may be worthless. On the other hand, a modest level of tamper-resistance may be both achievable and sufficient for many applications. First, attacks on hardware (unlike buffer-overflow attacks, for instance) are not in general subject to large-scale automation. Moreover, many nodes (and their SCBs) are in physical environments in which serious tampering is hard or would be easily detected—for example, in shared workspaces and data centers. In other environments, a key question is whether the people who are in a position to perform the tampering would benefit from it. Whenever the SCB works on behalf of users, defending them from viruses and other software attacks, we may not need to worry about protecting the SCB from the users.

Trust in an SCB may also be partly based on trust in its developer, its administrators, and other principals. For instance, if Acme makes chips with embedded secret keys, and issues certificates for the corresponding public keys, then the chips are reasonable trusted third parties only if Acme can be trusted to manage the secret keys appropriately. Thus, Acme is a trusted third party too. However, trust in Acme may be based on an open review, and may be further justified if Acme never has direct access to the secret keys.

On this basis, it seems reasonable or at least plausible that SCBs would be trusted third parties—and even trustworthy third parties—in specific contexts.

3. ASSUMPTIONS

We focus on NGSCB partly because of its practical importance, partly for the sake of concreteness, but most of the paper applies verbatim to other systems such as XOM; it may also apply to future versions of these systems, which continue to evolve. This section presents the main assumptions on which we rely.

We expect that the SCB in a system is able to communicate with other parts of the system, typically at a modest cost; in particular, this communication may be through local memory. In addition, we make the following assumptions:

- Authenticity: The capability of making assertions that can be verified by others (local or remote) as coming from this SCB, or from an SCB in a particular group. For instance, in a common design, the SCB holds a signature key that it can use for signing statements; a certification authority (perhaps operated by the SCB's manufacturer, owner, or a delegate) issues certificates for the corresponding public key, associating the public key with this SCB or with a group of trusted SCBs.

- Protection: Protection from interference from the rest of the system when performing local computations.

Two additional assumptions are not essential, but sometimes convenient:

- Persistent state: The SCB may keep some persistent state across runs. This state may be as simple as a monotonic counter. Using this monotonic counter, the SCB may implement mechanisms for maintaining more complex state. In particular, assuming that the SCB has a monotonic counter, it can maintain other state on untrusted storage, using digital signatures and encryption; the counter should be incremented, and its value attached to the state, whenever an update happens, thus offering protection against replay attacks.

- Weak timeliness: The SCB has secure means to know the time, to within the precision allowed by network and scheduling delays. In particular, the SCB may get the correct time signed by a trusted network time server TS for which it knows the public key. In each exchange with TS, the SCB would challenge TS with a fresh nonce (for example by applying a one-way hash function to a secret plus a monotonic counter). Network and scheduling delays may lead the SCB to accept an old value for the time, but never a future value. Without this assumption, the SCB can include nonces as proofs of timeliness for its assertions to on-line interlocutors. The nonces would be provided as challenges by those interlocutors. The assumption removes the need for the challenge messages.

4. VERIFIED COMMUNICATIONS WITH AN SCB

In this section we show how an SCB can serve as a trusted third party for checking and certifying communications. First, in section 4.1, we review examples of input verification, and their importance for security. Then, in section 4.2, we explain how these examples can rely on SCB support. Later sections are concerned with refining the examples, discussing benefits and drawbacks, and generalizing.

Throughout this paper, we emphasize communications that involve programs at their endpoints. Accordingly, we often refer to the sender as the caller and to the receiver as the callee. However, many of the ideas and techniques that we present do not require that the messages being exchanged are calls to program functions; they apply more broadly to arbitrary messages in a distributed setting.

4.1 Checking inputs

When a program receives data, it is prudent that it verify that the data has the expected properties before doing further computation with it (e.g., [Howard and LeBlanc, 2003]). These verifications may for example include:

- Checking that an argument is of the expected size, thus thwarting buffer-overflow attacks.
- Checking that a graph is acyclic, so as to avoid infinite loops in later graph manipulations.

- Checking that an argument is of the expected type or structure.
- Checking the validity of a "proof of work" (evidence that the sender has performed some moderately hard computation, of the kind suggested for discouraging spam; e.g., [Dwork and Naor, 1992; Jakobsson and Juels, 1999]).
- Checking that cryptographic parameters have particular properties (often number-theoretic properties) needed for security [Anderson and Needham, 1995, Principle 6].
- Checking that a set of credentials forms a chain and implies some expected conclusion, for example that the sender is a member of a group.

Further, interesting examples arise in cases where the data is code (or may include code):

- Checking that the data does not contain one of a set of known viruses.
- Checking that a piece of mobile code is well-typed. This mobile code might be written in a source language, an intermediate language, or in binary. As in Java Virtual Machines [Lindholm and Yellin, 1999] and the Common Language Runtime (CLR) [Box et al., 2002], the typing provides a base level of security. With some research type systems (e.g., [DeLine and Fahndrich, 2001; Myers, 1999]), the typing may ensure further properties, such as compliance with resource-usage rules and secure information-flow properties.
- Checking the legality of a logical proof that a piece of mobile code satisfies some property, for example an application-specific safety property, termination, or an information-flow property. Research on proof-carrying code [Necula, 1997] explores these ideas.
- More speculatively, checking that compiled mobile code is a correct implementation of a given source program (that is, that the compiler did not make a mistake in a particular run). Research on translation validation [Pnueli et al., 1998] explores these ideas.

As these and other examples illustrate, authenticating the origin of data is often essential, but further checking can be essential too. In particular, the checking can serve in preventing the spread of infections from senders to receivers.

Some checking may be done automatically by standard machinery in distributed systems; for example, remote procedure call (RPC) machinery can enforce simple typing properties before delivering arguments to remotely invoked

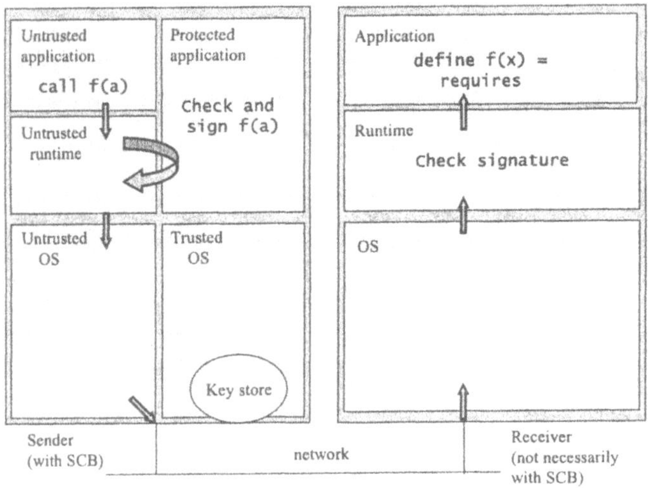

Figure 2. A verified input

procedures. Such automatic checking is particularly justified for generic properties that are easy to verify. On the other hand, application-specific properties and properties that are expensive to verify tend to be treated on a case-by-case basis.

4.2 Using an SCB

Suppose that a piece of code relies on a certain property of its inputs, and that therefore this property should be checked. The checking can happen at the code's boundary or deeper inside the code. It could also happen at the caller, though in general the caller may not know what property to ensure, and crucially the caller cannot always be trusted.

Having an SCB in the caller leads to a new possibility, depicted in Figure 2: the SCB can serve as a trusted third party that is responsible for the checking, and that certifies that the checking has succeeded.

This certification consists in a signed assertion that the call (including its arguments) satisfies a given property. The signed assertion should contain a proof of timeliness, such as a timestamp or a nonce. The signature may simply be a public-key digital signature. When the SCB and the consumer of the signature share a secret, on the other hand, the signature may be an inexpensive MAC (message authentication code). This MAC may be applied automatically if caller and callee communicate over an authenticated channel, such as can be implemented on top of the SSL and SSH protocols. This authenticated channel has another clear benefit: proving the identity of the caller to the callee.

When it receives a certificate, the callee should check that it matches the call, that it is timely, that it claims the expected property, and also that it is issued by a sufficiently trusted SCB. All these checks but the last should be straightforward. Checking that the certificate is issued by an appropriate SCB is a classical authorization problem (a matter of trust rather than of remote integrity verification). When the SCB is identified with a public key, the public key may be in a group of keys trusted for the purpose. On the other hand, the SCB may prove only that it is a proper SCB in a certain group, without revealing its exact identity; this case is more elaborate but does not introduce new difficulties.

There is no requirement that the callee have an SCB. However, an SCB at the callee can provide a secure environment in which to perform the checks just described; it can also serve for certifying properties of communications in the opposite direction, such as the result (if any) of the call.

There remains the problem of letting the caller's SCB know what property to check. This information may be hard-wired on a case-by-case basis. In general, it is attractive to envision that the property would be advertised along with the interface to the code being called. Much like the caller learns about the existence of the code entry point, and about the expected types and semantics of arguments, the caller should learn about the expected properties of these arguments.

Using an SCB for checking inputs has a number of desirable features, as well as some potentially problematic ones. Before we discuss them, however, it is useful to consider a few instantiations of the method for particular checks.

5. EXAMPLES

Next we consider four examples, both because of their intrinsic interest and in order to elucidate general features of the method described in section 4.2.

5.1 Typechecking

In the simplest example, the SCB of the caller typechecks the call, and writes a corresponding certificate.

For simple typing properties of small arguments, this example is wasteful. If the caller's SCB and the callee are not already communicating on an authenticated channel, then the callee may need to check some public-key certificates; when typechecking is simple and fast, trading it for a public-key operation is hardly attractive.

As arguments get larger, delegating the typechecking to the caller's SCB becomes more reasonable. For instance, suppose that the caller is uploading a large amount of data into the callee's database, and that this data is supposed to be in a particular format. In general, checking or imposing this format may require some processing and some buffering. If the caller's SCB can guarantee that the format is obeyed, then the callee may need to compute a message digest (relatively fast) and perform at most one public-key operation, independently of the size of the data, without any buffering.

Delegating the typechecking to the caller's SCB also becomes more reasonable for complex typing tasks. For instance, the callee may be relieved to avoid the task of checking that a piece of XML conforms to a particular schema, or that a piece of mobile code is well-typed. Indeed, the typechecking of mobile code can be fairly expensive, to the point where it is difficult or impossible on resource-constrained environments.

In a recent paper [Leroy, 2002], Leroy discusses the cost of traditional bytecode verification on Java cards, and also discusses alternatives. Leroy writes:

> bytecode verification as it is done for Web applets is a complex and expensive process, requiring large amounts of working memory, and therefore believed to be impossible to implement on a smart card.

The alternatives include both off-card verification and the combination of off-card code transformations with easier on-card verification. Leroy ingeniously develops this latter alternative. On the former alternative, Leroy writes:

> The drawback of this approach is to extend the trusted computing base to include off-card components. The cryptographic signature also raises delicate practical issues (how to deploy the signature keys?) and legal issues (who takes liability for a buggy applet produced by faulty off-card tools?).

Having the off-card verification done in the caller's SCB mitigates these concerns:

- Extending the trusted computing base to an SCB appears less problematic than extending it to an arbitrary machine with arbitrary software and arbitrary viruses.

- The deployment of SCBs should include the deployment of their keys and of certificates for those keys.

- The off-card verifier can be chosen by the consumer of the code, or a delegate, and the SCB can guarantee that it is this verifier that it runs. Therefore, the SCB would not be liable for a faulty verifier. (However, other parties would still have to be responsible for more fundamental infrastructure failures such as bugs in SCBs or leak of the master secret keys.)

Moreover, any work done in the caller's SCB needs to be done only once, while work done at the consumer needs to take place once per consumer (and even more often when consumers obliviously download the same piece of mobile code multiple times).

In addition to smart-cards, servers can also be resource constrained. In the design of busy servers that deal with many clients, one typically shifts as much work as possible to the clients. In our case, the client's SCB would be responsible for checking code uploaded to the server (servlets). For instance, when the server is a database, and its data cannot be sent to the client because of privacy considerations or sheer size, the client may upload code to run against the data; the client's SCB could ensure the safety of the code. More broadly, the client's SCB could also ensure that the code conforms to any server policies.

In short, although there exist clever alternatives, typechecking in the caller's SCB appears as a viable approach to an actual problem. Although it is not always advantageous, it does have some appealing properties, and it can be a good choice.

5.2 Proof checking

Research on proof-carrying code develops the idea that mobile code should be accompanied by proofs that establish that the code satisfies logical properties. As a special case, the properties may represent basic guarantees such as memory-safety, which can also be obtained by typechecking. However, proof-carrying code is considerably more general. As suggested above, the properties may include application-specific safety properties, termination, and information-flow security properties. For example, a proof may guarantee that the code uses only certain limited resources, or that it does not leak pieces of private user data. Such properties may be attractive whether the receiver of the code is a resource-constrained personal smart-card or a busy database server.

Although the verification of proofs is typically simpler than their construction, it is not a trivial task. It is roughly as hard as typechecking (discussed in section 5.1), and in fact proof checking can be formulated as a kind of typechecking. In addition, proofs can be bulky, creating communication overhead. For example, a recent paper [Henzinger et al., 2002] that treats device-driver properties includes proof sizes, for instance up to 156 KB of proof for a program of around 17 KLOC. Other proof encodings are possible (e.g., [Necula, 2001]), and may lead to a reduction of proof sizes by an order of magnitude. While these encodings are both insightful and effective, they can lead to slower proof checking, and in any case the proofs often remain much larger than signed statements. For example, a proof for the hotjava code takes 354 KB [Necula, 2001], substantially less than the code itself (2.75 MB), but more than a thousand times the size of a signature; checking the proof took close to one minute on a 400 MHz machine, much more than checking a signed statement.

Alternatively, with our approach, the SCB of the code producer could be responsible for checking the proof. The proof could be constructed outside the SCB, by whatever means, and given to the SCB with a cheap, local memory transfer, rather than network communication. The SCB could then transmit an assertion that the proof exists, in a certificate, rather than the proof itself. The consumer of the code would simply check the certificate rather than the proof. Leroy's concerns about off-card bytecode verification apply also to this scenario, though again the use of an SCB should mitigate them and offer some advantages.

To date, there is only limited experience in the deployment and use of proof-carrying code technology. Therefore any assessment of the use of SCBs in this context may remain rather speculative. Nevertheless, as for typechecking, this use of SCBs appears as a sensible and potentially attractive variant.

5.3 Certificate checking

For access control in distributed systems, the reference monitor that evaluates a request typically needs to consider digitally signed certificates and assemble evidence on whether the request should be granted. If the request comes from a source S and it is for an operation O on a target object T, the certificates may for example say that S is a member of a group G, that G is included in another group G', that all members of G' can perform O on objects owned by a principal P, and that P does in fact own T. Examples with chains of 5–6 certificates are not uncommon in some systems (e.g., [Clarke et al., 2001; DeTreville, 2002]). The certificates may be obtained by a variety of methods (pushed or pulled); selecting the relevant certificates and assembling them into a proof can be difficult. Therefore, several systems have, to various extents, shifted the work of providing proofs to the sources of requests [Wobber et al., 1994; Appel and Felten, 1999; Bauer et al., 2002]. Nevertheless, the checking of proofs remains in the reference monitor.

Using an SCB, we can go further: the source of a request need not present a pile of certificates or even a proof, but rather its SCB can provide a certificate that it has checked a proof. (In addition, the SCB should present certificates to establish its trustworthiness, and the reference monitor should check them, but these certificates may be trivial, and in any case they should not vary much from request to request.) Thus, the task of the reference monitor becomes simpler.

This approach could also have privacy advantages: the source's SCB need not reveal all the source's certificates—including the exact identity of the source and its group memberships—as those are processed locally. Private information about the source can thus be kept from the reference monitor, and also from any parties that somehow succeed in compromising the reference monitor, which may not have an SCB. Conversely, the reference monitor may be able to disclose its access-control policy to the source's SCB without making it public. (However, this disclosure is not essential: the SCB may provide only a partial proof if it does not know the access-control policy, so the approach applies in that case also.) Clearly, realizing this privacy advantage may require additional machinery, such as specifications of privacy properties that control the flow of certificates; the development of this machinery is perhaps interesting but beyond the scope of this paper.

While the explanation above concerns a reference monitor that evaluates a request, much the same applies to an on-line authority that issues certificates—for example, an authority that issues a certificate of membership in a group G to anyone who proves membership in another group G'.

More generally, the protected environment of an SCB appears as an appealing place for certificate processing and manufacturing. With some care, its weak timeliness properties should be adequate for this application.

5.4 Virus confinement and communications censorship?

Preventing the spread of viruses is an eminently worthy application of SCBs. Because viruses can in general attack anti-virus software, it is attractive to run that software under the protection of SCBs. In particular, when two nodes communicate, either or both can use their SCBs to check and certify the absence of known viruses in the data they exchange.

One may ask, however, whether any negative applications of SCBs might make them unattractive overall. In particular, the same infrastructure that blocks viruses could well be used for censoring other kinds of contents. Fortunately, communications censorship—at least in the form described here—can be avoided. First, there may be legal protections against it. Hardware attacks on SCBs may also defeat censorship, though they negate protection against viruses at the same time. Finally, censorship may be avoided at the software level, since communications between consenting nodes can circumvent SCBs. (We note however that there has been prior discussion of other forms of censorship, in which local files would be deleted [Anderson, 2003b].)

6. ASSESSMENT

In light of the preceding examples, we see that the shift of checking to the sender's SCB has a number of consequences, some of them rather attractive:

- The work is done at the sender, not the receiver. Therefore, we may not mind if there is quite a lot of work. In particular, we remove one opportunity for denial-of-service attacks on the receiver. This point is only significant if the work is substantial (more expensive than whatever signature verification is required). It may be particularly significant when the receiver is a resource-constrained device such as a smart-card or a server.

- Any auxiliary data needed for the checking is communicated only locally, not to the receiver across a network. This feature can result in simplifications and efficiency gains (as in the proof-carrying code example), and possibly also in privacy gains (as in the certificate-checking example).
- If the data is sent to multiple destinations, the checking of each property needs to be done only once at the sender, not once at every destination. (For example, the data might be mobile code being widely distributed, as discussed above.)
- The receiver should trust the sender's SCB. Specifically, if that SCB is somehow compromised (say, with a hardware attack), the checking may be circumvented. On the other hand, the receiver need not trust the rest of the sender, which may be incompetent, compromised, or malicious.

Some of these features are also obtained when the checking is done by a trusted third party placed at a firewall or at another machine managed by trusted system administrators. In comparison, using an SCB may increase concerns about hardware attacks. On the other hand, it may reduce any concerns about administrators, it saves communication, and it does not require special infrastructure.

7. AN EXAMPLE, STEP BY STEP

As a more concrete example, suppose that a server offers a generic computing service, initially with the following interface:

```
public void compute(p : Principal,
                    f : Code,
                    i : FileName,
                    o : FileName)
```

Here `f` is code to be executed (possibly in binary format), `i` a source of inputs for the code, `o` a destination for the outputs, and `p` the identity of the invoking principal. The secure-communication machinery can guarantee that `p` is not spoofed [Lampson et al., 1992]. Internally, the server may check `p` against access control lists, for example those for `i` and `o`. The server may also check that it is safe to run `f`, somehow—for example, by checking `f` for known viruses and also by relying on any types and other evidence of safety included with `f`.

With our approach, the interface may specify the requirements of the call, leaving their verification to the caller's SCB:

```
public void compute(p : Principal,
                    f : Code,
                    i : FileName,
                    o : FileName)
requires     p says safe(f),
             p says may-read(p,i),
             p says may-write(p,o),
             GoodSCB(p)
```

For simplicity, this interface identifies the principal p with its SCB. It is however easy to write versions in which the SCB need not put its full authority behind the call, in particular by requiring only that p be of the form "s quoting r" [Lampson et al., 1992], for some SCB s and some identity r:

```
public void compute(p : Principal,
                    f : Code,
                    i : FileName,
                    o : FileName)
requires for some r, s.
             p = (s quoting r),
             GoodSCB(s),
             s says safe(f),
             s says may-read(p,i),
             s says may-write(p,o)
```

Such requirements are particularly appropriate when r represents a piece of code at the client. Even when the SCB and its user are trustworthy (so in particular the user does not attempt hardware attacks on the SCB), some client code may not be.

When a client p imports this interface, it also learns about the requirements that calls should satisfy. When the client wishes to call `compute(p,f,i,o)`, it somehow finds proofs of `safe(f)`, `may-read(p,i)`, and `may-write(p,o)`. The proof of `safe(f)` may consist of a logical proof of some property of `f` and a certificate that associates the predicate name "`safe`" with this property. The proofs of `may-read(p,i)` and `may-write(p,o)` may be assertions signed by a trusted authority, perhaps by the server itself. In all cases, further certificates may be required, for instance certificates for the keys of the authorities in question, and certificates that place p, `f`, `i`, and o in particular groups.

The client provides this material to its SCB, along with the data for the call. The SCB can then verify and assert `safe(f)`; it can similarly assert `may-read(p,i)` and `may-read(p,o)`. The client should present these signed assertions along with its call, and with a certificate that its SCB is in the group `GoodSCB`. Upon receipt of the call, the server automatically verifies that the SCB's assertions match its requirements before launching the execution of `f`.

The server can be even more forthcoming on its expectations. In particular, it can provide some information on how `safe(f)`, `may-read(p,i)`, `may-write(p,o)`, and `GoodSCB(s)` may be established. For instance, the server could supply a piece of code that implements `safe`, and a rule that implies that (in its view) if `s` is a good SCB and `r` is a good program then `s quoting r` may read `i` and write `o`. These can also be attached to the interface that the client imports.

8. EXTENSIONS

In this section we briefly consider variants and extensions of the ideas described above.

A first, minor extension consists in taking into account auxiliary state that the SCB may keep. For instance, an SCB can certify network requests from its host up to some number (say, 1,000) per day. The requests may include calls on web services, such as search engines, and also requests to send e-mail (via SMTP) or to create free e-mail accounts. Of course, the requests can be broken into classes, with a different limit for each. Anyone that receives a non-certified request would have reason to suspect that it is generated by a program rather than a human user, and may disregard it or give it low priority. For this example, the SCB can simply rely on monotonic counters.

As this example shows, one advantage of performing checks at the caller's SCB is that the SCB can rely on any relevant auxiliary state it can keep. The state may not readily be available at the callee.

In further extensions, an SCB may do more than checking data: it can supply all or part of an input. The SCB can thus guarantee that the input is generated in a certain way. For example, the SCB can guarantee that the input is generated with a particular protocol stack; by running a particular compiler; with inlined safeguards that enforce a security policy, such as an inlined reference monitor [Erlingsson and Schneider, 2000]; with a particular application (for example, with a trusted tax-preparation package); by completing a particular form; or directly by a user, through secure I/O. Although these scenarios may be attractive, some of them may require running substantial pieces of code on the SCB. These scenarios often tend to fit into a fairly controlled approach to systems, which enforces not only what hosts say but also why they say it (what code they run).

In addition, the SCB can help when the input in question contains sensitive information (such as personal medical records). The SCB may be in charge of holding the sensitive information, and occasionally encrypting it and sending it to designated parties, or displaying it on a trusted output device. In such examples, the SCB is involved not in order to guarantee how the data is generated, but in order to protect its secrecy.

9. CONCLUSIONS

Trusted Computing gives rise to a new supply of potential trusted third parties. These trusted third parties may find a variety of applications in distributed systems—keeping sensitive personal information, preventing cheating in games, and possibly many more. In this paper we investigate the use of these trusted third parties for verified communications. We consider several instances of remote input checking, such as remote typechecking, proof checking, and certificate checking.

Despite the lively controversy on Trusted Computing, and despite the substantial progress in the development of its basic machinery, there remains much room for further thinking and experimentation. In particular, this thinking and experimentation should shed more light on the potential uses of this technology, which are important whether one prefers Trusted, Trustworthy, or Treacherous Computing.

Acknowledgements

This paper was written at Microsoft Research, Silicon Valley, and is partly based on discussions with Andrew Birrell, Mike Burrows, Luca Cardelli, John DeTreville, Cynthia Dwork, Ulfar Erlingsson, Cedric Fournet, Andy Gordon, Jim Gray, Butler Lampson, Paul Leach, Roy Levin, Greg Morrisett, Chuck Thacker, Chandu Thekkath, and Ted Wobber. Chandu Thekkath and Ted Wobber also suggested improvements to a draft of this paper. Thanks to all of them.

References

Anderson, R. (2003a). Cryptography and competition policy - issues with 'Trusted Computing'. On the Web at `www.ftp.cl.cam.ac.uk/ftp/users/rja14/tcpa.pdf`.

Anderson, R. (2003b). 'Trusted Computing' Frequently Asked Questions - TC / TCG / LaGrande / NGSCB / Longhorn / Palladium / TCPA. Version 1.1, on the Web at `www.cl.cam.ac.uk/~rja14/tcpa-faq.html`.

Anderson, R. and Needham, R. (1995). Robustness principles for public key protocols. In *Advances in Cryptology—CRYPTO '95*, pages 236–247. Springer.

Appel, A. W. and Felten, E. W. (1999). Proof-carrying authentication. In *Proceedings of the 5th ACM Conference on Computer and Communications Security*, pages 52–62.

Bauer, L., Schneider, M. A., and Felten, E. W. (2002). A general and flexible access-control system for the Web. In *Proceedings of the 11th USENIX Security Symposium 2002*, pages 93–108.

Box, D., Sells, C., and Pattison, T. (2002). *Essential .NET Volume I: The Common Language Runtime*. Addison Wesley.

Clarke, D., Elien, J.-E., Ellison, C., Fredette, M., Morcos, A., and Rivest, R. L. (2001). Certificate chain discovery in SPKI/SDSI. *Journal of Computer Security*, 9(4):285–322.

DeLine, R. and Fahndrich, M. (2001). Enforcing High-Level protocols in Low-Level software. In *Proceedings of the ACM SIGPLAN '01 Conference on Programming Language Design and Implementation (PLDI-01)*, volume 36.5 of *ACM SIGPLAN Notices*, pages 59–69.

DeTreville, J. (2002). Binder, a logic-based security language. In *Proceedings of the IEEE Symposium on Research in Security and Privacy*, pages 105–113.

Dwork, C. and Naor, M. (1992). Pricing via processing or combatting junk mail. In *Advances in Cryptology—CRYPTO '92*, pages 139–147. Springer.

England, P., Lampson, B., Manferdelli, J., Peinado, M., and Willman, B. (2003). A trusted open platform. *IEEE Computer*, 36(7):55–62.

Erlingsson, Ú. and Schneider, F. B. (2000). IRM enforcement of Java stack inspection. In *Proceedings of the 2000 IEEE Symposium on Security and Privacy*, pages 246–255. IEEE Computer Society Press.

Garfinkel, T., Pfaff, B., Chow, J., Rosenblum, M., and Boneh, D. (2003). Terra: A virtual machine-based platform for trusted computing. In *Proceedings of the 19th Symposium on Operating System Principles (SOSP 2003)*, pages 193–206.

Henzinger, T. A., Jhala, R., Majumdar, R., Necula, G. C., Sutre, G., and Weimer, W. (2002). Temporal-safety proofs for systems code. In *Proceedings of the 14th International Conference on Computer Aided Verification (CAV'02)*, pages 526–538. Springer.

Howard, M. and LeBlanc, D. (2003). *Writing Secure Code*. Microsoft Press, 2nd edition.

Jakobsson, M. and Juels, A. (1999). Proofs of work and bread pudding protocols. In *Proceedings of the IFIP TC6 and TC11 Joint Working Conference on Communications and Multimedia Security (CMS '99)*, pages 258–272. Kluwer.

Lampson, B., Abadi, M., Burrows, M., and Wobber, E. (1992). Authentication in distributed systems: Theory and practice. *ACM Transactions on Computer Systems*, 10(4):265–310.

Leroy, X. (2002). Bytecode verification on Java smart cards. *Software — Practice and Experience*, 32(4):319–340.

Lie, D., Thekkath, C., Mitchell, M., Lincoln, P., Boneh, D., Mitchell, J., and Horowitz, M. (2000). Architectural support for copy and tamper resistant software. In *Ninth International ACM Conference on Architectural Support for Programming Languages and Operating Systems (ASPLOS-IX)*, pages 168–177.

Lindholm, T. and Yellin, F. (1999). *The Java Virtual Machine Specification*. Addison Wesley, 2nd edition.

Myers, A. C. (1999). JFlow: Practical mostly-static information flow control. In *Conference Record of POPL '99: The 26th ACM SIGPLAN-SIGACT Symposium on Principles of Programming Languages*, pages 228–241.

Necula, G. C. (1997). Proof-carrying code. In *Conference Record of POPL '97: The 24TH ACM SIGPLAN-SIGACT Symposium on Principles of Programming Languages*, pages 106–119.

Necula, G. C. (2001). A scalable architecture for proof-carrying code. In *Functional and Logic Programming, 5th International Symposium, FLOPS 2001*, pages 21–39. Springer.

Pnueli, A., Siegel, M., and Singerman, E. (1998). Translation validation. In *Proceedings of the 25th International Colloquium on Automata, Languages, and Programming (ICALP 1998)*, volume 1384, pages 235–246. Springer.

Stallman, R. (2002). Can you trust your computer? On the Web at `www.gnu.org/philosophy/can-you-trust.html`.

Wobber, E., Abadi, M., Burrows, M., and Lampson, B. (1994). Authentication in the Taos operating system. *ACM Transactions on Computer Systems*, 12(1):3–32.

MAILLE AUTHENTICATION

A Novel Protocol for Distributed Authentication

Andrew Fritz and Jehan-François Pâris[1]

University of Houston, Department of Computer Science, Houston, Texas 77204-3010

Abstract: We present a decentralized solution to the distributed authentication problem. Unlike current schemes, Maille does not rely on a set of dedicated servers. Each participating node has a set of trusted peers that act as replicated repositories for its public key. Whenever a node A wants to contact another node B, it sends messages to its peers, which forward them to their peers, and so on until they reach B's peers. These peers then reply with B's public key and return them back through the paths the requests originally took. To guarantee the independence of replies, each node along each path forwards one reply only. Because of this, each Byzantine failure can introduce at most one false response. If the same key value is asserted by a qualified quorum of replies, A accepts it as the true key of B. Otherwise authentication fails.

Keywords: authentication, distributed security, peer-to-peer, Byzantine failure, denial of service, certificate authority

1. INTRODUCTION

> Maille: *... mesh, network, a coat of mail, ... A flexible fabric made of metal rings interlinked. It was used especially for defensive armor.*
> *-Webster's 1913 Dictionary*

In distributed environments, centralized services can become bottlenecks, and are often targets for attack, both electronic and political. As a result, any service that resides at some small fixed number of locations has the potential to limit the ultimate performance, availability and security of the system in question. Any critical portion of the distributed system controlled by one organization can effectively give control of the entire system to that organization.

[1] Supported in part by the National Science Foundation under grant CCR-9988390

Existing authentication systems require centralized management by a single organization or well-defined collaboration among organizations. This leads to a weakness. The Verisign problem—where Verisign changed the way many root DNS servers worked without the permission of the standards organizations or community—is a real-world example. Placing control of a central service required by many organizations into the hands of a single organization places the controlling organization in a monopoly position where they can use their management right to control the larger group. For political reasons, this type of structure may not be acceptable in many collaborations. In addition, it is inherently weak against legal and political attack. No known centralized authentication system can prevent this problem.

To handle normal failures, such as network outages, most systems replication authentication credential to a small fixed set of authentication authorities, or accept the limited scalability and uptime of a central authentication service. If all authentication server replicas happen to be on the same link, or even segment of the larger network, then a single failure may take them completely offline.

Furthermore, centralized authentication services provide a clear target for denial of service (DOS) attacks. By preventing authentication, an attacker can take an otherwise robust distributed system offline. If no one can be authenticated because all or most of the authentication authorities are unavailable, then the entire system is effectively taken offline.

Several researchers (e.g. Reiter and Stubblebine, 1997; Kahn, 1998) have noted that independent corroboration of credential is an effective way to remove this single point of failure. To our knowledge, no existing authentication system actually makes use of multiple independent sources of key information beyond a set of defined authentication servers. This paper presents the Maille authentication protocol, a new authentication protocol that eliminates authentication as a single point of failure. All nodes participate as equals to provide multiple verifiably independent sources of public keys. Each node has established trust relationships with some peers. Any node in the network may use its trusted peer relationships to obtain the public key of any other node in the network. When one node in the network wishes to authenticate another node, it attempts to acquire that nodes public key using its peer relationships. The protocol is structured so the requester can verify the independence of the sources of keys. It then uses a conventional nonce challenge against the node to establish one-way trust. If mutual trust is required, each node uses the protocol to establish the identity of the other, and a direct secure session can be established.

As a result, the Maille protocol provides a means to acquire the public key of an entity in a trusted manner without a trusted central authority. Since all nodes equally share authentication functions, there are no favored targets of attack in a Maille network. Maille acts as a completely distributed certificate authority.

The remainder of this paper is laid out as follows. Section 2 covers previous work. Section 3 defines the Maille-authentication protocol. Section 4 provides an analysis of Maille's vulnerability to attacks. Section 5 presents future work. Finally, section 6 has our conclusions.

2. PREVIOUS WORK

Kerberos (Steiner et al., 1988) is a centralized authentication system, designed to allow single-sign-on from trusted workstations. Kerberos based systems rely on a single or a small set of authentication servers. The Kerberos system uses a ticket scheme, which allows clients to authenticate against the Kerberos servers only once. Thereafter, for the lifetime of the ticket, no further authentication is required and services and other individuals can trust the ticket holder without having to know their key.

Kerberos does have several weaknesses. First, it is highly centralized, requiring one master server where all updates occur. Replication of the security information to other server will offload all authentication work, but cannot reduce the total amount of work the master server must do to update security information and to broadcast changes. Further, because Kerberos relies on a single master server for all changes, that server becomes a single point of failure from a hardware, software, security and political standpoint.

The KryptoKnight family of protocols (Bird et al., 1995) is designed for embedded devices and is optimized for speed and efficiency. It relies on a single, possibly replicated, authority to provide trusted keys and act as an intermediary during authentication for all clients. The main focus is on providing several protocols that allow the exchange of keys, challenges and responses to flow as efficiently as possible by allowing the use of information each of the parties may already have. The KryptoKnight protocol family does not address issues of scalability or how credentials are revoked. A Byzantine failure in an authority is catastrophic for all parties using that authority.

Public key infrastructure (PKI) (Adams and Lloyd, 1997) has become very popular for Internet commerce. It is also widely used in grid computing as the basis for the Globus Security Infrastructure (GSI) (Foster et al. 1998). PKI relies on a hierarchy of certificate authorities (CA) for scalability. At the top is the root CA, which signs certificates for servers in the second level and so on, until the lowest-level CAs are used to establish the identity of outside entities such as web servers. Revocations are handled through certificate expiration dates and revocation lists.

Replication of CA ensures that most authentications will not be affected by a single failure. However, the higher up the hierarchy an authentication is required to go, the more likely a single failure is to prevent successful authentication. Caching prevents most interactions from requiring the root CA and other high level CA servers. Nevertheless, a Byzantine failure at the root level will lead to a complete loss of security. Failures at lower levels will result in security breach for only part of the system. Politically, the root CA is a single point of failure.

PGP (Zimmermann, 1995) is a system designed to let many individuals authenticate each other without a central authority. It provides a method of creating and distributing keys among small clique of users and for deciding to trust a key acquired from a third party. How much trust can be placed in a public key is directly related to how many intermediaries it went through. A single failure in any intermediary can result in an incorrect key being used. Because of its decentralized nature, there is no global single point of failure since unless key repositories are used. PGP does not provide any way for two unrelated parties to authenticate each other than through trusted key repositories, which will then be single points of failure.

PathServer (Capkun et al., 2002) is an extension to PGP that fills the role of the trusted key repository suggested by the PGP users guide. The server models the network topology and tries to find mutually independent sources for the key, ensuring that agreement can be reach about their validity. In this way the normal PGP authentication system is strengthened against single failures in key sources and keys can be used that otherwise would not be acceptable.

However, PathServer itself represents a single point failure, as the requesting user still must trust a single entity for the key, namely the PathServer. The user is free to verify the key, but this results in the user's computer functioning as a PathServer. Further, the method used for finding independent keys is graph analysis. This technique requires much global knowledge and an approximation to escape the NP-HARD nature of the problem. PathServer does not assure that keys are actually independent, just likely to be that way.

Coca (Zhou et al., 2002) is a distributed certificate authority that makes use of threshold public key cryptography. Available servers transmit their portion of the certified key to a delegated server selected by the client. If the delegated server can collect enough parts, a complete key can be assembled. Coca is specifically designed to prevent DOS attacks. Further, if t is the minimum number of components required to assemble a public key and n the total number of servers, $t+1$ Byzantine failures must occur before the public key of a service is compromised and $(n-t)+1$ failures must occur to deny authentication.

Coca relies on a fixed set of cooperating servers that all have the public keys of all other servers. As the number of replicated servers increases to meet demand, the risk of at least one server being compromised increases. If the threshold used in the threshold cryptography is not increased to match the server pool size, then it become easier to compromise the keys Coca holds. Increasing the threshold for cryptography increases network and processor overhead as more servers and messages are required to collect enough key parts to perform authentication.

3. THE MAILLE PROTOCOL

We want to design a protocol that allows participating nodes to authenticate each other securely without any centralized point of failure, including management. Individual nodes must not need global knowledge of the network. The system should not be dramatically affected by multiple Byzantine failures and should experience graceful degradation when faced with DOS attacks or the loss of many participating nodes.

Maille has no centralized point of failure because it is a decentralized key distribution system made up of individual, equal and autonomous nodes. Each node in the network has a set of peer nodes. Each node's peers are explicitly trusted. Like in any large distributed authentication system, this trust is transitive. Each node trusts its peers, and all trust in other nodes is derived from this basic trust. A Maille node uses its trusted peer relationships to find a set of independent sources for key information. If enough completely independent sources agree on the key, then the acquired key can be used to challenge the other party to prove its identity via a nonce challenge.

Maille ensures that individual responses are independent by tracking the path (called a *response chain*) by which each response reached the requester. A response chain includes all nodes that relayed the message from the node holding a copy of the requested public key back to the requester. The protocol guarantees that non-independence chains will always be detected.

Maille also guarantees that each properly functioning node filters duplicates and extra responses. Thus, if *n* independent chains are found, the requester can be sure that *n* nodes independently returned chains with the same key. A single Byzantine failure can never introduce more than one false chain and one false key. In addition Maille requires a qualified quorum of independent agreeing chains. Hence many such failures in separate, but very specific parts of the network must occur to provide a false key.

3.1 Assumptions

1. Network transport is not secure unless Maille uses a secure channel it creates using keys it trusts.

2. Attackers know the network structure and can attack any machine in the network with equal effort.
3. Attackers know the protocol and can use that knowledge to find the most critical node(s) to any transaction.
4. Any node may fail or be compromised.
5. No single source of information can ever be trusted.
6. The underlying cryptography functions are strong. They cannot be broken easily without some other failure that provides clues.

3.2 Notations

In the following discussion, the notations listed below are used.

- A, B, C represent specific nodes in the network.
- X and Y represent any arbitrary node in the network.
- *f* (...) is a message, containing, amongst other, the parameters specified between the parentheses.
- *A* (in italics) represent the public key of node A.
- *Ø* is the special null key. It is treated exactly as any other key except where noted.

For example, A may send to B the message *kr*(A,C) to request the public key *C* for node C.

3.3 Node Structures

Each node in the network maintains the following:

- *kr_cache*: a cache of recent key requests storing the requester and a return seen flag (initially false).
- *trust_list*: a list containing the public keys of the peers of the node.
- *black_list*: a list of nodes that are blacklisted.

Note that *trust_list* and *black_list* must mutually disjoint.

3.4 Messages

- *kr*(A,C) denotes a key request that contains the ultimate source (A), the target (C). Other parameters include the remaining hop count and the immediate source (the peer that forwarded the message to the node).
- *trust*(*kr*(A,C),*C*) denotes a statement of trust by the sender to the receiver that *C* is the key requested by *kr*(A,C).
- *rkr*(...) denotes a return key request for the public key of C in the form: *trust*(*kr*(A,C),*C*) | *rkr*(B,*rkr*(...)) | *rkr*(B,*trust*(*kr*(A,C),*C*)), where B is the peer who sent the *rkr*(...).

3.5 Peer Relationships

Within the peer-to-peer network, each node has a set of trusted peers. Peer relationships are established offline from the point of view of the Maille protocol. However, there are several required features of peer relationships that the rest of the system relies on.

If two nodes A and B are peer nodes, then A has a trusted public key of B, and vice versa. These keys are only used to establish a bidirectional secure channel between A and B. For added security, each node has one key per peer so that the public key A uses to establish a secure channel with peer B is different than the public key A would use to establish a secure channel with peer C. The secure channel from A to B is called secure channel A-B.

How A and B exchange public keys initially (i.e. become peers), or how A and B chose each other as peers isn't the focus of this paper.

Upon startup, peers A and B mutually authenticate each other using the appropriate key pair and establish a secure channel for use later. This secure two-way channel is used to pass all messages from A to B and from B to A for the duration of the protocol. Thus, A can trust that any message arriving via the secure B-A channel came unaltered from B and only B and vice versa.

Each node A also maintains a key pair for third party authentications. The public key *A* is passed to the peers along with a version number. When a key request message arrives at a peer of A, it returns that key *A* to the node that requested it. An optional parameter *l* specifies the lifetime of the third party key pair. If additional protection against a combination of Byzantine failures and clear text cryptographic attacks is required, we may choose to require A to change its key pair from time to time.

3.6 Obtaining Keys

The Maille protocol is made up of three individual protocols, each executed in response to particular types of messages. The key request initiator protocol is used by nodes to find keys. The key request forwarding protocol controls the propagation of *kr*(…) messages in Maille and the return key request protocol governs the propagation of *rkr*(…) messages.

3.6.1 Key Request Initiator Protocol

When node A, which is part of the network, wishes to obtain a trusted public key for node C, which is also part of the network, it does the following:

1. Node A initiates a key search by sending to its peers *kr*(A,C).
2. It waits for *rkr*(...) responses from them.

3. When a response arrives, node A checks that the sending peer has not already returned a response. If it has, it drops this response and return to step 2.
4. Node A creates a new *rkr*(…): *rkr*(Z, *rkr*(…)) where Z is the peer that returned the original *rkr*(…).
5. Node A adds the new *rkr*(...) to the set of chains received and returns to step 2.

3.6.2 Key Request Forwarding Protocol

Any node X receiving a key request *kr*(A,C) from some peer node Y carries out the following procedure:

1. Node X verifies that the key request *kr*(A,C) is not in its *kr_cache*. If it is, it drops this instance of *kr*(A,C) and stops.
2. Node X checks whether it has in its *trust_list* a trusted public key *C* of node C in *trust_list* whose lease has not expired. If it has such key, it returns an *rkr*(*C*) to node Y in the form *trust*(*kr*(A,C),*C*) and stops.
3. Node X checks if there is an entry in *black_list* for C. If there is, it returns *rkr*(Ø) to Y in the form *trust*(*kr*(A,C),Ø)) and stops.
4. Node X subtracts one from the remaining hop count of *kr*(A,C). If the hop count is now 0, it drops *kr*(A,C) and stops.
5. Node X adds *kr*(A,C) to the node's *kr_cache*.
6. Node X sends *kr*(A,C) to all node's peers, except node Y.

3.6.3 Return Key Request Forwarding Protocol

Any node X receiving an *rkr*(...) message from a node Z does the following:

1. Node X checks whether *kr_cache* contain an entry that matches *kr*(A,C) of this *rkr*(...) message. If it does not, node X drops this *rkr*(…) message and stops.
2. If the *kr_cache* entry has the *seen flag* set, node X drops this *rkr*(…) message and stops. This ensures that node X will only return only one *rkr*(…) per *kr*(…). Otherwise two dependent chains would be returned from node X.
3. Node X marks the *kr_cache* entry for *kr*(A,C) of this *rkr*(...) message as *seen*.
4. Node X returns a new *rkr*(…) message to the immediate source of the *kr*(…) in *kr_cache*, node Y, in the form *rkr*(Z,*rkr*(...)) to verify that it received an *rkr*(…) message from Z over the secure Z-X channel.

3.6.4 Protocol Outcome

Carrying out this protocol will result in A receiving a set of *rkr*(…) messages (called chains).The *kr*(A,C) might cause A to receive a chain like:

rkr(A,*rkr*(B,*rkr*(X,*trust*(*kr*(A,C),*C*))))
which is more easily under stood in the form A>B>X>Y with key *C*

This means that A received an *rkr*(…) from B, who received an *rkr*(…) from X who received the key *C* from Y. Note that no node adds itself to the chain. It is added instead by the node to which it returns the *rkr*(…).

3.7 Picking a Winning Key

Whenever distinct conflicting keys are returned, we must decide which key to accept as the public key of C. This procedure is carried out even if only one distinct key is received. The score it produces is required in later steps.

Chains are separated into groups based on the public key they contain C_1, C_2, and so forth. Each unique key will receive a score that is the total of all the scores of all chains that contain that key. The score of each individual; chain is computed by raising a link trust factor $0 < w \leq 1$ to the power of the length of the chain $|rkr(...)|$:

$$w^{|rkr(\ldots)|}$$

The key with the highest total score is the only one considered.

3.8 Independence Analysis and Penalties

The wining key is not necessarily the correct key because rogue nodes could have flooded their peers with the incorrect key. Therefore an independence analysis is performed on all chains containing the winning key. The analysis must determine the number of collisions that occur. A collision is defined as the number of times any two chains contain the same node, or 1 minus the number of times each distinct node appears in any chain.

The number of collisions is used as a penalty. If all chains are independent, then the number of collisions is 0. Otherwise, the number of collisions is subtracted from the total score the key received in the previous step.

3.9 Determining if the Winning Key Should be Trusted

If the adjusted score of the winning key surpasses some threshold of trust *t*, that key is used. If that key is the null key (Ø) then C is added to the *black_list* of node A and authentication fails. Otherwise the key request fails and no authentication may occur. Node A may choose to retry the key request with a longer hop count.

3.10 Using Keys

Once node A has obtained a key *B* for node B, it will use this key to authenticate B by sending a one-way nonce challenge. If B wishes to authenticate A as well, it will need to acquire *A* and extend the challenge to a two-way challenge.

3.11 Tunable Parameters

Maille includes several tunable parameters. All the parameters of Maille are related in intuitive ways to the level of security those using Maille want.

3.11.1 Chain Scoring

The link trust factor *w* determines how much trust is placed on each hop a response takes. A value of *w* near 1 will result in longer chains being nearly as trusted as shorter chains. A value of 1 will result in all chains being trusted equally. A value below and near 1 is suggested.

The trust threshold *t* is used to determine if a key has amassed a qualified quorum of support. The value of *t* will need to be less than the minimum peer count *m*. If *w* is 1, that is, all chains are treated equally regardless of length, then one key with all *m* peers supporting it will receive a score of *m*. If *t* is above *m* the no key can satisfy the quorum.

Further, if *t* is too close to *m*, many keys may not be able to amass enough support to be trusted. Higher values of *t* will make incorrect authentication of rogue nodes less likely, but will also allow fewer rogue nodes to launch a successful DOS attack.

One good compromise would be to set *t* relative to the actual peer count a node has. It could be set to 2/3 of the total peer count, thus requiring 2/3 of all peers to return chains that support one key for the key to be trusted.

3.11.2 Peer Relationships

The minimum peer count *m* determines how strong the system is in general, assuming all nodes have not many more peers than *m*. If *m* is only 3 (and all nodes have exactly 3 peers), w is 1, and t is 2 then a minimum of only 2 Byzantine failures is required to cause a false key to win the competition. If m is raised to 30 and t to 20, 20 such failures are required for the false key to win, and 11 are required to deny authentication that should otherwise occur.

4. ANALYSIS

The Maille authentication protocol removes the centralized authority that normally acts to dispense trusted public keys. Maille's design also changes the system dynamics so that authentication experiences graceful failure, both in terms of availability and security. As nodes fail, or are compromised, trust is not completely lost. Because all nodes are operated separately by the participating organizations and participate equally and there is no hierarchy of nodes within the network and no single control point, no one organization is placed in a monopoly position.

4.1 Byzantine Failures and Impersonation

Consider that a node A wishing to authenticate some other node claiming to be C. To do this, it must acquire C's public key *C*. Should a node D try to impersonate C, it must either acquire C's private key, which only C has, or cause A to receive and trust its public key *D* instead of *C*. To prevent the first problem, C must protect its private key. We consider only the second case.

D may choose to operate alone or with others to provide incorrect keys. If it operates alone, it must provide enough responses to A that favor *D* instead of *C*. As in figure 1, the worst case occurs when D is peered with all of A's peers E, F, G and H. It could then fabricate an *rkr*(...) message that provided *D* and quickly send it to E, F, G and H. Because the fake *rkr*(...)s from D were received by A's peers before all other replies, they will be the only ones to reach A. If w is 1, *D* will receive a score of 4. However, during the independence analysis, 3 collisions would be found because E, F, G and H would all add D to the *rkr*(...) they forward. The key *D* will only receive a score of 1. If the trust threshold t is set appropriately, this will not be enough for A to trust *D*.

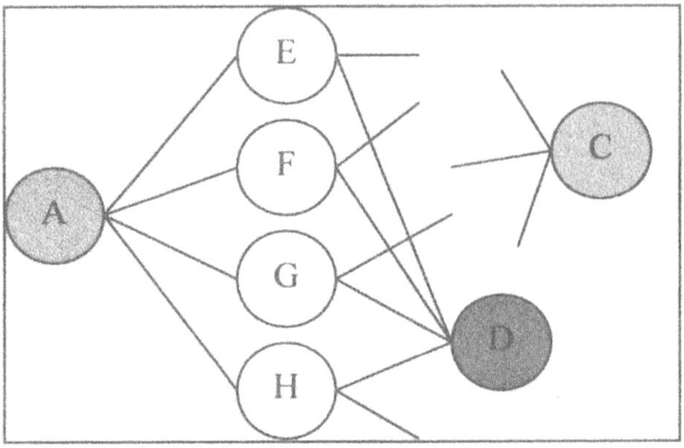

Figure 1. A is the requestor; C is the real target; D is the node that wishes to impersonate C; E, F, G and H are A's peers.

D may also choose to collaborate with other rogue nodes to trick A into accepting *D*.

Looking back at figure 1, we see that A may trust the wrong key if enough of its peers act as collaborating rogues. To trick A to accept the wrong public key for C, the number of peers f_a returning the same false key must satisfy the inequality

$$f_a \geq t/w$$

If w is 1, at least t of A's peers must be collaborating rogues for A to use the wrong key. For a small value of m, as in the example above, only a small number of rogues is required. The easiest counter measure is to select a value of $w<1$, which will increase minimum number of rouges required to fool A. Unfortunately, it will also reduce the likelihood a valid key will achieve a qualified quorum. In larger networks with larger values of m, t can be higher and many more collaborating rogues will be needed. If the rogues are not directly peered with A, more collaborating rogues may be required to fool A.

4.2 DOS Attacks

Rogue nodes may wish to exploit the Maille protocol to deny successful authentication. System without established peer relationships (i.e. computers that are not part of a Maille network) have no chance to exploit the protocol since all messages within the protocol flow over secure channels between peers. Outside nodes cannot insert messages.

Rogue nodes can produce messages that try to exploit flaws in the protocol. Consideration should be given to various possibilities. To prevent

A from obtaining a trusted public key for C, enough messages must be suppressed to prevent *C* from scoring higher than *t*. This would require a partial partitioning of the network such that there were less than *t* independent paths from A to C (assuming $w = 1$). However, as the number of paths decreases towards *t* it will become less likely due to the random routing used in Maille that A will receive enough independent replies to trust any key.

Rogue nodes could also exploit the independence analysis by providing a correct key with a fabricated chain. The artificial chain could be created to contain many of the nodes that might take part in the real chains. This would artificially penalize otherwise valid responses.

5. FUTURE WORK

We plan to extend the Maille protocol in several fashions. First, we plan to investigate techniques to automatically detect and blacklist rogue nodes. Second we want to develop protocols for adding and readmitting nodes to the network. Third we are designing an authorization protocol to work along side the Maille authentication protocol.

We are also currently building a simulation model that will allow us to explore the scalability of the system, routing, failure detection, network structure and restructuring.

6. CONCLUSIONS

The Maille authentication system is a completely distributed authentication system based on established asymmetric cryptography and nonce challenges. In some respects, it is similar to the certificate authorities of PKI because it acts as certificate authority to facilitate authentication. However, Maille authentication is designed to work in an environment where the assumptions made by most systems are unacceptable for political or security reasons. Maille is hardened against denial of service attacks, a reality of modern Internet computing. Because of its totally decentralized nature, Maille resists Byzantine failures, natural failures and organizational subversion. Its performance and security degrades gracefully in the face of failures or compromises instead of experiencing total collapse in the face of a small number of failures.

A Maille network is a peer-to-peer network where all nodes participate equally. Any node may use its peer relationships to acquire the trusted key of another node in the network. To do so, the Maille protocol is designed to find independent sources for the key of the target, and to maintain traceability of all who took part in finding the key. This combination allows the requesting node to know with a high degree of certainty that the key is authentic and for the target. Because the task of providing keys is not trusted to any one entity or small group of entities, one organization cannot use control of authentication to subvert the larger system.

REFERENCES

Adams, C., and Lloyd, S., 1997, Profiles and Protocols for the Internet public-key infrastructure, *Proc. 6th IEEE Workshop on Future Trends of Distributed Computing Systems*, pp. 220–224.

Bird, R., Gopan, I., Herzberg, A., Janson, Ph., Kutten, S., Molva, R., and Yung, M., 1995, The KryptoKnight family of light-weight protocols for authentication and key distribution, *ACM/IEEE Transactions on Networking*, **3**(1):31–41.

Capkun, S., Buttyan, L., and Hubaux, J.-P., 2003, Small worlds in security systems: an analysis of the PGP certificate graph, *Proc. 2003 ACM Workshop on New Security Paradigms*, pp. 28–35.

Foster, I., Kesselman, C., Tsudik, G., and Tuecke, S., 1998, A security architecture for computation grids, *Proc. 5th ACM Conference on Computer and Communication Security*, pp. 83–92.

Kahn, C., 1998, Tolerating penetrations and insider attacks by requiring independent corroboration, *Proc. 1998 ACM Workshop on New Security Paradigms*, pp. 122–133.

Reiter, M. K., and Stubblebine, S. G., 1997, Path independence for authentication in large-scale systems, *Proc. 4thACM Conference on Computer and Communications Security*, pp. 57–66.

Steiner, J, G., Neuman, C., and Schiller, J. I., 1988, Kerberos: an authentication service for open network systems, *Proc. 1988 Winter Usenix Conference*, pp. 191–201.

Zhou, L., Schneider, F. B., Van Renesse, R., 2002, COCA: A secure distributed online certification authority, *ACM Transactions on Computer Systems*, **20**(4):329-368.

Zimmermann, P., 1995, *The Official PGP User's Guide*. MIT Press, Cambridge, MA.

SUPPORTING END-TO-END SECURITY ACROSS PROXIES WITH MULTIPLE-CHANNEL SSL

Yong Song, Victor C.M. Leung, Konstantin Beznosov
Department of Electrical and Computer Engineering, University of British Columbia, 2356 Main Mall, Vancouver, BC V6T 1Z4, Canada

Abstract: Secure Socket Layer (SSL) has functional limitations that prevent end-to-end security in the presence of untrusted intermediary application proxies used by clients to communicate with servers. This paper introduces Multiple-Channel SSL (MC-SSL), an extension of SSL, and describes and analyzes the design of MC-SSL proxy channel protocol that enables the support for end-to-end security of client-server communications in the presence of application proxies. MC-SSL is able to securely negotiate multiple virtual channels with different security characteristics including application proxy and cipher suite.

Keywords: Network security; mobility; authentication; integrity; privacy.

1. INTRODUCTION

In this paper, we propose multiple-channel SSL (MC-SSL), a new protocol based on TLS/SSL [1] (for brevity, referred in this paper as SSL). It enjoys several advantages over current SSL variants. First, MC-SSL can significantly strengthen end-to-end security when application proxies or gateways are involved. Second, MC-SSL supports multiple cipher suites in the same connection so that different degrees of protection are available for client-server communications. Third, MC-SSL can flexibly satisfy various security requirements for content delivery because new factors, such as security policies, device capabilities, and security attributes of contents, are introduced into the security model.

MC-SSL is especially helpful for resource-constrained devices such as PDAs and cellular phones because they tend to need application proxies for functions such as content transformation or virus scanning. Such devices can also better utilize battery and other resources by selectively choosing the degree of cryptographic protection for data transmission. Moreover, because MC-SSL is a general protocol that can negotiate multiple security channels, it can flexibly meet the requirements from various terminals, servers, applications, and users. In this paper we mainly describe those parts of MC-SSL that are relevant to the support of end-to-end security in the presence of application proxies. For example, we do not present the concrete protocol related to multiple cipher suites. Instead, we focus on the high-level model issues and the detailed design of the proxy channel protocol, which is considered to be a prominent feature of MC-SSL. The MC-SSL architecture supports two types of channels between a client and a server: end-to-end channels and proxy channels. The proxy channel protocol is able to securely set up and use a proxy channel, and hence greatly improves the end-to-end security when a proxy is needed.

The remainder of the paper is organized as follows. Section 2 analyzes the functional limitations of SSL that triggered our work on MC-SSL. Section 3 presents the high-level description of MC-SSL. Section 4 discusses related work. Section 5 presents the proxy protocol, which is the focus of this paper. Section 6 discusses advantages and disadvantages of the proxy protocol. Section 7 concludes the paper, and describes future work.

2. PROBLEM MOTIVATION

Although it is currently a *de facto* security protocol at transport layer for Internet applications, SSL has several functional limitations. First, while SSL can provide an application with a secure point-to-point connection, it does not have facilities to securely deal with application proxies: if a proxy **P** is involved between a client **C** and a server **S**, **C** will normally set up a SSL connection with **P**, and then **P** will act as the delegate of **C** and set up another SSL connection with **S**. These kinds of proxies are needed for functional or performance reasons such as virus scanning, content transforming/filtering, or compression. For instance, in order to enable wireless terminals to access the Internet, WAP gateway architecture adopted the chain proxy model although the connection between **C** and **P** in the architecture is replaced with WTLS, a variant of TLS protocol. The SSL chain proxy model shown in the lower part of Figure 1, in which **P** can read and modify sensitive data at will, requires unconditional trust in **P** at least from one side, **S**'s or **C**'s. This can be satisfied only if **P** is administrated by the organization or individual that administrates **S** or **C** as well. In that case, the trust model is actually equivalent to the end-to-end trust model with two entities, and **P** can be merged with **S** or **C**. In other cases, **S** or **C** must take risks of information leakage and tampering at **P** because **P** is a third party in the trust model. Since communication through a third party application proxy requires **S** or/and **C** to trust **P** unconditionally with all the information passed through it, we consider the dilemma between passing all the data through **P** and not using a proxy at all as a limitation.

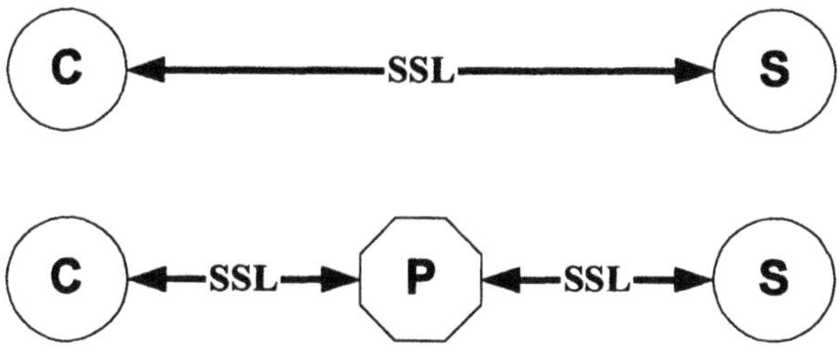

Figure 1. End-to-end model and chain proxy model

The second functional limitation of SSL is that it employs only one cipher suite at any given time. Although SSL allows re-negotiating the cipher suite of a connection, frequent re-negotiations are not practical because it is inefficient to change cipher suites back and forth using relatively expensive handshake protocol. Accordingly, lots of data is overly protected. For instance, a handheld user checks his corporate email inbox all day. He wants encryption for the id/password of his email account but he does not need strict confidentiality for his emails. It might be good enough for him to have good encryption for the id/password transmission, but no encryption for email contents. This way, the battery power will not be drained. However, the SSL architecture gives all or nothing dilemma. As another example, consider a handheld user who accesses the application server of her stock brokerage. She wants the stock prices to be accurate, but does not require them to be encrypted since they are publicly available; however, she requires the best security protection when she is temporarily transmitting id/password or doing transactions. In this case, using two different cipher suites are better than using only the strongest cipher suite because the battery power is greatly reserved. To summarize, the requirement for communication security does not entail the strongest cipher. Security is tightly related to other requirements. As pointed out by Abadi and Needham, encryption is not wholly cheap, and not asking precisely why it is being done can lead to redundancy [2]. The SSL's support for one cipher suite at a time combined with the relatively high cost of changing suites just in time makes it difficult for applications to optimize the strength of data protection according to the changes in the sensitivity of the data in the channel.

Further, to allow **S** to take various requirements into account and optimize a secure channel using power and other constraints, **C** may want to send **S** its terminal capabilities and the security policy configured for a particular application or server. For example, **C** could need to define whether proxies are allowed to be used for passing data with sensitivity below a certain level, and what types of data need proxies. Lack of negotiation support for proxies and multiple cipher suites is the third functional limitation of SSL, which directly results from the first and the second limitations. These functional limitations of SSL form an obvious gap between SSL and the requirements of real world applications and devices. When security-sensitive mobile applications become more popular, the gap will become more apparent.

3. HIGH LEVEL DESCRIPTION OF MC-SSL

Three key features of MC-SSL help us to address the above limitations of SSL. Going in reverse order, MC-SSL supports channel negotiation according to the parties' security policies, device capabilities, and security attributes of data.

To address second limitation (only one cipher suite at a time), MC-SSL provides a set of customizable secure channels in order to meet the practical requirements of different clients, servers, and applications. MC-SSL uses a multiple-channel model, in which each channel can possess its own characteristics including cipher suite and data flow direction.

To address the first limitation (the dilemma between an unconditionally trusted proxy and no proxy at all), we introduce a special type of channel: proxy channel, which enables MC-SSL to support partially trusted proxies, the focus of this paper. Figure 2 shows the conceptual proxy model of MC-SSL, in which three SSL connections form a triangle. **C-P-S** is a proxy channel, and **C-S** is an end-to-end channel. In this model, **C-P-S** is no longer an independent proxy channel that is shown in Figure 1. Instead, it relies on the **C-S** channel to control channel negotiation and application data delivery. Besides, **C** and **S** can deliberately choose **C-S** or **C-P-S** to deliver data according to the requirements for proxy and the sensitivity of data. As a result, sensitive data, such as id/password or credit card number, do not have to be exposed to **P**. The protocol for proxy channels is described in Section 5 and discussed in Section 6.

A MC-SSL session may negotiate zero or more proxy channels. Each of them and the corresponding end-to-end channel form a triangular relationship with the proxy as the third vertex. Theoretically, the maximum number of proxy channels solely depends on the available resources at **C** and **S**. However, a proxy channel currently supports only one proxy. We expect that for most practical client-server applications, one-hop proxy channels should suffice because multiple proxies can be transformed into a "proxy cluster" in which one proxy acts as the cluster head. On the other hand, we are planning to extend MC-SSL to support multi-hop proxy channels so as to further generalize the model of MC-SSL.

In SSL, a cipher suite consists of a key exchange algorithm, a cipher, and a hash algorithm, e.g., {RSA, 3DES_EDE_CBC/168, SHA-1}. The hash algorithm is used to compute Message Authentication Code (MAC). In MC-SSL, a cipher suite consists of only two elements: a cipher for data encryption/decryption, and a hash algorithm for MAC. We can define it as a structure as follows:

{*cipher and key size, hash algorithm for MAC*} (1)

A MC-SSL connection can have multiple cipher suites. We can characterize a point-to-point connection as follows: {*point 1, point 2, key exchange algorithm,* {*cipher suite 1, cipher suite 2, ...*}}, where each cipher suite forms a channel. Every MC-SSL connection must first negotiate a cipher suite strong enough to form the primary (or backbone) channel, which is employed to set up and control other channels, named as secondary channels. Figure 3 illustrates a sample connection between **A** and **B**, which is characterized by {**A**, **B**, RSA, {CS1, CS2, CS3, CS4}}, where RSA is the key exchange algorithm, and CS1 to CS4 are four different cipher suites. The primary channel is channel 1, for which CS1 is used.

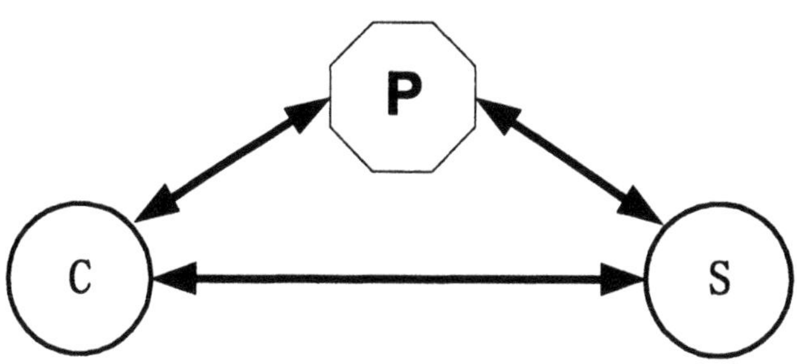

Figure 2. Triangular proxy model of MC-SSL

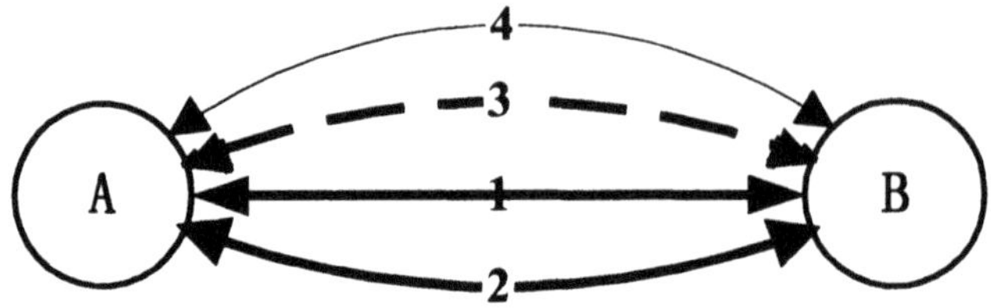

Figure 3. Multiple cipher suites inside a point-to-point connection

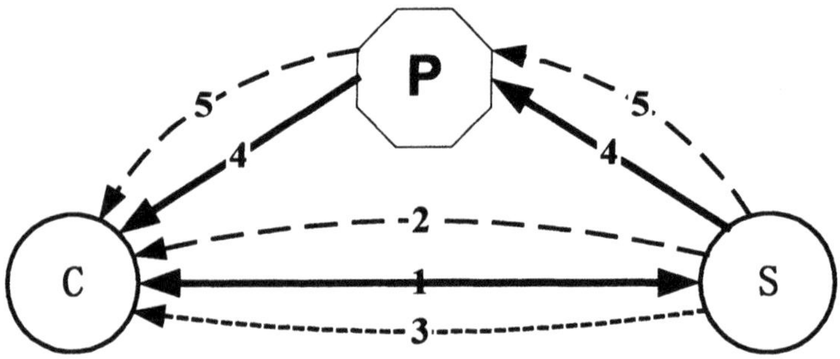

Figure 4. Multiple-channel model of MC-SSL

A combination of the proxy model and the multiple cipher suites produces the multiple-channel model shown in Figure 4. In MC-SSL, a channel can be defined as a virtual communication "pipe" with or without intermediate application proxies. Two MC-SSL endpoints communicate with each other through the pipe using a cipher suite. In addition, a channel can be either duplex, or simplex with a flow direction. We can characterize a MC-SSL channel with a set of attributes:

$$\{channel\ id,\ endpoint1,\ endpoint2,\ proxy,\ direction,\ cipher\ suite\} \quad (2)$$

Channel id is the identifier of a channel in a MC-SSL session context. Endpoint1 and endpoint2 are either DNS names or IP addresses of corresponding machines. Proxy attribute is null if a channel is an end-to-end channel; otherwise, it is the DNS name or IP address of the proxy in a proxy channel. Direction can be one of the following values: D, C, and S. D indicate a duplex channel; C or S indicates a simplex channel pointing to **C** or **S**, respectively. Cipher suite has been defined in expression (1). Figure 4 shows a sample MC-SSL session having five channels. Channel 1 and 4 are primary (or backbone) channels, and channel 2, 3, and 5 are secondary channels. In addition, only channel 1 is a duplex channel for application data; others are simplex channels from **S** to **C**. Typically, **C** can use channel 1 to send encrypted requests to S, and **S** can choose one of the five channels to send back the responses according to the contents. In addition, these channels are negotiated in the following order: channel 1 is the first; channel 2 and 3 are negotiated through channel 1; channel 4 is the first proxy channel, which is also negotiated through channel 1; channel 5 is negotiated through channel 4. Except for channel 1, other channels can be set up at any time. In Section 5 we present the protocol to negotiate and use primary channels, especially primary proxy channels.

4. RELATED WORK

We have not found other published research work that resembles MC-SSL protocol and this framework as a whole although there are other approaches that address the issues we are concerned about.

There are a few other approaches that address the end-to-end security for the case of chained proxies. One solution is to make application data unreadable to **P** by end-to-end encryption-based tunneling. For instance, the approach proposed by Kwon et al. [3] requires **C** to encrypt data twice: first for **S** using K_S, and then for **P** using K_P. Consequently, functions at the application layer, such as content transformation and virus scanning, cannot be performed by **P**.

Another solution is to simultaneously set up a SSL connection and a SSL chain between **C** and **S**, both shown in Figure 1. This approach is adopted by Kennedy [4]. To provide confidentiality, sensitive data is sent through the end-to-end connection instead of **P**. This is a typical approach, but it is insecure for the following reasons: most web/application servers still authenticate their clients using id/password. If **C** trusts **P** and gives its id/password to **P**, then **P** can impersonate **C** in unconstrained fashion. There are a number of solutions for **C** to avoid exposing id/password to **P** including sharing the master key or the symmetric session keys with **P**, or helping **P** sign the verification data. However, **P** can still impersonate **C** in a session and conduct person-in-the-middle attacks. MC-SSL is securer than this approach. In MC-SSL, every entity authenticates each other with their genuine identities; therefore, there is no impersonation in MC-SSL. Besides, every connection between any pair of **C**, **S**, and **P** has exclusive session keys.

A simple extension to SSL has been proposed by Portmann and Seneviratne [5] in order to get an extra cleartext channel without encryption and MAC protection; however, the security strength of this method is questionable because the cleartext channel is established without strict negotiation, and moreover, a malicious attacker can modify and inject cleartext data at will if there is no security protocol at the application layer. Besides, their approach is not capable of creating other types of channels except the primary end-to-end channel and the extra cleartext channel.

Finally, we would like to compare MC-SSL with XML security solutions including XML Security [6,7] and Web Services Security (WSS) [8,9]. XML Security is a set of core specifications that define XML syntaxes to represent encryption, hash, and digital signature. WSS is a framework that unites a number of existing and developing specifications for the purpose of constructing comprehensive security solutions for XML-based Web services. WSS is based on XML Security. Compared with XML Security and Web Services Security, MC-SSL is a complete and compact protocol under application layer, which is able to provide authentication, key exchange, and secure data transportation for client-server applications with or without the needs of proxies. On the other hand, both XML Security and WSS are not self-contained protocols, and they do not attempt to specify a fixed security protocol for authentication and key exchange so that they can have the extensibility and flexibilities to integrate existing or new security technologies at different layers. As with SSL, MC-SSL can be combined with XML Security, or adopted by WSS for securing Web service. For example, by combining XML Security with MC-SSL, an application can use MC-SSL to do authentication and key exchange for client, server, and proxies, and use XML Security to perform complex encryptions and/or digital signatures on application data.

5. PROXY CHANNEL PROTOCOL

In this section, we explain the design of proxy channel protocol in MC-SSL. We deliberately designed the proxy protocol as a protocol layer on top of SSL as shown in the right part of Figure 5. The left part shows the current Internet architecture. Such a design can keep underlying SSL protocol unchanged if a client needs only the MC-SSL proxy protocol without multiple cipher suites, or needs only SSL to access an SSL-based server without MC-SSL proxy protocol. The whole protocol described in this section deals with primary channels including the primary end-to-end channel and primary proxy channels, such as channel 1 and 4 shown in Figure 4.

The proxy protocol consists of three sub-protocols: handshake, application data, and alert protocols. Handshake protocol sets up the proxy channel, application data protocol defines messages for transporting application data, and alert protocol conveys warnings and fatal errors. The alert protocol of MC-SSL is similar to that of SSL. We do not describe the alert protocol in this paper. Please refer to RFC 2246 [1].

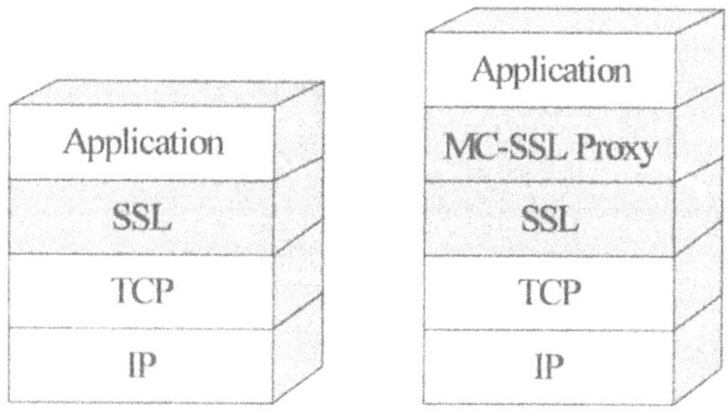

Figure 5. SSL and MC-SSL proxy protocol in the Internet layered architecture

5.1 Handshake protocol

A full handshake is shown in Figure 6. There are four stages: **C-S** handshake, **C-P** handshake, **P-S** handshakes, and confirmations of success. The proxy protocol is based on SSL. Three SSL connections are established to provide basic authentication, confidentiality, and data integrity between peer points. In addition, the **P-S** connection may be replaced by a permanent SSL or an IPSec connection.

After setting up a SSL connection, **C** and **S** exchange a pair of hello messages to initiate a MC-SSL session. Both hello messages, i.e. MC_CLIENT_HELLO and MC_SERVER_HELLO, have the following fields: 1) protocol version; 2) session id, which is generated by **S** to identify a MC-SSL session; 3) MAC key, which is used for the MAC in APP_DATA_CONTROL_PROXY messages; 4) the hash algorithm for MAC.

C then sends two messages to **S**: CLIENT_SECURITY_POLICY and CLIENT_CAPABILITIES. They tell **S** about security policy and device capabilities of **C**. Security policy may define whether a proxy is allowed to deliver a certain type of information. Device capabilities include hardware and software information such as screen resolution, power, CPU, memory, OS, browser capabilities, virus scanning capability, etc.

After a MC-SSL session is started, **S** or **C** can start negotiating a proxy channel at any time when necessary. In Figure 6, **S** sends **C** a PROXY_SUGGESTION_S2C message, which contains information such as the purpose of the proxy, the channel direction (simplex or duplex and so on), the DNS name and the certificate of the proxy. **C** sends

PROXY_REQUEST_C2S back to S. It has similar fields as the previous message. S responds with PROXY_REQUEST_RESPONSE_S2C to give the final decision.

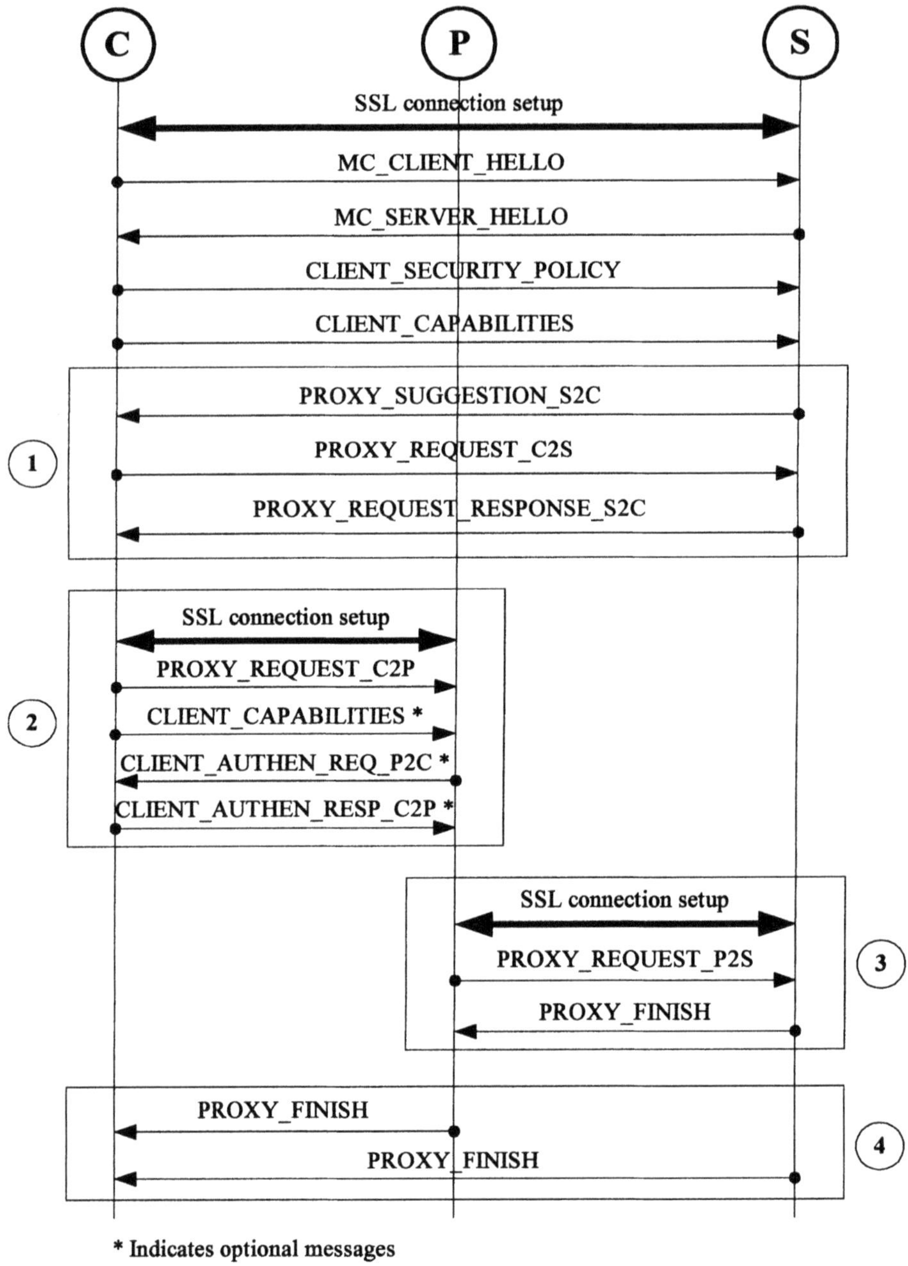

Figure 6. Message flow for a full handshake

The **C-P** handshake also starts with a SSL handshake, and then **C** sends **P** a PROXY_REQUEST_C2P message to inform **P** the session id, processing needed, channel direction, preferred authentication methods, handshake type, and the IP address and port number of **S**. In addition, a flag indicates if CLIENT_CAPABILITIES will follow. CLIENT_AUTHEN_REQ_P2C and CLIENT_AUTHEN_RESP_C2P is a pair of messages for **P** to authenticate **C**. The former tells **C** the authentication method, such as user id/password, challenge/answer, or PKI certificate, and the latter returns authentication data.

If the authentication is passed, **P** will start the handshake with **S**. Note that during the SSL handshake **S** and **P** should authenticate each other using their certificates. PROXY_REQUEST_P2S in Figure 6 carries a session id for **S** to bind the proxy channel with the corresponding end-to-end channel. The last three messages return the final result of negotiating a primary proxy channel.

As SSL does, MC-SSL supports the resumption of a cached session, which results in an abbreviated handshake. Some messages in Figure 6, such as the client and server hello messages, the **C-P** and **P-S** proxy request messages, and the final proxy finish message, are still necessary, but others are omitted in an abbreviated handshake.

5.2 Application data protocol

The purpose of the application data protocol is to transport application data between **C** and **S**. There are two ways to do this: one is through the end-to-end channel as shown in the A part of Figure 7; the other is through the proxy channel under the control of the end-to-end channel, as shown in the B part of Figure 7. Figure 7 only shows that **S** sends data to **C**. The transmission in the opposite direction uses the same messages.

To use the end-to-end channel, the sender encapsulates data into an APP_DATA_DIRECT message and sends to the receiver through the end-to-end channel. To transport a piece of content through the proxy channel, three messages are involved: APP_DATA_TO_PROXY, APP_DATA_FROM_PROXY, and APP_DATA_CONTROL_PROXY. They have the same sequence number. The sender wraps the content into APP_DATA_TO_PROXY, and sends it to **P**. After processing the content, **P** generates APP_DATA_FROM_PROXY, and forwards the result to the receiver. Besides, the sender will directly send the receiver an APP_DATA_CONTROL_PROXY message in order to control the behavior of **P**. The rest of this section briefly describes these three messages.

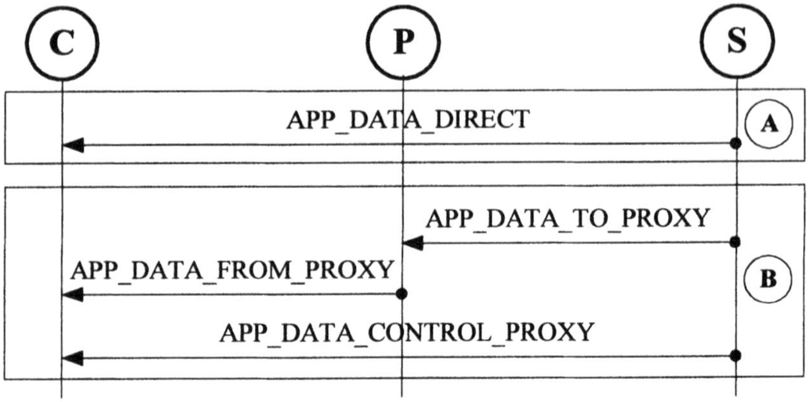

Figure 7. Message flow for transporting application data

APP_DATA_TO_PROXY contains such fields as content, processing request, and change restriction. Field content can be a complete or fragmental piece of content. Processing request tells **P** how to process the content. Change restriction indicates if the content is unchangeable, modifiable, or discardable. APP_DATA_FROM_PROXY includes such fields as content and result. The former is the processed content. The latter outlines the processing result. APP_DATA_CONTROL_PROXY conveys information such as proxy channel id, content attributes, change restriction, and MAC. Proxy channel id indicates which proxy channel the corresponding APP_DATA_TO_PROXY has gone through. Field content attribute is a compound string that describes some attributes of content, e.g. content types. The benefit to know about content types is that the receiver can test if **P** has injected new types of potentially dangerous code. MAC is used for verifying the integrity of unmodifiable content, and is calculated with HMAC hash function described in RFC 2246 [1]. MAC keys and the hash algorithm are negotiated in the initial hello messages.

6. DISCUSSION OF PROXY PROTOCOL

As pointed out in section 2, when an application proxy is needed, some applications allow or force **C** to use a SSL chain to access **S** through a proxy such as a WAP gateway. Consequently, the end-to-end security completely depends on the degree of trustworthiness of **P**. However, it is rare to have an unconditionally trusted **P** administrated by a third party. MC-SSL lets **C** and **S** themselves decide whether or not to have a proxy channel and how to use it after having the secure end-to-end channel in place. Besides other advantages, this feature enables making such a decision after both parties have determined the degree of trust in each other, which could be a function of the corresponding credentials.

The possible misuse of a proxy is another problem of the SSL chain model. For instance, a user may mistakenly keep using the same proxy configured in his browser even when a proxy or gateway that is hosted (and therefore trusted) by a bank is available for online banking. To avoid such a misuse, when starting a secure session, **C** should always first connect to **S** rather than **P**, then negotiate a proxy with **S**, and finally connect to **P** to set up the proxy channel. That is the order in which MC-SSL proceeds.

The following is our informal discussion of why MC-SSL proxy protocol can enhance end-to-end security in the presence of proxies.

1. **P** is authenticated by **S** as a proxy instead of a client; therefore, **P** cannot impersonate **C**. **P** is not only authenticated by its certificate, but also by the session id received from **C** as a security token especially if session id is implemented to be a cryptographically random string. Moreover, **S** has already obtained the certificate of **P** before verification.
2. As described earlier, because **C** first connects and negotiates with **S** rather than **P**, and the proxy is decided after **S** already has the security policy and terminal capabilities of **C**, the risk of misusing proxies is reduced. **S** can decide if **C** needs a proxy and what type of proxy is needed, and even suggest a proxy.
3. Sensitive data such as id/password and credit card information can be transported through the end-to-end channel, while relatively non-sensitive data including Web pages, software, and email attachments can go through proxies for content scanning or transformation. Note that content scanning can increase system security. Neither does an end-to-end SSL connection nor a SSL proxy chain provide both benefits.
4. A proxy channel can be explicitly negotiated as a one-way channel, which eliminates the chance that **P** turns a response channel into a request channel. For instance, we can have a one-way proxy channel that only allows responses from **S** to **C**; all requests from **C** to **S** have to go through the secure end-to-end channel. As a result, **P** cannot send any fake request to **S** to trigger a transaction.
5. When unmodifiable data is delivered through **P**, the MAC of the data is calculated and sent to the receiver through the end-to-end channel, and hence **P** cannot modify the data without being detected.
6. When modification is permitted, content attributes such as content types are sent to the receiver through the end-to-end channel. Information of content types can prevent **P** from injecting dangerous code or contents.

We also need to analyze security issues still unsettled in MC-SSL proxy protocol. Since the proxy protocol is on top of SSL, the strength of communication security between two peer points is no weaker than that of SSL. In addition, the cryptographic technique of the end-to-end MAC is also inherited from SSL. Therefore, we do not search for vulnerabilities related to the cryptographic techniques, and we mainly analyze the security issues at the protocol level.

1. It is not easy for **S** and **C** to decide on using which channel to deliver data: end-to-end channel or proxy channel. There are a number of questions to answer such as the following: What is the sensitivity of the data? Is the data too sensitive to be delivered through the proxy? What if sensitive data needs a proxy? What is the worst consequence if **P** modifies the data? What if insensitive data mixes with sensitive data? And so on. For **S**, the security attributes of contents such as sensitivity level need to be defined beforehand. Moreover, **S** needs to get the security policies to help answer those questions. For **C**, it is normally much harder to answer them. A conservative strategy is for **C** to choose the end-to-end channel whenever the answer is unclear. Fortunately, in typical Internet applications, **S** does not have to use a proxy to process the data from **C**. To summarize, **S** and **C** must define their security policies, security attributes of data or contents, and device capabilities. How to define them is beyond the scope of this paper.
2. The proxy protocol alone cannot guarantee that **P** has correctly performed its task. When data is modifiable, there are mainly two types of threats: first, **P** can modify requests or responses between **C** and **S**; second, **P** could inject code if the original content has some embedded code. The strategy to thwart the first threat is neither to send sensitive data through an untrustworthy proxy, nor to use any data from it as sensitive data. To mitigate the second threat, **S** may deliver contents without embedded code, or with embedded code which type is harmless to **C**.

For some of the above problems, XML Security [6,7] is a good solution if both **C** and **S** support it because XML is very flexible to describe data and its attributes. We can use different keys to protect different parts of a XML document. For instance, **S** can use the key for the end-to-end channel to encrypt a XML element that requires end-to-end confidentiality before **S** wraps the XML document into an APP_DATA_TO_PROXY message. As mentioned in Section 4, XML Security can be well combined with MC-SSL to further enhance end-to-end security.

7. CONCLUSIONS AND FUTURE WORK

In this paper, we present Multiple-Channel SSL, a new security protocol extended from SSL. The multiple-channel model of MC-SSL is more flexible and general than SSL, and hence it is able to satisfy diverse requirements for different applications, especially for emerging mobile or wireless applications. MC-SSL exhibits three advantages: first, it improves end-to-end security in the presence of application proxies; second, MC-SSL supports multiple cipher suites in the same connection so that appropriate communication security can be selectively applied to different data or contents; third, MC-SSL supports channel negotiation according to security policies, device capabilities, and security attributes of contents.

We further present the proxy channel protocol including handshake protocol and application data protocol. The handshake protocol completes negotiation, authentication, key exchange, and channel binding, while the application data protocol supports delivering application data through a proxy channel. Besides analyzing the improvement in end-to-end security, our discussion examines some unsettled issues and also suggests solutions to address them.

MC-SSL is our ongoing research work, which we divide into multiple phases. In the first phase, we developed the model and the proxy protocol. In the second phase, we develop the protocol to support multiple cipher suites, and also extend MC-SSL to support multi-hop proxy channels.

REFERENCE

[1] T. Dierks and C. Allen, "The TLS Protocol Version 1.0," RFC 2246, Jan. 1999

[2] M. Abadi and R. Needham, "Prudent Engineering Practice for Cryptographic Protocols," *IEEE Trans. on Software Engineering*, Vol. 22, No. 1, pp. 6-15, Jan. 1996

[3] E.K. Kwon, Y.G. Cho, and K.J. Chae, "Integrated Transport Layer Security: End-to-End Security Model between WTLS and TLS," *Proc. IEEE 15th Int. Conf. on Information Networking*, Jan. 2001

[4] D. J. Kennedy, "An Architecture for Secure, Client-Driven Deployment of Application-Specific Proxies," Master Thesis in Computer Science, University of Waterloo, 2000

[5] M. Portmann and A. Seneviratne, "Selective Security for TLS," *Proc. IEEE 9th Int. Conf. on Networks*, pp. 216-221, Oct. 2001

[6] W3C, *XML Signature Recommendations*, http://www.w3.org/Signature/, Feb. 2002

[7] W3C, *XML Encryption Recommendations*, http://www.w3.org/Encryption/, Dec. 2002

[8] IBM Corp. and Microsoft Corp., "Security in a Web Services World: A Proposed Architecture and Roadmap," http://www-106.ibm.com/developerworks/webservices /library/ws-secmap/, Apr. 2002

[9] OASIS Open, "Web Services Security: SOAP Message Security," http://www.oasis-open.org/committees/documents.php?wg_abbrev=wss, working draft 17, Aug. 2003

A CONTENT-PROTECTION SCHEME FOR MULTI-LAYERED RESELLING STRUCTURES

Pei-Ling Yu, Pan-Lung Tsai, and Chin-Laung Lei
Department of Electrical Engineering, National Taiwan University

Abstract: Since the idea of digital watermarks was proposed, various watermarking schemes have been developed to protect digital contents in electronic transactions. However, most schemes limit themselves to the simplified buyer-seller model that can be applied to only a small number of cases rather than common scenarios in real life. In this paper, a more practical watermarking scheme is proposed to protect digital contents in real-world transactions using PKI (Public-Key Infrastructure). The proposed scheme is considered more practical in the sense that one or more reselling agents may exist between buyers and the original seller, and the seller can package the merchandise in advance without the involvement of buyers. The packaged merchandize can hence be made publicly accessible (e.g., on top of an open shelf) for the purpose of demonstration and circulates in the market. The proposed scheme also preserves the anonymity of the buyers while guaranteeing that the distributor of any illegal copy will be unambiguously identified.

Keywords: Digital watermark, Public-Key Infrastructure, content protection, anonymity, privacy protection

1. INTRODUCTION

In modern times, a wide variety of materials have been digitized to facilitate the processing of information. Unfortunately, the convenience of data manipulation encourages many kinds of software piracy in that the efforts of retrieving and duplicating contents have been greatly reduced as well. To alleviate the problem of piracy, various content-protection schemes have been proposed to restrict the access and use of digital contents to legitimate users only. Content-protection schemes in electronic commerce generally consist of two important parts, one for ensuring that only those users who have paid are granted access to digital contents, and the other for tracking down the unauthorized content distributors who should be responsible for the pirated copies found in the market.

In the content-protection schemes [1, 2, 3, 4, 5] that make use of the digital watermarking technology [6, 7, 8, 9, 10], a watermark uniquely associated with each transaction is inserted into the copy of the digital content to be sold. In order to achieve the goals of secure and fairness, the seller and the buyer are required to perform the watermark insertion collaboratively using cryptographic techniques when carrying out the transaction. Such schemes guarantee that neither the seller nor the buyer alone is able to remove the watermark embedded in the copy of the digital content. When a pirated copy of the digital contents is found sometime later, the watermark can then be extracted and/or detected to identity a particular transaction and the corresponding buyer, who should now be responsible for the piracy.

Such watermark-enabled content-protection schemes have several major drawbacks. Firstly, these schemes are primarily designed to deal with purely electronic transactions where the entire digital content is going to be transferred across the Internet. This can be prohibitive or unreliable for large-volume digital contents such as digital videos. Secondly, the preparation of the merchandize requires the involvement of buyers so that buyer-specific information can be used to generate effective watermarks. This prevents the original seller of the digital content to pre-package the merchandize without knowing the actually buyers. Thirdly, most schemes focus on the direct-sale business model and consider only the interactions between the seller and the buyer in the transaction. In real-life scenarios, it is common to have intermediary reselling agents, and these schemes simply fail on such multi-layered reselling structures. To sum up, these schemes are not practical and can barely used in real-world transactions.

Taking ISO (International Organization for Standardization) Store [11], the on-line shopping services for ISO publications, as an example will make previous discussions more clear. In order to protect its copyright, ISO Store requires a buyer to register at first, and then dynamically inserts the buyer's name and affiliation into the documents sold in the transaction. Such approach is hardly applicable to merchandize that is high-volume or accompanied by non-digital materials, and can only be realized in purely electronic transactions. It works exclusively for direct-sale model and does not allow the merchandize to be pre-packaged. As a result, the circulation of the merchandize through reselling agents is prohibited.

Another important goal of a practical content-protection scheme is to preserve the privacy of buyers. Although the substantial objective of content-protection schemes is to track down the unauthorized distributors, we cannot simply presume the guilt of buyers by treating them as potential distributors of pirated copies of digital contents and hence deprive buyers of their privacy. Besides, the rights to shopping anonymously are essential to current practice of commercial transactions.

In this paper, a practical content-protection scheme is proposed to achieve the goal of protecting digital contents in real-world transactions. The proposed scheme integrates the digital watermarking technology with Public-Key Infrastructure (PKI) [12] and allows the existence of intermediary reselling agents, in addition to the basic direct-sale business model. It also permits the original seller to pre-package the merchandize without the involvement of buyers (i.e., without knowing who the buyers are during merchandize preparation). The privacy of buyers and the rights of reselling agents are also considered. The rest of the paper is organized as follows. Section 2 describes the proposed scheme in detail, and section 3 presents the security analysis along with further discussions. Section 4 finally concludes the paper.

2. THE PROPOSED SCHEME

In this section, a content-protection scheme using digital watermarks and PKI is presented with each phase in an individual subsection. The goals to be achieved are:

a) Full access to a single copy of the digital content to be sold is granted only after the buyer successfully registers himself to the original seller for that copy. On the other hand, reselling agents may decide to present samples with lower quality (e.g., thumbnails, movie trailers, etc.) or in non-digital forms (e.g., posters) to attract potential buyers, as in real-world commercial transactions.
b) When a pirated copy is found in the market, the original seller can always track down the real identity of the responsible distributor, who must be a buyer in certain transaction happened previously.
c) The privacy of buyers is well protected. As long as a buyer does not commit piracy, he is guaranteed to remain anonymous during transactions and arbitrations.

The proposed scheme is based on the following assumptions:

1. At least one robust watermarking insertion, extraction, and/or detection algorithm exists for the digital content to be sold. Watermarking schemes

for certain types of digital contents have been well studied, as mentioned in section 4, while others may still need more research efforts. Nevertheless, as more and more researchers devote themselves to this area, this assumption will definitely hold for more and more types of digital contents.
2. A PKI implementation has been deployed to provide adequate authentication and non-repudiation services.
3. The original seller does not intentionally frame a buyer by distributing illegal copies herself and then accusing the buyer of piracy. In subsection 3.3, a possible extension to the proposed scheme dealing with malicious sellers will be described and hence further relax this assumption.

Several roles are used when explaining the details of the proposed scheme. They are listed and briefly described as follows.

S: The seller, who is the rightful owner of the original digital content and wants to make a profit on the sales of it.
B: The buyer, who wants to purchase a copy of the digital content from **S** via intermediary reselling agents.
CA: A trusted certification authority, who is responsible for issuing anonymous certificates.
ARB: A decent arbiter, who will make righteous judgments on disputes according to the evidences presented.

Some notations used throughout the rest of the paper are defined as:

$U \oplus V$: The watermarked/fingerprinted copy of digital content U. The binary operator $\oplus$ denotes the operation of watermark insertion, and V is the chosen watermark to be inserted.
$Priv_{\mathbf{I}}$: The private key of identity **I**.
$Pub_{\mathbf{I}}$: The public key of identity **I**.
$Cert_{\mathbf{I}}$: The digital certificate issued to subject **I** by **CA**. By using X.509-compliant digital certificates, $Priv_{\mathbf{I}}$ and $Pub_{\mathbf{I}}$ can be derived from $Cert_{\mathbf{I}}$.
$Sign_{Priv\mathbf{I}}(M)$: The signature of message M signed by **I** with $Priv_{\mathbf{I}}$.
$E_{Pub\mathbf{I}}(M)$: The ciphertext of message M encrypted with $Pub_{\mathbf{I}}$. The encryption can be performed by anyone.
$D_{Priv\mathbf{I}}(C)$: The original message (plaintext) of ciphertext C decrypted by **I** using $Priv_{\mathbf{I}}$.
$Enc_J(M)$: The ciphertext of message M encrypted with symmetric key J. The encryption can be performed by anyone that knows J.

$Dec_J(C)$: The original message of ciphertext C decrypted with symmetric key J. The decryption can be performed by anyone that knows J.

2.1 Initialization

In order to remain anonymous during transactions and (possibly) unsolicited arbitrations, **B** first applies to **CA** for an anonymous certificate. An anonymous certificate is a special type of digital certificates with the corresponding subject field being a pseudonym rather than the real identity of the applicant. Let **P** denote the pseudonym used by **B**. The information of binding the issued anonymous certificate, $Cert_{\mathbf{P}}$, to **B** will then be kept safely by **CA**. Upon receiving the anonymous certification, **B** saves it in his personal computer, and stores another copy in a specialized smart card so that the anonymous certificate can be easily used whenever needed.

B may use a single anonymous certificate in multiple transactions, or he may request a distinct anonymous certificate for each individual transaction. Using the same anonymous certificate in many transactions to the same seller may allow the seller to gain extra knowledge about the purchasing behavior of **B**. However, it is undoubtedly more convenient and more efficient for **B** to reuse his anonymous certificates. The issue of reusing anonymous certificates will be revisited in subsection 3.2.

2.2 Merchandize Preparation

For each copy of digital content X, **S** randomly generates a watermark W and a one-time key K for symmetric encryption. **S** then performs watermark insertion to get the watermarked/fingerprinted copy Y:

$$Y = X \oplus W \tag{1}$$

Then **S** encrypts Y with K to get the ready-to-ship form Z:

$$Z = Enc_K(Y) \tag{2}$$

S also derives a product key, PK, from K:

$$PK = Hash(K) \tag{3}$$

Note that *Hash* is a public one-way hash function. It has the following property that everyone who knows K can easily calculate PK, while it is considered impossible to do the reverse (i.e., figuring out K from PK).

Finally, **S** computes the digital signature SIG:

$$SIG = Sign_{PrivS}(PK, Z) \quad (4)$$

After the computation, the merchandize is packaged as (PK, Z, SIG). **S** then stores the tuple (PK, K, W) in her sales records and distributes (PK, Z, SIG), along with any accompanied materials (e.g., posters, free gifts, etc.), to one of the reselling agents.

Note that **S** does not need to know the actual buyers in the phase of merchandize preparation, and the merchandize is allowed to freely circulate in the market once it is packaged. This is exactly how most real-world commercial transactions proceed, in contrast to the common approach used in purely electronic transactions, which usually asks a buyer to authenticate himself beforehand so that buyer-specific information can be blended into the merchandize during preparation.

2.3 Merchandize Sale

The reselling agent who obtains sets of packaged merchandize from **S** may decide to resell them to next-level reselling agents, or directly to the real customers. If a reselling agent does not trust whom he gets the merchandize from, he may verifies the digital signature, SIG, included in the merchandize to ensure that the content has not been altered after the initial packaging. Subsequently, the agent may make the merchandize publicly accessible for the purpose of demonstration, as how he does to other non-digital merchandize. Once a buyer purchases a set of packaged merchandize, he then follows the registration and activation processes stated in the next two subsections, respectively.

2.4 Merchandize Registration

When **B** decides to buy a set of packaged merchandize, he performs the registration of the merchandize to **S** as part of the transaction. First **B** opens the package and extracts the product key, PK, of the merchandize. Then he inserts the smart card holding his anonymous certificate, $Cert_P$, into the card reader provided by the reselling agent, and types PK in the input device associated with the card reader. Upon receiving $Cert_P$ and PK, the program stored in the smart card is instructed to run in "registration mode" and begins to compute:

$$PKS = Sign_{Priv_P}(PK) \tag{5}$$

Recall that $Priv_P$ is the private key of pseudonym **P** and can be obtained directly from $Cert_P$. Then the program sends ($Cert_P$, PK, PKS) to **S**. When **S** gets ($Cert_P$, PK, PKS), she verifies the signature, PKS, and rejects the request if it is invalid. If the request contains a valid signature, **S** then searches her sales records for an entry associated with PK. If no such entry is found, **S** rejects the request. Otherwise, **S** further checks the status of the merchandize associated with PK, and also rejects the request if the merchandize turns out to have been registered. If everything goes fine, **S** extracts the corresponding K from the entry and computes:

$$AK = E_{Pub_I}(K) \tag{6}$$

Then **S** adds ($Cert_p$, PK, PKS) to the entry so that the merchandize is marked as registered, and sends the encrypted activation key, AK, back to the program. When the program running on the smart card successfully receives AK from **S**, it stores AK in the smart card and completes the transaction. If the request is rejected by **S** in the middle way, the program notifies **B** and aborts the transaction.

2.5 Merchandize Activation

After a successful transaction, **B** takes his merchandize and smart card home. Before **B** can make use of the digital content he just bought, he has to carry out one last action, activating the digital content. **B** starts by inserting the digital media (contained in the packaged merchandize) and his smart card into the corresponding readers. Then the program stored in the smart card is instructed to run in "activation mode" and invokes an auxiliary program running on **B**'s personal computer associated with the readers. The first thing the auxiliary program does is to verify that SIG is a valid signature of (PK, Z), followed by computing:

$$K = D_{Priv_I}(AK) \tag{7}$$

The auxiliary program proceeds to check the validity of K by evaluating equation (3) stated above. Finally **B** gets the watermarked/fingerprinted copy of the digital content by calculating:

$$Y = Dec_K(Z) \tag{8}$$

Note that an auxiliary program is needed here for the sake of computation efficiency. Although it is possible to perform the cryptographic operations mentioned in this subsection totally on the smart card, the restricted computation power of the smart card will result in intolerably long processing time when dealing with high-volume digital contents.

If anything goes wrong during activation, **B** may simply return the broken merchandize and ask for refunding. The proposed scheme ensures that **B** will be able to present concrete evidences to show that the merchandize is broken and he does not have to be responsible for it.

2.6 Arbitration

Disputes may arise when a pirated copy is found in the market. The arbitration proceeds as follows. By running an appropriate watermark extraction and/or detection algorithm on the pirated copy, **S** can easily find out the distributor, who must be a buyer in some previous transaction, from her sales records. **S** then presents corresponding ($Cert_P$, PK, PKS, W) to **ARB**. Upon receiving the request from **S**, **ARB** checks whether PKS is a valid signature of PK signed by **P** and W actually exists in the pirated copy. If both conditions are true, **ARB** sends $Cert_P$ to **CA** and asks **CA** to reveal the real identity of pseudonym **P**. Once knowing who the buyer is, **ARB** makes the final judgment.

3. DISCUSSIONS

In this section, we first present the security analysis of the proposed scheme, followed by pointing out the issue that arises from reusing anonymous certificates. In the last subsection, we describe a possible extension to the proposed scheme so that buyers can be further protected from malicious sellers, and hence the assumption that there is no ill-intentioned seller can be relaxed.

3.1 Security Analysis

The security of the proposed scheme primarily comes from the underlying watermarking scheme and PKI. As long as the watermarking scheme adopted is robust, the watermark embedded in a watermarked/ fingerprinted copy can always be extracted and/or detected. In addition, the authentication and non-repudiation services provided by PKI guarantee that nobody can lay his sins on others.

Based on the properties stated above, the proposed scheme is able to successfully protect the copyright of digital contents. Once a pirated copy is found in the market, **S** can unambiguously identify the anonymous certificate, $Cert_{\mathbf{P}}$, based on the watermark embedded in that copy. Besides, the evidences supplied by **S** during the arbitration are able to prove that **P** was indeed involved in certain previous transaction because nobody except **P** would be able to generate signature *PKS*.

The rights of **B** are also protected. On one hand, his real identity remains unexposed unless he is proven to have committed piracy. On the other hand, it is not possible for some malicious agents to sell an already-registered copy to **B** because **B** will be aware of the failure on his registration during the transaction. Furthermore, there is no chance for any reselling agent to peek into or alter the content of the merchandize since the merchandize is packaged in its encrypted form, *Z*, and has a digital signature, *SIG*, on it.

The digital signature, *SIG*, also contributes to the preservation of reselling agents' rights. It helps a reselling agent to ensure that the packaged merchandize has not been altered by his predecessors. Moreover, an honest reselling agent does not have to worry about having purchased already-registered copies because once a copy is found to have been registered, it is guarantee that the guilty one will always be unambiguously identified.

3.2 Reusing Anonymous Certificates

The anonymity of the proposed scheme basically relies on the introduction of a trusted third party, **CA**. Under such circumstances, if **B** uses a different anonymous certificate for each individual transaction, his real identity will never get exposed as long as **CA** is not compromised. However, this is both inconvenient and inefficient because **B** will then have to manage a large number of his certificates and **CA** will need tremendous storage for even more certificates of all buyers. An intuitive solution will be allowing **B** to reuse his anonymous certificate.

However, reusing a single anonymous certificate in many transactions to the same seller will make these transactions linkable and allow that particular seller to infer personal information about **B** via data-mining techniques. It is a potential threat because the seller may become able to take advantage of **B** without knowing **B**'s real identity. A simple way for **B** to alleviate the risk is to apply for several anonymous certificates and randomly choose one before a transaction takes place because doing so will increase the efforts for seller to infer useful information. As a matter of fact, the policy for reusing anonymous certificates depends on how **B** concerns his privacy. There is a tradeoff between the complication of certificate management and the degree of privacy preserved.

3.3 Protecting Buyers from Malicious Sellers

In the proposed scheme, **S** is assumed to have no bad intentions. This is necessary because **S** has direct access to every watermarked/fingerprinted copy that is going to be sold, and hence she can easily frame **B** by waiting for **B**'s registration and then distributes the registered copy herself. Following the arbitration process, **S** can successfully charge **B** with piracy, which in fact he never did, and make him be mistakenly judged as the guilty one, who should be responsible for the illegal distribution.

This issue is known as *customer's right problem* [4]. Some schemes [2, 5] have been developed to deal with this particular issue as follows. In order to prevent **S** from framing **B**, a watermark certification authority (**WCA**) is introduced to perform watermark generation in the encrypted domain. By encrypting the generated watermark with **B**'s public key, WCA inhibits **S** from knowing the actual watermark. **S** is still responsible for inserting the watermark to the digital content, but the insertion operation is performed in the encrypted domain and hence **S** has no access to the (decrypted) watermarked/fingerprinted copy that **B** will finally get.

Here we describe general guidelines about how the proposed scheme can be extended to overcome *customer's right problem* by following the spirits of the schemes mentioned above. We also introduce the existence of **WCA** for watermark generation, but, since we don't want the involvement of buyers in the phase of merchandize preparation, the schemes mentioned above cannot be applied directly. Instead, **WCA** generates an extra key pair for each request from **S**, and uses the public key of the key pair to encrypt the watermark. During the transaction, **B** is required to make an additional contact to **WCA** to get the corresponding private key so that he has enough information to activate the watermarked/fingerprinted copy later. As a result, this extended scheme successfully achieves the goal of protecting buyers from malicious sellers.

4. CONCLUSIONS

In this paper, a practical watermarking scheme is proposed to protect digital contents in real-world transactions using PKI (Public-Key Infrastructure). The proposed scheme enables the original seller to pre-package the merchandize without the involvement of buyers, and permits the circulation of packaged merchandize in the market by allowing one or more reselling agents to exist between the original seller and buyers. The digital watermarking technology is seamlessly integrated with PKI to make the proposed scheme easily applicable to real-world transactions.

In the proposed scheme, the copyright of the digital content belonging to the original seller is well protected, as the distributor of any pirated copy is guaranteed to be unambiguously identified. On the other hand, the privacy of buyers is preserved since a buyer can always stay anonymous unless he is proven to have committed piracy. The rights of reselling agents are not forgotten. Mechanisms are provided to detect broken merchandize during resale, and the seller's traceable records of registration also entitle reselling agents the immunity of accusations.

REFERENCES

1. J. Doming-Ferrer and J. Herrera-Joancomarti, "Efficient Smart-Card Based Anonymous Fingerprinting," Proceedings of International Conference on Smart Card Research and Applications (CARDIS '98), LNCS 1820, pp. 221-228, September 1998.
2. N. Memon and P. W. Wong, "A Buyer-Seller Watermarking Protocol," IEEE Transactions on Image Processing, Vol. 10, No. 4, pp. 643-649. April 2001.
3. B. Pfitzmann and A.-R. Sadeghi, "Coin-Based Anonymous Fingerprinting," Proceedings of 17th Annual IACR Eurocrypt Conference (EUROCRYPT '99), LNCS 1592, pp. 150-164, May 1999.
4. L.Qiao and K. Nahrstedt, "Watermarking Schemes and Protocols for Protecting Rightful Ownership and Customer's Rights," Journal of Visual Communication and Image Representation, Vol. 9, No. 3, pp. 194-210, September 1998.
5. C.-L. Lei and M.-H. Chan, "An Efficient Anonymous Buyer-Seller Watermarking Protocol," 2002 International Computer Symposium, Workshop on Cryptology and Information Security, December 2002.
6. G. Depovere, T. Kalker, and J.-P. Linnartz, "Improved Watermark Detection Reliability Using Filtering Before Correlation," Proceedings of IEEE International Conference on Image Processing (ICIP '98), Vol. 1, pp. 430-434, October 1998.
7. I. J. Cox, J. Kilian, F. T. Leighton, and T. Shamoon, "Secure Spread Spectrum Watermarking for Mulimedia," IEEE Transactions on Image Processing, Vol. 6, No. 12, pp. 1673-1687, December 1997.
8. J. R. Hernandez, F. Perez-Gonzalez, J. M. Rodriguez, and G. Nieto, "Performance Analysis of a 2-D-Multipulse Amplitude Modulation Scheme for Data Hiding and Watermarking of Still Images," IEEE Journal on Selected Areas in Communications, Vol. 16, No. 4, pp. 510-524, May 1998.
9. A. Herrigel, J. O. Ruanaidh, H. Petersen, S. Pereira, and T. Pun, "Secure Copyright Protection Techniques for Digital Images," Proceedings of Second International Workshop on Information Hiding (IH '98), LNCS 1525, pp. 169-190, April 1998.
10. N. Memon and P. W. Wong, "Protecting Digital Media Content," Communications of the ACM, Vol. 41, No. 7, pp. 35-43, July 1998.
11. ISO Store, http://www.iso.org/iso/en/prods-services/ISOstore/store.html.
12. IETF Public-Key Infrastructure (X.509) Working Group. http://www.ietf.org/html.charters/pkix-charter.html

AN ASYMMETRIC CRYPTOGRAPHY SECURE CHANNEL PROTOCOL FOR SMART CARDS

Konstantinos Rantos and Konstantinos Markantonakis
Information Security Group, Royal Holloway, University of London, Egham, Surrey, TW20 0EX, United Kingdom, k.rantos@rhul.ac.uk
Information Security Group Smart card Centre, Royal Holloway, University of London, Egham, Surrey, TW20 0EX, United Kingdom, k.markantonakis@rhul.ac.uk

Abstract: Smart card secure channel protocols based on public key cryptography are not widely utilised mainly due to processing overheads introduced in the underlying smart card microprocessors and the complexities introduced by the operation of a PKI infrastructure. In this paper we analyse the significance of public key secure channel protocols in multi application smart cards. We believe that multi application smart card technology (e.g. the GlobalPlatform smart card specification) should benefit more from the advantages of public key cryptography specifically for the initiation and maintenance of a secure channel. This paper introduces a public key based cryptographic protocol for secure entity authentication, data integrity and data confidentiality. The proposed secure channel protocol uses a combination of public key, secret key and the main idea behind the Diffie-Hellmann key establishment protocols in order to achieve the desired goals.

Key words: Secure Channel Protocol, Public Key Cryptography, Diffie-Hellmann, GlobalPlatform, Java Card, Multi-application smart cards

1. INTRODUCTION

In the recent years with the introduction of multi-application smart cards it became possible to securely host multiple applications, dynamically and securely download or delete them at any point during the card's lifecycle. As a result, the complexity of the smart card operating system

(SCOS) increased exponentially. Similarly, the complexity of the terminal applications increased significantly as new architectures [1, 2] emerged. Likewise, as smart card technology evolves (the performance of smart card cryptographic algorithms improves) and as new smart card applications are invented the benefits of public key cryptography are widely scrutinised.

Multi-application smart card technology can benefit from the use of public key cryptography both at the application level and in the SCOS level e.g. with the provision of secure channel protocols based on Public Key Infrastructures (PKI). Current versions of secure multi-application smart card standards [6] do not fully take advantage of the benefits of public key cryptography, specifically for the provision of a secure channel mechanism. The reasons range from the increased prices due to the additional processing power, up to the potentially limited performance of public key cryptographic primitives in the current generation of smart card microprocessors, or simply because there is no immediate need for such functionality.

The advantages and disadvantages of public key cryptography are widely documented in the academic literature [3, 4, 5]. In this paper we propose a public key secure channel protocol for smart cards. The protocol is based on the well known Diffie-Hellman key exchange protocol and it was designed by taking into account the processing and storage restrictions of current smart card microprocessors. Alongside with the protocol description we also provide a discussion on the operation and security requirements for its successful and efficient operation. We believe that as the number of smart card applications increases and the nature of smart card applications changes along with the differentiations on the operational requirements (e.g. dynamic application downloading and deletion), the demand for efficient smart card PKI will potentially increase.

The remainder of this paper is organised as follows. Firstly, we set up the scenery by elaborating more on motivation behind the paper along with providing an overview of the main characteristics of a multi-application smart card standard, namely GlobalPlatform [6]. Subsequently, we highlight the main characteristics of the supporting public key infrastructure required for the successful operation of the protocol. Moving to the core idea of this paper we present the protocol details and architectural design. In order to provide a more complete coverage of the issues surrounding the implementation and operation of the proposed architecture we also provide a discussion around the security properties of the protocol by highlighting practical issues that imposed certain design decisions. Finally, we discuss several practical issues and provide directions for further research.

2. PUBLIC KEY SMART CARD SECURE CHANNEL PROTOCOLS AND THE REAL WORLD

In the following sections we provide an overview of limiting factors along with the driving forces behind the adoption of public key cryptography in multi application smart card platforms. Similarly, we highlight the main characteristics of a widely used multi application smart card standard in order to provide a reference point to the specifics of an existing architecture along supporting the case for the existence of such a protocol.

2.1 Motivation

The advantages and disadvantages of public key cryptography have been a topic of discussion for many years. The significance of public key cryptography in smart cards, impose certain restrictions and complexities that are unique to smart card microprocessors and the nature of the infrastructures they operate.

A few years ago the main prohibiting factor for the utilization of public key cryptography in smart card microprocessors was the limited processing power of the underlying technology. However, following a number of significant improvements both at the hardware [24] and software level [20, 21, 22] the operation and performance of public key cryptography in smart card microprocessors has improved significantly.

The nature of smart card applications is also changing. There are instances, in which public key cryptography specifically for the establishment of a secure channel might be considered beneficial e.g. when two unknown parties want to establish keys and protect subsequent communications. Another example can be drawn by the nature of the fact that secure channels are normally used for personalization or in order to protect post issuance operations (e.g. application/card management functions [6], protection of application or smart card operating system (SCOS) data [25]).

Although the significance of public key cryptography in a smart card environment cannot be underestimated at the same time the drawbacks are negligible. For example, a secure channel protocol designed specifically for smart cards has to be as lightweight as possible, depending of course on the underlying security and operational requirements. Furthermore, in order to improve the required performance and fulfill the security objectives, a combination of cryptographic primitives and algorithms might be used. Finally, further constraints arise from the fact that a public key based architecture requires the existence of a public key infrastructure (PKI) [26] for the management keys and certificates.

Our proposed protocol aims to fulfil some of the aforementioned requirements. It is designed by keeping in mind the performance requirements and operational characteristics of smart card microprocessors. Although there is a plethora of public key cryptography secure channel protocols [33], most of them are not specifically designed by taking into account the specific characteristics of smart cards. For example, smart card microprocessors have limited communication buffers, often ranging between 190-255bytes. Therefore, if a protocol requires a large number of messages to be exchanged between the card and an off-card entity this will add to the communication and processing overheads [32]. Furthermore, the nature of a public key infrastructure requires the existence of cryptographic key certificates. For example, if a protocol requires regular checks in order to identify whether certificates are revoked or expired this might add to overall protocol security but on the other hand it will potentially complicate its mitigation in smart card environment.

The proposed solution does not claim to introduce a protocol based on new cryptographic techniques. Instead it is an implementation adaptation of existing cryptographic primitives and techniques which are carefully selected in order to be used in a smart card environment. Before moving into the details of the proposed architecture, we highlight the main characteristics of a multi aplication smart card platform.

2.2 An Overview of GlobalPlatform Card Specification

In this section we highlight the main characteristics and the core components of the GlobalPlatform (GP) card specification [6], as a typical example of a multi application smart card architecture that could benefit from the utilization of the proposed protocol. Please note that among the main reasons behind the description of the GlobalPlatform architecture is that it provides the necessary functionality (e.g. secure storage of keys, key management, etc.) required by the protocol. However there are no restrictions or prerequisite for a specific type of smart card technology as the protocol could be utilised and implemented either at the application or at the (SCOS) [7, 8] level irrespectively of the characteristics of the underlying smart card microprocessor.

The GlobalPlatform smart card architecture is comprised of a number of on-card components that offer secure multi-application card management functionality at any given point during the card's lifecycle. Furthermore, the GlobalPlatform smart card architecture is closely coupled with the Java card

[9] technology although there are no restrictions on it's portability to other smart card platforms [10, 11].

The functionality provided by the underline card management system includes the necessary mechanisms (e.g. secure channels [12]) that enable secure communication with the outside world. A secure channel is a mechanism that allows a card and an off-card entity to authenticate each other and establish session keys in order to protect the integrity and confidentiality of subsequent communications.

The GlobalPlatform card specification defines two protocols which are used to establish a secure channel. SCP01 is defined in Appendix D of the GlobalPlatform card specification as a symmetric key protocol that provides three levels of security (i.e. mutual authentication, integrity and data origin authentication, confidentiality). The details of the other secure channel protocol (SCP02) can be found in Appendix E of the GlobalPlatform card specification. The two protocols use symmetric key cryptography for the authentication, establishment of session keys and protection of subsequent communication between the card and the outside world. Although, the existing protocols are mainly used for card content management purposes they can also be used by applications for secure communications. For example, secure communication between a card and an off-card entity is considered necessary whenever a sensitive operation (e.g. during cryptographic key exchanges) is about to be performed.

Another main component of GlobalPlatform is the notion of security domains. GlobalPlatform security domains are the on-card representatives of the card Issuer or an application provider. It is the security domains that allow Issuers to share control, over selected portions of their card, with approved partners. Additionally, security domains are responsible for cryptographic functions and key handling/separation functionality. In terms of communicating with the off-card entity in a secure way the security domains implement different secure channel protocols, as aforementioned. For the purpose of this paper we will be using the notion of a security domain as a mechanism that will securely store keys and control access to the secure channel mechanisms.

The GlobalPlatform smart card specification is becoming the de-facto mechanism for secure application handling especially for Java cards [9] used in the GSM [28] and finance sectors [27]. There are currently on going discussions in order to enhance the functionality offered with the provision of additional secure channel protocols based on public key cryptography. In the following sections we present the main characteristics of the proposed protocol.

3. THE PROPOSED PUBLIC KEY ARCHITECTURE

In a multi application smart card usage scenario the entities that are likely to get involved in a communication session with the card are the Issuer and any Application Provider who has a business relationship with the Issuer. Given the number of the entities involved, there is clearly a need for a Public-Key Infrastructure (PKI) [29, 30] that assists these entities in managing their keys and supports the security functions of the proposed protocol. The supporting functions of a PKI include key certification, authorisation of participating entities, and the ability of a participating entity to have multiple keys.

In this section we highlight the main characteristics of a PKI for supporting the protocol described in this paper. For simplicity and in order to sustain the practicality of the overall architecture the description of the proposed infrastructure will provide examples linked with the GP architecture as described above. Furthermore, we also assume that adequate key and entity management procedures are in place.

According to the proposed infrastructure, each participating off-card entity (being an Issuer or an Application Provider) has a key pair (namely certification key pair) which is used for the certification of other keys. The public key of this key pair is securely loaded on the card (e.g. in a security domain that represents the off-card entity on the card). The corresponding private key is used for the certification of RSA public encryption keys (which are used for the establishment of a secure channel). These certificates bind the included public key to the entity that is authorised to use this public key encryption key during the establishment of a secure session. As an alternative, the certification key pair might belong to a Certification Authority, which has a business relationship with the off-card entity.

Secure loading and replacement of these keys can take place by establishing a secure channel that will enable the secure transfer of keys to the card (e.g. by using the Put Key command as described in the GP specifications). Initial keys for the Issuer can be hard-coded (e.g. masked in ROM) and used, during the personalisation phase, for the loading of the public certification keys. Loading of the public keys for Application Providers has to be done in a secure way (e.g. during the loading of the corresponding GP security domains or during the personalisation of these security domains). Following the loading of these certification keys, any public encryption key that belongs to an entity recognised by the security domain and is certified using the certification private key can be used for the establishment of a secure channel.

Given the proposed infrastructure, the card (or a security domain) is able to tell whether the key presented to it belongs to an entity that is authorised to establish a secure channel by verifying the certificate. For instance, if the certified key belongs to an Application Provider and is certified using the certification key loaded on the Application Provider's logical space in the card (e.g. a security domain) then the off-card entity is authorised to establish a secure session with one of the applications belonging to this Provider.

To strengthen the security provided by this scheme and considering that the off-card entity might use the certification key pair to certify keys not used by this protocol, certificates have to explicitly state that the certified keys are authorised to be used for the establishment of a secure channel. This explicit authorisation is granted when specified in one of the certificate extensions. Given an Issuer who would typically have many certified keys for different purposes, there is clearly a need to protect the card from accidental or deliberate misuse of a key that is not authorised for this purpose. Therefore, the card should only use those keys that explicitly state in a dedicated extension that they can be used for communications with the card, and more specifically, for establishing a secure channel.

Apart from the off-card entities' RSA public encryption keys, the proposed protocol requires each card to have one or more Diffie-Hellman keys [12]. There are two options for the certification of these keys; either the card has a single key pair which is certified by the Issuer and shared among applications (or security domains) that exist on the card, or each application (or security domain) has its own key pair certified by the entity it belongs to. The second option provides more flexibility as it allows the corresponding entity to specify the format based on their applications requirements. Given that none of these approaches introduce any risks to the security of the protocol it is up to the issuer's discretion to adopt either of these options. Please note that the infrastructure required for supporting the certification and verification of these keys by the participating off-card entities is beyond the scope of this paper.

4. A PUBLIC KEY SECURE CHANNEL ESTABLISHMENT PROTOCOL

In this section we present a protocol that utilizes well-established public key techniques for mutual authentication and key establishment between a smart card and an off-card entity based on the Diffie-Hellman key agreement protocol and a combination of symmetric and asymmetric cryptography. The established session keys are used for providing integrity and confidentiality on the exchanged messages. For this protocol the following requirements must be satisfied:

1. All entities share public values p and a, where p is a large prime number and a is an element of large prime multiplicative order modulo p. We will write a^x for (a^x mod p) throughout.
2. Each card has a Diffie-Hellman key agreement key pair. More specifically, card C has private key agreement key y with corresponding public key a^y. The card's key pair can be either generated off-card by the issuer or the application provider and subsequently loaded onto the card, or it can be generated on-card (if the functionality is provided by the card). In either case the public key has to be certified by the corresponding off-card entity, i.e. the issuer or an application provider.
3. The host has an RSA public encryption key (*HPEK*), which is certified by the corresponding certification authority.
4. The card and the host share a symmetric cryptosystem and two different key generation functions (e.g. a one-way function) *f1* and *f2*.
5. The card is capable of generating pseudorandom numbers.
6. Each security domain on the card has a trusted copy of its owner's (issuer or application provider) public certification key whose corresponding private key is used by the off-card entity for issuing certificates.

The proposed protocol, which involves a host (off-card entity) H and a card C, consists of the following steps (please note that messages in brackets are considered optional):

1. The host initiates the protocol by generating a random secret x, $1 \leq x \leq p-2$, and computes α^x. The host sends (e.g. as part of the INITIALIZE UPDATE command) the computed value α^x, together with its public encryption key certificate *Cert(HPEK)*:

$$\text{H} \rightarrow \text{C:} \quad \alpha^x, \text{Cert(HPEK)}, \{\text{options}\}$$

   where *options* is used by the host to inform the card on certain communication requirements (e.g. protecting certain card details and whether the card has to return to the host the certificate on its Diffie-Hellman public key or just the certificate's identification number).

2. On receipt of the message the card verifies the certificate *Cert(HPEK)* using the preloaded public certification key of the corresponding off-card entity. If the certificate verification is successful and the entity is pre-

authorised (e.g. the entity possesses a security domain) then the card generates the following: a random value *randC* and if the *options* field sent by the host requires anonymity the card generates a session key *K1* and encrypts its Diffie-Hellman certificate (*Cert(C-DH)*) using this key (encryption is done using a symmetric cryptosystem). Note that, if neither the card, nor the host require anonymity the card still has to send its certificate or the certificate's serial number (CSN) but this time in clear. The card, using the host's public encryption key HPEK, encrypts the random number *randC*, the *CSN*, and if encryption of *Cert(C-DH)* is required, the key *K1*, and sends the following response to the host:

$$C \rightarrow H: \quad E_{HPEK}(randC, \{K1\}, CSN), E_{K1}(Cert(C\text{-}DH))$$

where $E_K(M)$ denotes encryption of message *M* using key *K*, and *CSN* is the card's unique identifier. *K1* is the output of a key generation function *f1* whose input is the shared Diffie-Hellman key α^{xy}, i.e. $K1 = f1\ (\alpha^{xy})$.

3. On receipt of the card's response the host decrypts the first encrypted message using its private decryption key and extracts the random value *randC*, the optional key value *K1* and the *CSN*. If anonymity was a requirement, which implies that the card's certificate is encrypted, the host decrypts the received certificate and checks its validity. Following that, the host generates the shared key α^{xy} using the card's public Diffie-Hellman key and the secret value *x* that the host generated, and verifies that the key *K1* used to encrypt the certificate was the output of the key generation function *f1*. The calculation of the shared secret and the correct verification of K1 enables the host to authenticate the card. The host sends (e.g. with the EXTERNAL AUTHENTICATE command) the following encrypted, using a symmetric cryptosystem, message that consists of the random value *randC* and the optional *session keys*:

$$H \rightarrow C: \quad E_{K2}(randC, \{session\ keys\})$$

where *K2* denotes the output of a key generating function *f2* which takes as input the shared Diffie-Hellman key α^{xy}, i.e. $K2 = f2\ (\alpha^{xy})$. The optional *session keys*, if sent to the card, will be used as the session keys for the established session. This is useful during card personalisation and card updates where the off-card system has pre-computed the messages to speed up the process. Note that if the off-card entity does not send *session keys*, a similar key generating function can be utilised for the generation of session keys (which will be used to provide integrity and confidentiality for the exchanged messages).

4. On receipt of the above message the card generates the temporary key *K2*, decrypts the received message, and checks whether the random number included in the message matches the number that the card sent to the host. This check proves to the card that the message is not a replay of an old message.

If all the steps are successful, the host and the card will use the established session keys (or the keys provided by the host in step three of the protocol) for the protection of exchanged messages throughout this session.

5. PROPERTIES AND SECURITY ANALYSIS

The proposed protocol provides mutual authentication and session key establishment between the communicating entities, i.e. an off-card entity and the card. The established session keys can be used to optionally provide integrity and message authentication as well as confidentiality on subsequent communications. Although the protocol is based on public key techniques it takes into account the restricted computing resources offered by a smart card (as briefly described in the previous sections). Therefore, the number of expensive computations (like the ones required by public key cryptography) are minimised to avoid processing overheads.

One of the factors that could affect the number of expensive computations was the choice of the Diffie-Hellman keys. Diffie-Hellman keys can be of two flavours; either long term, preferably certified, keys or just short term keys that are typically used for a single session. The card's Diffie-Hellman key pair is fixed so that to avoid the computational overhead required for the generation of a new key pair (a relatively computationally expensive operation for a smart card given that the card has this capability) for each session. However, it is assumed that the host possesses the computational resources for computing and storing a large number of key pairs. For that particular reason it uses a new key pair (for each communication), as opposed to a fixed certified one, so that to avoid one more certificate verification on the card. Note that the host can generate these keys in advance to avoid delays introduced by the generation of these keys during the establishment of a secure session.

5.1 Compromise of Cryptographic Keys

Among the main issues surrounding the deployment and operation of a security protocol is the compromise of the scheme's private keys. If a card's

Diffie-Hellman key pair is compromised it is the Issuer's decision whether to terminate or block this card, or simply update this card's Diffie-Hellman key pair. In the GP analogy if the key belongs to an Application Provider's security domain the Application Provider has to simply update this key by using the Put Key command.

If an off-card entity's RSA encryption key pair is compromised, the off-card entity has to perform the following actions in order to prevent further use of the compromised key by a malicious user:

1. The off-card entity has to generate a new certification key pair, which will replace the one used to certify the compromised key.
2. The off-card entity has to generate a new RSA encryption key pair and certify the public key of this key pair using the new private certification key. Note that if the issuer has issued multiple certification keys, it then has the option not to generate a newly created key pair but use an existing one.
3. All the cards that carry the old public certification key have to be updated with the new public key. As soon as the cards obtain the new certification key they will be able to reject certificates that were created using the compromised key.

Replacement of the certification key pair is also deemed necessary when RSA public encryption key certificates are due to expire to ensure that a key is not used beyond its expiration date. The off-card entity can use the above method to replace these keys.

An off-card entity, being the Issuer or an Application Provider, can have multiple RSA encryption key pairs to avoid unnecessary exposure of a single key. Given that the public key of this key pair is certified by a certification private key whose public counterpart is loaded on the card, the card will be able to verify this key and use it for the establishment of the secure channel. Off-card entities can also use multiple certification keys. In that case, however, the off-card entity has to have access to information that will assist it in the choice of the correct public encryption key certificate, prior to initiating the establishment of a secure channel. In the GP analogy this information can be part of the security domain management data provided to the host as a response to a SELECT command as defined in [6].

5.2 Protocol Efficiency

It can be argued that protocol is relatively heavy, especially when compared with corresponding symmetric key protocols. However, the advantages that public key cryptography has to offer will have to be balanced with the expected processing and architectural overheads. A direct comparison with similar smart card public key protocol (addressing analogous security concerns and with comparable functionality) is not possible due to the fact that most of them are not publicly available. Most of the publicly available smart card secure channel protocols are based on symmetric cryptography techniques.

However, by taking into account the performance of cryptographic algorithms as defined in [31, 32] we can provide some indicative estimates on the performance of our cryptographic protocol, please refer to Table 1.

From the above table we can observe that cryptographic operations of the protocol can be completed in less than a second. Please note that this figure does not include the time spent by the SCOS to form the messages according to the protocol requirements and also move any data from EEPROM to RAM and vice versa. Furthermore, it does not include any performance measurements for the transmission of APDUs as required in each step in the protocol. However, they give an indication as to how much time is spent in the cryptographic part of the protocol.

Table 1. Performance of the Cryptographic Protocol According to Theoretical Timmings.

Operations	Approximate Timings (ms)
• An RSA signature verification for the host certificate verification	~160
• A random generation (RandC)	~ 30
• An RSA encryption for encrypting the RandC, the card CSN and the key K1.	~250
• A DH computation of a shared secret value g^xy	~300
• A secret key encryption for the encryption of the card certificate cert(C-DH)	~10
• One computation for f1	~10
• One computation for f2	~10
	~770ms

Furthermore, in order to successfully verify the actual performance details of the protocol we are currently, experimenting with its development in a Java card Ver. 2.1 [18] and GP 2.1 platform [19]. We believe that in the final version of the paper we will also have obtained the required performance measurements which will be included as another section (i.e.

performance measurements from a Java card implementation of the protocol) in the paper.

6. CONCLUSIONS

In this paper we have outlined the necessity and importance of using public key based cryptographic protocols for the establishment of secure channels in a multi application smart card environment. Although public key protocols were not widely used in smart card microprocessors due to their limitations in processing power, recent technological improvements [14, 15] along with improvements in the operation of cryptographic algorithms [16, 17], make the whole idea more attractive and more feasible.

The core of this paper is dedicated in the development of secure channel establishment protocol that uses standardised public-key techniques (e.g. Diffie-Hellman) in order to provide mutual authentication and key establishment. The supporting infrastructure required to sustain the protocol's cryptographic operations is also defined. The proposed protocol, which benefits from the advantageous key management functionality provided by public key cryptography, can be utilised in any smart card microprocessor. It can be used both by the underlying SCOS and by smart card applications. More importantly, it can also be smoothly integrated in the architecture of existing multi application smart card technologies as in the case of GP.

The future demands for public key smart card protocols will increase taking into account the needs and architectural/business models of various security sensitive applications. We are currently experimenting with the theoretical and practical implementation details around the design of public key secure channel protocols (e.g. based on elliptic curve cryptography) and also compare their performance with other existing protocols.

REFERENCES

1. PC/SC Workgroup, "Specifications for PC-ICC Interoperability", www. smart cardsys .com
2. OpenCard Consortium, "OpenCard Framework Specification OCF", www. opencard.org
3. B. Schneier, Applied Cryptography, Second Edition, John Wiley & Sons, 1996.
4. W.Rankl, W. Effing, "Smart Card Handbook ", John Willey and Sons, 1997.
5. A. Menezes, P. van Oorschot, S. Vanstone, "Handbook of Applied Cryptography", Boca Raton CRC Press, 1997.

6. GlobalPlatform, "Open Platform Card Specification", Version 2.1. June 2001. http://www.globalplatform.org.
7. P. H. Hartel and E. K. de Jong Frz. "Smart cards and card operating systems". In J. Bartlett, editor, *UNIFORUM'96*, pages 725--730, San-Francisco, California, Feb 1996. Uniforum, Santa Clara, California.
8. Constantinos Markantonakis, "The Case for a Secure Multi-Application Smart Card Operating System", Springer-Verlag Lecture Notes in Computer Science Vol. 1396.
9. Javasoft, "Java cardPlatform Specifications", Version 2.2, September 2002, http://java.sun.com/products/javacard/specs.html
10. Microsoft, "Windows for Smart Card", http://www.microsoft.com/HWDEV/TECH /input/smartcard/
11. MAOSCO, "MULTOS Reference Manual Ver 1.2", http://www.multos.com/
12. International Organization for Standardization, Genève, Switzerland. ISO/IEC 7816–4, Information technology—Identification cards—Integrated circuit(s) cards with contacts—Part 4: Interindustry commands for interchange, 1995.
13. W. Diffie and M. Hellman, "New directions in cryptography", IEEE Transactions on Information Theory, 22:644-654, 1976.
14. R. Ferreira, R. Malzahn, P. Marissen, J.-J. Quisquater and T. Wille, "FAME: A 3rd generation coprocessor for optimising public key cryptosystems in smart card applications",Smart card Research and Advanced Applications -- Cardis '96, Publ. Stichting Mathematisch Centrum, pp. 59-72, 1996.
15. T. Boogaerts, "Implementation of Elliptic curves cryptosystems for smart cards", *CARDIS 1998*, 14-16th September 1998.
16. Helena Handschuch, Pascal Paillier, "Smart Card Cryptoprocessors for Public Key Cryptography", In Springer Verlag. Third Smart Card Research and Advanced Application Conference- CARDIS'98, September 1998.
17. Guillou L.C, Ugon M, Quisquater J.J, 1991, "The Smart card (A standardised Security Device Dedicated to public Cryptology)", Contemporary Cryptology: The science of Information Integrity, ISBN 0879422777.
18. Gemplus, GemXpressoRAD, Gemplus, 2003.
19. Giesecke & Devrient, StarSIM Developer Suite, G&D 2003.
20. Jean-Sébastien Coron, Marc Joye, David Naccache, and Pascal Paillier , "Universal padding schemes for RSA" in Yung, Ed., Advances in Cryptology -- CRYPTO 2002, vol. 2442 of Lecture Notes in Computer Science, pp. 226-241, Springer-Verlag, 2002
21. Sébastien Coron, David M'Raïhi, and Christophe Tymen, "Fast generation of pairs (k,[k]P) for Koblitz elliptic curves" by Jean In S. Vaudenay and A.M. Youssef, Eds., Selected Areas in Cryptography, vol. 2259 of Lecture Notes in Computer Science, pp. 151-164, Springer-Verlag, 2001
22. Marc Joye, Pascal Paillier, and Serge Vaudenay "Efficient generation of prime numbers" In Ç.K. Koç and C. Paar, Eds., Cryptographic Hardware and Embedded Systems - CHES 2000, vol. 1965 of Lecture Notes in Computer Science, pp. 340-354, Springer-Verlag, 2000
23. R. Ferreira, R. Malzahn, P. Marissen, Jean-Jacques Quisquater, T. Wille, FAME: A 3rd generation coprocessor for optimising public key cryptosystems in smart card applications, In P. H. Hartel et al., editor(s), Smart card Research and Advanced Applications -- Cardis '96, pages 59-72, 1996
24. UCL, "A Smarter Chip for Smart cards", www.dice.ucl.ac.be/cascade, 1996

25. Konstantinos Markantonakis, "Secure Log File Download Mechanisms for Smart Cards", Third Smart card Research and Advanced Application Conference (CARDIS'98) , September 14-16 1998, UCL Louvain-La-Neuve-Belgium, Final proceedings to be published by Lecture Notes in Computer Science, volume 1820.
26. ISO/IEC 11770-3, "Information technology -- Security techniques -- Key management -- Part 3: Mechanisms using asymmetric techniques", ISO 1999.
27. Laura Geele Wang, "Smart Visa and Java Technology", http://java.sun.com/features/2001/06/visa.html , June 04, 2001
28. 3GPP, "GSM 03.48 Digital cellular telecommunications system, SIM Toolkit Secure Messaging", http://www.3gpp.org/ftp/tsg_cn/WG4_protocollars/Temp/SMG%2323/TDocs/P-97-790.pdf
29. ISO/IEC 11770-1, "Information technology - Security techniques – Key management - Part 1: Framework", 1996
30. ITU-T X.509, "The directory – Public key and attribute certificate frameworks", (2000)
31. Helena Handschuh, Pascal Paillier, "Smart Card Crypto-Coprocessors for Public-Key Cryptography", The Technical Newsletter RSA Laboratories, Vol 1, Number 1, Summer 1998
32. Konstantinos Markantonakis, "Is the Performance of the Cryptographic Functions the Real Bottleneck?", IFIP TC11 16th International Conference on Information Security (IFIP/SEC'01), June 11-13, 2001, Paris, France, *In "Trusted Information: The New Decade Challenge" , Kluwer Academic Publishers* , ISBN 0-7923-7389-8, pages 77-92
33. Colin Boyd, Anish Mathuria, "Protocols for Authentication and Key Establishment", Springer Verlag in Information Security and Cryptography, June 15, 2003

IPSEC CLUSTERING

Antti Nuopponen,[1] Sami Vaarala,[2] and Teemupekka Virtanen[3]

[1] *Emic Networks*
antti.nuopponen@iki.fi

[2] *Stinghorn*
sami.vaarala@iki.fi

[3] *Helsinki University of Technology*
tpv@tml.hut.fi

Abstract IPsec based VPNs are widely used to secure connections in the Internet. As the bandwidth in the Internet grows there is a need to create more powerful fault tolerant IPsec systems. We present an IPsec clustering model based on sharing the processing of each IPsec connection to all cluster nodes using a master node. This clustering model offers good scalability, fine graded load balancing, and fault tolerance while maintaining all IPsec security features. We also present a test implementation of the clustering model with test results.

Keywords: IPsec, high availability, clustering

INTRODUCTION

The Internet Protocol security architecture (IPsec)[2] is commonly used to secure connection in the Internet. In the Internet the bandwidth is growing all the time, and the users are taking advantage of it. This sets higher and higher performance requirements to all components of the Internet, IPsec systems included.

One approach to the improve performance of an IPsec implementation is to use high powered machines with dedicated cryptographic hardware. Another approach to increase the capacity of an IPsec implementation is to group multiple machines in a way that they act as one IPsec node. This technique is called clustering. In addition to better scalability the clustering also provides fault tolerance, and an ability to dynamically add and remove members from the system without affecting the cluster operation.

A straightforward way to cluster IPsec is to divide each IPsec connection to a dedicated node in the cluster. This approach however suffer form several disadvantages. The load balancing can only be made in the connection level. If there is for example a cluster with three nodes and there are two connections into the cluster there is no way to take advantage of all cluster nodes.

The most serious disadvantage is the lack of ability to survive member failure situations transparently to the IPsec clients without compromising the security of the IPsec. The IPsec security associations (SA) in the cluster could be synchronized between cluster nodes, and in a failure situation a new member could start to serve connections of the member that failed. This however creates a security vulnerability because the replay protection information of the cluster member that failed is lost. Because the IPsec replay protection information has to be updated after every packet it can not be kept synchronized between cluster nodes with a reasonable overhead. After the cluster member has failed there is no way of telling which packets it already had received.

In order to create an IPsec clustering model that maintains all IPsec security features and is transparent to the IPsec clients a new approach is taken. The clustering presented in this paper is based on sharing the processing of each IPsec connection (SA) to all cluster nodes using a master node. The model assumes that the cluster members do not have an ability to keep IPsec security associations synchronized real-time and that every member of the cluster can potentially fail in any point of time.

The paper is organized as fallows. In section 1 existing clustering models are presented. Section 2 describes the clustering model presented in this paper. Section 4 describes a test implementation and the conclusions are given in section 5.

1. EXISTING CLUSTERING MODELS

There are some commercial IPsec clustering products available, but technical details of them are hard to obtain.

One IPsec clustering model is based on broadcasting all packets to all cluster members in layer 2 and then selecting the processing node based on some information on the packet. In this model there is no master node in the cluster and all cluster nodes make the choice whether to process a packet or not individually. Because each cluster node perform the IPsec processing independent of each other there is no way to keep the IPsec SAs synchronized, and thus the cluster can no guarantee the replay protection in a failure situation.

Other method to cluster IPsec is to dedicate a member of the cluster to each IPsec connection when the connection is initiated. In this model the same cluster member does all the processing of a given SA, and if the member fails then the SA has to be renegotiated. To avoid the renegotiation the cluster could synchronize each SA to more than one slave and then survive failure situation.

Unfortunately the SA replay protection information becomes the problem as in the previous model.

2. CLUSTERING ARCHITECTURE

The IPsec clustering architecture presented in this paper is based on sharing the same IP addresses between all cluster nodes. Each member of the cluster has at least two network interfaces, which are named public and private interface. Since the cluster is a security gateway between the IPsec client and the correspondent node only the IPsec tunnel mode is supported[2].

All packets sent to the cluster are first received by the cluster master, which is one of the cluster nodes and is automatically selected among the cluster members to handle the packet distribution. The master then forwards each packet to a cluster node for processing. Other members of the cluster are called slaves.

This kind of clustering technique makes the whole cluster to appear as a single IP network node, and no changes to the existing IPsec clients are required as long as the cluster does not lose the SA information.

It is assumed that an IPsec client can set up an IPsec SA with the cluster using Internet key exchange (IKE)[5]. Detailed description about how that is done is out of the scope of this paper, but for example IKE negotiations can be handled in the cluster master node.

Forwarding model

The forwarding in the master node is done packet-by-packet basis using a load sharing function that takes in the packet and produces an id that points to a member of the cluster to which to forward the packet for processing. The master itself does not take part to the IPsec processing because it would make it impossible to survive a master failure without compromising the replay protection.

The advantage of this kind of load sharing is that differences between IPsec connections do not affect the load differences of individual cluster members. If an IPsec connection creates a traffic peak the effect is a load peak in all cluster members not just in one member. Other big advantage is the ability to maintain IPsec security in failure situations.

With this kind of forwarding model two issues arise; first is the IPsec replay protection when receiving packets and the second is traffic based SA lifetime. It is unrealistic to keep the replay protection information synchronized real-time and therefore a cluster member A can not know what sequence number a member B has already received. If a packet with a sequence number n is forwarded to the cluster member A for processing and the packet is re-sent to the cluster and forwarded to the cluster member B, IPsec replay protection does not work.

Next subsections describe how these problems can be solved.

Load sharing function

In order to maintain the replay protection feature of the IPsec when receiving IPsec packets the load sharing function must always map the same sequence number to the same authentication context, or to drop it if the authentication context no longer exists. To achieve this, the load sharing function uses the IPsec sequence number field as input and performs deterministic mapping from the sequence number to a cluster member id.

The sequence number field must always be present on an IPsec packet, and the sender must add incremental sequence number to every packet[4, 3]. If the receiver explicitly tells the sender not to use reply prevention the sender can stop adding the sequence number into each IPsec packet. This does not limit the use of the sequence number in the cluster solution, because the cluster is the receiver and the cluster does not send the notification to the IPsec client. This ensures that the client must add a sequence number to each packet.

The IPsec ESP[4] transform allows operation without packet authentication. In this mode the sequence number is not authenticated and the replay prevention is not used. The clustering solution can handle this mode in a same way than other modes by treating all packets as if they passed the replay protection check. As described in [6] the use of ESP without authentication opens security vulnerabilities, and thus it is better to use authentication with ESP.

If the ESP with authentication is used all IPsec packets that are secured using the same SA have a unique sequence number value that can not be modified. Because sequence number field is present in every IPsec packet, it can not be modified and it is unique in context of one SA, it is a good basis for multiplexing the connections between cluster nodes.

Because a cluster member will most likely lose its SA information in a failure situation the authentication context is linked into the cluster member in a way that every time a member leaves the cluster and joins it again a new authentication context is created.

Outbound packets do not have the sequence number when they are received by the master. In order to use the same load sharing model the cluster master first allocates a sequence number, places it to the packet and then uses the load sharing function to decide to which member to send the packet for the actual IPsec processing.

Inbound IP traffic processing

Inbound IP traffic that does not have an IPsec protection are handled by the cluster master as specified in [2].

When an IPsec protected packet is received by the master it first performs an SA lookup to find the correct SA for the packet. If no such SA is found the packet is discarded. After the master has located the correct SA it uses the load sharing function of that SA to get the cluster member id to which to forward the packet. The function may also produce a null result which indicates that the packet must be discarded. Ones the slave receives the packet it performs the same SA lookup to find the correct SA to use. If the slave can not locate the SA it discards the packet and send notification about missing SA to the master.

If the master receives notification about missing SA it send the SA to all cluster members.

The cluster master also maintains additional information about packets it forwards. This information includes the highest seen sequence number and highest authenticated sequence numbers. The master uses this information to survive from failure situations. The highest seen sequence number is the larges seen sequence number for each SA that the master has forwarded to a cluster node. The highest authenticated sequence number is the highest sequence number that the cluster nodes have reported correctly authenticated.

Outbound IP traffic processing

In outbound packet processing the cluster master first checks the correct action from the IPsec Security Policy Database (SPD)[2]. If the action is to apply IPsec and use replay protection the master then allocates next sequence number to the packet and uses a load sharing function to decide to which cluster member to send the packet to. The master attaches the sequence number into the packet so that the slave knows which sequence number to use.

Ones the slave receives the packet from the master it locates the correct SA from the SPD and then applies the IPsec protection specified in the SA using the sequence number assigned by the master. After packet is protected the slave sends it to the other endpoint of the IPsec connection.

Handling failure situations

There are two kind of failures that can occur; first is the master failure and the second is the slave failure. These are handled differently, and as long as they do not occur simultaneously the cluster can survive them transparently to the IPsec clients and without compromising security features of the IPsec.

In a slave failure situation the cluster master performs a load re-assignment that removes the failed slave from the cluster. This is done by defining a new load sharing function starting from the next sequence number after the highest seen sequence number. Because the master forwards all packets to the cluster members for processing this guarantees that no replay attack is possible.

When the master fails a new master is automatically selected. The new master does not forward any packets until it has asked the highest authenticated sequence number for each SA from all cluster members. Ones the new master has this information it creates a new load sharing function for each SA. These functions only allow packets that have higher sequence number than already authenticated. Because the master does not perform IPsec processing the cluster slave members have all the replay protection information. If the master would participate into the IPsec processing there would be no way to survive from the master failure without compromising the replay protection.

In a situation were the master and at least on of the slaves fail simultaneously, the cluster has to drop all SAs, because it does not have enough information to recover without compromising the security.

Changing the mapping functions

When the state of the cluster changes in some way, for example a new member is added or a member fails, the cluster must update the load sharing function. The update must be made in a way that the mapping of an already seen sequence number does not change. This can be done by defining the new load sharing function in parts. When a new sharing scheme is taken into use the function is redefined starting from a sequence number larger than the largest seen by the master so far. For example the load sharing function could be defined as follows:

$$F(s) = \begin{cases} F1(s) & \text{,for } s \leq 100 \\ F2(s) & \text{,for } 100 < s < 500 \\ F3(s) & \text{,for } s \geq 500 \end{cases}$$

In this example packet with sequence number smaller or equal than 100 are mapped using function $F1$, packet with sequence numbers between 101 and 499 are mapped using function $F2$, and packet with sequence number larger or equal of 500 are mapped using function $F3$. The function $F1$ could for example map packet to slaves one, two, and three. The function $F2$ could represent a failure of a slave when packet are mapped to slaves one and two. Ones the slave three has rebooted it can rejoin the cluster and a new function $F3$ taken into use.

IPsec specification[2] defines the replay protection in a way that packets with sequence number smaller by the size of the replay window that the highest seen sequence number are automatically dropped. This allows the clustering implementation to forget the previous definitions of the load sharing functions after packet with larger sequence numbers are received. The required receive window size is 32, and recommended is 64. In the above example when the clustering has seen and authenticated a packet with sequence number larger than 164 it can forget the function definition $F1$. This feature allows the clustering to maintain only two function definition as long as it does not redefine the function before is has seen and authenticated packet with larger enough sequence number in order to forget the previous definition.

Replay protection information synchronization

In order to properly handle slave failure situation, the cluster needs to synchronize the replay protection information periodically. Synchronization has to be done in a way that it prevents potential denial of service attack that could be made using fake packets that have a very high sequence number. This attack is explained in more detail in section 3.

Figure 1 illustrates the sequence number synchronization. A message named "packet (xx)" is an IPsec protected packet with a sequence number of xx. At first the master receives two packets with sequence numbers 10 and 11. The master forwards these packets to slaves 1 and 2 and marks the highest sequence number seen to 11 ($s_h = 11$). Then the master receives a forged packet with sequence number 50 and forwards it to the slave 2. Now the highest seen sequence number is 50 ($s_h = 50$), even though the slave 2 dropped the packet because it was not authenticated correctly.

Figure 1. Periodic synchronization of sequence number counters

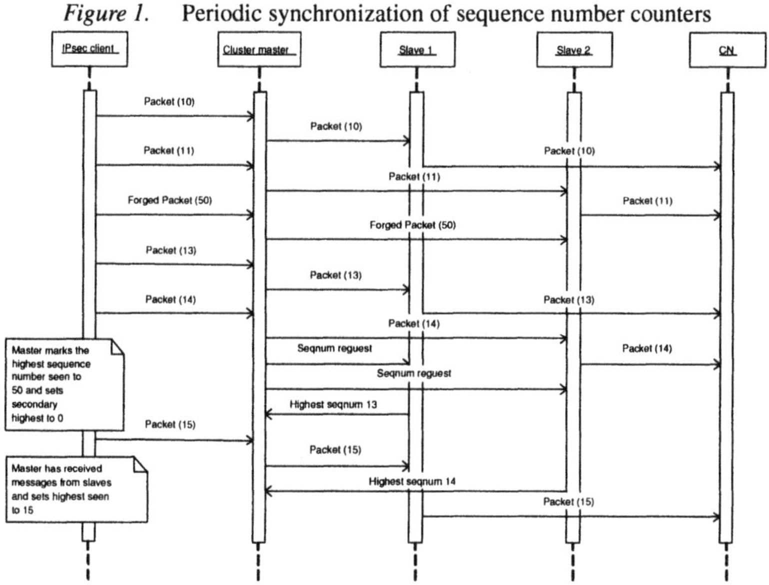

After the packet with sequence number 14, the master sends the periodic sequence number synchronization request to all slaves. When the master sends

the request it saves the highest seen sequence number ($s_s = 50$), and set the current value of the highest seen sequence number to zero ($s_h = 0$). Before all slaves have responded to the request the master has forwarded a packet with the sequence number 15 ($s_h = 15$) to the slave 1. After all slaves have replied to the request, the master checks the highest correctly authenticated sequence number from all slaves. Then the master sets the highest seen sequence number to the maximum of the current highest seen by the master ($s_h = 15$) and the highest authenticated by slaves ($s_a = 14$).

If the packet with sequence number 50 would have been correctly authenticated the slave 2 would have sent this in its sequence number synchronization reply, and the highest seen sequence number would have ended up to be 50.

The sequence number synchronization is done in this way to prevent a denial of service attack using forged sequence numbers. In this attack a malicious node sends fake packets to the cluster. These packets have a valid SPI value and a maximum sequence number. These packets cause the highest seen sequence number seen in master to be the same as in attack packets. If the cluster did not have a mechanism to roll back the highest seen sequence number a single packet would be enough to cause the value to remain high. This would cause a infinitely long delay when a slave fails, and the load sharing must be reassigned. Details how the slave failure situation is handled is explained in section 2.

3. ANALYSIS

Security association lifetimes

An IPsec security association has two different lifetimes: first is the time based lifetime defining how long an SA can be used, and the second is the traffic amount based lifetime defining how much data can be transferred using an SA.

Each cluster member can monitor the time based lifetime individually, and the only requirement is that cluster members keep their clocks synchronized. This can be done for example using the network time protocol[1].

The traffic amount based lifetime is more difficult to handle, because each cluster member that participates to the IPsec processing processes only a portion of the actual traffic. Fortunately traffic lifetime can be handled by being little over cautious. This is done in a way that each cluster member sends periodically the amount of traffic it has received to the cluster master. The master then send the total amount of traffic received back to each cluster member. In addition to this the cluster master also counts the traffic it forwards to the cluster members. If the traffic lifetime would be exceeded by adding the forwarded amount of traffic the master stops forwarding traffic of this SA and send request for data counters to all cluster members. Now the master knows the actual amount of traffic received using this SA. If it does not exceed the traffic lifetime the master will continue forwarding traffic. Ones the lifetime would be again exceeded with the forwarded amount the master repeat the procedure. When the lifetime is actually exceeded the master will remove the SA from its database and optionally sends IKE notification to the IPsec client.

In a typical IPsec setup a new SA is negotiated when the old one is about to exceed its lifetimes, and that is why the cluster does not have to perform this kind of procedure in practice.

In a slave failure situation the master must relay on the amount it has forwarded to the clients and update the traffic counters according to it. This does not cause any security issues since the forwarded amount of traffic is always at least as large as the actual amount of traffic received using an SA.

Security

This section gives an analysis of the differences of the clustering method presented in this paper and a single node IPsec implementation. This analysis show that the security of the system presented is as good as single node IPsec implementation.

In a steady state the packet processing is logically done exactly as in a single node IPsec implementation as long as the load sharing function guarantees that all IPsec packets protected using the same SA with the same sequence number are forwarded to the same authentication context.

Because the changing of the load sharing function is done by defining a new function starting from a given sequence number there is no way that a packet could be forwarded using two different mapping functions.

The master failure situation does not generate any security weaknesses as long as only the master fails because other members of the cluster have the whole replay protection information available and the new master only forwards packets with larger sequence numbers that already received before the old master failed.

The slave failure situation is the only situation were there are a change to attack the cluster. In a situation were a slave fails a malicious node can perform a denial of service attack by sending fake IPsec packet that have very high sequence number. This attack is carried out by constantly sending these packets, and ones a slave fails the cluster master has forwarded a very high sequence number to that slave. Because the master does not know whether the packet was correctly authenticated it has to assume that it was, and only define a new load sharing function after this sequence number. This generates a situation were the cluster is unable to take a new load sharing function into use and the portion of the traffic previously handled by the failed node is discarded.

To make this attack less serious the cluster constantly synchronizes the replay protection information as specified in 2. Because of this synchronization the attacker must be actively sending packets when a slave fails. If it's not the highest seen sequence number is rolled back during the periodic synchronization.

There is a possibility to initiate a targeted denial of service attack against a single cluster node. This attack is performed by generating forged packets with valid SPI and high sequence number. When these packets are sent to the cluster they will end up to the same cluster node, since they have the same sequence number. When this kind of attack occurs the cluster will re-balance the load and the attack will automatically move to the next node. As long as the attack is ongoing the cluster will keep re-balancing the load over and over again. This will cause disruption to the cluster, but compared to single node implementation the situation is better since the whole cluster can not be taken down.

4. IMPLEMENTATION

We made an test implementation of the clustering model presented in this paper using an existing clustering implementation running on Linux. The existing clustering worked on a connection level in a way that each IPsec connection is assigned into a single cluster node. In a case of member failure the existing clustering requires a new IKE negotiation.

Load sharing function

The load sharing function $F(s) = id$ was implemented using a sharing table that contains 256 entries. Each of these entries contains a pointer to a member in the cluster. The table is filled in pseudo random way so that desired portion of the table entries point to a desired node. This way the packet forwarding code can just use last eight bits of the sequence number as an index into the table to get the id of the correct member to which to forward the packet.

The pseudo random function used to fill the sharing table needs not to be very good. Just a simple integer hash that divides sequence numbers in a pseudo random way works fine. The main point about the function is that there would not be long sequence of same ID in the share table to avoid bursts of traffic.

Figure 2 shows an example of a share table from a cluster that has three nodes that participate to the IPsec processing.

Figure 2. Example share table

1	2	1	3	1	2	1	3	1	2	1	3	1	2	1	3

Performance testing

The performance testing was done using four Pentium 4 1,8GHz machines as a cluster, two Pentium 4 2,4GHz as IPsec clients and own Pentium 3 machine as a server that runs a tool called iperf. The iperf is a tool that measures network throughput using UDP or TCP.

Throughput using the single node IPsec was measured to be 52Mbit/s in both directions. At this speed the processor load of the IPsec node handling the IPsec client was 100%, and the processor load of the IPsec client was about 40%. The machines used as an IPsec client and gateway were identical. The difference between processor loads is caused be the different software architecture of the Linux and Windows IPsec implementations.

The performance of the implementation made in this paper was found to highly depend on the amount of simultaneous TCP connections used to measure the performance. Analysis of the implementation showed that this effect was caused by reordering of packets when they were put through the cluster. Reordering caused TCP to drop window size and triggered the TCP fast retransmission that wasted the bandwidth. Overall the TCP behaves badly when reordering occurred. When the amount of concurrent TCP connections was increased the reordering effect became smaller for a single TCP stream and the performance improved significantly. More details how TCP reacts on reordering of packets is explained in [7].

The reordering effect sets an upper limit to throughput of a single TCP stream in this kind of cluster. When this limit is reached the TCP starts to suffer from the reordering. Estimation about this limit can be calculated using the network latency between cluster members. The amount of reordering can be calculated from the network throughput and extra delay in the route through a slave. The amount means that how many packets can be passed through the master before a packet sent through a slave arrives to the receiver.

In order to remove the re-ordering problem the cluster could be modified to route all packets through a ordering node that would use upper layer protocol information such as TCP sequence numbers to restore the right order. In principle this node could be the cluster master. However routing packets twice through the master would degrees the performance of the cluster.

Table 1 summarizes the results of the performance measurements using a single TCP stream. Upload means the traffic throughput from the IPsec client to the server, and the download means the traffic throughput from the server to the IPsec client. In a two member setup the traffic was divided in a way that the master handled 40% of the traffic and slave handled 60%. In a three member setup the master did not handle traffic at all and both slaves handled 50% each. In four member setup the master handled 1% of the traffic and each slave handled 33%.

Table 1. Results of the performance measurements

	Download	Upload
Single	53 Mbit/s	53 Mbit/s
1 member	47 Mbit/s	46 Mbit/s
2 members	41 Mbit/s	28 Mbit/s
3 members	68 Mbit/s	43 Mbit/s
4 members	30 Mbit/s	43 Mbit/s

As the table shows the three member setup provided the best performance of the new cluster. This was caused by the fact that the master did not handle the traffic at all and traffic routes through both slaves were almost equal in term of latency. In two and four member setups the traffic going through the master caused more reordering than in three member setup, because the route through the master was faster than the route through the slaves. Later measurements showed that even in a setup were the master does not handle traffic at all reordering seems to happen, because the throughput increase in this setup was also significant when traffic was divided to multiple TCP streams.

Fail-over testing

The fail-over testing was done with a cluster of two machines.

The first test measured how a failure of a slave affects the traffic throughput. The slave was removed from the cluster when the IPsec client was downloading through the cluster. The throughput was measured as an average in 5 second intervals using 5 simultaneous TCP streams. The traffic was divided between cluster members in a way that the master handled 40% of the traffic and the slave handled 60%.

Throughput in the first interval before the slave removal was 70Mbit/s. In the second interval were the slave was removed the throughput was 42Mbit/s, and in the last interval were the master handled the whole traffic the throughput was 47Mbit/s.

When the slave was removed it took 600ms from the master to notice that slave was removed. During this time slave's portion of the traffic was lost. After the master noticed that the slave was missing it re-assigned the load share, and the packet loss ended.

As can be seen from the throughput in interval were the slave was removed the impact was small. The throughput was only 5% smaller than it was using single node with no packet loss. Of course if the measurement interval would be smaller the affect would be larger. The 5 second interval gives impression about what kind of impact a slave failure would cause in a real situation. If the total traffic through the cluster is less than 47Mbit/s when a slave fails there is no real affect to clients. If the total throughput is larger then all connections will be slowed down after the failure.

5. CONCLUSIONS

The IPsec clustering method presented in this paper provides high availability and scalability while maintaining all security features of the IPsec.

If the architecture presented in this paper would be taken into commercial use, more extensive test should be made to see how the cluster reacts to different kind of load peaks, and how the SA synchronization work in setup were the amount of SAs is in order of thousands. For use in very high speed connections such as site-to-site VPNs the packet re-ordering issues must be solved.

References

[1] David L. Mills: Network Time Protocol (Version 3). Request For Comments 1305, March 1992.

[2] Randall Atkinson, Stephen Kent: Security Architecture for IP. Request For Comments 2401, November 1998.

[3] Randall Atkinson, Stephen Kent: IP authentication header (AH). Request For Comments 2402, November 1998.

[4] Randall Atkinson, Stephen Kent: IP Encapsulating Security Payload (ESP). Request For Comments 2406, November 1998.

[5] Dave Carrel, Dan Harkins: The Internet Key Exchange (IKE). Request For Comments 2409, November 1998.

[6] Steven M. Bellovin: Problem Areas for the IP Security Protocols. Proceedings of the Sixth Usenix UNIX Security Symposium, July 1996.

[7] Ethan Blanton, Mark Allman: On making TCP more robust to packet reordering. ACM Computer Communication Review, 32(1), January 2002.

IMPROVING SECURE DEVICE INSERTION IN HOME AD HOC NETWORKS

Olivier Heen, Jean-Pierre Andreaux, and Nicolas Prigent
Thomson R&D, Security Laboratory, Rennes, France
{Olivier.Heen; Jean-Pierre.Andreaux; Nicolas.Prigent}@thomson.net

Abstract Home ad hoc networks are sets of devices that self-configure and interact to offer enhanced services to their users. These networks are heterogeneous, dynamic and fully decentralized. Moreover, they generally lack a skilled administrator. These properties dramatically reduce the efficiency of classical security approaches: even defining the boundaries of such networks can be difficult. A way to solve this problem has been recently proposed, using the concept of secure long-term communities. This solution relies on one critical operation: the secure insertion of a device in the home ad hoc network. In this article, we propose two ways to improve this operation, using store-and-forward techniques. The first improvement deals with the ability to achieve insertion under loose connectivity circumstances. The other improvement deals with the ability for the user to use any trusted device in order to realize the insertion, especially in heterogeneous networks.

Keywords: Home Networks, Ad Hoc Networks, Security, Key-management.

INTRODUCTION

Communities are sets of principals linked by a trust relation. They can be encountered in many fields, from social interactions and commerce to corporate networks and home ad hoc networks.

Home ad hoc networks can be viewed as communities where principals are home devices (*figure 1*) and where trust relations are preconditions for device communication. A home ad hoc network may evolve along time: some devices could be inserted, while some others could be removed. Nevertheless, the home ad hoc network should still ensure consistency (a device is in the home ad hoc network or not), as well as correct trust management (a device can securely check if another one belongs to the home ad hoc network).

Due to their nature and environment, home ad hoc networks present additional constraints:

High heterogeneity

Devices have various computing powers, battery lifetimes, operating systems, communication protocols. Generally, not all the devices that constitute a home ad hoc network are able to communicate directly: see for instance the digital TV set and the modem in figure 1.

Erratic connectivity

A home ad hoc network may physically split and merge arbitrarily, according to the user's moves and devices availability. As showed in figure 2, the MP3 player, the digital assistant and the digital camera can for instance be temporarily disconnected from the rest of the network.

Poor administration

For wide public acceptance, home ad hoc networks should not be administration intensive for home users. Users may not have skills, motivation or time to perform advanced administrative tasks.

No central device

Users will acquire devices according to their needs. Therefore the existence of a specific device in the network cannot be assumed. In figure 1 for instance, no device is intended to interconnect all the other devices. Moreover, due to erratic connectivity, no device can be considered always available.

No central information

As some devices may be lost or stolen, trust management information should not rely on central information [9]. Consequently, solutions based on pre-shared secret data or centralized certification authority cannot be used.

One important point to ensure the security of a home network is first to define its boundary. For instance, technologies such as virtual private networks or firewalls suppose some notion of boundary or at least of an "inside" to be protected and an "outside" to be protected from. Consequently, before protecting the home ad hoc network, it is important to securely define which devices belong to it.

To achieve this, existing approaches such as [2] [5] [8] [13] [15] [16] rely on a secure insertion operation, generally with user participation. Other approaches, such as [17] do not require user participation, at the price of

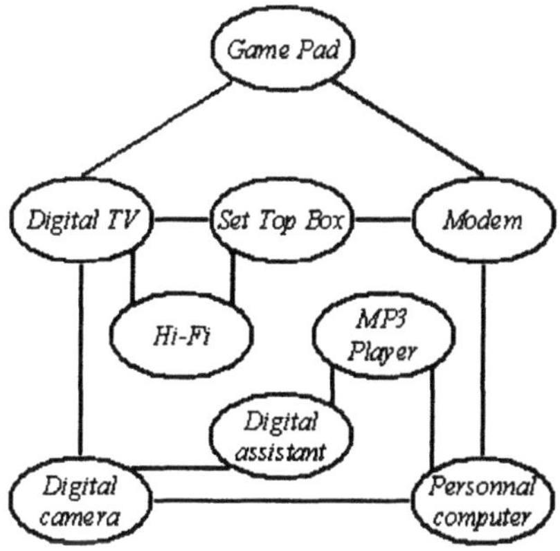

Figure 1. A home ad hoc network.

possible erroneous insertions. In this paper we propose generic improvements for robustness and handiness of the insertion operations with user participation.

In section 1, we indicate notations and recall from [13] a set of basic community operations, as well as their consequences on trust relations.

In section 2, we show how store-and-forward techniques can improve the robustness of the insertion operation.

In section 3, we show how store-and-forward techniques can improve the handiness of the insertion operation.

1. PRELIMINARIES

1.1 Notations

We mainly adopt notations from [13] and add explicit trust relations. Individual devices are denoted by letters. Trust relations are denoted by arrows: $x \rightarrow y$ means that device x trusts device y as being in its home ad hoc network. Mutual trust is denoted $x \leftrightarrow y$, meaning that $x \rightarrow y$ and $y \rightarrow x$ simultaneously hold.

Note that, in home ad hoc networks, trust relations between devices are transitive. Even if all the devices of the home ad hoc network do not necessarily belong to one and the same person, all the users of the same household share the same interest in interconnecting their devices securely. Consequently, they share the same policy about the devices that belong to the home ad hoc network, and all the devices that belong to it follow the same policy. In other words, when a device is inserted in a home ad hoc network, all the users agree on this insertion. Moreover, the devices that belong to the same home ad hoc network can reasonably trust each others: when x trusts y as being in its home ad hoc network, all the devices of this home ad hoc network should also consider y as so. Transitive trust is denoted $x \rightarrow^* y$, which means that x trusts y as being in its home ad hoc network because a device z, that is trusted by x ($x \rightarrow z$), trusts or transitively trusts y ($z \rightarrow^* y$) as being in the home ad hoc network.

The set of devices that can bidirectionally communicate with x at a given time is denoted $\Phi(x)$.

The local community of x is denoted $\Lambda(x)$ and defined by $\Lambda(x) = \{y \mid x \rightarrow y\}$.

A community defined by its elements is denoted $\mathcal{C} = \{x_1 \dots x_n\}$ with $\forall x_i, x_j \in \mathcal{C}, x_i \rightarrow^* x_j$.

1.2 Basic operations

We now define the main operations on communities in home ad hoc networks. In this section, operations are defined supposing there is total connectivity between all the devices of the community. We show in section 2 how the insertion operation can be approximated under more realistic circumstances.

Initialization turns an isolated device into a community that contains only this device. Typically, initialization is done when acquiring a new device and turning it on for the first time.
$\mathit{init} : x \mapsto \{x\}$ with $\Lambda(x) = \{x\}$

Insertion adds a device to an existing community. For convenience, we suppose that the new device has been initialized just before insertion starts.
$\mathit{insert} : \{x_1 \dots x_n\}, \{y\} \mapsto \mathcal{C} = \{x_1 \dots x_n, y\}$ with $\Lambda(y) = \Lambda(x_i) = \mathcal{C}$

Merge is the extension of the insertion operation to communities.
$\mathit{merge} : \{x_1 \dots x_n\}, \{y_1 \dots y_m\} \mapsto \mathcal{C} = \{x_1 \dots x_n, y_1 \dots y_m\}$ with $\Lambda(x_i) = \Lambda(y_j) = \mathcal{C}$

Removal happens when a device has to leave the home ad hoc network, typically when it is sold or given. This device is considered available at the time of removal.
$\mathit{remove} : \{x_1 \dots x_n, y\}, y \mapsto \mathcal{C} = \{x_1 \dots x_n\}$ with $\Lambda(x_i) = \mathcal{C}$ and $\Lambda(y) = \{y\}$

Banishment is another form of removal, that occurs when a device that has to be removed from the community is not available, e.g. because it was lost or stolen.

$banish : \{x_1 \ldots x_n, y\}, y \mapsto \mathcal{C} = \{x_1 \ldots x_n\}$ with $\Lambda(x_i) = \mathcal{C}$ and $\forall i, x_i \not\leftrightarrow y$

2. ROBUST INSERTION

2.1 Realistic insertion conditions

In this section, we address the case of insertion under realistic circumstances, that is when perfect connectivity is not ensured. A home ad hoc network may physically split. In this case, it is made of physically separated partitions, and the devices of each partition cannot communicate with the devices of the others. Nevertheless, they still are in the same community. This happens for instance when the user temporarily leaves his or her home, taking some portable devices with him or her (*figure 2*). When she or he is back, all the devices are able to communicate altogether again.

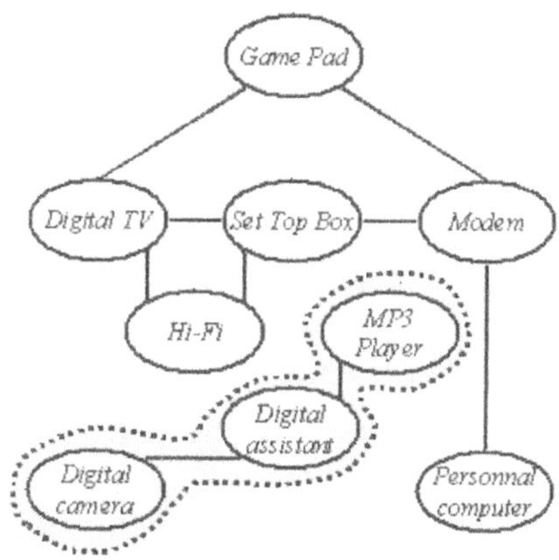

Figure 2. A partitioned home ad hoc network.

A partitioned community can be defined as: $\mathcal{C} = \{x_1 \ldots x_n, y_1 \ldots y_m\}$ with $\Phi(x_1) = \ldots = \Phi(x_n)$ and $\Phi(x_i) \cap \Phi(y_j) = \emptyset$. The partition $\{x_1 \ldots x_n\}$ is hereafter called the first partition. The rest of the community is hereafter called the other partitions.

When a user inserts a new device in the first partition of the community, the ideal insertion operation defined in section 1.2 is not possible. Only the devices in the first partition can receive the information that a new device is inserted. Devices from the other partitions are not reachable.

We propose to minimize this effect by using a secured and fully distributed store-and-forward mechanism of trust management information.

2.2 First stage: gaining trust

In our proposal, each device owns a public/private key pair. This key pair can for instance be set-up during the initialization operation. A device knows the other devices of its community through their public key, or a secure hash of their public key, as proposed in SUCV [10] and CAM [11]. We hereafter call the public key or the secure hash of a device its "provable identity".

We also choose that trust relation $\rightarrow$ is expressed by certificates. $x \rightarrow y$ stands for $Cert_x(y)$, meaning that the device x has issued a certificate for the device y's public key. Trust relation $\rightarrow^*$ is expressed by chains of certificates. $x \rightarrow^* y$ holds for the chain of certificates $1 \leq i \leq n, Cert_{x_i}(x_{i+1})$ with $x_1 = x$ and $x_n = y$.

Each device periodically broadcasts its provable identity to inform the other devices of its physical availability.

We now indicate the algorithm for inserting y in the community of x.

1 x detects the incoming device y by receiving its provable identity.

2 x notifies the user, displays the provable identity of y using a secure and user friendly representation such as a random art visualization [12], and waits for user confirmation.

3 The user checks that the provable identity displayed on x's screen is really y's provable identity (for instance by asking y to display a random art visualization of its own provable identity). If so, the user confirms to x that y belongs to x's community. x then generates a certificate $x \rightarrow y$, and stores it.

4 x sends the certificate $x \rightarrow y$ to y.

5 y stores the certificate $x \rightarrow y$ if and only if it already knows $y \rightarrow x$.

6 y answers by a certificate $y \rightarrow x$ if and only if it already knows $y \rightarrow x$ or $y \rightarrow^* x$.

7 If x received the answer, x stores the certificate $y \rightarrow x$.

8 If x knows the certificate $y \rightarrow x$, then x notifies the user that the insertion of y is complete.

Here is the explanation that this algorithm correctly achieves insertion, provided that the user confirms to each device that the provable identity it displays matches the one to be inserted.

During the insertion, both devices apply the same algorithm. Without loss of generality, we suppose that the user first logs on device x. He or she sees the notification that a new device y is waiting for insertion. The user checks the random art representation of y's provable identity. Consequently, no other public key than y's can be inserted. The user then confirms. At this time, y has no information about x. In particular, y does not know that $y \rightarrow x$, and will not reply to any message from x. Then, at the end of step 4, x only learned $x \rightarrow y$, and y learned nothing.

Now, the user logs on y to complete the insertion. y executes the same algorithm, with x knowing that $x \rightarrow y$. Thus, x answers to y message on step 6, providing y more information on trust relation. At the end of step 8, x learned $y \rightarrow x$ (x already knew $x \rightarrow y$) and y learned $y \rightarrow x$ and $x \rightarrow y$. In other words, both x and y learned that $x \leftrightarrow y$. □

At this stage, robustness comes from the fact that a device never assumes that sent messages are received. Instead, the user validates insertion by giving confirmation to the first device and then to the second device.

2.3 Second stage: spreading trust

In order to reduce user's involvement and to spread trust, the devices of the same community exchange information about the other devices that belong to it.

On a regular basis, a device x sends to each device y for which it has the certificate $x \rightarrow y$ and the other one $y \rightarrow x$ all the certificates it has about the devices z_i that x trusts as being in the community (i.e. the certificates $x \rightarrow z_i$). It also sends the certificates or the chains of certificates it has that prove that these z_i trust x (i.e. $z_i \rightarrow x$ or $z_i \rightarrow^* x$), as explained in the following algorithm:

1 x detects y, for which it has a mutual trust relation with, i.e. for which it has both certificates $x \rightarrow y$ and $y \rightarrow x$.

2 x sends to y all the certificates it has about the devices z_i that it trusts as being in the community, i.e. the certificates $x \rightarrow z_i$, as well as the certificates or the chains of certificates it has that prove that these z_i trust x, i.e. $z_i \rightarrow x$ or $z_i \rightarrow^* x$.

3 If y knows $y \rightarrow x$ and $x \rightarrow y$, for each z_i that y does not know as $y \rightarrow z_i$, y generates the certificate $y \rightarrow z_i$ and stores it. It also appends the certificate $x \rightarrow y$ to the certificate $z_i \rightarrow x$ or the chain

of certificates $z_i \rightarrow^* x$. y stores this chain of certificates $z_i \rightarrow^* y$. y now knows that z_i belongs to the community, and can prove it the next time y and z_i meet, as explained later.

4 If y knows $y \rightarrow x$ and $x \rightarrow y$, for each z_i that y does know as $y \rightarrow z_i$, it checks the length of the certificate chain obtained by appending the certificates $z_i \rightarrow x$ or the chains of certificates $z_i \rightarrow^* x$ to the certificate $x \rightarrow y$. If it is shorter than the chain of certificates it already has for z_i, it replaces it by the new one.

At the end of the algorithm, y knows all the devices z_i that x trusts as being in the community. Moreover, y has all the chains of certificates to prove z_i that they belong to the same community.

When y detects an incoming device z_i that it trusts, i.e. for which is has a certificate $y \rightarrow z_i$, and for which it does not have the certificate $z_i \rightarrow y$ but a chain of certificates $z_i \rightarrow^* y$, y sends to z_i the chain of certificates $z_i \rightarrow^* y$ to prove to z_i that they belong to the same community, as explained in the following algorithm:

1 y detects an incoming device z_i, y knows that $y \rightarrow z_i$, and y has a chain of certificate $z_i \rightarrow^* y$.

2 y sends to z_i the chain of certificate $z_i \rightarrow^* y$.

3 If the received information $z_i \rightarrow^* y$ is correct, i.e. if the chain of certificates is valid, z_i computes the certificate $z_i \rightarrow y$, stores it, and sends it to y.

At the end of the algorithm, both y and z_i know they are in the same community, and each of them knows that the other one knows it.

2.4 Related Work

Some proposals have already been made for secure group management in ad hoc networks. Zhou and Haas [18] have proposed to use threshold cryptography to certify public keys in ad hoc networks. Some special devices, known as servers, have shares of a private key corresponding to a public key known by all the devices belonging to the group. A device is inserted when a minimum number of servers generates a partial signature of its public key. This proposal requires a certain number of servers to be present to insert a new device, and consequently does not perfectly fit with the requirements we stated for home ad hoc networks.

Stajano [14] proposes to manage groups of devices by uploading policy rules during the imprinting phase of the Resurrecting Duckling [15] about the devices that should be trusted. However, the knowledge of the trusted devices is obtained only at the time of imprinting, and nothing is proposed to update it. Some other proposals [6] [1] use and extend [15] to manage groups, but none of them proposes updates of the groups that fit with the requirements we stated for home ad hoc networks.

Other works have also proposed to use web of trust in order to set up trust relations between principals in an ad hoc network. Because there is no always-available-PKI in ad hoc networks, Capkun, Buttyan and Hubaux [3] [7] [4] propose that each device stores a subset of the available certificate. Two devices that want to securely communicate share the certificates they know and try to find a chain of certificates that could certify each other's public key. These proposals differ from ours because the goals are different. While Capkun, Buttyan and Hubaux intend to create pairwise secure relations between devices, we intend to securely manage groups. Consequently, the way each device manages its certificates is different.

3. HANDY INSERTION

Aside from robustness, another essential criterion in home ad hoc networks is handiness. If insertion is not handy enough, users will eventually try to circumvent boring security mechanisms related to it. In section 3.1, we recall how the user is involved in most existing insertion methods. In section 3.2, we modify the proposed mechanism to make the insertion less demanding to the user. In section 3.3, we discuss some consequences and extensions to the *merge* operation.

3.1 State-of-the-art

Different proposals [2] [5] [8] [13] [16] have been made to securely insert a device in a group of devices. Because they come along with their own notion of community and their own definition of "secure insertion", all these proposals are hardly comparable. Moreover, not all these proposals fit home ad hoc networks, first because devices are not always available in home ad hoc networks, and also because devices cannot always communicate by pair, mostly due to the fact that the devices in home ad hoc networks may use heterogeneous communication protocols and media. Nevertheless, some common properties can be drawn out:

User participation

Generally, both the insert*ed* device and the insert*ing* device require user actions such as entering a PIN code or a password, pressing a button, generating a key and inserting it in the right devices, etc.

Auto-detection

To reduce user participation, incoming devices announce themselves to, or are detected by, at least one device in the community.

Need for secure channel

Most insertion methods require transmission of secret information, either between devices or between their users, e.g. whispering a password.

It also appears that, whenever auto-detection is used, the user is not given the possibility to choose the inserting device.

Instead, the detector may become the inserting device (*figure 3*). This case could happen for instance with the Resurrecting Duckling [15] and with 802.11 [8] in ad hoc mode.

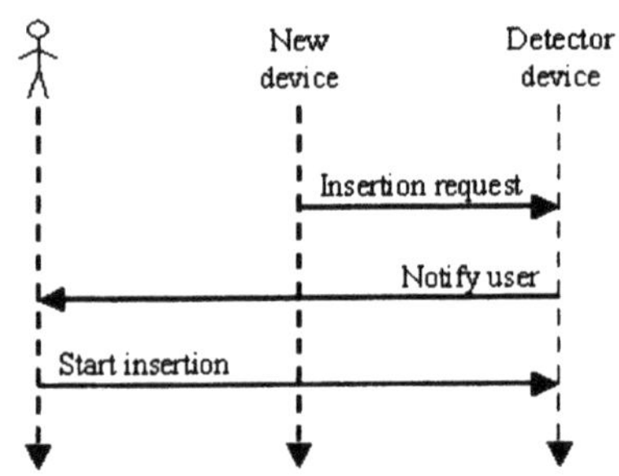

Figure 3. The detecting device immediately becomes the inserting device.

Another option is that the detector forwards information to a predetermined controller that becomes the inserting device (*figure 4*). This case could happen for instance with Zigbee [5] [9], UPnP [16] and 802.11 [8] in infrastructure mode when the access point is used as a controller. Controller based solutions generally correspond to centralized solutions, that do not meet home ad hoc networks requirements.

Auto-detection should not imply forced choice of the inserting device. First of all, this complicates the user's understanding of the process when he or she is surprised by the automatic choice. Moreover, this forbids insertion when the user does not have enough rights to log on the detector or controller device, even though he or she possesses enough rights to log on other devices of the community that could have endorsed the role of inserting device.

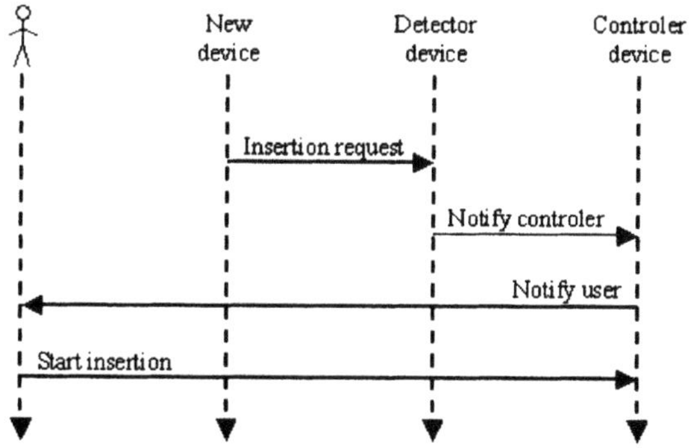

Figure 4. The detecting device forwards to a controller, that becomes the inserting device.

3.2 Free choice of the inserting device

We now modify the beginning of the insertion method described in section 2 so that the user freely chooses the inserting device. In order to do so, we propose to use a store-and-forward technique: devices of the community store insertion requests instead of immediately serving them. The user freely chooses a device in the community, and asks to start insertion. This device asks all the reachable devices of its community for their stored pending requests, and then endorses the next steps of the insertion process, on behalf of the formerly detector devices.

More precisely, if the user chooses x, then x sends a message to all the devices in $\Lambda(x)$ that are reachable, including itself. Each of these devices that has pending requests sends them back to s. If at least one pending request is collected, the chosen device can execute its usual insertion protocol, playing the role of the inserting device. When modifying the beginning of the insertion process, one should care about some technical details:

- Once a detector has sent its pending requests, it should forget them immediately in order to avoid retransmission of formerly honored requests.

- When the chosen device endorses the role of inserting device, there is no need to notify the user since he or she is already logged on, as done in step 2 of the algorithm in section 2.2.

The beginning of the algorithm explained in section 2.2 has to be modified as follows:

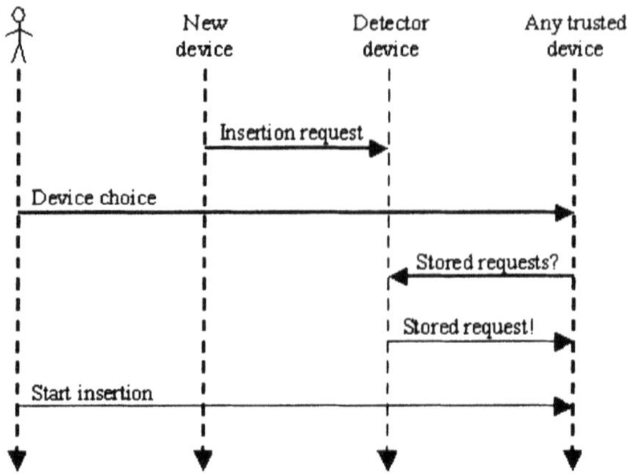

Figure 5. A new method for starting device insertion.

1 The user chooses a device x of the community, logs in, and asks for extended insertion.

2 x securely sends to all the reachable devices $\{y_0 \ldots y_n\}$ of its community a request for pending request.

3 Each device y_i broadcasts the provable identity of x and listens for provable identities they do not know as being in the community.

4 Each device y_i sends to x the list of provable identities it has stored.

5 The algorithm described in section 2.2 continues normally at step 2. The only difference is that x sends the messages to the device that has returned to it the provable identity to be inserted. This device forwards the messages between x and the device to be inserted.

It is important that each device y_i broadcasts the provable identity of x. Indeed, if it does not, the device to be inserted in the community can not receive the provable identity of x, and consequently can not insert x in its own community.

3.3 Consequences

This proposal can be applied to many insertion method, since only the beginning of the process needs to be modified.

The user finds more comfort in choosing his or her favorite device for performing insertion. Moreover, insertion will benefit from rich user interactions such as checking the random art visualization [12] of a provable identity instead of its hexadecimal string, key generation could benefit from mouse moves, etc. If at least one device in the community can perform such functions and make its life easier, the user will intuitively choose it.

An interesting point is that this extension of the insertion of a device in a community can also be used for the merge of two communities. Consequently, none of the detecting devices need to be used, (*figure 6*) and the user can freely choose the inserting device in both the communities to be merged. Technical details of this extension are not discussed in this paper.

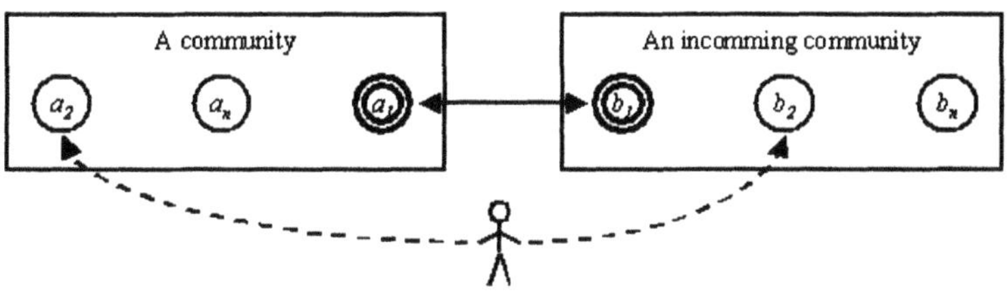

Figure 6. Unconstrained device choice when merging communities.

CONCLUSION

Home networks are a very demanding kind of communities. Most difficulties come from the mix of security needs and poor administration. In this paper, we've proposed ways to improve robustness and handiness of one critical home ad hoc network operation: the secure insertion of a device. We believe that using these proposals in home devices will allow better security acceptance from the user. Further work in the same direction should now consider extensions such as: allowing insertion of guest devices within the community or allowing temporary merging of communities.

Acknowledgments

The authors would like to thank Refik Molva and the anonymous reviewers for their insightful remarks and comments. They also would like to thank Alain Durand and Jean-Louis Diascorn for their valuable remarks and support.

References

[1] D. Balfanz, D. Smetters, P. Stewart, and H. Wong. Talking to strangers: Authentication in adhoc wireless networks. In *Proceedings of the ISOC Network and Distributed Systems Security Symposium*, Feb. 2002.

[2] Bluetooth Inc. Bluetooth core specification 1.1, 2001.

[3] S. Capkun, L. Buttyan, and J.-P. Hubaux. Self-organized public-key management for mobile ad hoc networks. *IEEE Transactions on Mobile Computing*, 2(1), January-March 2003.

[4] S. Capkun, J. P. Hubaux, and L. Buttyan. Mobility helps security in ad hoc networks. In *Proceedings of the Fourth International Symposium on Mobile Ad Hoc Networking and Computing*, 2003.

[5] E. Callaway et al. Home networking with IEEE 802.15.4 : a developing standard for low rate WPAN. *IEEE Communications Magazine*, pages 70–76, Aug. 2002.

[6] L. Feeney, B. Ahlgren, and A. Westerlund. Spontaneous networking: an application-oriented approach to ad hoc networking. *IEEE Communications Magazine*, June 2001.

[7] J. P. Hubaux, L. Buttyan, and S. Capkun. The quest for security in mobile ad hoc networks. In *Proceedings of ACM Symposium on Mobile Ad Hoc Networking and Computing (MobiHOC)*, Long Beach, CA, October 2001.

[8] IEEE Standard Department. IEEE 802.11 Standard for Wireless LAN Medium Access Control (MAC) and Physical Layer (PHY), 1999.

[9] T. S. Messerges, J. Cukier, T. A. Kevenaar, L. Puhl, R. Struik, and E. Callaway. A security design for a general purpose, self-organizing, multihop ad hoc wireless network. In *1st ACM Workshop on Security of Ad Hoc and Sensor Networks (SASN)*, Oct. 2003.

[10] C. Montenegro and C. Castelluccia. Statistically Unique and Cryptographically Verifiable (SUCV) identifiers and addresses. In *NDSS'02*, Feb. 2002.

[11] G. O'Shea and M. Roe. Child-proof authentication for MIPv6 (CAM). *ACM SIGCOMM Computer Communication Review*, 31(2):4–8, 2001.

[12] A. Perrig and D. Song. Hash visualization: a new technique to improve real-world security. In *International Workshop on Cryptographic Techniques and E-Commerce (CrypTEC '99)*, pages 131–138, 1999.

[13] N. Prigent, J.-P. Andreaux, C. Bidan, and O. Heen. Secure long term communities in ad hoc networks. In *1st ACM Workshop on Security of Ad Hoc and Sensor Networks (SASN)*, Oct. 2003.

[14] F. Stajano. The Resurrecting Duckling – what next? *Lecture Notes in Computer Science*, 2133:204–211, 2001.

[15] F. Stajano and R. Anderson. The Resurrecting Duckling: Security issues for ad-hoc wireless networks. In *7th International Workshop on Security Protocols*, pages 172–194, 1999.

[16] The UPnP Initiative. The UPnP Forum, *www.upnp.org*, 2004.

[17] V. Cahill et al. Using trust for secure collaboration in uncertain environments. *Pervasive Computing Mobile And Ubiquitous Computing*, 2(3), July-September 2003.

[18] L. Zhou and Z. J. Haas. Securing ad hoc networks. *IEEE Network*, 13(6):24–30, 1999.

SPAM FILTER ANALYSIS*

Flavio D. Garcia[1], Jaap-Henk Hoepman[1], and Jeroen van Nieuwenhuizen[2]
[1] *Department of Computer Science, University of Nijmegen, the Netherlands.*
{flaviog,jjh}@cs.kun.nl

[2] *Department of Computer Science, University of Twente, the Netherlands*

Abstract Unsolicited bulk email (aka. *spam*) is a major problem on the Internet. To counter spam, several techniques, ranging from spam filters to mail protocol extensions like *hashcash*, have been proposed. In this paper we investigate the effectiveness of several spam filtering techniques and technologies. Our analysis was performed by simulating email traffic under different conditions. We show that genetic algorithm based spam filters perform best at server level and naïve Bayesian filters are the most appropriate for filtering at user level.

Keywords: Spam, unsolicited email, email abuse, security, networking, Bayesian filtering, genetic algorithms, text classification.

1. INTRODUCTION

Spam [16], officially called unsolicited bulk email (UBE) or unsolicited commercial email (UCE), is rapidly becoming a major problem on the Internet. At the end of 2002, as much as 40% of all email traffic consisted of spam[1,2], and recent reports estimate that this amount has risen to more than 50%[3].

To handle this increasing load of junk email [17], several spam filtering techniques exist to automatically classify incoming email as spam, and to reject or discard email classified as such [20]. In this paper we investigate the effectiveness of these spam filtering techniques and technologies.

For users, receiving spam is quite a nuisance and costs money. In a recent study of the European Community [4], it was estimated that the cost for receiving spam for an average Internet user is in the order of 30 euro a year. But the costs

---

*Id: spam-filter.tex,v 1.31 2004/04/28 09:57:35 flaviog Exp

[1] `http://zdnet.com.com/2100-1106-955842.html`

[2] `http://www.linuxsecurity.com/articles/privacy_article-6369.html`

[3] `http://zdnet.com.com/2100-1105_2-1019528.html`

[4] `http://europa.eu.int/comm/internal_market/privacy/docs/studies/spamstudy_en.pdf`

of spam goes well beyond the total costs of all recipients. Each ISP pays for each email message received, because it must be stored in a mail box and it takes up a certain amount of bandwidth. The total cost has been estimated in the order of 10 billion euro a year [10].

A second problem with spam is the impact it has on the Internet backbone. Spam sent over the Internet backbone causes delays for all Internet users. Furthermore, because most spammers use mailing lists that have outdated addresses on them, many messages are rejected ("bounced"). This mandates the operator of the intended destination to send a return response, wasting even more bandwidth [13].

Bulk mailers use several different techniques to send their spam. Often, bulk mailers misuse the SMTP protocol or use badly configured MTAs (so-called open-relays) to hide their tracks [10] [15]. We describe these techniques in detail in section 2.1.

There are at least three fundamentally different ways to counter spammers [5].First, bulk mailers can be prevented to send spam by blocking or limiting access to mail servers. Another method is make spamming less profitable, for example by incurring a cost on every email message sent [7]. A third method aims to detect and remove all spam once it is sent by applying different types of filtering techniques that use the special characteristics of spam to recognise it [2] [3][9] [12] [19]. These techniques are discussed in section 2.2.

Our analysis of countermeasures against spam focuses on filtering techniques. We are interested in measuring the accuracy level of these filters in practice. Some of the filtering techniques not only look at the content of each message, but also consider the email traffic at large (e.g., methods that try to detect duplicate mail messages, or checksum schemes that match incoming messages with a database of known spam messages). To faithfully analyse such spam filters, we built a simulator to generate realistic email traffic and test the filters with it. In section 3 we give a description of our analysis method. We have analysed the performance of the filters in two settings: while being used at the server level and while being used by the end user directly. While running at server level, the filter might use the information about connections from the server. On the other hand, while running at user level, the filter is able to be trained or customised as a function of user specific characteristics. Moreover, we have measured the different behaviour of filters depending on the type of bulk mailer used to generate the spam. We refer to section 3 for details.

To summarise our results: we have found that filters based on genetic algorithms, perform best at ISP level and naïve Bayesian filters perform best at user level. We discuss the results of our analysis on a per filter basis in section 4.

2. SPAM: PRODUCERS AND COUNTERMEASURES

In this section we describe the most common techniques used by bulk mailers. We also describe current proposals for countering spam, with focus on filtering techniques.

2.1 Bulk mailing techniques

Spammers use so-called bulk mailers to send spam. These bulk mailers are capable of sending huge volumes of email without going through a specific mail server or a particular ISP. Some bulk mailers are capable of sending approximately 250,000 messages an hour over a 28.8kb/s modem line [10]. This enormous amount of messages is attained by contacting more than one mail server at the same time and misusing the resources of the ISPs.

The bulk mailers used by spammers have several features to hide their tracks. Most bulk mailers do not use the mail server of their ISP, but instead connect to the destination mail server directly or use a so-called open relay. This way, the spammer avoids to be detected by his ISP. An open relay is a SMTP or ESMTP server that allows everyone to use that server to relay mail. To make the tracking even harder when an open relay is used, most bulk mailers add so-called bogus received headers to the spam message (in front of the *real* received headers added by the SMTP protocol). By adding these bogus headers they hope to redirect any tracking to a site in the fake header.

Bulk mailers also include features which try to outsmart spam filters. The most processor consuming feature is to personalise every message for a recipient. This personalisation of messages can be classified into two types. In the first type, the spammer only uses the victim's mail address as the recipients address instead of using the `Bcc:` headers to send the message to recipients. In the second type, he also uses the name or mail address of the victim to personalise the body of the message. Less processor consuming techniques include the randomisation of the `Subject:` field and the `From:` address line. Some bulk mailers also forge the Message-ID and/or do not send the `To:` header in the SMTP session. Another technique especially developed to confuse Bayesian filters, is to add extraneous text in the body of a message. This text is usually a set of randomly selected words from a dictionary, or just some paragraphs from news or books.

2.2 Countermeasures

There are two fundamentally different methods to counter spam. The first method tries to prevent bulk mailers to send spam, e.g., by incurring a cost on every email message sent, or by blocking or limiting access to mail servers for spammers. The second method aims to detect and remove all spam once it is sent, by applying different types of filtering techniques.

2.2.1 Spam prevention. The most direct way to prevent spam is to close all open relays on the Internet, and to strengthen the SMTP protocol to disallow bogus received headers and require sender authentication, to facilitate bulk mailer tracking. This forces bulk mailers to send spam through their own ISP, but relies on these ISPs to block their accounts. More fundamentally, this approach goes against the open philosophy of the Internet and poses yet another threat to privacy on the Internet. Moreover closing all open relays would not be sufficient, as spammers are now shifting to the use of open proxies to hide their tracks[5], or even use hacked computers.

One interesting and perhaps more feasible approach proposed to stop junk email is by using economic based solutions. The principal attractiveness of spam is that sending large amounts of small email messages is relatively cheap compared to other direct marketing techniques. The idea behind such economic based methods is to make the sending of email more expensive, thereby making it less attractive to send huge amounts of mail. The two main categories of economic solutions are computing time based systems (that force the spammer to spend considerable amounts of his computing resources to send a single spam message) and money based systems (that charge a small amount of money for every email sent).

Computing time based systems. In computing time based economic solution, the sender of a message is required to compute a moderately expensive function. This function is called a *pricing function* [21] [4]. The idea is that for a legitimate sender it is not too expensive in computer time to send a message to a recipient, but for a spammer, who has to send many messages, it is. This expensive use of computer time makes it much harder for a spammer to send large volumes of mail within an acceptable time.

The question whether such a system will work in practice is hard to answer. First of all, this feature has to be incorporated into the Internet, and this may not be easy (although, admittedly, it does not require changes to the SMTP protocol and could simply be enforced by mail user agents). Second, there is the problem of hardware backward compatibility. A user using an old computer must be able to send an email in a reasonable amount of time. This rules out the use of too costly pricing functions. But then for a spammer using modern hardware, the cost in time to send a message may become almost equal to zero. It seems impossible to find a pricing function that suits both needs (although methods based on memory bandwidth limitations appear a little more promising [1]).

[5] `http://zdnet.com.com/2100-1106-958847.html`

Money based systems. Money based systems are based on channelised email systems where users require payment before reading message arriving on certain channels [21]. This payment can be in the form of electronic cash to automate the process. This approach would make it more costly to send Junk-E-mail, which makes it less attractive.

Problems with money based systems include (among others) the adoption of the system by users, and the absence of a global electronic cash system. It is therefore hard, if not impossible, in practice to introduce a money based spam prevention system.

2.2.2 Spam Filters. Filter based countermeasures against spam can be divided into two main categories: cooperative filtering and heuristic filtering. The main principle of the former is that there can be cooperation between spam originators and spam recipients. Such a cooperative filtering system requires a network-wide implementation of, and adherence to, a set of standards for identifying spam. Because most spammers try to hide to track such an implementation is not likely to appear, hence we will focus on heuristic filtering.

Heuristic filters work on the assumption that it is possible to distinguish between normal mail and spam by applying heuristic rules to a message. We can distinguish three types of heuristic rules: origin based filtering, filtering based on traffic analysis and content-based filtering.

Origin based filtering. Origin based filtering happens before a message is fully received by the computer of the recipient. The most prominent method in this class uses the so-called blacklists (e.g., the blacklist from ordb.org[6]). These blacklist can be used to refuse IP or TCP connections from spam originators, but also to refuse mail if the domain name given at the `FROM:` command is on the blacklist. This method can be circumvented by relaying mail through the SMTP servers of legitimate originators that are not on the blacklist. Another disadvantage is that this sites are frequently under denial of service attacks (e.g., the blacklist of osirusoft.com).

A second approach is to configure the SMTP server to perform a reverse DNS lookup to find the IP-address associated with the domain name given at the `MAIL FROM` command. If this IP-address is not equal to the IP-address of the TCP-connection, the SMTP server can refuse to handle the message. This method can also be circumvented by using relay hosts.

Whitelists are another origin based filter method. They contain the senders (or domains) from which incoming mail is automatically accepted for delivery. All other mail is refused by default. To enable legitimate senders to reach the recipient, a whitelist based system will return a request for confirmation to the

[6] `http://www.ordb.org/`

sender, who should reply to this message within a short period of time. When a whitelist is used it is almost certain that no spam will reach the inbox of the user. The disadvantage, however, is that a whitelist has a very high false positive rate, which may seriously confuse or irritate unknown but legitimate senders. Moreover, bulk mailers increasingly match the from header in their mails to the domain of the recipient, using known (and probably trusted) senders in the same domain.

Filtering based on traffic analysis. Traffic analysis based filtering can be used at the mail server of the ISP. Here the log files of the SMTP server can be used to detect anomalies in the normal traffic stream. Anomalies that can appear and indicate the spam are anomalies in the connection time to the server, anomalies in the amount of mail coming from a certain host, the fact that a message is send to more recipients as normal from a certain host or the fact that mail is relayed.

Content-based filtering. Content-based filtering happens after a message is fully received (including the body of the message). In this case, filtering can also be based on known keywords in the subject and body of the message, common features of spam and the use of signature/checksums from databases on the internet.

Naïve Bayesian filtering is a new content-based mechanism for spam filtering [19] [2] [11]. Before it can be used, a Bayesian filter must be trained with a set of spam and a set of legitimate emails (aka. *ham*) that have been previously classified. For each word w in the training sets the filter estimates the probability that it occurs in a spam message ($C = \mathbf{S}$) or in a ham message ($C = \mathbf{H}$) using

$$P(W = w|C = \mathbf{S}) = \frac{S(w)/N_S}{S(w)/N_S + H(w)/N_H},$$

where $S(x)$ is the number of occurrences of word x in the spam set, $H(x)$ is the number of occurrences of word x in the ham set, and N_S and N_H are the sizes of the spam and ham training sets respectively.

When a new message $M = \{w_1, ..., w_N\}$ arrives, the filter determines the n most *interesting* words $\{\dot{w}_1 \dots \dot{w}_n\} \subseteq M$, where interesting means $P(\dot{w}_i) \approx 1$ or $P(\dot{w}_i) \approx 0$). Using these 'interesting' words, the filter then computes the probability that M is spam using Bayes rule [8] [3]

$$P(C = \mathbf{S}|\overrightarrow{W} = M) = \frac{P(C = \mathbf{S}) \prod_i P(W = \dot{w}_i|C = \mathbf{S})}{\sum_{k \in \{\mathbf{S},\mathbf{H}\}} P(C = k) \prod_i P(W = \dot{w}_i|C = k)}$$

Then if this probability is greater than a given threshold T, M is classified as spam.

A third method to filter on content is the use of *genetic algorithms*[7]. Genetic algorithms use so-called feature detectors to score an email message. In practice, these feature detectors are a set of empirical rules that apply to the message and return a numeric value. A genetic program is then represented as trees and has associated a training set and a fitness function. The evolutionary mechanism is accomplished by two basic operations, namely "crossover" and "mutation". This process intends to find a minimum in the fitness function. This score can then be used to classify a message as spam or ham. A more detailed explanation of spam filtering by genetic programming can be found in [14].

Another approach which has the ability to learn is the use of *neural networks*. Like Bayesian filters, neural networks must be trained first on a set of spam and non-spam messages. After this training the neural network can be used to classify incoming mail message based on common features in email messages [6].

It is also possible to classify mail based on the content by the use of *signature/checksum schemes*[8] in a cooperative system. When a mail message arrives a signature/checksum is calculated for this message and compared to the values in special spam databases on the Internet. If the checksum matches any of the values in the database, the message is regarded as spam.

3. METHOD OF ANALYSIS

In this section we detail the method of our analysis. Our goal is to measure the effectiveness of several spam-filtering techniques on realistic email traffic patterns, both at ISP and at the user level. This traffic should reflect the fact that different users have different traffic patterns and their emails also differ in their contents. In an ideal situation we would perform these measurements on real email traffic, but due to privacy reasons, that is infeasible (unless we restrict attention to a very small and atypical consenting sample like staff at a university). Instead we perform this analysis by generating both normal and spam email traffic using a simulator, and measuring the accuracy level of each analysed spam filter.

3.1 Mechanism of the analysis

Our analysis aims to measure how good a spam filter is at preventing spam from being delivered to the end user, while still allowing legitimate emails to pass through unblocked. To this end, we compute for each spam filter, the false acceptance and rejection rate. Moreover, we propose a single measure of the

---

[7] `http://spamassassin.org/`

[8] `http://www.rhyolite.com/anti-spam/dcc/`

filter's wrongness as a function of its false acceptance and rejection rates. Using this measure, we can rank the accuracy of the evaluated filters. We also study how the spam personalisation and new techniques like the inclusion of random words affects this performance.

For each message, a spam filter does one of four things:

$S \to S$: it correctly classifies a spam message as spam,

$S \to H$: it falsely accepts a spam message as ham,

$H \to H$: it correctly classifies a ham message as legitimate, or

$H \to S$: it falsely rejects a ham messages as spam.

When running the spam filter on n messages, n_S of which are spam messages and the remaining n_H are legitimate messages, we write $n_{S\to S}$ for the number of correctly classified spam messages, and define $n_{S\to H}$, $n_{H\to S}$ and $n_{H\to H}$ analogously. Then the false acceptance rate (FAR) and the false rejection rate (FRR) are defined as

$$FAR = \frac{n_{S\to H}}{n_S} \qquad FRR = \frac{n_{H\to S}}{n_H} \, .$$

Clearly, the false acceptance rate can be artificially decreased to 0 by blocking all messages. This increases the false rejection rate though. The same happens with the false rejection rate: it can be decreased to 0 by not blocking any messages at all. A good spam filter has a low FAR as well as a low FRR. We want to be able to rank the performance of spam filters, as expressed by their accept and reject ratios, using a single scalar value. At first, it would seem natural to define the *wrongness* W of a spam filter as the distance to the origin when plotting the performance of the filter in the FAR/FRR plane. However we do not consider these errors to be symmetric, given that it is much worse to falsely reject a legitimate message than to accept a spam message. We can think of a false positive as an error, and a false negative as an effectiveness indicator. Another way to approach this, is to say that one is willing to tolerate a small increase in the false reject ratio for a significant reduction of the false accept ratio. Trying to represent that we propose the function

$$W(FAR, FRR) = (FRR + \epsilon)^2 (FAR + \epsilon) \, , \tag{1}$$

where ϵ is a small constant (i.e., $\epsilon = .01$).

3.2 Modelling of the normal email traffic

The simulator needs to generate email traffic that faithfully emulates the behaviour of a normal user, to test the filters with. In order to do that we need content for the bodies of these emails, and an effective method to generate normal email traffic patterns.

Ideally, the bodies of the messages should come from real mail sent by different kinds of users. However, due to privacy considerations, this is not possible. Instead, we use bodies from a wide variety of USENET (news) messages. This variety is necessary to avoid creating email messages with a limited set of subjects and vocabulary. We do note however that USENET messages do not contain HTML code (whereas some of the spam messages considered (see section 3.3) do).

One important issue to model is the fact that the contents of the messages sent by different people varies. A person's vocabulary varies as a function of his occupation, education, hobbies, etc. In order to reflect that in our model, each sender of an email belongs to a specific topic group chosen randomly. Each topic group only takes bodies from a specific news group.

Another important issue to model is that in normal email traffic, people mostly get mail from people they know. To model this behaviour we enumerate senders and receivers and use a normal distribution to find people close to the sender of the message. The use of the normal distribution means that most mail will be send to people close to the sender in the list, but that there is still a chance of sending mail to people less close to the sender.

Mailing lists also are considered as a special case, because in some aspects they behaves like spammers, sending multiple copies of the same content to many different users, so this behaviour may confuse some filters. In our model, each mailing list has a email address database which is initialised at the beginning of the simulation. When it sends an email, it iterates over its database sending the same message to all the users in it. When the end of the database is reached, another message is chosen and the process starts again.

3.3 Modelling of the spam traffic

To generate realistic spam email-traffic, we have analysed the behaviour of bulk-mailers. An important characteristic of spam is that it arrives in bulk. Whenever a bulk mailer start sending, it keeps sending for some period of time, until the spam message is delivered to all the users in it's database.

An issue when generating the spam traffic is to make sure that the percentage of spam of the total email is realistic. At the end of 2002, 40%[9,10] of the total email traffic consist of spam, and is rapidly growing, accounting for more than 50 percent on June, 2003[11].

---

[9] `http://zdnet.com.com/2100-1106-955842.html`

[10] `http://www.linuxsecurity.com/articles/privacy_article-6369.html`

[11] `http://news.zdnet.co.uk/internet/ecommerce/0,39020372,39118223,00.htm`

In our model of spammers, each spammer has a database with email addresses. When a spammer starts sending, it continues sending as fast as possible, until the message is delivered to the whole database. A spammer can send personalised spam. In this case, each spam message has only one target address, say login@server, and the line "Dear login," is added to the body of the message.

To generate spam mail we also need content for the bodies of the spam messages. We use a variety of spam archives to extract these bodies[12]. In a separate simulation, we also include a collection of very recent spam messages in order to show the impact of the spam evolution in the filtering techniques.

3.4 The simulator

The simulator must generate authentic-looking, email traffic and bulk mail patterns. We also believe that the simulator should be as general as possible. That is, it should be easy to use with existing filters and with filters that might be invented in the future. In addition it should also be easy to add new features of bulk-mailers.

The architecture chosen to fulfil these requirements is to split the simulator into four parts (see Fig. 1). The first part is the traffic generator. This is the main

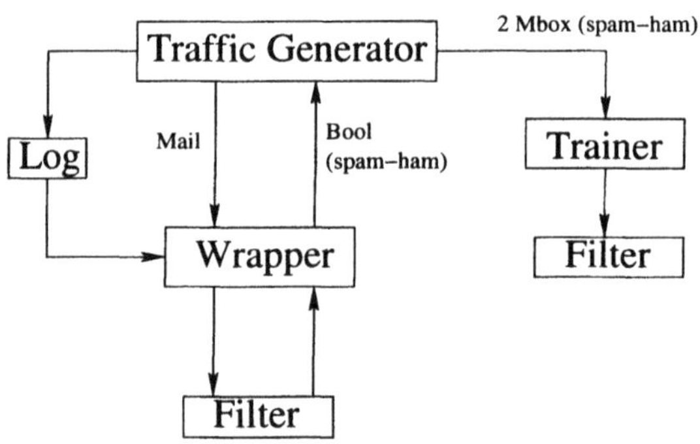

Figure 1. The design of the simulator

module, responsible for generating email traffic according to its configuration and deliver it. The traffic generator has a set of senders and receivers. Each sender has a probability to send an email every one step of the simulation and the traffic generator iterates over the set of senders in a round-robin fashion. After the sender is selected, if it is a normal user, the traffic generator first generates the number of targets this email will have, in which field (i.e., `To:`, `Cc:`, `Bcc:`)

[12] `http://www.spamarchive.org/`

and then selects the receivers. In order to select the receivers, we first take the index of the sender (in the list of senders) and then we generate a offset using the normal distribution. The index plus the offset yields the receiver. In other words, the smaller the difference between your index and the index somebody else, the closer that person is to you, and the more likely it is that you exchange messages. In the case of a spammer, it changes to state sending, and it will keep sending until the spam message is delivered to all the users in it's database. In case of a mailing list, it behaves like a normal user except for the fact that their targets are selected from its database and the same body is sent to the whole database.

The second part is the wrapper, a program that receives an email on its standard input and is responsible for running the filter, take its classification (spam or ham) and send the response back to the traffic generator. If the filter is at the server level, the traffic generator will log every simulated connection to a log file, emulating the log file of the mail server. This file is accessible for the wrapper as well. Each filter may have different input formats, so the wrapper program is filter-specific.

The third part is the trainer. Given that some filters (e.g., Bayesian) need to be trained before they can be used, and the email traffic received for each user has different patterns, the simulator is able to generate a certain amount of correctly classified traffic, and send it to two different files, one for ham and one for spam. After that, the trainer is executed, and it should to be able to train the filter with this data.

3.5 The analysed filters

Some of the aforementioned filtering techniques are not included in our analysis for several reasons. Some filters are highly dependent on unknown constants (e.g., the number of spammers who are not in a blacklist). By setting those constants to an arbitrary value, we would basically be setting the results of those simulations. We also restrict our analysis to open source filters. We therefore choose to test the following set of popular spam filters: *Filtering based on traffic analysis*: Mail volume-based filter, and *Content Based Filters*: Distributed Checksum Clearinghouse (DCC), Genetic algorithm based spam filter (SpamAssassin), and Naïve Bayesian Filters (Bogofilter, Spamprove, Bmf).

4. SPAM FILTER COMPARISON

In this section we give for each analysed filter a short description of the results of the simulation. For each of the filters, we have measured their effectiveness both when applied at user level and at server level, against the following types of spam traffic: (1) non-personalised spam, (2) personalised spam, and (3) non-personalised recent spam. The qualitative results (wrongness, FAR and FRR)

Table 1. Filter performance: non-personalised spam.

Filter	*Level*	*FRR*		*FAR*		$W * 10^5$
		mean	$conf * 10^2$	*mean*	$conf * 10^2$	
Bogofilter	U	.0000	.005	.144	1.750	1.56
Bmf	U	.0117	.518	.029	2.041	1.86
Mail volume	S	.0000	.000	.633	2.854	6.44
SpamAssassin	U/S	.0071	.093	.213	0.928	6.57
DCC	U/S	.0005	.034	.624	1.466	7.11
Bmf	S	.0004	.016	.656	3.817	7.20
Bogofilter	S	.0000	.010	.709	2.838	7.27
Spamprove	U	.0085	.476	.215	5.442	7.80
Spamprove	S	.0024	.239	.695	3.729	10.9

Table 2. Filter performance: personalized spam.

Filter	*Level*	*FRR*		*FAR*		$W * 10^5$
		mean	$conf * 10^2$	*mean*	$conf * 10^2$	
Mail volume	S	.0000	.000	.196	1.174	2.07
Bogofilter	U	.0000	.000	.233	4.819	2.43
Bmf	U	.0037	.401	.186	10.11	3.70
Spamprove	U	.0005	.033	.325	5.585	3.76
SpamAssassin	U/S	.0070	.059	.217	1.555	6.60
DCC	U/S	.0028	.071	.710	1.816	11.9

Table 3. Filter performance: non-personalised recent spam.

Filter	*Level*	*FRR*		*FAR*		$W * 10^5$
		mean	$conf * 10^2$	*mean*	$conf * 10^2$	
Bmf	U	.0030	.159	.070	4.630	1.37
Bogofilter	U	.0000	.006	.129	2.835	1.41
Spamprove	U	.0040	.386	.138	3.013	2.94
SpamAssassin	U/S	.0074	.080	.179	1.356	5.74
Mail volume	S	.0000	.005	.642	3.188	6.58
DCC	U/S	.0005	.034	.786	1.624	8.80

are summarised (together with 95% confidence intervals) in several graphs and tables. The tables contain the performance of the filters both at user and at server level, indicated in the *Level* column by U and S respectively. If the performance of the filter did not depend on its level, U/S is used to indicate this fact. Note that in case a Bayesian filter runs at server level, it did not get a per-

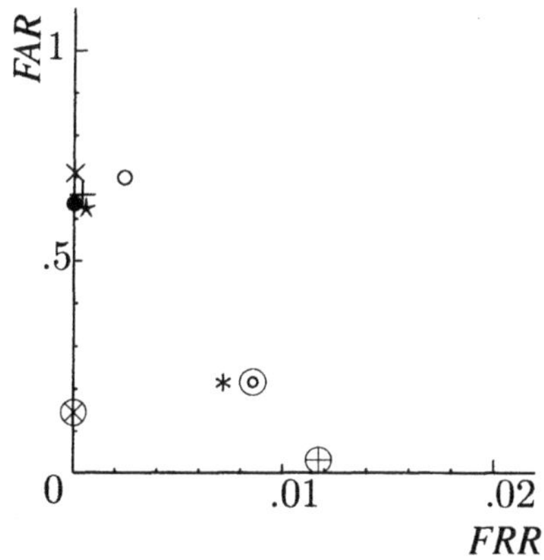

Figure 2. Non-personalised spam.

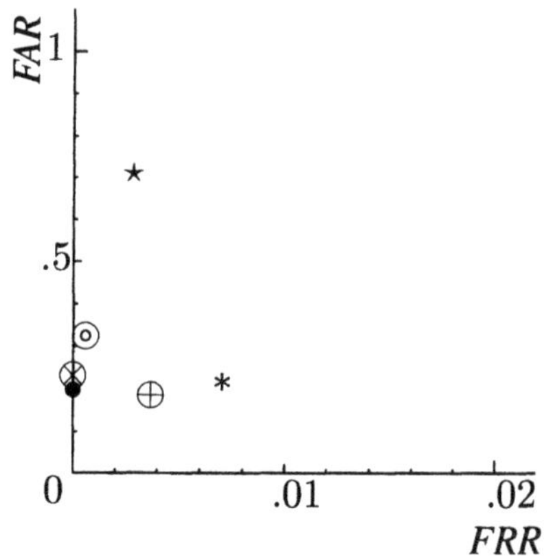

Figure 3. Personalised spam.

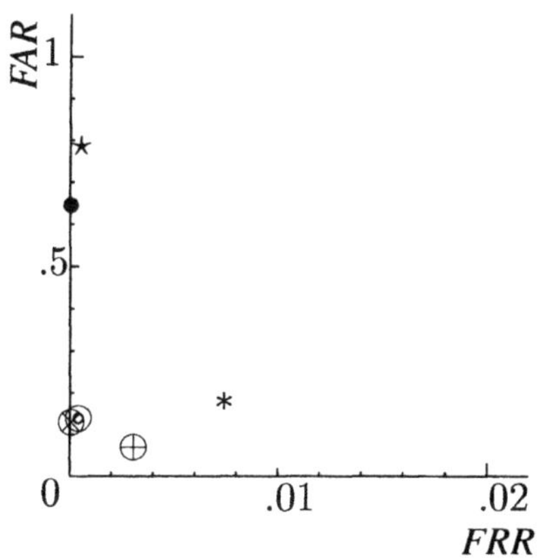

Figure 4. Non-personalised recent spam.

Filter	*Level*	*Symbol*
Bogofilter	S	×
Bogofilter	U	⊗
Bmf	S	+
Bmf	U	⊕
Spamprove	S	∘
Spamprove	U	⊚
Mail volume	S	•
SpamAssassin	U/S	∗
DCC	U/S	⋆

user training but a general one. Table 1 contains the results for non-personalised spam. Table 2 shows the results of the simulation when personalised spam was used. Note that in this case the filters were only tested when running at user level. Finally, for Table 3 non-personalised spam was used but this time the spam mails were collected during the last two months. The purpose of this is to show the impact of the new spamming techniques (e.g., like including random words in the body of the message), especially on the Bayesian filters. The performance of the filters is also shown graphically in Fig. 2–4 (excluding the confidence intervals, because they are quite narrow).

In the remainder of this section we discuss the performance and behaviour of each of the tested filters individually, in separate subsections.

4.1 Mail volume-based filter

The volume-based filter uses an algorithm that checks how much email is received from a specific host during the last connections (the last 1500 lines from the log file in the simulation, which is the same as used by Kai's Spamshield[13]). If the amount of mail received is greater then a certain threshold, then the mail is classified as spam. This filter was able to correctly classify all the legitimate emails, for a high enough threshold. The drawback of this filter is that the FAR achieved is in general very high.

When personalisation is used, this filter performs very well. We consider this to be a side effect, because for personalised mail the bulk mailer has to establish many more connections in order to deliver the same amount of emails. That makes it easier to detect with mail volume analysis. Detection is easily avoided however using multiple open-relays at the same time.

4.2 Distributed Checksum Clearinghouse

The Distributed Checksum Clearinghouse filter tested is a modification of the standard DCC filter version 1.2.14[14]. The modification disables the report function to the Internet server. Instead, a local database was used and the filter reports the emails to this database.

The percentage of false positives of the DCC filter is small, but the performance filtering spam is small as well. This performance can be improved by a less conservative threshold, but this has a direct impact on the FRR.

When personalised spam is used the accuracy of this filter lowers slightly.

4.3 Genetic algorithm based spam filter

The genetic trained filter that we used in our simulations was SpamAssassin version 2.60[15]. The default filter configuration was used in our simulation.

This filter performs very well, achieving one of the best performances of the evaluated filters at ISP level, with the non-personalised spam. This accuracy is not largely affected when personalised spam is used. One drawback of this filter is that it is the most computationally expensive of the evaluated filters.

4.4 Naïve Bayesian Filters

Several implementations of Bayesian filters were evaluated and there were important performance differences between them. An outstanding performance was achieved by Bogofilter when it runes at user level, it had almost no false

---

[13] `http://spamshield.conti.nu/`

[14] `http://www.rhyolite.com/anti-spam/dcc/`

[15] `http://spamassassin.org/`

positives and more than 75% of the spam was filtered in every simulation. We suspect that this good performance is due to the Fisher's method, see Robinson [18]. Bmf has also a very good performance, comparable to Bogofilter, but less conservative. It archives lowest FAR of the evaluated filters, but the number of false positives in some circumstances is certainly high. Spamprove has a low efficacy compared with other Bayesian filters. We suspect this is related to the fact that Spamprove ignores HTML code.

A general characteristic of Bayesian filters is that their performance is lowered drastically when they run at ISP level. Another general characteristic of Bayesian filters is that most of the wrongly classified emails are very short. We suspect that in these cases there is not enough information to perform statistical analysis.

Personalisation of the spam does not have a big impact on the accuracy of Bayesian filters. If anything, it improves the accuracy of some of them. We suspect this improvement is related to the fact that keywords like "login" in the body of the message becomes a good spam indicator.

5. CONCLUSIONS

Filtering at the ISP level. The most efficient way for filtering at the ISP level seems to be using a genetic algorithm. This require a big amount of processing power, but when that is available it is certain a good option. A mail volume-based filter can be established as a first line of defence even when its performance is low because for a high enough threshold they give no false rejects and are computationally cheap.

Filtering at the user level. The best way for a user to filter spam seems to be a naïve Bayesian filter with the Fisher's method like Bogofilter. Depending on how important it is for the user to loose email, Bmf also can be considered as a good option.

The simulator is in the public domain, and can be downloaded from `http://www.cs.kun.nl/~flaviog/spam-filter/`.

References

[1] Martín Abadi, Mike Burrows, Mark Manasse, and Ted Wobber. Moderately Hard, Memory-bound Functions. In *Proc. of the 10th Annual Network and Distributed System Security Symposium*, pages 25–39, February 2003.

[2] I. Androutsopoulos, J. Koutsias, K. Chandrinos, G. Paliouras, and C. Spyropoulos. An Evaluation of Naive Bayesian Anti-Spam Filtering. In G. Potamias, V. Moustakis, and M. van Someren, editors, *Proc. of the workshop on Machine Learning in the New Information Age*, pages 9–17. 11th European Conference on Machine Learning, 2000.

[3] I. Androutsopoulos, G. Paliouras, V. Karkaletsis, G. Sakkis, C. Spyropoulos, and P. Stamatopoulos. Learning to filter spam e-mail: A comparison of a naive bayesian and a

memory-based approach. In H. Zaragoza, P. Gallinari, and M. Rajman, editors, *Proc. of the workshop "Machine Learning and Textual Information Access", 4th European Conference on Principles and Practice of Knowledge Discovery in Databases (PKDD)*, pages 1–13, 2000.

[4] Adam Back. Hashcash. `http://www.cypherspace.org/hashcash`, March 1997.

[5] L.F. Cranor and B.A. LaMacchia. SPAM! *Communications of the ACM*, 41(8):74 – 83, 1998.

[6] Rich Drewes. An artificial neural network spam classifier. `http://www.interstice.com/drewes/cs676/spam-nn/spam-nn.html`, August 2002.

[7] Cynthia Dwork and Moni Naor. Pricing via processing or combatting junk mail. In *Proceedings of the 12th Annual International Cryptology Conference on Advances in Cryptology*, pages 139–147. Springer-Verlag, 1993.

[8] William Feller. *An Introduction to Probability Theory and Its Applications, Volume 1.* Wiley Text Books, third edition, 1968.

[9] Eran Gabber, Markus Jakobsson, Yossi Matias, and Alain J. Mayer. Curbing Junk E-Mail via Secure Classification. In *Financial Cryptography*, pages 198–213, February 1998.

[10] Serge Gauthronet and Etienne Drouard. Unsolicited commercial communications and data protection, 2001. Commission of the European Communities.

[11] Paul Graham. Better Bayesian Filtering. `http://paulgraham.com/better.html`, January 2003.

[12] Hall, R.J. Channels: Avoiding unwanted electronic mail. In *Proc. 1996 DIMACS Symposium on Network Threats*, pages 85–102. DIMACS, 1996.

[13] S. Hambridge and A. Lunde. DON'T SPEW, A Set Of Guidelines For Mass Unsolicited Mailings and Postings (spam). RFC 2635 `http://www.rfc-editor.org/rfc/rfc2635.txt`, 1999.

[14] Hooman Katirai. Filtering Junk E-Mail: A Performance Comparison between Genetic Programming and Naive Bayes. `http://members.rogers.com/hoomank/papers/katirai99filtering.pdf`, October 1999.

[15] G. Lindberg. Anti-Spam Recommendations for SMTP MTAs. RFC 2505 `http://www.rfc-editor.org/rfc/rfc2505.txt`, 1999.

[16] Monty Python's Flying Circus. *Just the words*, volume 2, chapter 25, pages 27–28. Methuen, London, 1989.

[17] Jon Postel. On The Junk Mail Problem. RFC 706 `http://www.rfc-editor.org/rfc/rfc706.txt`, November 1975.

[18] Gary Robinson. A Statistical Approach to the Spam Problem. `http://radio.weblogs.com/0101454/stories/2002/09/16/spamDetection.html`, 2003.

[19] Mehran Sahami, Susan Dumais, David Heckerman, and Eric Horvitz. A Bayesian Approach to Filtering Junk E-Mail. In *Learning for Text Categorization: Papers from the 1998 Workshop*, Madison, Wisconsin, 1998. AAAI Technical Report WS-98-05.

[20] Alan Schwartz and Simson Garfinkel. *Stopping Spam, stamping out unwanted email & news postings*. O'Reilly, 1998.

[21] S.Hird. Technical Solutions for Controlling Spam. In *proceedings of AUUG2002*, Melbourne, September 2002.

COLLECTIVE SIGNATURE FOR EFFICIENT AUTHENTICATION OF XML DOCUMENTS

Indrajit Ray and Eunjong Kim
Department of Computer Science, Colorado State University
Fort Collins, CO – 80523, USA.
{indrajit, kimeu}@cs.colostate.edu

Abstract: XML (eXtensible Markup Language) is the de-facto standard for document representation and exchange on the Web. Researchers have previously proposed access control models and schemes for XML documents that allow one to disseminate selectively, portions of an XML document to the user community based on different policies. Such selective dissemination of an XML document creates a new problem, namely, how to authenticate portions of the XML document independent of other portions and/or the complete document. In this paper, we present a novel scheme based on one-way accumulator functions that allows the user to have a guarantee that a portion of an XML document does indeed belong to the original document.

Key words: authentication, XML documents, message digests, digital signatures

1. INTRODUCTION

Extensible Markup Language (XML) is fast replacing Hyper Text Markup Language (HTML) as the de-facto standard for information representation and exchange on the Web. A major advantage of XML over HTML, which contributes to the former's popularity, is the provision for defining tags, nested document structures, and document types; these features together not only allows efficient representation of document content but also document structure, thus allowing increased control of the granularity of information that can be disseminated over the Web.

As the information size in XML format grows, we can expect these documents to be too large to be retrieved at one time. In fact, almost always will portions of an XML document be retrieved at any given instance. Consider for example an XML document that contains the most recent information about international trading practices and customs regulations for various countries. Such a document may, very well be extended over several thousand of web pages. Suppose now a client wants information about a specific import regulation at the New York harbor for imports from France. The client will obviously want only the relevant portions delivered to him or her and not the entire document. Thus, the server that hosts the information (or is responsible for delivering the information) will only send a small portion of the XML document to the client. However, the client, no doubt, will like to have the guarantee that the information delivered to him or her is from the original document containing the regulations. This will enable the client to have the most up-to-date information from the right sources.

There are other reasons also why an XML document often needs to be delivered in bits and pieces with each piece being delivered independent of other pieces. As pointed out by Bertino and Ferrari [7] it is often the case that in large organizations the same XML document is often shared among many users. Such an XML document may contain information of different sensitivity degrees and thus the document may need to be selectively disseminated among these users. Different access control policies are applied to different components within the same XML document to retrieve the various portions, which are then independently pushed to individual users.

For all these cases, the client gets only portions of the original XML document at a time. This mode of content delivery has some serious integrity issues particularly in critical areas such as government, health, finance, law, etc. The integrity of these individual portions needs, no doubt, to be ensured in transit. This is easily achieved by cryptographic techniques such as message authentication codes and digital signatures. The bigger problem is ensuring the authenticity of each individual portion. How can a recipient have the guarantee that a portion of an XML document that is received belongs to the original document?

The central goal of this paper is to allow verification of answers to queries of a small portion from a large XML document, without searching for or worrying about the remaining portions of the document. As discussed later in section 2, this problem has been previously addressed by other researchers. In this paper, we propose a new approach to signing XML documents based on the notion of one-way accumulator functions. We introduce the concept of collective signatures to sign each portion of the document. This technique allows the recipient of a portion of the XML document to verify the integrity of the portion as well as its authenticity in an efficient manner.

The rest of the paper is organized as follows. Section 2 provides some background information about the technologies that are need. Section 3 discusses some of the more important related works. Section 4 presents our approach. We conclude in Section 5 by discussing our future work.

2. BACKGROUND

Some of the well-recognized benefits of using XML as a data container are its simplicity, richness of the data structure, and excellent handling of international characters [13].The basic concept of an XML document is that an element can be nested to any depth and can contain other sub-elements. An element contains a portion of the document delimited either by a pair of start and end tags (e.g., <tag-name> and </tag-name>). A document is represented as an ordered, node-labeled tree. The conventional terminology for XML document is found in W3C proposals such as, XML Information Set [8] or Xpath [9].

We adopt the example of a typical XML document from [7]; the example is shown in Figure 1. The document is a bulletin that provides information on trading legalities all over the world. It contains a special section, that is, the Import-Export, for international trading legislation. For each law, the document provides information on the country that issues the law, on the law topic, and on related laws.

The start tag of each element can specify a list of attributes providing additional information for the element. Attributes are of the form, *name* = *attvalue*, where *name* is a label and *attvalue* is a string delimited by quotes. Attributes can have different types allowing one to specify the element identifier, additional information about the element, or links to other elements of the document. To be referenceable, an element must have an attribute of type ID whose value provides a unique identifier. Each element can be empty and empty elements are characterized only by a start tag and do not contain data content or sub-elements.

An XML document is often represented as a labeled graph where nodes represent elements and attributes, and edges represent relationships between them. A node representing an element contains the element identifier (*id*). An element identifier can be the ID attribute value associated with the element, or can be automatically generated by the system, if no attribute of type ID is defined. A node representing an attribute contains its associated value. For simplicity, data content of an element is represented as a particular attribute whose name is content and whose value is the element data content itself. The graph contains edges representing the element-attribute and the element-sub-element relationships, and link edges representing links between elements. Edges are labeled with the tag of the destination node and are represented by solid lines, whereas link edges are labeled with the name of the corresponding attribute and are represented by dashed lines.

```
<InternationalLaw Date=”11/20/2003”>
    <Law Country=”USA” RelatedLaws=”LK75”>
        <Topic> Taxation</Topic>
        <Summary>....</Summary>
    </Law>
    <Law Id=”LK75” Country=”USA”>
        <Topic> Import-Export </Topic>
        <Summary>.....</Summary>
    </Law>
    <BluePageReport>
        <Section GeoArea=”Europe”>
            <Law Country =”Germany”>
                <Topic> Guns </Topic>
                <Summary>.....</Summary>
            </Law>
        .......
        </Section>
    </BluePageReport>
</InternationalLaw>
```

Figure 1. An example XML Document

Figure 2 shows the graph representation of the XML document in Figure 1. In the figure, nodes representing elements are denoted by white ovals, whereas nodes representing attributes are denoted by gray ovals. Nodes representing elements have the corresponding *id*. For meaningful exchange of XML documents and formal description of admissible structures of XML documents, a document type definition (DTD) is typically associated with a collection of XML documents. DTD specifies the rules that XML documents must follow.

A DTD consists of declarations for elements, attributes, notations, and their content. It is composed of two parts: the element declarations and the attribute list declarations. The element declarations part specifies the structure of the elements contained in the document. The attribute list declarations part specifies, for each element, the list of its attribute names, types, optional clauses, and default values. Path queries are used to address subtree structures of an XML document using regular expressions over XML element names. DTDs can be represented as a graph using a representation similar to those of XML documents. An XML source is a set of XML documents and DTDs.

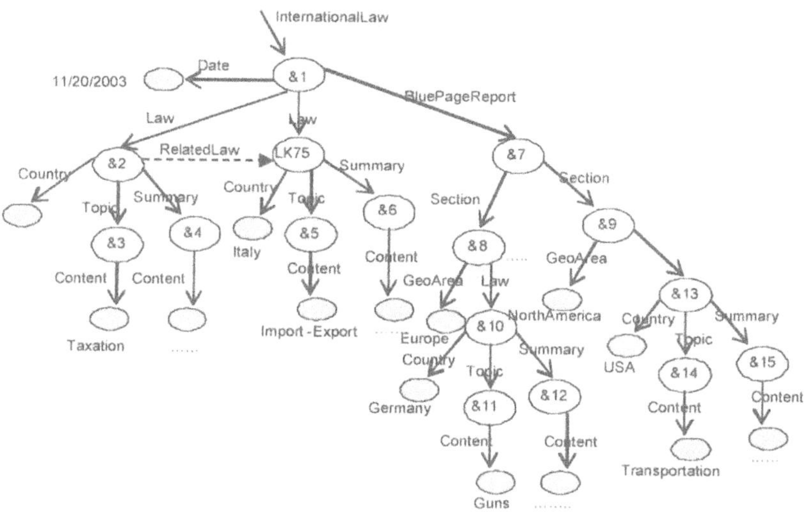

Figure 2. Graph representation of the XML document from Figure 1

3. RELATED WORK

As described in section 2, XML documents have a simple data model based on trees. DOM [6], which is a tree-like representation, is a standardized application programming interface that defines how XML documents are to be accessed by programs. DOM specification includes a hashing procedure very similar to Merkle-hashing. DOMHASH provides a concrete definition of the hash value calculation. The procedure "hashes the leaves of the document, and recursively proceeds up the document tree, hashing both the element types as well as the elements within the document" [1].

The DOMHASH process works as follows. A Merkle hash tree [10] associates hash values with nodes in a tree. The leaves get the hash of the values, and each interior node gets the hash of all the values of all its children, combined in a suitable way. Once the root digest is authenticated, anyone can certify answers to queries, using only the hash values to provide evidence of a correctly conducted search. If the root hash of an entire document *D* is known to a party *P*, it is possible to provide evidence to *P* that any subtree τ of the document occurs under *D* without revealing all of *D*. For this, *P* can DOMHASH the subtree τ to get the root hash of τ. Then *P* can be given just the hash value of the siblings of τ and the siblings of all its parents and *P* can recompute the root hash of *D*. Figure 3 shows how it works. *P* is given the hash value of τ (for example, 23). If *P* has the hash values of sibling of τ (in this case, *h*(34)) and the siblings of all its parents (*h*(*h*(312), *h*(1123)) and the value *hl (hash of the left sub-tree)*, *P* can recompute the root hash of *D* and thus be sure that the value 23 did occur in the tree. *P* can be assured the hash values cannot be forged because the hash function *h* is one-way. Once a client has a secure way of obtaining the root hash of this structure, it is also possible to prove credibly that the answers to the selection queries are complete.

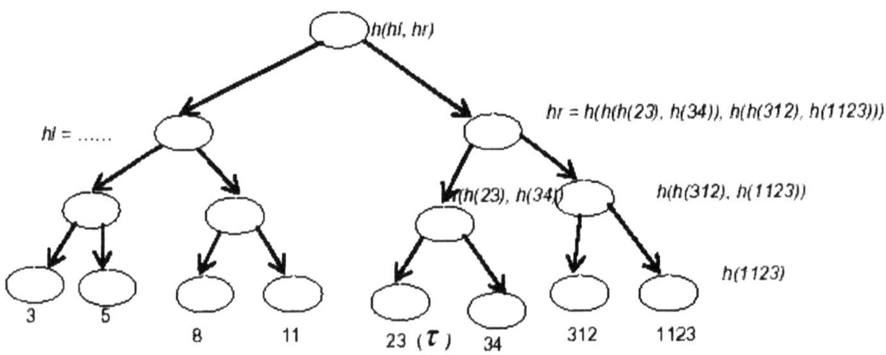

Figure 3. Merkle hash tree associates hash values with nodes in a tree

These techniques are useful in the context of XML documents. However, note that XML data is not always as structured, and requires some additional mechanism to certify answers to the queries. Another drawback of the DOMHASH technique is that it requires additional information about a node's siblings and parents.

Devanbu [1] proposed an approach to signing XML documents, which allow untrusted servers to answer certain types of path queries and selection queries over XML documents without the need for trusted on-line signing keys. This approach enhances both the security and scalability of publishing information in XML format over the Internet. In addition, it provides flexibility in authentication parts of XML documents. However, in this approach a server and a client have to keep track of all path information to get the answer for each path query as above. Although the authors suggest a new data structure, an *xtrie*, to store the different possible answers to path queries, this is a cumbersome procedure and in some cases, it is a very inefficient way to access the data.

Steinfeld and Bull [11] introduced the Content Extraction Signature (CES) as a new type of digital signature. The use of CES with documents enable a user to handle and process these statements in a more selective manner which enable selective disclosure of verifiable content by providing dynamic loading of the transforms for signing and verifying documents. It enables the user to embrace a minimal information model whereby all of the extra details that are not required can be omitted from the transaction. This approach gives some ideas for using access control model with one-way accumulator.

4. OUR APPROACH

We begin the discussion of our approach to generating signatures for XML documents by introducing the notion of one-way accumulators. A one-way accumulator function is a special type of hash function. Hash functions [3] are generally used to reduce their arguments to a pre-determined size. Therefore, it consumes arbitrary length of input and generates a fixed length of the output. *One-way hash functions* requires that collisions of the form $h(x, y) = h(x', y')$ for given x, y, and y' are hard to find. This means, given, x, y, y', it is generally hard to find a pair $(x', y') \in X \times Y$ such that $h(x, y) = h(x', y')$.

One-way accumulator [2] is a simple one-way hash function which satisfies the *quasi-commutative* property. A function h: $X \times Y \rightarrow X$ is said to be quasi-commutative if for all $x \in X$ and for all y_1, $y_2 \in Y$, $h(h(x,y_1),y_2) = h(h(x,y_2),y_1)$.

As defined by Benaloh and Mare [2], a family of one-way accumulators is a family of one-way hash functions each of which is *quasi-commutative*. From above property, if one starts with an initial value $x \in X$, and a set of values y_1, $y_2 \ldots$ y_m $\in Y$, then the accumulated hash $z = h\left(h\left(h\left(\text{K } h\left(h\left(h(x,y_1),y_2\right),y_3\right),\text{K },y_{m-2}\right),y_{m-1}\right),y_m\right)$ would be unchanged if the order of the y_i were permuted. This forms the intuitive basis of our approach.

Suppose an XML document stored at a server has a set of values $Q = (q_1, q_2, \text{K }, q_n)$. The server chooses an intial value x_0. Using an one-way accumulator function $h(x,q)$, the server can compute $x_i = h(x_{i-1}, q_i)$, where $i \in [1,n]$. Owing to the quasi-commutative property, the result $A = x_n$ is independent of the order of the q_i's . The server calculates the signature of the XML document as $\left[A, S_{prv}\right]$, where S_{prv} is the private key of the server. For any value q_i, the server can compute a *witness* A_i given by: $A_i = h\left(h\left(\ldots h\left(x_0, q_1\right), \ldots, q_{i-1}, q_{i+1}\right) \ldots, q_n\right)$.

When the server needs to send q_i to a client such that the client wants to verify the authenticity of q_i, the server basically sends A_i, q_i and the signature of the entire XML document, $\left[A, S_{prv}\right]$. The receiver computes $A' = h(A_i, q_i)$ and compares it with the value A that is obtained by decrypting the signature $\left[A, S_{prv}\right]$ using the server's public key.

We are now ready to provide complete details of our approach for collective signatures.

4.1 Collective Signature Generation

We begin by providing the details regarding generating the accumulated hash using a one-way accumulator function. The most well-known one-way accumulator function to date is based on modular exponentiation techniques similar to RSA public key cryptography. Benaloh and Mare were the first to introduce this technique [2]; it was later improved by Nyberg [13], Baric and Pfitzmann [14], Gennaro, Halevi and Rabin [15], Sander [16] and Sander, Ta-Shma and Yung [17]. We use an approach similar to the one proposed by Benaloh and Mare which uses the exponential accumulator, $\exp(y, x) = y^x \bmod N$, for suitably chosen values of y_0 and the modulus N. In particular, choosing $N = p \bullet q$ for two primes p and q, makes the exponential accumulator function as difficult to break as the RSA cryptosystem. However, as pointed out by Goodrich et al. [18], the exponential accumulator is not associative; any updates require significant recomputations. Goodrich et al. provide an optimized solution based on the notion of *universal-2 hash functions*, which we adopt. Briefly, a family $H = \{h : A \to B\}$ of functions is *universal-2* if for all $a_1, a_2 \in A$, $a_1 \neq a_2$ and for a randomly chosen function h from H

$$\Pr_{h \in H}\left\{h\left(a_1\right) = h\left(a_2\right)\right\} \leq \frac{1}{|B|}$$

The following steps are involved in the computation of an RSA based one-way accumulator. Let the source (that wants to generate the one-way accumulator) contain data set S consisting of n k-bit elements; that is $S = \{e_1, e_2, \ldots, e_n\}$.

1. The source chooses two strong primes p and q such that $p, q > 2^{3/2 k}$.

2. The source also chooses a large base a that is relatively prime to $N = p \times q$.
3. Finally the source chooses a random function h from a universal-2 hash family and publishes a, N and h. The source keeps p and q secret.
4. For each element e_i of S, the representative x_i is computed such that $h(x_i) = e_i$ and x_i is a prime.
5. The accumulation is then computed as $A = a^{x_1 x_2 K\, x_n} \bmod N$. The maximum size of A is the size of N, which is at least $3k$ bits long.
6. For each element e_i the *witness* A_i is given by $A_i = a^{x_1 x_2 K\, x_{i-1} x_{i+1} K\, x_n} \bmod N$.

To verify if a given e_i belongs to S the receiver needs to know the corresponding x_i, A_i and A that is signed by the source. It computes $A' = \left(A_i^{x_i} \bmod N\right)$. If $A' = A$ then e_i is verified to belong to S.

The only problem to this scheme that prevents its use in XML document verification is the assumption that each element e_i is of equal length and fixed in size. However, this is usually not the case. To solve this we replace the XML document element e_i by its corresponding digest created through a well-known cryptographically strong hash algorithm like SHA or MD5. The complete algorithms are given in the following section.

ALGORITHM 1: Accumulated signature generation

<u>Input</u>: An XML document D which consists of sub-documents d_i.
A cryptographically strong hash function h' to generate message digests.
A key pair $\left(K_{prv}, K_{pub}\right)$ which represents the public-private key pair for the server.

<u>Output</u>: A universal hash function h.
A table τ with the tuples (sub-document d_i, hashed value e_i, pointer to representative value x_i)
A binary tree τ' whose leaf nodes are the pairs (e_i, x_i, A_i), where x_i is the representative for e_i, and A_i is the corresponding witness for x_i.
An accumulated hash value A for the entire document.

<u>Steps:</u>

1. For each leaf node d_i of the XML tree for D, compute a fixed size (say 128 bit or 160 bit) message digest $e_i = h'(d_i)$ where h' is a cryptographically strong hash function like SHA or MD5. Let the

size of each e_i be *k* bits. Store the result in a table τ, sorted against the value of e_i. Choose values *a*, *p*, *q* and *N* as discussed earlier. Choose also a universal-2 hash function, *h*. Publish *a*, *N* and *h*.

2. For each hashed value e_i generated in step 1, compute the representative value x_i of e_i.
3. Construct a complete binary tree τ', so that each leaf node of the tree contains a representative value x_i for e_i. Store the value (e_i, x_i) in the leaf node. Leave the internal nodes empty for the time being. Adjust the corresponding pointer in the table τ to point to point to this leaf node.
4. Perform a post-order traversal of τ' visiting each node after having visited all its children. Stop when the root node is traversed. When a node *u* is reached in the traversal process compute a value $\Omega(u)$ as follows:
   a. If *u* is a leaf node and *u* has a value x_i compute $\Omega(u) = x_i \bmod \phi(N)$ where $\phi(\bullet)$ represents the Euler totient function. Store $\Omega(u)$ temporarily in the location for A_i in the node.
   b. If *u* is not a leaf node and *v* and *w* are its children, compute $\Omega(u) = \Omega(v)\Omega(w) \bmod \phi(N)$. Store the value $\Omega(u)$ in the internal node.
5. Perform a pre-order traversal of the tree τ'. For each node *u* that is traversed, compute the accumulation $A(u)$ of all values that are not stored in any descendant of *u* as follows:
   a. If *r* is the root, $A(r) = a$. Store the value signed by the server's private key K_{prv}.
   b. If $u \neq r$, $A(u) = A(g)^{\Omega(w)} \bmod N$, where *g* is the parent node of *u* and *w* is its sibling. Replace the previous value $\Omega(u)$ by $A(u)$. It can be shown trivially that each $A(u)$ that is generated is the witness A_i for x_i.

Stop when A_i is computed for all leaf nodes.

ALGORITHM 2: Query Processing

Input: The XML document *D*, table τ and binary tree τ' generated in algorithm 1, and a query *q* for sub-document d_i from a client

Output: A message *M* which consists of the document d_i resulting from the query, the hash value $h'(d_i) = e_i$ corresponding to d_i, the representative value x_i, the witness A_i corresponding to e_i and the signed accumulated hash value $\left[A, K_{prv}\right]$.

Steps:

1. Retrive the sub-document d_i from the XML document *D*, based on the query *q*.
2. Compute the message digest $h'(d_i) = e_i$ for the subdocument d_i just retrieved.
3. Perform a binary search on the table τ to locate entry e_i. Let the entry be in row *r*.
4. Follow pointer in row *r* to get to a leaf node in the binary tree τ'.
5. Retrieve the values (e_i, x_i, A_i).
6. Visit the root of the tree τ' to get the signed accumulated value $\left[A, K_{prv}\right]$.
7. Concatenate the values retrieved in steps 1-6 and create a message *M*.
8. Generate a message digest M_d for this whole message *M*.
9. Send *M* and M_d as answer to the query.

ALGORITHM 3: Answer Verification

Input: A message $M = \left(d_i, e_i, x_i, A_i, \left[A, K_{prv}\right]\right)$ which is the result of a query on a document *D*, together with a digest M_d of *M*.
Values *N*, functions *h* (the universal-2 hash function) and h' (to compute simple message digests) and K_{pub}, the server's public key.

Output: A yes/no answer depending on whether or not the sub-document d_i belongs to the document *D*.

Steps:

1. Compute the hash value $h'(M) = M'_d$ of *M*. If $M'_d = M_d$ proceed to the next step, else return with an answer no.
2. Retrieve and decrypt $\left[A, K_{prv}\right]$ from the message *M* using the public key K_{pub} of the server. If the signature verifies successfully (that is the decryption is successful) proceed to next step; else retrun with no.
3. Compute $A' = \left(A_i^{x_i} \bmod N\right)$. If $A' = A$ return with answer yes; else return with answer no.

Note that if the XML document is static, the query process incurs very little overhead compared to a standard query, which does not require such features. The table τ and the binary tree τ' can be computed once during generation of the XML document itself. Providing the values is independent of the size of the XML document.

5. CONCLUSION AND FUTURE WORK

The main goal of this paper is to suggest an efficient way to disseminate a portion of an XML document so that the receiver is assured about the origination of the sub-document and its authenticity and at the same time has the assurance that the document is actually a portion of the original document. The information that is provided to the user to achieve this is only what is related to the user's request and nothing else. This is unlike in previous schemes. The protocol uses one-way accumulator for accumulated hash value, which is formed from a given set of values in entire document. This protocol is based on two properties of one-way accumulator; *quasi-commutative* and *one-wayness.*

At this stage of our work, we have assumed that the XML document is static. Thus, all necessary values can be precomputed. In future, we plan to investigate algorithms to efficiently compute these values when the XML document is changing. We also plan to investigate faster algorithms to compute the accumulator functions.

ACKNOWLEDGMENT

This work was partly supported by the U.S. National Science Foundation under award number IIS-0242258. Any opinion, findings and conclusions or recommendations expressed in this publication are those of the authors and do not necessarily reflect the views of the NSF. The authors would like to thank the anonymous reviewers for their useful comments and suggestions.

REFERENCE:

[1] P. Devanbu, M. Gertz, A. Kwong, C.Martel, G.Nuckolls, S. G. Stubblebine. "Flexible Authentication of XML Document," *Proceedings. of the 8th ACM conference on Computer and Communications Security,* pp. 136-145, November 2001.

[2] J.C. Benaloh and M. de Mare. "One-way accumulators: A Decentralized Alternative to Digital Signatures," *Proceedings of*

EUROCRYPT'93: Advances in Cryptology, Lecture Notes in Computer Science, vol. 765, pp 274-285, Springer-Verlag, 1993.

[3] R. Merkle, "A Digital Signature Based on a Conventional Encryption Function," *Proceedings of Crypto '87: Advances in Cryptology,* Lecture Notes in Computer Science, vol. 293, pp.369-378, Springer-Verlag, 1987.

[4] D. Eastlake, J.Reagle and D. Solo, "XML-Signature Syntax and processing", W3C Recommendation, February 2002.

[5] A. Shamir, "On the Generation of Cryptographically Strong Pseudo-Random Sequences." *Proceedings of the 8th Colloquium on Automata, Languages and Programming ICALP '81*, Lecture Notes in Computer Science, vol. 115, pp 544-550, Springer-Verlag, 1981.

[6] Digest Values for DOM (DOMHASH). RFC 2803 Internet Society, available from http://www.faqs.org/rfcs/rfc2803.html, April 2000

[7] Elisa Bertino and Elena Ferrari, "Secure and Selective Dissemination of XML Documents," *ACM Transactions on Information and System Security,* Vol. 5, No. 3, August 2002, pp. 290-331

[8] J. Cowan, "XML Information Set". W3C Recommendation, October 2001.

[9] J. Clark and S. DeRose, "SML Path Language (XPath)". W3C Recommendation, November 1999.

[10] R. Merkle, "A Certified Digital Signature". *In Advances in Cryptology-CRYPTO'89.* Lecture Notes in Computer Science, vol. 435, pp 234-246, Springer-Verlag 1989.

[11] R. Steinfeld, L. Bull and Y. Zheng, "Content Extraction Signatures", *Proceedings of the 4th International Conference on Information Security and Cryptology (ICISC 2001)*, Lecture Notes in Computer Science, vol. 2288, pp. 285-304, Springer-Verlag 2001.

[12] H. Maruyama, K. Tamura, and N. Uramoto, "XML and Java, Developing Web Applications," Addison Wesley 1999.

[13] R. Nyberg, "Fast Accumulated Hashing", *Proceedings of the 3rd International Workshop on Fast Software Encryption.* Cambridge, U.K. 1996. Published as D. Gollman editor, Lecture Notes in Computer Science, vol. 1039, pp 83-87, Springer-Verlag 1996.

[14] N. Baric and B. Pfitzmann, "Collision-free accumulators and fail-stop signature schemes without trees", *Proceedings of EUROCRYPT '97: Advances in Cryptology.* Lecture Notes in Computer Science vol. 1233, pp 480-494. Springer-Verlag 1997.

[15] R. Gennaro, S. Halevi and T. Rabin, "Secure Hash-and-Sign Signatures Without the Random Oracle" *Proceedings of EUROCRYPT '99: Advances in Cryptology.* Lecture Notes in Computer Science, vol. 1592, pp 123-139, Springer-Verlag 1999.

[16] T. Sander, "Efficient Accumulators Without Trapdoor", *Proceedings of the 2nd International Conference on Information and Communications Security 1999 (ICICS 1999)*. Lecture Notes in Computer Science, vol. 1726, pp 252-262, Springer-Verlag 1999.

[17] T. Sander, A. T-Shma and M. Yung, "Blind, Auditable Membership Proofs", *Proceedings of Financial Cryptography '00, 2000.* Lecture Notes in Computer Science, vol. 1962, pp 53-71, Springer-Verlag 2000.

[18] M. T. Goodrich, R. Tamassia and J. Hasic "An Efficient, Dynamic and Distributed Cryptographic Accumulator", *Proceedings of the 5th International Conference on Information Security, 2002.* Lecture Notes in Computer Science, vol. 2433, pp 372-388, Springer-Verlag 2002.

UPDATING ENCRYPTED XML DOCUMENTS ON UNTRUSTED MACHINES

Prakash Reddy, Robert N. Mayo, Eamonn O'Brien-Strain, Jim Rowson, and Yuhong Xiong
Hewlett-Packard Labs

Abstract: With XML and other data types becoming increasingly used in distributed systems, we have a need to update this data in a way that preserves privacy and integrity. Prior work has developed ways of encrypting XML documents for privacy, and adding integrity codes to ensure that the data is not tampered with. In this paper we present an algorithm that allows XML documents, or other tree-structured data, to be updated without decrypting them. In our model of a distributed system, several trusted machines have access to the decrypted form of a document and may request changes to it. These change requests are encrypted and sent to an untrusted update machine for processing. The update machine is able to take the original encrypted document, apply the encrypted changes, and produce an updated encrypted document. In addition, an integrity code is produced that proves the untrusted machine performed the update correctly. In practice, our algorithm allows trusted machines in a distributed system to send incremental updates to a storage server, even if that server is not allowed access to the clear text.

Key words: XML, Security, Incremental Cryptography, XOR MACs, Incremental change support

1. INTRODUCTION

Distributed systems often have the need to cache data at many locations, and then propagate incremental changes to the data [8]. For instance, consider the case of several mobile computers that each cache a shared calendar of meetings, with the master copy of the shared calendar stored at an Internet storage provider that should not be allowed to inspect the calendar. When a user of a mobile computer changes a meeting time in his copy of the calendar, it needs to tell the storage server to update the master calendar appropriately.

In our model, we consider the storage server to be untrusted, in that it is not allowed access to the clear text meeting times. The mobile computers, however, are trusted as they are part of the workgroup that has access to the calendar. We would like to store our data in encrypted form, to prevent snooping, on the storage server, and add an integrity code to verify that the storage server hasn't modified the data inappropriately.

However, there needs to be a way for the storage server to update the documents it holds. Our algorithm presents a way for the mobile computers to send encrypted update requests, called *deltas*, to the storage server. Deltas are an application-generated set of consistent changes. Deltas are encoded such that the storage server can perform the updates without needing access to the clear text. This is done using a form of incremental cryptography [2] [3]. Incremental cryptography is a set of algorithms that can re-compute cryptographic functions in time proportional to the amount of change in the document, rather than in proportion to the overall size of the document. Our technique, however, is novel in that the updates can be processed by an *update server* without knowledge of the key.

2. BASIC CONCEPTS OF XML

XML, or eXtensible Markup Language [5], is a popular way of representing hierarchical data using a textual representation. The standardization of the syntax and the textual representation make it a useful and neutral interchange format between systems and organizations.

Each XML document contains a single tree of nodes, each node called an element. Each element begins and ends with a tag, as in "<employee> ... </employee>. Data for the element may be associated with the element tag as name-value pairs called attributes, and/or it may be included as free-form text between the tags. Child elements may also appear between the tags.

In the example below, `web="lime.org"` is an attribute containing data for the element `base`. The text `OK rock.` is free-form text associated with the `base` element, while `color` and `value` are child elements.

```
<base web="lime.org">
      Ok rock.
      <color>White</color>
      <value basic="cheap">
            <dollars>29.33</dollars>
      </value>
</base>
```

3. OUR SYSTEM MODEL

In our model of the system, we have trusted *update generators*, untrusted *update processors*, and trusted *verifiers*. In the introduction, we gave an example where the update generator was a mobile computer with a user's calendar, while the update processor was a storage server. In addition, in that example the mobile computers are also trusted verifiers, in that they verify the integrity of a document when they receive it.

Update generators request modifications by preparing *deltas*, or lists of changes. These are sent to update processors. For example, our storage servers store the documents safely. But, acting as update processors, they also incrementally update the documents with deltas from the mobile computers. Trusted verifiers verify that the updates were done properly.

4. ALGORITHMS

XML documents in our system are encoded, encrypted, and then appended with a Message Authentication Code (MAC). We present algorithms for each of these three steps. In addition, documents can be updated by applying deltas. In the process of applying deltas, it is possible for conflicts to occur. We discuss options and algorithms for handling these.

4.1 Encoding and encrypting the XML

XML documents are tree structures. Our encoding and encryption scheme is designed to protect the content as well as the hierarchy. Hiding the hierarchy prevents unauthorized persons from inferring the structure of the information from the encrypted document.

We will describe our process in two phases: encoding and encryption. In the encoding phase, the tree-structured document is flattened into an unordered list of elements, or nodes. The encryption phase then encrypts these elements.

In XML, every node has some data (tags, attributes, and text), possibly some child pointers, and a parent pointer (unless it is the root node). An example document is shown below:

```
<base web="lime.org">
      Ok rock.
      <color>White</color>
      <value basic="cheap">
            <dollars>29.33</dollars>
```

```
        </value>
    </base>
```

We will describe the encoding process using approximate textual descriptions. An implementation of the algorithm would bypass these textual descriptions, working directly on the data structures instead.

The first step in our encoding is to assign an arbitrary ID to each element. This is shown below in brackets:

```
    <base[109] web="lime.org>
        Ok rock.
        <color[558]>White</color>
        <value[971] basic="cheap">
            <dollars[623]>29.33</dollars>
        </value>
    </base>
```

Now that elements have IDs, we can flatten the hierarchy into an unordered list of nodes, in no particular order. Each line contains its own node ID, the node ID of its parent, and then the contents of the node with IDs substituted for the children:

```
[971] 109, <value basic="cheap" [623]>
[623] 971, <dollars>29.33</dollars>
[109] null, <base web="lime.org>Ok rock.
[558][971]</base>
[558] 109, <color[558]>White</color>
```

To encrypt the document, we encrypt each line independently. We leave the initial node ID in clear text, and add a random number to the start of the line's encrypted data. Thus, using encryption function *E ()*, we have:

```
[971] E(10290304, 109, <value basic="cheap" [623]>)
[623] E(98740123, 971, <dollars>29.33</dollars>)
[109] E(57093489, null, <base web="lime.org>Ok
rock.[558][971]</base>)
[558] E (95347892, 109, <color [558]>White</color>)
```

The random number strengthens the encryption. This is needed because XML documents have certain properties: a document often has multiple identical nodes, and the amount of data in a node may be small. Adding the random number allows each node to be unique, and also lengthens the amount of data to be encrypted. The choice of the encryption algorithm and the length of the random number are not discussed here. They should be chosen to provide sufficient security for the application, while still meeting any size and performance constraints imposed by the application. We will rely on the uniqueness of each node, provided by the random number, in the conflict resolution algorithm, presented later.

4.2 Encoding the Deltas

Given a document, it is possible for a trusted update generator to request modifications to the original document by preparing a *delta*, which is a list of commands, each of which is one of the following:

- **ADD *id*, E(*txt*)**
  This adds a new node to the XML document. The encrypted data contains the same data used to represent a node, explained previously.

- **DELETE *id***
  This simply deletes the specified node from the document.

- **REPLACE *id*, H(E(*txt_old*)), E(*txt*)**
  The specified node is replaced with the specified data. The ID of the node is not changed. We include a hash of the old encrypted value of the node, to allow the update processor to uniquely identify the data in the node. This command is similar to a DELETE of the node, followed by an ADD using the same node ID.

Addition and deletion will affect the parent nodes, in that their child pointers change. Therefore ADD and DELETE commands will be followed by REPLACE commands to replace the parent nodes with correct ones. Since the node ID of the parent does not change, no further propagation of the changes is needed. Since we require that deltas be consistent, it is assumed that when an intermediate node is deleted then the command list also contains commands to delete all the descendent nodes.

While the delta containing these commands is generated on a trusted machine, they can be applied to the encrypted document on an untrusted update machine, which is the primary benefit of our system.

4.3 Applying the Deltas

Applying the delta involves executing the ADD, DELETE, and REPLACE commands. For ADD, the untrusted machine simply adds the ID and encrypted text into the document as a new line. For DELETE, it simply deletes the line with the specified ID. REPLACE substitutes the specified data for the data that is found in the file for that node. If the hash of the data that is found in the document does not match the value specified in the REPLACE command, we consider it a conflict (described later).

4.4 Document Integrity

Since our documents are modified by untrusted machines, we need to verify that the document was properly processed. MACs have been used in other systems for a similar purpose: verifying that the document was not altered in transit. In these systems, a MAC is generated by a sender using the document and a shared key, and the document and MAC are transmitted to the receiver. The receiver generates its own MAC from the received document and the shared key, and then verifies that it matches the transmitted MAC.

A traditional MAC must be regenerated using the shared key whenever the document is changed. Our system, however, allows an untrusted machine to update a document, and this machine does not have access to the key. Furthermore, we would like the amount of computation done by the machine to be proportional to the size of the delta, not the size of the document.

To meet these requirements, we use what we call an *integrity code (IC)* that is similar to a MAC. Like a MAC, a shared key is needed to generate an IC and to verify it. Each document and each delta contains an IC, produced using the shared key. The ICs are designed such that they can be combined when applying a delta, without knowledge of the shared key. The resulting IC matches what would be produced by computing an IC by a pass over the entire updated document, allowing verification at a later time.

4.4.1 Integrity code generation for a complete document

To compute the *integrity code (IC)* for a document, we compute the IC for each line of the document separately and XOR these together to produce the IC for the overall document. As a special case, if the document has no lines then its IC is 0.

To compute the IC for a line, we decrypt the line using our shared key and then apply a one-way hash function, H (). For instance, to compute the IC for this line in the encrypted document:

```
[558] E (95347892, 109, <color[558]>White</color>)
```

We compute

```
IC = H ([558] 95347892, 109, <color[558]>White</color>)
```

We do not require the use of a specific hash function. All we require is that it be a one-way function (that is, no H^{-1} exists such that $H^{-1}(H(x)) = x$) and that the function be largely collision-free (that is, H(x) is unlikely to equal H(y)). In practice, MD5 or a member of the SHA family will work fine.

4.4.2 Verification of integrity codes

It is simple for a trusted machine to verify the IC of a document. It simply recomputes the IC of the document and compares it with the IC attached to the document. If they match, the document is valid. A mismatch indicates improper processing.

4.4.3 Integrity code generation for incremental changes

In order to support incremental changes to documents, we need to allow the incremental update of integrity codes. We do this by attaching to each delta, or set of update commands, an *incremental integrity code (IIC)* that can be applied to a document's IC to reflect the updates. Neither generating nor applying the IIC should require access to the entire document, nor should it involve re-generation of the IC of the entire document.

In our method, the bitwise XOR function is used to apply the IIC. That is, when the delta has been applied to the document by an untrusted machine, that machine may update the integrity code for the entire document by XORing the document's IC with the delta's IIC, producing a new IC for the document.

To compute the IIC for a delta, we compute an IIC for each line in the delta and XOR them together, producing the IIC for the overall delta. Deltas contain a list of commands, and we generate an IIC for each one depending upon its type:

- **ADD *id*, E(*txt*)**
  This adds a new node to the XML document. The IIC for this change is just H (*txt*). *Txt* is the clear text for a node. As described previously, this contains a random number, the ID of the parent and children, and the text data for the node. When the untrusted machine adds this line to the document and XORs in this IIC, the new IC will match the updated document.

- **DELETE *id***
  This deletes the specified node from the document. The IIC for this is simply the IC of the old node. To compute this, we find the node

to be deleted in the existing file and compute the IC for that line. When the untrusted machine deletes this line from the document and XORs in this IIC, the new IC will match the updated document. This is because the IIC, when XORed, effectively "backs out" the old node's contribution to the overall IC. We are relying on the property that if x = a XOR b, then we can "back out" b with another XOR, as in a = x XOR b.

- **REPLACE *id*, H(E(*txt_old*)), E(*txt*)**
  The specified node is replaced with the specified data. The ID of the node is not changed. We compute the IIC as if this was a DELETE command followed by an ADD command, XORing those two IICs together. In other words, the IIC consists of the XOR of two values: the IC of the line to be deleted, and the IC of the new line.

Once the IICs for each line of the delta have been computed, we XOR them together to produce the IIC for the overall delta. The individual line IICs are not saved. The overall IIC is attached to the delta and sent to the machine that will process the delta. When the untrusted machine performs the update, it executes each line in the delta and then XORs the IIC of the delta with the IC of the document, producing a new IC that matches the contents of the updated document.

4.5 Data freshness and conflicts

Since we allow multiple applications to make incremental modifications to XML documents we have to address the issue of data freshness. Our primary goal is to prevent applications from updating data based on stale copies. In systems that allow multiple applications to make changes to a document the granularity of freshness is often the entire document. In our case the granularity is at the level of a node. This finer level of granularity allows us to support changes to documents which would normally be considered stale. For example if two applications (A and B) cache the same document and if application "A" makes a change and updates it, the copy held by application would be considered stale in the traditional sense. In our case it is possible for application "B" to make changes to the document as long as these changes do not conflict with the changes made by application A. Applications in our system are free to make changes and any conflicting changes would be recognized by the update processors.

Conflicts are when a set of changes made to a document do not agree with another set of changes made to the same document. Conflicts can be one of the following two types, *real* or *false*. For example if a document has an attribute called color and if Bob changes it to green, while at the same time Jim changes it Red, the system will not be able to decide which of the changes to accept. This is a real conflict. The other type of conflict is false; in this case the two sets of changes may appear to be inconsistent because of the granularity of the encoding. However a system that understands the semantics of the document may be able to automatically resolve this type of conflict. For example if we have a node P that has two children b and c. If Bob inserts a new child before child b and Jim inserts another child after c, both changes may be acceptable, however our particular encoding method would view this as a conflict in the children of node P.

In any system that supports incremental updates by multiple applications, conflict can occur and is not specific to encrypted documents. An example of a system that supports incremental updates is a source code control system used by a group of developers. Such systems need to detect when conflicts occur and resolve them. The systems that understand the semantics of the data deal with false conflicts automatically and resort to policies or human intervention for resolving real conflicts. In our case since the update processors deal with encrypted data they will not be able to distinguish between real and false conflicts.

In this paper we are primarily concerned with detecting conflicts. The procedure for resolving conflicts is beyond the scope of this paper, but would most likely enlist application-specific interpretation of the data to enable a meaningful conflict resolution.

4.5.1 Conflicts from multiple applications

Several applications may need to access and update the data in an XML document, which results in multiple deltas being sent to the update processor. This may lead to conflicts. It would be possible to avoid conflicts by restricting data access to one application at a time. However this would be an impractical policy. Typically applications tend to operate on different parts of the data and their updates are non-overlapping. We offer a solution that is optimized for this common case but we handle the exception situation.

4.5.2 Detecting conflicts

When applications make updates to documents, they generate deltas which are sent to the update processor, which is responsible for applying the deltas. It detects a conflict if a delta replaces a node that has been replaced by a previous delta. Add and delete may also cause conflicts, but these will be caught by the associated replace command. Recall a replace command includes a hash of the node's original encrypted data, plus the encrypted version of the modified data. When processing a replace command, the update processor finds the node data associated with the node to be replaced and compares its hash with the hash sent as part of replace and, if they match, replaces it with the new data. If there is a mismatch it marks it as a conflict and appends the delta to the document. The update processor, given a delta, computes if there are any conflicts before applying the delta. A single conflict will disable the merge of this delta, and instead the delta is simply appended to the end of the document.

5. OTHER WORK

There is considerable body of work that is related to our proposal. Most of this work is complementary to what we are proposing. Our solution encompasses the areas of XML encryption, incremental change support for XML, incremental cryptography and XOR based MACs. We give a brief overview of some of the work in these areas.

5.1 XML Encryption

Since XML documents are widely used to store important data, security is a major concern. Hirsch [6] gives a broad overview of the issues associated with XML security. Current work in this area includes two developing standards, XML-Encryption [9] [1] and XML-Signature [10] [1].

XML-Encryption and XML-Signature are going through the standardization process as part of W3C. In brief, the XML-Encryption standard aims to specify a process for encrypting data and representing the result in XML. The XML-Encryption supports encryption of arbitrary data, an XML element, or an XML element's content. The specification defines how the encrypted data is to be represented and does not specify what to encrypt. On the other hand our solution specifies how an XML document is encrypted and also how incremental updates can be performed. The syntax proposed by the XML-Encryption can be adopted to represent our encryption scheme.

XML-Signature provides integrity, message authentication and signer authentication services for any type of data. XML Signature is a method of associating a key with a block of data, and representing it as XML. While this specification is an important component of secure XML applications, it is not sufficient to address all application security and incremental update issues.

Our solution generates MACs that are used to verify the integrity of the document and also support incremental updates to the document as well as the MAC itself. Since XML-Signatures map one XML document into another, they have no effect on our processing of documents.

5.2 Incremental change support for XML

Increasingly, distributed systems are being built to allow multiple users to update to a single data set simultaneously [7]. The key issue in supporting this is the ability to merge changes as they are made. The proposal made by Fontaine [7] aims to support merging of updates at a much finer level of granularity (attributes) than our solution of merging at the element level. However, if attribute-level of merging is required, our method could be adapted to provide it. The motivation behind our decision to support a coarser level of granularity is based on the assumption that applications may operate on a single data set simultaneously but they tend to be non-overlapping. Our solution works well under this assumption. Another difference of our work is that it operates on encrypted versions of the data rather than the clear text.

The key issue in a merge process is the node matching algorithm. Fontaine et al [7] proposes a tree matching algorithm. The algorithm walks through the corresponding nodes in each input XML tree and performs a tree structured comparison. This tends to be time consuming and requires work not proportional to the amount of change. In our model, identification is a trivial algorithm that involves a simple look up of the node's unique ID. In our system a node is assigned a unique ID and that is associated with the node for the node's lifetime.

The fact that there are several commercial entities developing tools to support concurrent changes to XML data validates the importance of this field. Since there is no clear solution that addresses incremental support combined with security, this is a promising and relevant field of study,

5.3 Incremental Cryptography

Using cryptographic algorithms to protect the security of data is a well known technique, however the algorithms used tend to operate on the entire data and any time the data is modified the entire document would have to be re-encrypted. In systems that support incremental changes and simultaneous updates, this technique would not be practical. The goal of incremental cryptography is to design cryptographic algorithms with the property that having applied the algorithm to a document, it is possible to quickly update the result of the algorithm for a modified document, rather than having to re-compute it from scratch. Incremental cryptography can provide large efficiency gains when small changes are frequently made to large documents [2].

Our solution to support incremental changes to XML documents and securing them lends itself very well to incremental cryptographic techniques. Incremental cryptography and its application to virus protection were proposed by Bellare et al [3] and they base their techniques on previous work on XOR MACs [4]. Our solution proposes a technique for incrementally encrypting changes specific to XML documents and more generally to hierarchical oriented or record oriented data structures. The Bellare proposal is aimed at supporting incremental changes to documents by breaking them up into smaller units and encrypting them. It does not specify the exact process by which a document is broken down and what happens if the change crosses the unit boundary. However they highlight the same issues we address. We came to the same conclusion that they have reached in that we cannot hope to support incremental encryption and at the same time hide the amount of difference between the two documents. Our solution also supports concurrent changes and merging of these changes on untrusted servers.

5.4 XOR MACS

The XOR MAC scheme was also originally described Bellare et al [4]. This scheme was designed to support incremental changes, however these changes were limited to just text replacement operations. This XOR scheme was subsequently enhanced to support add and delete operations [3]. The enhancement included the introduction of the chaining technique. The algorithm involves three steps. 1) Computing a hash for each subsection of the document. 2) Combining adjacent hash values using another hash, and then accumulating these results with the XOR function. Combining adjacent values ensures that the order of the subsection is reflected in the final MAC. 3) Generating the MAC of the document by hashing the result of the XOR accumulation.

Incremental updates including deletions and additions involve computing the hashes of the subsections being added or deleted, and adjusting the accumulated value to reflect these additions or deletions. Finally the accumulated value is hashed to produce the MAC.

Our initial MAC computation is similar to this scheme, however the definition of subsections is not defined, and they do not specify how changes across subsections would be handled. In our model, the blocks for which we compute the integrity code are based on the structure of the document. Computing incremental MACs in our proposal is fairly straight forward. We do not need the chaining since our encoding of the document is immune to the reordering of subsections. Also, in the above mentioned scheme it is assumed that the final document's MAC can be computed only on trusted machines and is assumed not to contain conflicts. It therefore cannot be used to support merging of changes on untrusted machines.

5.5 Threats

The proposed system could be subjected to several kinds of threats. We have grouped these into three main categories. We will briefly identify the threats and address some of the possible solutions

5.5.1 Encoding related

Since we encode the XML document on a per node basis any party that has access to the encrypted document could easily infer the number of nodes in the document. If this is information is deemed important it is possible to add additional dummy nodes which can be easily filtered out during document decryption phase.

The choice of the encryption algorithm is not specified and the security of the document depends on the specific encryption scheme. Users have the flexibility in picking the appropriate encryption scheme to ensure the level of security.

5.5.2 Update related

As incremental updates are supported through command lists, it is possible for an adversary to reconstruct partial hierarchies by monitoring the command lists. This will reveal the structure of the document but not the contents of the document. One way to deal with this would be to generate additional commands which would make the reconstruction difficult but not impossible. This threat can be prevented by using a secure channel to transfer the command lists.

Unauthorized parties can generate false updates which would be accepted by the un-trusted server and this may result in preventing real trusted updates. This technique can be used to launch a denial of service attack. Again a secure channel can prevent this threat. This threat does not affect the integrity of the document as changes to documents are accepted only after the changes have been validated.

5.5.3 Standard network related

The common network related threats can be launched and most of these will lead to some sort of denial of service. None of these attacks can be used to make a legitimate modification to the document. If the update command lists are sent over public networks, they can intercepted and modified, however the integrity code checking would catch any such violation. Hosts in the middle can do traffic analysis, determine the size and frequency of changes, perhaps compute the size of the document etc. They may even alter the commands list to remove some valid changes. We offer no specific solutions to these threats. If these threats are considered important, they can be prevented by using a secure channel to transfer the command lists.

6. STATUS AND FUTURE WORK

We have implemented most of these algorithms in support of an application we are building, but the conflict detection and resolution parts are not yet finished.

We are investigating the application of this technology to other (non-XML) representations. In particular, we believe a variant of our method could be used to do similar operations on directed graphs, which are more general than trees.

In order to avoid the need for trusted update generators to have access to the entire document, we are working on methods of retrieving subtrees of documents and computing integrity codes on them. This would allow even more general incremental operation.

We are experimenting with different granularities of encodings, in order to support conflicts better. In particular, we are experimenting with refining our encoding so it doesn't encode a single node's data as one encoded node. Rather, each attribute (name-value pair) and each block of free-form text would be encoded separately. This should reduce the number of false conflicts.

More ambitiously, we are investigating the ability to do queries on encrypted data.

7. CONCLUSIONS

We have applied past work and created new algorithms to solve an important distributed systems problem: the incremental modification of data on untrusted machines. We have structured our system to work on the popular XML format, although the underlying techniques apply to other representations.

The problem solved has wide applicability to distributed systems. Our model of an untrusted storage server is just one example. There are other patterns of data flow in other distributed systems that could benefit from our algorithms.

8. REFERENCES

[1] IBM Corporation. alphaWorks XML Security Suite, *As described at http://www.alphaworks.ibm.com/tech/xmlsecuritysuite, December*, 2003.

[2] M. Bellare, O. Goldreich, S. Goldwasser, Incremental Cryptography: The case of Hashing and Signing – Crypto 94 Proceedings, Vol 839, Springer-Verlag.

[3] M. Bellare, O. Goldreich, S. Goldwasser. Incremental Cryptography and Application to Virus Protection, Proceedings of the 27th ACM Symposium on the Theory of Computing, May 1995.

[4] M. Bellare, R. Guerin, P. Rogaway. XOR MACs: New Methods for message authentication Using Finite Pseudorandom Functions, Oct 1995.

[5] Tim Bray, Jean Paoli, C. M. Sperberg-McQueen, Eve Maler. *Extensible Markup Language (XML) 1.0 (Second Edition)* , W3C Recommendation, October 6, 2000.

[6] Fredrick J. Hirsch. Getting Started With XML Security, *As described at http://www.fjhirsch.com/xml/xmlsec/starting-xml-security.html* .

[7] Robin L. Fontaine. Merging XML files: a new approach providing intelligent merge of XML data sets, *As described at http://www.deltaxml.com/pdf/merging-xml-files.pdf* .

[8] Andrew S. Tanenbaum, Maarten van Steen, Distributed Systems: Principles and Paradigms, Prentice Hall, 2002.

[9] World Wide Web Consortium. XML Encryption Syntax and Processing. *As described at http://www.w3.org/TR/xmlenc-core/*, December 10, 2002.

[10] World Wide Web Consortium. XML-Signature Syntax and Processing. *As described at http://www.w3.org/TR/xmldsig-core/*, February 12, 2002.

EFFICIENT SIMULTANEOUS CONTRACT SIGNING

Ľubica Líšková[1] and Martin Stanek[2]

[1] *Department of Mathematics*
Faculty of Civil Engineering
Slovak University of Technology
Radlinského 11, 813 68 Bratislava, Slovak Republic
liskova@math.sk

[2] *Department of Computer Science*
Faculty of Mathematics, Physics and Informatics
Comenius University
Mlynská dolina, 842 48 Bratislava, Slovak Republic
stanek@dcs.fmph.uniba.sk

Abstract Simultaneous contract signing is a two-party cryptographic protocol: two mutually suspicious parties wish to exchange signatures on a contract. We propose novel and efficient protocol for contract signing based on a construction by Even, Goldreich, and Lempel. We focus on the reduction of on-line computational complexity of the protocol. A significant part of the most time-consuming operations can be pre-computed. An important component used in our protocol is an efficient oblivious transfer, which can be of interest per se.

Keywords: contract signing, cryptographic protocols, oblivious transfer

1. INTRODUCTION

Simultaneous contract signing is a two-party cryptographic protocol: two mutually suspicious parties, let us denote them A and B, wish to exchange signatures on a contract. Intuitively, a fair exchange of signatures is one that avoids a situation where A can obtain B's signature while B cannot obtain A's signature and vice-versa. Contract signing protocols can be partitioned into two categories: protocols that use trusted third party either on-line or off-line (so called optimistic protocols) [1, 12, 19], and protocols without trusted third party [5, 9–11, 13]. We are interested in protocols from the second category.

Protocols without trusted third party are based on gradual and verifiable release of information. Hence, if one participant stops the protocol prematurely, both participants have roughly the same computational task in order to find the other participant's signature. A security problem can arise in a setting where A has much more hardware power than B. Then the actual time needed for finishing the computation would be unbalanced favorably for A. Timed protocols solve this problem by employing task for which parallel computation has roughly the same complexity as a sequential one [5, 11].

The motivation for our work is the following problem: a server has to participate in many contract signing protocols, possibly with large number of clients. A natural requirement is the time-efficiency of the protocol. An efficient protocol improves the throughput of the server and contributes to the overload protection. A server load is not uniform and thus it makes sense to pre-compute significant part of the protocol beforehand.

Our contribution. We present a contract signing protocol where substantial amount of computation can be done in advance. Our aims are the simplest possible construction and overall efficiency of the protocol. Therefore we based the protocol on construction by Even, Goldreich, and Lempel [10].

A novel oblivious transfer, based on RSA, is used as a key component of the protocol. The security of the contract signing protocol highly depends on the properties of oblivious transfer. We prove the security of oblivious transfer in the random oracle model. To address efficiency issues we show computation vs. communication tradeoff for our oblivious transfer based on ideas from [18].

Let us mention that also ElGamal based oblivious transfer, e.g. [18], can be used in our contract signing protocol, instead of RSA based one. Employing RSA allows further reduction of computation overhead, since public exponent can be made small.

Related work. There are several protocols for contract signing without trusted third party. These protocols have many appealing properties, for example, they are statistical zero-knowledge [9] or resist parallel attack [5, 11, 13], and allow an exchange of standard signatures. On the other hand, strong security properties lead to protocols that employ computationally demanding components, such as zero-knowledge proofs. Although our protocol does not enjoy these properties, it is more efficient than other protocols.

Many cryptographic applications (protocols) make intensive use of oblivious transfer. Hence, efficiency of oblivious transfer influences the overall efficiency of these protocols. Our oblivious transfer is a modification of the protocol by Juels and Szydlo [16]. The problem of amortizing the cost of multiple oblivious transfers has been considered by Naor and Pinkas [18]. They based their construction on computational Diffie-Hellman assumption, while we start with RSA assumption. Both constructions can also be used for reducing the cost of recent protocol for extending few oblivious transfers into many [15].

Our security proofs for oblivious transfers make use of random oracles. The application of random oracles in the security analysis of cryptographic protocols was introduced by Bellare and Rogaway [4]. Security proofs in a random oracle model substitute a hash function with ideal, truly random function. This approach has been applied to many practical systems, where the ideal function must be instantiated (usually as a cryptographically strong hash function). Recently, several papers pointed out that the instantiation can break the security of a protocol [3, 7]. On the other hand, the counterarguments are specially designed protocols, unsafe when instantiation takes place.

Organization of the paper. We present and analyze oblivious transfer protocols in Section 2. Section 3 provides an exposition of our contract signing protocol. We evaluate benefits and drawbacks of the protocol in Section 4.

2. OBLIVIOUS TRANSFER

Oblivious Transfer (OT) protocol, more specifically OT_1^N protocol, allows two parties (sender and chooser) to solve the following problem. The sender has N strings $m_0, \ldots, m_{N-1}$ and wishes to transfer one of them to the chooser in a secure manner:

- the chooser can select particular m_b which he wishes to obtain;
- the chooser does not learn any other string except m_b;
- the sender does not know which m_b was transferred.

Oblivious transfer is used as a key component in many cryptographic applications, such as electronic auctions [11, 17], contract signing [10], and general multiparty secure computations [2, 14]. Many of these applications make intensive use of oblivious transfer.

We modify and extend construction of RSA-based OT_1^2 protocol from [16]. Most oblivious transfer protocols employ some kind of ElGamal encryption. This results in increased computational overhead as the chooser must perform at least one modular exponentiation. Using special RSA-based oblivious transfer allows to reduce the chooser's complexity. Presented protocols make use of hash function H, modelled as random oracle (truly random function) [4], to facilitate security proofs. First, let us overview the construction from [16].

Notation. We use the following notation through the entire Section 2. All protocols in this paper use RSA public-key system. Let $n = pq$ be an RSA public modulus, where p and q are primes. Let $Z_n = \{0, 1, \ldots, n-1\}$ and $Z_n^* = \{a \in Z_n \mid \gcd(a, n) = 1\}$. An RSA public key consists of modulus n and public exponent e such that $\gcd(e, \varphi(n)) = 1$. Recall that $\varphi(n) = (p-1)(q-1)$ is the Euler function. The corresponding private key (exponent) d satisfies condition $ed \equiv 1 \pmod{n}$. The protocols involve exclusive knowledge of private key d by the sender. A bitwise exclusive OR operation is denoted by $\oplus$. All other computations are in Z_n. The hash function, denoted by H, is modelled as a truly random function (random oracle) in the security analysis. For simplicity of notation we write $H(a_1, \ldots, a_l)$ for the hash function applied to the concatenation of l-tuple $(a_1, \ldots, a_l)$.

2.1 JS protocol

The protocol assumes that $e = 3$. It proceeds as follows:

1 The sender selects randomly and uniformly an integer $C \in Z_n^*$ as well as keys $k_0, k_1 \in Z_n^*$. Strings m_0, m_1 are encrypted using these keys – resulting in $(H(k_0^3), H(k_0) \oplus m_0)$, $(H(k_1^3), H(k_1) \oplus m_1)$. The sender sends encrypted strings together with C to the chooser.

2 The chooser, wishing to receive m_b for $b \in \{0, 1\}$, selects a random element $x \in Z_n^*$. The chooser computes two values (x_0, x_1):

$$\begin{aligned} (x_0, x_1) &= (x^3, Cx^3) \quad \text{if } b = 0; \\ (x_0, x_1) &= (x^3/C, x^3) \quad \text{if } b = 1; \end{aligned}$$

and sends them to the sender.

3 The sender verifies that $x_1/x_0 = C$ and uses the private key d to compute a reply for the chooser $(z_0, z_1) = (x_0^d k_0, x_1^d k_1)$.

4 The chooser then makes use of x to extract k_b in the obvious fashion: $k_b = z_b/x$. Given k_b, the chooser can extract m_b from $(H(k_b^3), H(k_b) \oplus m_b)$.

The protocol is used as a building block of verifiable proxy oblivious transfer, see [16]. The authors state (without explicit proof) that in the random oracle model the value m_{1-b} is hidden from the chooser in a semantically secure manner, assuming the validity of RSA assumption. The value b is hidden from the sender in an information-theoretic sense.

2.2 Efficient oblivious transfer

The JS protocol can be simplified by observing that x_0 and x_1 always satisfy the equation $x_1/x_0 = C$. Hence sending just one of them is sufficient. We change the relation between x_0 and x_1 to $x_0 = x_1 C$, to free the chooser from the computation of modular inversion. Moreover, we exclude the keys k_0, k_1 from the protocol, since they are not needed.

We want to prove sender's security by comparison with the ideal implementation (model). The ideal model uses a trusted third party that receives m_0

and m_1 from the sender and b from the chooser, and tells the chooser m_b. We require that for every distribution on the inputs (m_0, m_1) and any probabilistic polynomial adversary A substituting the chooser there exists a simulator S_A such that:

1. S_A is a probabilistic polynomial-time machine substituting the chooser in the ideal model;

2. outputs of A and S_A are computationally indistinguishable.

This results in the sender's security, since the ideal model hides the value m_{1-b} perfectly. For extensive study of various definitions of protocol security in the ideal model see [6].

It is not clear how to prove the security of the JS protocol with respect to the ideal model. Therefore we modify the construction of encrypted strings to facilitate the proof.

OT_1^2 protocol. The protocol will run a number of times with some its parameters pre-computed. Therefore we introduce a random string R to differentiate the instances of the protocols.

1. The sender selects randomly and uniformly an integer $C \in Z_n^*$ and sends it to the chooser.

2. The chooser, wishing to receive m_b for $b \in \{0, 1\}$, selects a random element $x \in Z_n$. The chooser computes the value $x' = x^e C^b$ for the sender.

3. The sender selects a sufficiently long random string (or counter) R. Then the sender sets $x_0 = x'$ and computes $x_1 = x_0 C^{-1}$. The sender encrypts strings m_0, m_1 into ciphertexts E_0, E_1:

   $$\begin{aligned} E_0 &= H(R, x_0^d, 0) \oplus m_0; \\ E_1 &= H(R, x_1^d, 1) \oplus m_1. \end{aligned}$$

   The reply for the chooser consists of values R, E_0, E_1.

4. The chooser then makes use of x to decrypt m_b from E_b: $m_b = E_b \oplus H(R, x, b)$.

The value x' (i.e. x_0) is uniformly distributed in Z_n. For any element x' and any $b \in \{0, 1\}$ there exists an x such that $x' = x^e C^b$. The sender can compute x by $x = (x' \cdot C^{-b})^d$. Hence, the chooser's security is protected in an information-theoretic sense – the sender cannot determine b, even with infinite computational power.

The sender's security is protected computationally. We prove it in the random oracle model, where the hash function H is modelled as a random function. Let A be an adversary. The simulator S_A simulates both the sender and A. First, S_A picks a random value C and sends it to A. When A sends x_0 the simulator computes $x_1 = x_0C^{-1}$. It selects random strings R, E_0, E_1 and sends them in response (as if they were the sender's answers). S_A continues the simulation of A and monitors all its queries to H. All queries not containing a valid triple (R, x, i), for $i \in \{0, 1\}$, are answered at random. We say that a triple (R', x, i) is valid if $x_i = x^e$ and $R = R'$. If A asks for $H(R, x, i)$, where the argument is a valid triple, then S_A asks a trusted third party in the ideal model for m_i. The simulator sets $H(R, x, i) = E_i \oplus m_i$ to allow A to decrypt correctly. Whatever A outputs, so does S_A.

The distribution of simulated communication with A is identical to the distribution of real communication between the sender and A. Hence the output of S_A cannot be distinguished from the output of A in the real communication with the sender. The only exception is the case when A asks for both valid triples: $H(R, x, 0)$ and $H(R, x^*, 1)$. In this case, from validity of the triples it follows that $x_0 = x^e$ and $x_1 = (x^*)^e$. Then the value $x \cdot (x^*)^{-1}$ is the decryption of C:

$$(x \cdot (x^*)^{-1})^e = x^e \cdot (x^*)^{-e} = x_0 \cdot (x_1)^{-1} = C.$$

Since $C \in Z_n^*$ is random, we cannot distinguish the two distributions with non-negligible probability, assuming the RSA assumption holds.

The input of H contains a random string R to ensure the inputs in different invocations of the protocol are different. We have shown that the knowledge of valid triples is equivalent to breaking of RSA. Hence we can run the protocol with the same value C polynomially many times.

Computation overhead. The protocol allows both parties to pre-compute some values beforehand, since they depend neither on actual values m_0, m_1, b, nor communication between parties. The sender can compute C^{-d} and the chooser can compute x^e. When choosing $e = 3$, the chooser needs two modular multiplications instead of one (expensive) exponentiation. The sender computes x_1^d directly: $x_1^d = x_0^d C^{-d}$, thus performing only single exponentiation x_0^d on-line. Comparison of computation overhead, for $e = 3$ and T instances of the protocol, in a pre-computed and fully on-line implementation, is in Table 1. Notice the saving of exponentiations, which is the most time-consuming operation.

***OT_1^N* protocol.** We extend the OT_1^2 protocol to the case of N strings $m_0, \ldots, m_{N-1}$ employing ideas from [18]. In order to simplify the description of the protocol we set $C_0 = 1$. Although the OT_1^N protocol is not needed for the contract signing protocol directly, it is useful for improving efficiency of the OT_1^2 protocol by means of computation/communication tradeoff. Moreover, this protocol can be of interest per se, since the chooser does not compute any

	on-line		*pre-computed*	
	sender	*chooser*	*sender*	*chooser*
pre-computation				
exponentiations			1	0
multiplications			0	$2T$
on-line				
exponentiations	$2T$	0	T	0
multiplications	T	$3T$	T	T
H evaluations	$2T$	T	$2T$	T

Table 1. OT_1^2 computation overhead: on-line and pre-computed implementation

exponentiation (for small e). The protocol makes use of a random string R (again) to differentiate among multiple invocations.

1 The sender selects randomly and uniformly integers $C_1, \ldots, C_{N-1} \in Z_n^*$ and sends them to the chooser.

2 The chooser, wishing to receive m_b for $0 \leq b < N$, selects a random element $x \in Z_n$. The chooser computes value $x' = x^e C_b$ and sends it to the sender.

3 The sender selects a sufficiently long random string (or counter) R. The sender computes $x_u = x' C_u^{-1}$, for all $0 \leq u < N$, and encrypts strings m_u – resulting in ciphertexts E_u:

$$E_u = H(R, x_u^d, u) \oplus m_u.$$

The reply for the chooser consists of $E_0, \ldots, E_{N-1}$, and R.

4 The chooser decrypts m_b from E_b: $m_b = E_b \oplus H(R, x, b)$.

The protocol has security properties similar to the properties of the OT_1^2 protocol. Since the value x' is uniformly distributed in Z_n, the chooser's security is protected information-theoretically. From the sender's perspective x' can be arbitrary for any b and any $C_1, \ldots, C_{N-1}$ – just set $x = (x' C_b^{-1})^d$.

The sender's security is protected computationally (again). Let A be an adversary. We describe a simulator S_A in the ideal model, using A as a black-box. The simulator and the adversary have computationally indistinguishable outputs in the random oracle model, i.e. assuming a truly random hash function H. The simulator starts with choosing $C_1, \ldots, C_{N-1}$ randomly. S_A "sends" them to A and simulates it to obtain x'. The simulator computes $x_u = x' C_u^{-1}$, for all $0 \leq u < N - 1$. Then, in reply, S_A selects all the values $\{E_u\}_{0 \leq u < N}$ at random. The simulator monitors all queries to (random oracle) H and checks their validity. We say that the triple (R', x, b) is valid if an only if $x_b = x^e$, $R = R'$ and $0 \leq b < N$. If a triple is not valid, S_A answers the query randomly. In case of valid triple S_A asks a trusted third party in the ideal model for m_b. Then it sets $H(R, x_b, b) = m_b \oplus E_b$ to allow A to decrypt the string correctly. The simulator's output is whatever A outputs.

It is easy to verify that the distribution of simulated communication with A is identical to the distribution of real communication between the sender and A. Hence the output of S_A cannot be distinguished from the output of A in real communication. The only exception is when A queries H with two distinct valid triples, say (R, x, b) and (R, x^*, w), since S_A is unable to get both m_b and m_w from a trusted third party. If A can find two valid triples, s(he) can decrypt the following ciphertext:

$$C_w^{-1} C_b = (x_w (x')^{-1})(x' x_b^{-1}) = x_w x_b^{-1} = (x^* x^{-1})^e.$$

Assuming validity of RSA assumption and the randomness of $C_1, \ldots, C_{N-1}$, the adversary can find two valid triples with negligible probability as long as N is polynomially bounded.

Computation overhead. The Naor-Pinkas OT_1^N protocol from [18] uses computational Diffie-Hellman assumption to guarantee computational security for the sender. Using RSA allows us to reduce chooser's computational overhead, since the public exponent e can be made small.

Similarly to OT_1^2 we can run the protocol polynomially many times with the same values $C_1, \ldots C_{N-1}$. In order to reduce complexity, some of the values can be pre-computed: $C_1^{-d}, \ldots, C_{N-1}^{-d}$ (sender), x^e (chooser). Then the on-line requirements, when running T instances of OT_1^N, are T exponentiations for the sender and no exponentiation for the chooser ($e = 3$).

Remark. The protocol can be easily accommodated in environments where pre-computation is not possible. We sketch briefly a way to reduce the overall number of exponentiations in such cases. The sender selects only t integers $C_0, \ldots, C_{t-1} \in Z_n$, where $t = \lceil \lg N \rceil$. The chooser changes the computation of x' to $x' = x^e C_0^{b_0} \cdot \ldots \cdot C_{t-1}^{b_{t-1}}$, where $(b_0, \ldots, b_{t-1})$ is the binary representation of b. The senders computes $x_u = x' C_0^{-u_0} \cdot \cdots \cdot C_{t-1}^{-u_{t-1}}$, for all $0 \leq u < N$, where $(u_0, \ldots, u_{t-1})$ is the binary representation of u. The security properties of the modified protocol are comparable with the original OT_1^N.

Computation vs. communication tradeoff for OT_1^2. It is quite common that network bandwidth "outperforms" CPU power, i.e. computation is slower than the ability to transfer data over communication lines. The idea of decreasing computation overhead while increasing communication complexity was cleverly used in [18]. The approach can be applied in our protocols as well. The situation is as follows: The chooser and the sender need to perform many

OT_1^2 protocols. They group them in blocks of l protocols (the exact value of l will be specified later). One block is implemented as follows. The sender has l pairs of strings $(m_{1,0}, m_{1,1}), \ldots, (m_{l,0}, m_{l,1})$. Instead of using l instances of the OT_1^2 protocol, the sender creates 2^l strings in the form $m_{1,i_1} m_{2,i_2} \ldots m_{l,i_l}$, where $i_1, \ldots, i_l \in \{0, 1\}$. These strings are the sender's inputs in OT_1^N protocol for $N = 2^l$. The strings can be too long to be encrypted as described in the OT_1^N protocol. Therefore, the sender encrypts them separately and transfer the keys via oblivious transfer. The encryptions are sent off-line.

Optimal performance is achieved when the time spent by computation is equal to the time spent by communication. We evaluate the situation for the sender who performs exponentiation (RSA decryption). Exponentiation is the most computationally intensive operation. Thus, for simplicity, we will ignore other operations. Let t_e (sec) be a time required for one exponentiation. We denote by v (bits/sec) the bandwidth between the chooser and the sender. Communication complexity of the on-line part of the protocol is $2^l k + r$, where r is the length of R and k is the length of encryption key. The optimal value l is then $l = \lg((t_e v - r)/k)$.

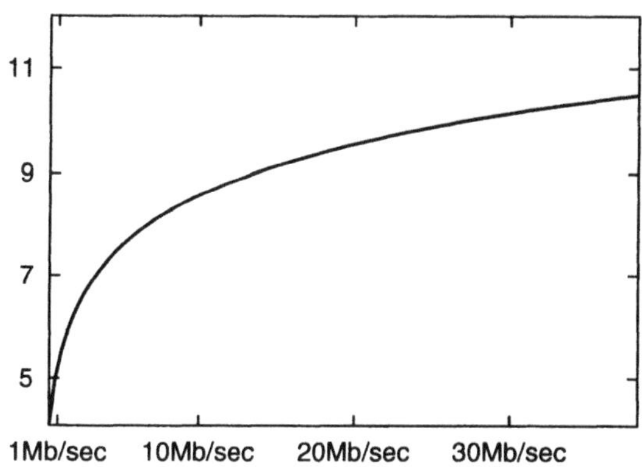

Figure 1. Optimal l for given bandwidth

To show relevance of the described tradeoff, let us give an illustration in the real world settings. Let $k = r = 128$. Let n (RSA modulus) be 1024 bits long. Taking benchmarks from [8], one RSA decryption takes 4.63ms on a 2.1GHz Pentium 4 processor. The optimal value of l as a function of the bandwidth is depicted in Figure 1.

3. CONTRACT SIGNING

The protocol for contract signing from [10] uses an oblivious transfer as a basic component. Our protocol is inspired by this construction but differs from it in the following aspects:

- a criterion when the contract is considered binding (the original protocol uses threshold acceptance);
- an efficient oblivious transfer, described in Section 2.2, allows to compute a significant part of the operations in advance.

These properties guarantee small on-line computation overhead.

Protocols for simultaneous contract signing usually consist of two interlaced protocols. Both participants are in symmetric situations – each of them wants to transfer its own signature in exchange for the other participant's signature. Our description employs both exchanges.

Let us denote by $Sig_A(m)$ a digital signature of a message m created by the participant A. Our protocol is independent of chosen digital signature algorithm. Let k be a security parameter. For the purposes of contract signing, we define *C-signature* (or $CSig$) of a message m. It is a triple:

$$CSig_A(m) = (Sig_A(m, R), Sig_A(R, i, 0), Sig_A(R, i, 1)),$$

for arbitrary $i \in \{1, \ldots, k\}$ and a random binary string $R \in \{0,1\}^k$ long enough to avoid collisions among instances of the protocol. A C-signature is valid if and only if all its parts are formed correctly and have valid signatures.

3.1 The protocol

Alice and Bob simultaneously transfer C-signatures of contract M. We will use A and B as shortcuts for Alice and Bob, respectively. The protocol employs symmetric keys. We assume the length of a symmetric key is a multiple of the security parameter k and not shorter than the length of a signature. Other settings are discussed in Section 3.2. We denote by $A \leftrightarrow B : OT_1^2(m_0, m_1)$ the instance of an oblivious transfer protocol with A playing the role of the sender (possessing two strings m_0, m_1), and B playing the role of the chooser.

1 Alice chooses $R_A \in \{0,1\}^k$ and Bob chooses $R_B \in \{0,1\}^k$. Alice and Bob exchange the first parts of their C-signatures together with R_A, R_B:

   $A \rightarrow B$: $R_A, Sig_A(M, R_A)$,
   $B \rightarrow A$: $R_B, Sig_B(M, R_B)$.

2 Alice and Bob validate the received signatures. They compute the values $Sig_{A/B}(R, i, b)$, for all $i \in \{1, \ldots, k\}$ and $b \in \{0,1\}$. Alice chooses

random symmetric keys $K_{A,i,b}$. Similarly, Bob chooses keys $K_{B,i,b}$. The keys are used for encryption of corresponding signatures. Alice and Bob exchange encrypted signatures, for all $i = 1, \ldots, k$ and $b = 0, 1$:

$A \rightarrow B$: $K_{A,i,b} \oplus Sig_A(R_A, i, b)$,
$B \rightarrow A$: $K_{B,i,b} \oplus Sig_B(R_B, i, b)$.

3 Alice uses an oblivious transfer protocol to exchange exactly one value $K_{A,i,b}$ from each pair of symmetric keys (a pair consists of keys with equal i and different b values):

$A \leftrightarrow B$: $OT_1^2(K_{A,i,0}, K_{A,i,1})$, for $i = 1, \ldots, k$.

Bob selects the key which he wants to receive, the first or the second one for each pair, randomly.

4 Analogously, Bob uses an oblivious transfer protocol to exchange exactly one value $K_{B,i,b}$ from each pair of his keys (Alice's choice for each oblivious transfer is random):

$B \leftrightarrow A$: $OT_1^2(K_{B,i,0}, K_{B,i,1})$, for $i = 1, \ldots, k$.

5 Alice and Bob decrypt half of the signatures using the received keys and validate the signatures. They divide every symmetric key into k pieces of equal length (recall that the length of keys is multiple of k):

$$K_{A,i,b} = K^1_{A,i,b} \mid\mid K^2_{A,i,b} \mid\mid \cdots \mid\mid K^k_{A,i,b},$$
$$K_{B,i,b} = K^1_{B,i,b} \mid\mid K^2_{B,i,b} \mid\mid \cdots \mid\mid K^k_{B,i,b},$$

where the operator '||' denotes concatenation of strings. Alice and Bob gradually exchange the pieces of keys, i.e. for $w = 1, \ldots, k$:

$A \rightarrow B$: $K^w_{A,1,0}, K^w_{A,1,1}, \ldots, K^w_{A,k,0}, K^w_{A,k,1}$,
$B \rightarrow A$: $K^w_{B,1,0}, K^w_{B,1,1}, \ldots, K^w_{B,k,0}, K^w_{B,k,1}$.

Transfers are interlaced, so both parties send the pieces for $w + 1$ only when they already received the pieces for w. Alice and Bob check after each transfer that the half of received pieces is equal to the corresponding pieces of the keys obtained via oblivious transfers. They continue the protocol only if the check is successful.

Security. A valid C-signature consists of three parts. The first part is transferred in step 1. Other parts are exchanged gradually from step 2 to step 5. Security of the protocol heavily depends on the properties of oblivious transfer. They guarantee that Bob learns exactly one part of each signature pair $Sig_A(R_A, i, 0)$, $Sig_A(R_A, i, 1)$ in step 3. Moreover, Alice does not know which of them has been obtained by Bob. Similar situation arises for Bob's signature pairs in step 4. Hence, neither party knows the C-signature before the last step. Pieces of keys needed to decrypt the remaining signatures are exchanged in step 5. The pieces of known keys are used to check for possible cheating.

The probability of successful cheating in the protocol depends on the security parameter k. Notice that any signature pair completes the C-signature. If a participant wants to cheat (i.e. not providing valid C-signature for the other), it must lie in every pair. Hence the probability of successful cheating is equal to the probability of guessing which keys the other participant obtained in oblivious transfers, i.e. 2^{-k}.

Performance. We analyze on-line and off-line computation overhead of the protocol. Most of the operations can be pre-computed, i.e. computed off-line. It follows from the properties of our OT_1^2 protocol and from the fact that the strings transferred by OT_1^2 do not depend on the contract. As the roles of both participants in the protocol are symmetric, computational and communication needs are equal. We calculate the number of signing operations and exponentiations performed in one instance of the protocol. Other operations, such as signature verification, multiplication, hash function evaluation etc. are neglected in the analysis since they are considerably easier to compute.

Signatures in the first step of the protocol must be computed on-line – they depend on the contract. On the contrary, all of the second step (employing $2k$ signatures) can be pre-computed in advance. Oblivious transfers in steps 3 and 4 require k on-line and 1 off-line exponentiations. The last step involves neither exponentiation nor signing. Table 2 summarizes these facts.

	off-line	*on-line*
signatures	$2k$	1
exponentiations	1	k

Table 2. Computation overhead of a participant

The on-line computation overhead can be further reduced by employing technique of computation/communication tradeoff for OT_1^2 described in Section 2.2.

3.2 Implementation issues

The description of the protocol left open some issues which can be addressed in an actual implementation:

- The security parameter k influences the efficiency of the protocol and the resistance to cheating. An appropriate value of k should balance these requirements, and can be, for example 100 or 128.

- The signatures in step 2 are encrypted using one-time pad. Employing a symmetric encryption algorithm allows signatures much longer than the length of symmetric keys.

- The hash function H is used in OT_1^2 protocol. Although modelled as a truly random function in our analysis, it is instantiated as a cryptographic strong hash function (such as SHA-1 or RIPEMD-160) in the implementation.

- Further simplification removes the hash function H from OT_1^2 protocol altogether. We substitute it with a simple combination of its arguments, for example XOR. This modification breaks the security of OT_1^2 in the random oracle model. On the other hand, we are not aware of any practical attack on our contract signing protocol when employing this modification or modified OT_1^2 protocol itself.

Finally, one can easily change the oblivious transfer protocol in the implementation. It is possible to use ElGamal based oblivious transfer, such as [18]. Nonetheless, employing our RSA based construction with small public exponent e saves exponentiations on the choosers side of the oblivious transfer, see Section 2.2.

4. BENEFITS AND DRAWBACKS

Let us sum up the properties of our contract signing protocol. On the positive side, the protocol is fast and easy to implement. Moreover, it allows any signature algorithm to be used in signing the parts of a C-signature. A precomputation and a computation vs. communication tradeoff can be employed for reducing the on-line complexity of the protocol. We believe the protocol is suitable for a situation where a party (server) has to participate in many contract signing protocols constantly.

When comparing the protocol with other constructions one can spot its drawbacks. Firstly, the protocol does not provide a protection in the case of unbalanced computing power. A participant with greater computing power possesses an advantage over the second one. S(he) can terminate the protocol prematurely and compute the remaining data faster than the other party. This can be reduced by unbalanced release of keys in step 5 if the difference is known in advance.

Additionally, the security proofs for oblivious transfers are based on random oracles. A necessity to instantiate the hash function leaves the possibility of unsafe implementations. On the other hand, we found no unsafe instantiations of the hash function at all.

5. CONCLUSION

We have presented a contract signing protocol with emphasis on performance. We see a possibility for subsequent research in improving the on-line complexity of the protocol further. An attractive objective is to remove the use of random oracles in our RSA based oblivious transfer while preserving its performance.

Acknowledgments

The first author acknowledges partial support from the APVT grant No. 023302. The second author was partially supported by VEGA grant 1/0131/03.

References

[1] N. Asokan, V. Shoup, M. Waidner: Optimistic fair exchange of digital signatures, In *Advances in Cryptology – EuroCrypt '98*, LNCS 1403, 591–606, Springer-Verlag, 2000.

[2] D. Beaver: Minimal-latency secure function evaluation, In *Advances in Cryptology – EuroCrypt '00*, LNCS 1807, 335–350, Springer-Verlag, 2000.

[3] M. Bellare, A. Boldyreva, A. Palacio: An Un-Instantiable Random-Oracle-Model Scheme for a Hybrid-Encryption Problem, *Cryptology ePrint Archive, Report 2003/077*, Available at `eprint.iacr.org`, 2003.

[4] M. Bellare, P. Rogaway: Random Oracles are Practical: a Paradigm for Designing Efficient Protocols, In *1st ACM Conference on Computer and Communication Security*, 62–73, ACM Press, 1993.

[5] D. Boneh, M. Naor: Timed Commitments, In *Advances in Cryptology – Crypto '00*, LNCS 1880, 236–254, Springer-Verlag, 2000.

[6] R. Canetti: Security and Composition of Multiparty Cryptographic Protocols, *Journal of Cryptology*, Vol. 13, No. 1, 143–202, 2000.

[7] R. Canetti, G. Goldreich, S. Halevi: The Random Oracle Methodology, Revisited, In *Proceedings of the 30th ACM Symposium on Theory of Computing*, 209–218, ACM Press, 1998.

[8] W. Dai: Crypto++ 5.1 Benchmarks, Available at `www.eskimo.com/~weidai/benchmarks.html`.

[9] I.B. Damgård: Practical and provably secure release of a secret and exchange of signatures, In *Advances in Cryptology – EuroCrypt '93*, LNCS 765, 200–217, Springer-Verlag, 1993.

[10] S. Even, O. Goldreich, A. Lempel: A Randomized Protocol for Signing Contracts, In *Advances in Cryptology: Proceedings of Crypto '82*, 205–210, Plenum Publishing, 1982.

[11] J.A. Garay, M. Jakobsson: Timed Release of Standard Digital Signatures, In *Financial Cryptography '02*, LNCS 2537, 168–182, Springer-Verlag, 2002.

[12] J.A. Garay, M. Jakobsson, P.D. MacKenzie: Abuse-Free Optimistic Contract Signing, In *Advances in Cryptology: Proceedings of Crypto '99*, LNCS 1666, 449–466, Springer-Verlag, 1999.

[13] J.A. Garay, C. Pomerance: Timed Fair Exchange of Standard Signatures, In *Financial Cryptography '03*, LNCS 2742, 190–207, Springer-Verlag, 2003.

[14] O. Goldreich, S. Micali, A. Wigderson: How to play any mental game – a completeness theorem for protocols with honest majority, In *19th ACM Symposium on the Theory of Computing*, 218-229, ACM Press, 1987.

[15] Y. Ishai, J. Kilian, K. Nissim, E. Petrank: Extending Oblivious Transfers Efficiently, *Advances in Cryptology – Crypto 2003*, LNCS 2729, 145–161, Springer-Verlag, 2003.

[16] A. Juels, M. Szydlo: An Two-Server Auction Protocol, In *Financial Cryptography '02*, LNCS 2537, Springer-Verlag, 2002.

[17] M. Naor, B. Pinkas, R. Sumner: Privacy preserving auctions and mechanism design, In *1st ACM Conference on Electronic Commerce*, 129–139, ACM Press, 1999.

[18] M. Naor, B. Pinkas: Efficient oblivious transfer protocols, In *12th Annual ACM-SIAM Symposium on Discrete Algorithms*, 448–457, 2001.

[19] B. Pfitzmann, M. Schunter, M. Waidner: Optimal Efficiency of Optimistic Contract Signing, In *Symposium on Principles of Distributed Computing*, 113-122, 1998.

DHCP AUTHENTICATION USING CERTIFICATES

Jacques Demerjian and Ahmed Serhrouchni
Ecole Nationale Supérieure des Télécommunications – LTCI-UMR 5141 CNRS, France
{demerjia, ahmed}@enst.fr

Abstract: In this paper, we describe several methods of DHCP authentication. We propose an extension to DHCP protocol in order to allow a strict control on equipments by using a strong authentication. This extension, called E-DHCP *(Extended-Dynamic Host Configuration Protocol)* is based on two principles. The first one is the definition of a new DHCP option that provides simultaneously the authentication of entities (client/server) and DHCP messages. The technique used by this option is based mainly on the use of asymmetric keys encryption RSA, X.509 identity certificates and attribute certificates. The second principle is the attribution of PMI (*Privilege Management Infrastructure*) attribute authority server functionalities to DHCP server. This server creates an attribute certificate to the client, which ensures the relation between the identity certificate of the client and the allocated IP address. This attribute certificate will be then used in the access control.

Key words: Access Control, Attribute Certificate, Authentication, DHCP, X.509 Identity Certificate.

1. INTRODUCTION

A protocol of dynamic attribution of Internet addresses is necessary for the functioning of a considerable number of networks and this, for two reasons. The first is the lack of Internet addresses, which does not allow static attribution of addresses. The second is that the mobility of the equipment is adapted to dynamic addressing. DHCP *(Dynamic Host Protocol Configuration)* [1], is thus in the centre of networks architectures, this protocol provides a framework for passing configuration information to hosts on a TCP/IP network. DHCP is based on the *Bootstrap Protocol*

'*BOOTP*' [3], adding the capability of automatic allocation of reusable network addresses and additional configuration options.

DHCP is built directly on UDP [12] and IP [7], which are as yet inherently insecure. There was no attempt in the design of DHCP to protect against malicious Internet hosts, and consequently the protocol is vulnerable to a variety of attacks [6]. Unauthorized DHCP servers may be easily set up. Such servers can then send false and potentially disruptive information to clients such as incorrect or duplicate IP addresses, incorrect routing information, incorrect domain name server addresses, and so on. Clearly, once this seed information is in place, an attacker can further compromise affected systems. Malicious DHCP clients could masquerade as legitimate clients and retrieve information intended for those legitimate clients. Where dynamic allocation of resources is used, a malicious client could claim all resources for itself, thereby denying resources to legitimate clients. Therefore, DHCP currently provides no authentication or security mechanisms.

For all of these problems, a solution would be to authenticate both the client and the server. Many different contributions [8][13] regarding how DHCP should be authenticated already exist. Some contributions include cryptography, some do not.

In this paper, we propose an extension [2] to DHCP protocol in order to allow a strict control on the equipments by strong authentication. This extension ensures on one hand, the authentication of the entities and DHCP messages and, on the other hand, the access control in DHCP system.

The remainder of the paper is structured as follows: Section 2 introduces the DHCP basic operations; section 3 presents the DHCP importance; section 4 explores the DHCP shortcomings and vulnerabilities. Section 5 presents some existing contributions that define how authentication should be handled in DHCP, and exposes their limits. Section 6 illustrates our proposed extension, and finally section 7 concludes the paper and identifies possible future work.

2. BASIC DHCP OPERATIONS

The DHCP provides a way to automate and manage the network configuration of desktop computers and other network devices that use the TCP/IP protocol. DHCP, like *BOOTP* uses client-server model and is set on UDP, utilizing ports 67 and 68. It uses the same packet format as *BOOTP* for its compatibility [5]. DHCP relays messages between the client and the server. In DHCP a client initiates all interactions, and a server replies.

By using DHCP, dynamically configuring the host on the network is done by a simple handshake. The process to be followed to get configuration data from DHCP server can be divided into two phases.

In the first step, the client broadcasts a *DHCPDiscover* message to collect proposals from servers. The client may specify preference of a lease and/or an IP address.

A DHCP server receiving the *DHCPDiscover* message may or not return *DHCPOffer* message (Many servers may receive the *DHCPDiscover* message). If a server decides to respond, it offers a selection of configuration parameters and puts an available address into *yiaddr* field and broadcasts the *DHCPOffer* to the client. At this point, there is no agreement of an assignment between the server and the client.

In the second step, the client gets one or more *DHCPOffer* and chooses one server from them. The client puts the IP address of the chosen server into the '*Server identifier*' option of a *DHCPRequest* and broadcasts it over the network. Each server checks the '*Server identifier*' option. If it does not match its own address, the server considers it as an implicit decline. The selected server sends the *DHCPAck* (if its address is available) or the *DHCPNak* (for example, the address is already assigned to another client).

The client which gets the *DHCPAck* starts using the IP address. If it gets DHCPNak, it restarts to broadcast a *DHCPDiscover* message. If the client finds a problem with the assigned address of *DHCPAck*, it sends *DHCPDecline* to the server, and broadcasts a new *DHCPDiscover*. The client can release the address before its lease expires by *DHCPRelease* [21].

3. IMPORTANCE OF DHCP

The introduction of DHCP alleviated the problems associated with manually assigning TCP/IP client addresses. Network administrators have quickly appreciated the importance, flexibility and ease-of-use offered in DHCP. DHCP has several major advantages over manual configurations. No manual reconfiguration is required at all. DHCP reduces the amount of work required to administer large IP networks by eliminating the need to individually assign, configure, and manage a permanent IP address for every machine. Each computer gets its configuration from a "*pool*" of available numbers automatically for a specific time period. When a computer has finished with the address, it is released for another computer to use. Configuration information can be administered from a single point. Major network resource changes (e.g. a router changing address), requires only the DHCP server be updated with the new information, rather than every system.

4. DHCP SECURITY

In this section, we illustrate major DHCP shortcomings and vulnerabilities.

4.1 DHCP shortcomings

DHCP suffers from some significant shortcomings, which include [22]:

1. The lack of robust administrative tools: DHCP lacks such administrative capabilities as the ability to associate an address with a user name.
2. The lack of intelligence: The DHCP server is simply a server, distributing IP addresses as requested and collecting them when they expire according to a set of simple rules. It doesn't ask questions about the user or track information other than the IP address and lease parameters. It cannot perform more sophisticated tasks to effectively manage the IP address asset.
3. Limited security: DHCP servers provide only limited facilities, such as log files, to audit IP address distribution or maintain a record of IP address usage. After manually combing the file, the administrator comes up with a MAC address, which still doesn't identify the machine to which the address has been assigned. Yet, such identification is critical when trying to track network activity at a given time. As a result, DHCP provides no effective way to identify conflicts or track down rogue addresses without manually examining reams of log files.

It is clear that DHCP was never intended as a full IP address management solution. Rather, DHCP was designed to perform one critical task very well: the automatic assignment of temporary IP addresses and maintenance of an IP address pool.

4.2 DHCP vulnerabilities

There was no attempt in the design of DHCP to protect against malicious Internet hosts, and consequently the protocol is vulnerable to a variety of attacks. Since the DHCP server doesn't do any authentication of client *DHCPDiscover* requests, any intruder can effectively impersonate the identity of any client that divulges its identification information [6]. Likewise, an intruder can impersonate a DHCP server, and send erroneous information to any local DHCP client. When connecting to the network with the DHCP service, even an illegitimate user without the right to use the network can obtain an IP address from the DHCP server.

DHCP itself does not have an access control, illegitimate users inside of a network segment can easily abuse inside or outside the network. To solve this problem, introduction of a MAC address authentication scheme has been proposed, whereby, the MAC address of the equipment must be registered on the DHCP server before accessing the network. When an IP address is requested, the server authenticates the equipment by the MAC address.

Using authentication by MAC address constrains the user to use the IP address affected by the DHCP server on the terminal with the same MAC address. In this mechanism, the DHCP server authenticates the terminal through its MAC address rather than the client. However, since only registered terminals can use an IP address, as it stands, the MAC authentication is inconvenient. Moreover, illegitimate users who fabricate a MAC address can easily deceive the DHCP server and obtain an IP address [19]. Therefore, DHCP in its current form is quite insecure. Hence, for all of these problems, we need stricter new authentication mechanisms, which can provide both entity (client/server DHCP) authentication and content authentication of DHCP messages.

5. EXISTING CONTRIBUTIONS

Several different contributions regarding how DHCP should be authenticated already exist. Among them:

1. **DHCP Authentication via Kerberos V** [13]: This authentication method authenticates the client only, and involves communication with the Kerberos server, in addition to the DHCP standard communication.
2. **Token Authentication** [8]: This involves sending a token such as a plaintext password from the client to the server to identify the client. This protocol provides only weak entity authentication and no message authentication. This mechanism is vulnerable to interception and provides only the most rudimentary protection against inadvertently instantiated DHCP servers.
3. **Delayed Authentication** [8]: This requires a shared secret key for each client on each DHCP server with which that client may wish to use the DHCP protocol. Each secret key has a unique identifier that can be used by a receiver to determine which secret was used to generate the MAC *(Message Authentication Code)* in the DHCP message. The authenticity of DHCP messages is confirmed using both an index to that key and a hash of the packet using that key. Delayed Authentication is the most secure and interesting contribution for DHCP Authentication, which has been more formally designed and accepted than many of the others. For this reason, we shall introduce its related issues.

5.1 Delayed Authentication issues

The main issues of this option are key distribution and key flexibility. None of these affect the security of the protocol, but both have potential to affect its applicability in practice. The protocol relies on shared cryptographic keys being already known by the client and the server.

The first issue is one of the major drawbacks to the use of shared keys [18]. Their distribution is complicated. The technical specification of Delayed Authentication itself attempts to remedy this and suggests using a master server key with multiple client keys to simplify the key distribution, but this can decrease system security.

The second issue (flexibility) to using shared keys becomes apparent when the client switches between networks. Different networks should require different keys, and this introduces a new issue with shared key management: the key chain. Management of multiple shared secret keys can quickly become cumbersome. A real digital signature mechanism such as RSA [9], would provide a better security.

The delayed authentication option is exposed to additional drawbacks:

1. Delayed Authentication is vulnerable to a denial of service attack through flooding with *DHCPDiscover* messages, which are not authenticated by this protocol. Such an attack may overwhelm the computer on which the DHCP server is running and may exhaust the addresses available for assignment by the DHCP server.
2. Delayed authentication does not support interdomain authentication.
3. Delayed authentication may also be vulnerable to a denial of service attack through flooding with authenticated messages, which may overwhelm the computer on which the DHCP server is running as the authentication keys for the incoming messages are computed.

6. E-DHCP

Because of the inherent vulnerabilities of the current authentication mechanisms, it proves to be necessary to find solutions answering effectively this legitimate security preoccupation.

We propose an extension to DHCP protocol called E-DHCP *(Extended-Dynamic Host Configuration Protocol)* in order to stricter control of the equipment user through a stronger authentication process. This extension ensures on the one hand, the authentication of the entities and DHCP messages and, on the other hand, the access control to a DHCP system.

The following sections provide an overview of E-DHCP, and a scenario demonstrating the use of E-DHCP for obtaining an IP address and the set of configuration parameters, then a scenario to show the way our method can be used to access resources and services within a network. Finally in this section, we present E-DHCP advantages.

6.1 E-DHCP Overview

The E-DHCP solution is based mainly on the certificate concept; so it is useful to briefly define some certificate related concepts before going further.

6.1.1 Essential background and concepts

In order to access a resource, both authentication and authorization are needed. PKI (*Public Key Infrastructure*) can provide a strong authentication support for a system by using PKCs (*Public Key Certificate*), while PMI (*Privilege Management Infrastructure*) can provide authorization support for a system by using ACs (*Attribute Certificate*). The use of public-key certificates proves the identity of the certificate holders. X.509 certificate is widely accepted as the appropriate format for public key certificates [23].

Similar to PKC, an AC binds the attributes such as group membership, roles, or other authorization information associated with the AC holder to that entity through the signature of a so-called AA (*Attribute Authority*).

As outlined in RFC 3281, an AC may consist of the following fields:

- ***Version***: This field indicates the version (1 or 2) of the AC format in use.
- ***Holder***: This field is used to bind an attribute certificate to an X.509 PKC. The *Holder* field identifies the client with which the attributes are being associated. Identification can be either by name or by reference to an X.509 PKC. This field is a *SEQUENCE* allowing three different syntaxes: *baseCertID*, *EntityName* and *objectDigestInfo*. Only one option should be present. For any environment where the AC is passed in an authenticated message or session and where the authentication is based on the use of X.509 PKC, the *holder* field should use the *baseCertificateID*. With the *baseCertificateID* option, the holder's PKC *serialNumber* and *issuer* must be identical to the AC holder field.
- ***Issuer***: This field identifies the AA that issued the AC.
- ***Signature***: This field indicates the AC digital signature algorithm.
- ***Serial Number***: This field contains a unique AC serial number.
- ***Validity Period***: This field contains a time period during which the AC is assumed to be valid.
- ***Attributes***: This field contains information (*SEQUENCE OF Attribute*) concerning the AC Holder. Each Attribute may contain a set of values.

- ***Issuer Unique Identifier***: This field is used to make the name of the issuing AA unambiguous, in the case where the same name was reassigned to different authorities through time. This field is optional.
- ***Extensions***: This field allows the addition of new fields to the AC. The extensions defined for ACs provide methods for associating additional attributes with holders. This profile also allows communities to define private extensions to carry information unique to those communities.

6.1.2 E-DHCP Principles

E-DHCP is based on two principles:

The first is the definition of a new DHCP option [17] "*E-DHCP Authentication*" that provides simultaneously the authentication of entities and DHCP messages. The technique used by this option is based mainly on the use of *asymmetric* keys encryption RSA, X.509 identity certificates [10] and attribute certificates [11]. The definition of new DHCP options is possible because the options field envisages the implementation of new options [2].

The second is the attribution of the AA *(Attribute Authority)* server functionalities of a PMI *(Privilege Management Infrastructure)* [11] to the DHCP server. This server creates the client attribute certificate 'AC'. This certificate ensures the relation between the client identity certificate and the allocated IP address. This AC will be then used in the access control.

6.1.3 E-DHCP Requirements

1. The client must hold a valid X509 identity certificate delivered by a trusted CA (*Certification Authority*).
2. The server must hold a valid X509 identity certificate delivered by a trusted CA (*Certification Authority*).

6.1.4 E-DHCP Architecture

Figure 1 depicts the different components of E-DHCP architecture which are described as follows:

1. ***E-DHCP Client***: An 'E-DHCP Client' or 'client' is an equipment using DHCP to obtain configuration parameters.
2. ***E-DHCP Server***: An 'E-DHCP Server' or 'server' is a server that:

a) Returns configuration parameters and IP address to E-DHCP clients.
b) Creates a client AC, which contains the IP address allocated.

3. ***X.509 Identity Certificate Database***: Is a Database where entities (client or server) X.509 Identity Certificates are saved.

4. ***Attribute Certificate Database***: Is a Database where clients ACs are saved.

Details about the interaction between these elements will be presented in 6.2 and 6.3.

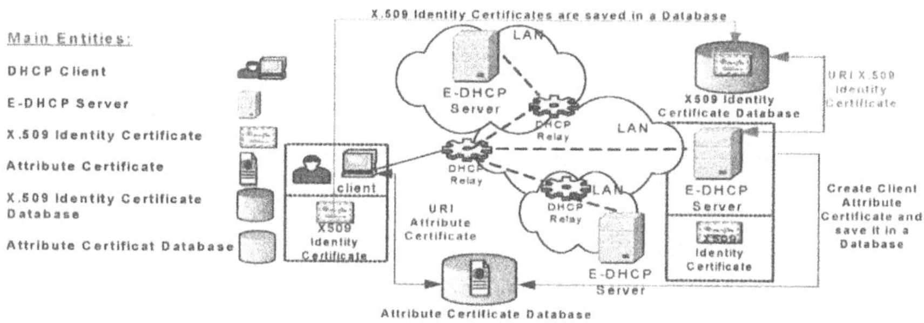

Figure 1. E-DHCP Architecture

6.1.5 E-DHCP Authentication option structure

This sub-section presents the "*E-DHCP Authentication*" option structure (see Fig. 2), which constitutes the first E-DHCP principle.

Bits:
0 1 2 3 4 5 6 7 8 9 10 11 12 13 14 15 16

Code	Length	Flag	URIIdentityCertificate	URIAttributeCertificate
AuthenticationInformation				

Figure 2. E-DHCP Authentication option structure

- ***Code***: Indicates the option code (To Be Determined).
- ***Length***: Indicates the entire option length.
- ***Flag*** (0 or 1): Indicates if the client used the server public key or alternatively if the server used the client public key to encrypt the content of the field "*AuthenticationInformation*" (if key used, flag=1, otherwise, flag=0. Default value is 1). This field allows the message receiver to know if their public key was used by the message sender.
- ***URIIdentityCertificate***: Defines the X.509 identity certificate URI *(Uniform Resource Identifiers)* [4] of the message sender (client or server).
- ***URIAttributeCertificate***: Defines the client attribute certificate URI. This certificate is created by the E-DHCP Server.
- ***AuthenticationInformation***: Contains the signature value if Flag=0. The signature is applied to the whole DHCP message including the header and the options except 'hops' and 'giaddr'. This signature is created

using the message sender's private key. The sender may then encrypt this signature using the receiver public key, and put the resulting value in the *AuthenticationInformation* field, which means Flag=1.

This double action signature/encryption requiring the client to be in possession of the server public key, would avoid a denial of service attack through flooding with non-authenticated *DHCPDiscover* messages. The same technique is used by the server to sign the message and encrypt its signature if the server is in possession of the client public key.

6.2 E-DHCP Scenario

The E-DHCP works the same way as DHCP Delayed Authentication. That is, the client and server send authentication information in an option within each DHCP packet and the DHCP protocol itself remains unchanged.

The following steps present a scenario of successful DHCP exchanges between clients and servers:

1. The client broadcasts a *DHCPDiscover* message (see Fig.3) on its local physical subnet. This message may include options, and among them, those which suggest values for the network address and lease duration or, that which provides both entity and message authentication "*E-DHCP Authentication*"option, etc.

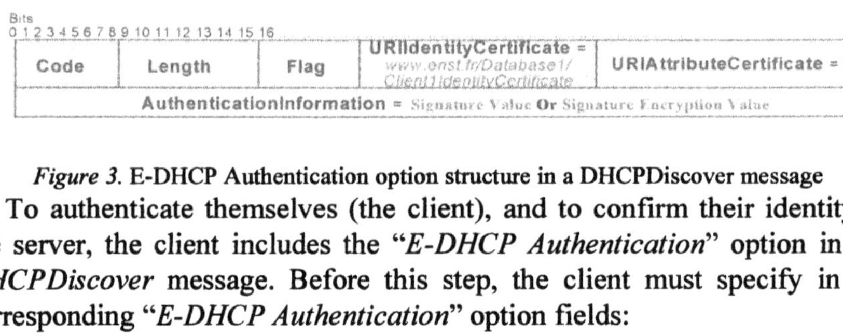

Figure 3. E-DHCP Authentication option structure in a DHCPDiscover message

To authenticate themselves (the client), and to confirm their identity to the server, the client includes the "*E-DHCP Authentication*" option in the *DHCPDiscover* message. Before this step, the client must specify in the corresponding "*E-DHCP Authentication*" option fields:

- '***URIIdentityCertificate***': Defines the URI (Uniform Resource Identifiers) of his X.509 identity certificate.
- '***URIAttributeCertificate***': Puts the value 0 in this field.
- '***Flag***': Indicates if they used the E-DHCP Server public key to encrypt the content of the field "*AuthenticationInformation*" (if key used, Flag=1, otherwise, Flag=0).
- '***AuthenticationInformation***': Contains the signature value if Flag=0. The signature is applied to the whole *DHCPDiscover* message including the header and the options except 'hops' and 'giaddr'. This signature is created using the client private key. The client may then encrypt this

signature using the server public key and put the resulting value in the *AuthenticationInformation* field, which means Flag=1.

N.B: The use of Digital signatures provides authenticity and integrity of transmitted data, and the use of encryption guarantees confidence in sensitive data.

2. To validate the authentication of the client and the incoming message '*DHCPDiscover*', the E-DHCP Server:

a) Uses the URI of the client X.509 identity certificate, contained in the '*URIIdentityCertificate*' field of *DHCPDiscover* message to extract the client X.509 identity certificate, and then to extract the client public key from this identity certificate.

b) Verifies the value contained in the 'Flag' field:

- If the field value is equal to 0, the server uses the client public key (extracted from the client X.509 identity certificate) in the verification of the validity of the signature (contained in the '*AuthenticationInformation*' field).
- If the field value is equal to 1, the server uses its private key to decrypt the value contained in the '*AuthenticationInformation*' field. The result of this decryption is the 'signature'. The server uses the client public key (extracted from the client X.509 identity certificate) in the verification of this signature.

If the authentication of the message and the client is validated, the server prepares an offer to send to the client. Otherwise, the server discards the message.

c) May choose to accept unauthenticated *DHCPDiscover* messages, or only accept authenticated *DHCPDiscover* messages based on its policy.

d) Responds with a *DHCPOffer* message (see Fig. 4) that may include "*E-DHCP Authentication*" option.

To authenticate itself (server), and to confirm its identity to the client, the server includes the "*E-DHCP Authentication*" option in the *DHCPOffer* message. Before this step, the server must specify in the corresponding "*E-DHCP Authentication*" option fields:

- ***'URIIdentityCertificate'***: Defines the URI *(Uniform Resource Identifiers)* of the server X.509 identity certificate.
- ***'Flag'***: Indicates if the sever used the client public key to encrypt the content of the field "*AuthenticationInformation*" (if key used, Flag=1, otherwise, Flag=0).
- ***'URIAttributeCertificate'***: Puts the value 0 in this field.
- ***'AuthenticationInformation'***: Contains the signature value if Flag=0. The signature is applied to the whole *DHCPOffer* message including the header and the options except 'hops' and 'giaddr'. This signature is created using the server private key. The server may then encrypt this

signature using the client public key and put the resulting value in the *AuthenticationInformation* field, which means Flag=1.

Figure 4. E-DHCP Authentication option structure in a DHCPOffer message

3. To validate the authentication of the server and the incoming message 'DHCPOffer', the client:

a) Uses the URI of the server X.509 identity certificate, contained in the '*URIIdentityCertificate*' field of *DHCPOffer* message to extract the server X.509 identity certificate, and then to extract the server public key from this identity certificate.

b) Verifies the value contained in the 'Flag' field:

- If the field value is equal to 0, the client uses the server public key (extracted from the server X.509 identity certificate) in the verification of the validity of the signature (contained in the '*AuthenticationInformation*' field).
- If the field value is equal to 1, the client uses its private key to decrypt the value contained in the '*AuthenticationInformation*' field. The result of this decryption is the 'signature'. The client uses the server public key (extracted from the server X.509 identity certificate) in the verification of this signature.

If the message or the server fail to pass the authentication validation or, if the offer suggested by the E-DHCP Server is unacceptable to the client; the latter discards the message, and sends a new *DHCPDiscover* message.

Otherwise, if the message and the server pass successfully the authentication validation and, if the offer suggested to the E-DHCP Server is acceptable by the client, the latter sends a *DHCPRequest* message (see Fig.5) to the server (including "*E-DHCP Authentication*" option): (1) requesting offered parameters from the selected server and implicitly declining offers from all others, (2) confirming the correctness of the previously allocated address after, e.g., system reboot, or (3) extending the lease on a particular network address.

The client follows the same steps to specify the "*E-DHCP Authentication*" option field's value followed-up in the *DHCPDiscover* message detailed in 1, with only one possible difference which is to put, in the '*URIAttributeCertificate*' field, the URI attribute certificate (In this case the client sends the *DHCPRequest* message to extend the lease on a particular network address).

Bits: 0 1 2 3 4 5 6 7 8 9 10 11 12 13 14 15 16

Code	Length	Flag	URIIdentityCertificate = www.enst.fr/Database2/ServerE-DHCPIdentityCertificate	URIAttributeCertificate = www.enst.fr/Database1/AttributeCertificateClient1
AuthenticationInformation = Signature Value Or Signature Encryption Value				

Figure 5. E-DHCP Authentication option structure in a DHCPRequest, DHCPRelease or DHCPDecline message

4. In the same way (as described in 2.a and 2.b) the server validates the authentication of the client and the *DHCPRequest* message.

If the authentication of the message and the client is validated or, if the server can't satisfy the client request, the server discards the message and sends a *DHCPNack* message to the client.

Otherwise, if the message and client pass successfully the authentication validation, the server verifies the value contained in the *'URIAttributeCertificate'* field:

a) If the value is equal to 0. The server creates an AC (version 2) for the client and saves it in the 'Attributes Certificate Database'. The issuer of this AC (E-DHCP Server) specifies the allocated IP address associated with the DHCP client in the AC *'Attributes'* field. In addition, the server can (optionally) specify in the *'Extensions'* field: (1) The MAC address of the equipment or, (2) The configuration parameters affected by the E-DHCP Server. The *'Holder'* field identifies the DHCP client that possesses the IP address. The validity period of this AC will correspond to the lease validity period. As from the moment when the validity period of the lease expires, the client can no longer exceed the access control server.

b) If the value is equal to an URI (Which means that the client has already an AC) then the server uses this URI to extract the AC from the Database. The server checks the certificate validity. If it is not expired, he can renew it. Otherwise, the server creates a new client attribute certificate and saves it in the 'Attributes Certificate Database'.

E-DHCP Server sends a *DHCPAck* message (see Fig.6) to the client containing "*E-DHCP Authentication*" option.

Bits: 0 1 2 3 4 5 6 7 8 9 10 11 12 13 14 15 16

Code	Length	Flag	URIIdentityCertificate = www.enst.fr/Database2/ServerE-DHCPIdentityCertificate	URIAttributeCertificate = www.enst.fr/Database1/AttributeCertificateClient1
AuthenticationInformation = Signature Value Or Signature Encryption Value				

Figure 6. E-DHCP Authentication option structure in a DHCPAck message

E-DHCP Server follows the same steps to specify the "*E-DHCP Authentication*" option field's value as in *DHCPOffer* message in 2, with only one difference that the server puts, the URI client AC (renewed or created) in the *'URIAttributeCertificate'* field.

5. In the same way (as described in 3.a and 3.b), the client validates the authentication of the server and the message *DHCPAck*.

If the authentication of the message and the server is validated, the client uses the configuration parameters in the lease, the IP address that is affected to him/her, then extracts (following the given URI), and uses its attributes certificate containing the Internet address allotted dynamically.

6.3 Service access scenario

E-DHCP was proposed in order to allow a strict control on equipment by using a strong authentication. The final objective is to allocate to the equipment an AC containing the Internet address dynamically allocated. This certificate ensures the relation between the client identity certificate and the allocated IP address. This AC will be then used in the access control. For their (equipment) authentication within network architectures, the equipment can prove its address by presenting its identification certificate and the AC.

As soon as the client receives their IP address and attributes certificate, it becomes possible to reach the services offered beyond the access control server. A scenario of access control is illustrated in Figure 7.

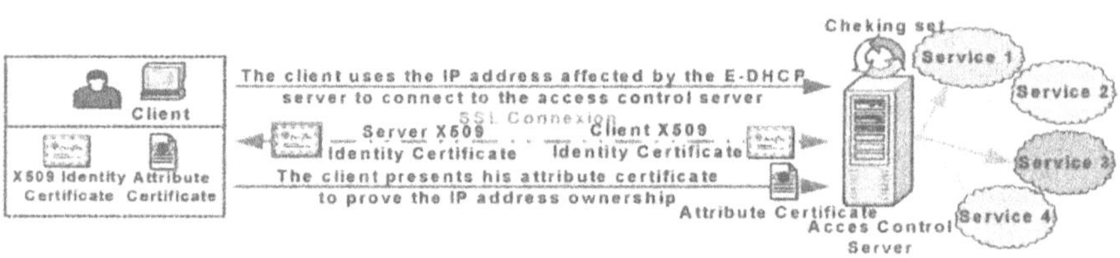

Figure 7. Scenario of access control

The steps to be followed are:

1. The client uses the IP address allocated by the E-DHCP Server to establish a connection with the access control server.
2. The client and the access control server use 'SSL client authentication' and 'SSL server authentication' [14] which allow:
   a) A server to confirm a client identity.
   b) A client to confirm a server identity.
3. The client presents his attributes certificate to the access control server
4. The access control server verifies:
   a) Identity certificate (Validity period, certification chain, etc.)
   b) AC (Validity period, allocated IP, authorized service, etc.)
   c) Validity of link between the X509 identity certificate and the AC.
   d) The link between the identity of the client and the IP address.
5. If the verification in the preceding part is successful, the access control server allows the client to be connected to the authorized service.

6.4 E-DHCP advantages

These are the E-DHCP advantages, listed in order of importance:

1. E-DHCP avoids changing the current DHCP protocol.
2. It provides simultaneously the authentication of entities (client/server) and the authentication of DHCP messages. The technique used is based mainly on the use of asymmetric keys encryption, X.509 identity certificates and attribute certificates.
3. It uses the RSA digital signature mechanism, which provides a better security than symmetric encryption. The use of this mechanism eliminates key distribution and key flexibility problems existing in the use of shared keys.
4. It allows a strict control over the equipment by using a strong authentication (using X.509 Identity and Attributes Certificates 'AC').
5. *DHCPDiscover* messages are authenticated by this protocol, which makes the protocol invulnerable to denial of service attack through flooding with unauthenticated *DHCPDiscover* messages.
6. Is invulnerable to message interception.
7. It supports inter-domain authentication.
8. The use of AC confirms the client IP address ownership.

7. CONCLUSION AND FUTURE WORK

This paper has presented an extension to DHCP protocol. This extension, called E-DHCP *(Extended-Dynamic Host Configuration Protocol)*, uses asymmetric keys encryption RSA mechanism, X.509 identity certificates and, attribute certificates 'AC' to provide simultaneously the authentication of entities (client/server) and the authentication of DHCP messages. In E-DHCP, DHCP server asks on an Attribute Authority server to create a client AC, which ensures the relation between the client identity certificate and the allocated IP address. This AC is used in the access control.

We have implemented E-DHCP by modifying the open source and free DHCP code base, developed by the Internet Software Consortium [20], then by the development of an attribute authority, to which the DHCP server is attached. We point out that the keys management protocol ISAKMP [15] supports the attributes certificates. This is why we believe that E-DHCP perfectly articulates and interoperates with IPSec [16] protocol using the certificates. A future direction of our research is to validate the interoperability of our proposition with IPsec through real scale developments and tests.

8. ACKNOWLEDGEMENTS

The authors would like to thank Mr. Salim Ferraz who has provided detailed reviews and much helped to produce this paper.

REFERENCES

1. R. Droms “Dynamic Host Configuration Protocol”, IETF, RFC 2131, Mar. 1997.
2. R. Droms and S. Alexander, "DHCP Options and BOOTP Vendor Extensions", IETF, RFC 2132, Mar. 1997.
3. B. Croft and J. Gilmore, "BOOTSTRAP PROTOCOL (BOOTP) ", IETF, RFC 951, Sep. 1985.
4. T. Berners-Lee, R. Fielding and L. Masinter, "Uniform Resource Identifiers (URI): Generic Syntax", IETF, RFC 2396, Aug. 1998.
5. R. Droms, "Interoperation Between DHCP and BOOTP", IETF, RFC 1534, Oct. 1993.
6. C. Perkins and K. Luo, "Using DHCP with computers that move", Wireless Networks, Mar. 1995.
7. M. del Rey, "INTERNET PROTOCOL", IETF, RFC 791, Sep. 1981.
8. R. Droms and W. Arbaugh, "Authentication for DHCP Messages", IETF, RFC 3118, Jun 2001.
9. J. Jonsson and B. Kaliski,"Public-Key Cryptography Standards (PKCS) #1: RSA Cryptography Specifications Version 2.1", IETF, RFC 3447, Feb. 2003.
10. Information technology-Open Systems Interconnection-The Directory: Authentication framework, ITU-T Recommendation X.509, 1997.
11. Information technology-Open Systems Interconnection-The Directory: "Public-key and attribute certificate frameworks", ITU-T Recommendation X.509, 2000.
12. J. Postel, "User Datagram Protocol", IETF, RFC 768, Aout 1980.
13. Hornstein and al., "DHCP Authentication via Kerberos V", Internet Draft, Nov. 2000.
14. A. Freier, P. Karlton and P. Kocher, "The SSL Protocol, Version 3.0", Netscape Communications Corp., November 1996. Standards Information Base, The Open Group.
15. D. Maughan, M. Schertler, M. Schneider and J. Turner, "Internet Security Association and Key Management Protocol (ISAKMP) ", IETF, RFC 2408, Nov. 1998.
16. S. Kent and R. Atkinson,"Security Architecture for the Internet Protocol", IETF, RFC 2401, Nov. 1998.
17. R. Droms, Procedure for Defining New DHCP Options, IETF, RFC 2489, Jan. 1999.
18. G. Glazer and al., "Certificate-Based Authentication for DHCP", Mar. 2003.
19. T. Komori and T. Saito, "The secure DHCP System with User Authentucation", LCN’02, 27th Annual IEEE Conference on Local Computer Networks, Nov. 2002.
20. Internet Software Consortium. Dynamic Host Configuration Protocol Distribution.
21. A. Tominaga, O. Nakamura, F. Teraoka and J. Marai, "Problems and solutions of DHCP", INET'95, The 5th Annual Conference of the Internet Society, Apr. 1995.
22. Threshold Networks, "RAZZO IP server appliance – Integrated solution for management of IP, DNS and DHCP", White paper, Apr. 2001.
23. J. Demejian, A. Serhrouchni and F. Tastet, "Why certificates don’t meet e-business needs?", SSGRR’03W, International Conference on Advances in infrastructure for e-Electronic, e-Business, e-Education, e-Science, e-Medicine on the Internet, Jan. 2003.

RECURSIVE SANDBOXES: EXTENDING SYSTRACE TO EMPOWER APPLICATIONS

Aleksey Kurchuk and Angelos D. Keromytis
Columbia University
ak2097@columbia.edu; angelos@cs.columbia.edu

Abstract The *systrace* system-call interposition mechanism has become a popular method for containing untrusted code through program-specific policies enforced by user-level daemons. We describe our extensions to *systrace* that allow sandboxed processes to further limit their children processes by issuing *dynamically* constructed policies. We discuss our extensions to the *systrace* daemon and the OpenBSD kernel, as well as a simple API for constructing simple policies. We present two separate implementations of our scheme, and compare their performance with the base *systrace* system. We show how our extensions can be used by processes such as *ftpd, sendmail,* and *sshd.*

Keywords: Access control, process sandboxing, operating systems

1. INTRODUCTION

Systrace [Provos, 2003] has evolved to a powerful and flexible mechanism for containing untrusted code by examining system-call sequences issued by monitored processes. A user-level process assists the operating system kernel in determining whether a system call (and its associated arguments) are permitted by some externally-defined policy. This scheme allows unmodified programs to be effectively sandboxed and, combined with other convenient features (such as automatic policy generation and a graphical user interface), has made *systrace* widely used by OpenBSD [ope,] and NetBSD. However, the strength of the *systrace* approach — *i.e.,* keeping applications unaware of the sandbox — also makes the system less flexible than it could otherwise be.

Consider the case of a simplified *sendmail* daemon, which listens on the SMTP port for incoming connections, parses received messages, and delivers them to the recipient's mailbox. Typically, sendmail will invoke a separate program to locally deliver a newly received email message. Different delivery programs may be invoked, based on parameters such as the recipient, *etc.*

In a straightforward application of *systrace,* we would specify a policy that would limit filesystem accesses (*e.g.,* the *open()* system call) to files in the system mail-spool directory (*e.g., /var/spool/mail/*) and the sendmail configuration file. Compromise of the sendmail process, *e.g.,* through a buffer overflow attack, would restrict damage to these files and directories. However, if the local delivery program is susceptible to a buffer overflow attack when delivering to a particular user (*i.e.,* when using a particular local-delivery program), the subverted child process will be able to read or write to any user's mailbox and the configuration file.

More generally, static *systrace* policies do not allow the system to further limit its privileges based on information it has learned during its execution. A roughly similar example of such behavior is use of *seteuid(),* or of POSIX capabilities, where processes that do not need specific privileges can simply revoke them on themselves or their children processes.

We present an extension to *systrace* that allows sandbox-aware processes to communicate with their controlling daemon and specify further *restrictions* that can be placed on children processes of theirs. These policies can only **refine** the privileges already bestowed upon the sandboxed process — no "rights amplification" is possible. Processes that are aware of the sandbox can use a simple API to specify common restrictions they want to apply to their offspring. More elaborate restrictions can be specified by directly issuing *systrace* policies.

To ensure the monotonicity of privileges (*i.e.,* the fact that added policies do not expand the privileges of sandboxed processes), we explored two implementations. In the first approach, we used the KeyNote trust-management system [Blaze et al., 1999] to specify, parse, and evaluate policies in the *systrace* daemon. We modified the kernel to allow for indirect communication between the sandboxed process and the controlling *systrace* daemon, and modified the latter to generate and evaluate KeyNote policies. In the second approach, we used a recursive policy-evaluation model that emulates the KeyNote assertion evaluation process, which allowed the use of unmodified *systrace* policies at the cost of increased implementation complexity. Our benchmarks show that the added performance penalty of our extensions is negligible.

Paper Organization. The remainder of this paper is organized as follows. Section 2 discusses related work in process sandboxing and Section 3 gives an overview of the Systrace mechanism and KeyNote. Section 4 describes our extensions to Systrace, while Section 5 presents a preliminary performance evaluation of our extensions. Section 6 discusses our plans for future work, and Section 7 concludes this paper.

2. RELATED WORK

The Flask system [Spencer et al., 2000] extends the idea of capabilities and access control lists by the more generic notion of a *security policy*. The Flask micro kernel system relies on a security server for policy decisions and on an object server for enforcement. Every object in the system has an associated security identifier, requests coming from objects are bound by the permissions associated with their security identifier. However Flask does not address the issue of cooperation amongst clients, servers and networks to deliver reliable and secure services to clients. Its notion of the security identifier is very limiting, in our system we require any number of conditions to hold before we provide a service, for example user identification might not be enough to grant access to a service, the user might also be required to access the service over a secure channel. As a minor issue, we have demonstrated that our prototype can be implemented as part of a widely used, commodity operating system, as opposed to a more fluid experimental micro-kernel.

System call interception, as used by systems such as TRON [Berman et al., 1995], MAPbox [Acharya and Raje, 2000], Software Wrappers [Fraser et al., 1999] and Janus [Goldberg et al., 1996] is an approach to restricting applications. TRON and Software Wrappers enforce capabilities by using system call wrappers compiled into the operating system kernel. The syscall table is modified to route control to the appropriate TRON wrapper for each system call. The wrappers are responsible for ensuring that the process that invoked the system call has the necessary permissions. The Janus and MAPbox systems implement a user-level system call interception mechanism. It is aimed at confining helper applications (such as those launched by Web browsers) so that they are restricted in their use of system calls. To accomplish this they use *ptrace(2)* and the */proc* file system, which allows their tracer to register a call-back that is executed whenever the tracee issues a system call. Other similar systems include Consh [Alexandrov et al., 1998], Mediating Connectors [Balzer and Goldman, 1999], SubDomain [Cowan et al., 2000] and others [Fraser et al., 1999, Ghormley et al., 1998, Walker et al., 1996, Mitchem et al., 1997].

Capabilities and access control lists are the most common mechanisms operating systems use for access control. Such mechanisms expand the UNIX security model and are implemented in several popular operating systems, such as Solaris and Windows NT [Custer, 1993]. The Hydra capability based operating system [Levin et al., 1975] separated its access control mechanisms from the definition of its security policy. Follow up operating system such as KeyKOS [Hardy, 1985, Rajunas et al., 1986] and EROS [Shapiro et al., 1999] divide a secure system into compartments. Communication between compartments is mediated by a reference monitor.

The methods that we mentioned so far rely on the operating system to provide a mechanism to enforce security. There are, however, approaches that rely on safe languages [Levy et al., 1998, Tardo and Valente, 1996, Leroy, 1995, Hicks et al., 1998], the most common example being Java [McGraw and Felten, 1997]. In Java applets, all accesses to unsafe operations must be approved by the security manager. The default restrictions prevent accesses to the disk and network connections to computers other than the server the applet was down-loaded from.

3. OVERVIEW OF *systrace* AND KeyNote

3.1 *systrace*

Systrace is a utility for monitoring and controlling an application's behavior by means of intercepting its system calls. Systrace provides facilities for confining multiple applications, interactive policy generation, intrusion detection and prevention, and can be used to generate audit logs. For a full account of how Systrace works, refer to [Provos, 2003]. Here, we give a brief review of the basic design of Systrace.

Taking a hybrid kernel/user-space approach, Systrace consists of a user-level daemon and a small addition to the OS kernel. Once a system call is issued by the monitored application, it is intercepted by the kernel part of Systrace, and is matched against an in-kernel policy. The in-kernel policy is a simple table of system calls and responses. If an entry for that system call exists, the result – allow or deny – is used in performing the operation. If no entry exists, such 'fast path' for the system call is impossible, and a user-level daemon is asked for a policy decision.

When the policy daemon receives a system call to be evaluated, it looks up the policy associated with the process, translates the system call arguments, and checks if the policy allows such a call.

Allowing for interactive policy generation, Systrace will prompt the human user for input on system calls not already described in the policy. If interactive policy generation is turned off, a system call that is not explicitly described in the policy will be denied.

3.2 KeyNote

KeyNote is a simple trust-management system and language developed to support a variety of applications. Although it is beyond the scope of this paper to give a complete tutorial or reference on KeyNote syntax and semantics (for which the reader is referred to [Blaze et al., 1999]), we review a few basic concepts to give the reader a taste of what is going on.

The basic service provided by the KeyNote system is *compliance checking;* that is, checking whether a proposed *action* conforms to local *policy.* Actions in KeyNote are specified as a set of name-value pairs, called an *Action Attribute Set.* Policies are written in the KeyNote *assertion language* and either accept or reject action attribute sets presented to it. Multiple assertions can be combined in specifying a policy, by constructing a delegation-graph, with each issuer of an assertion specifying the conditions under which the recipient of the assertion can perform some action. Ultimately, for a request to be approved, an assertion graph must be constructed between one or more POLICY assertions and one or more requesters. Because of the evaluation model, an assertion located somewhere in a delegation graph can effectively only refine (or pass on) the authorizations conferred on it by the previous assertions in the graph.

Each service that needs to mediate access, queries its local compliance checker on a per-request basis (what constitutes a "request" depends on the specific service and protocol). The compliance checker can be implemented as a library that is linked against every service or as a daemon that serves all processes in a host.

4. EXTENDING *systrace*

In this section, we present two extensions to Systrace: nested policies and dynamic policy generation. Nested policies allow for a composition of several policies to be recursively applied to an application, and dynamic policy generation allows a Systrace-aware application to set an appropriate policy for itself and even to generate such policy at runtime.

4.1 Nested Policies

The motivation behind nested policies is that priveleges, granted to an application should be no greater than priveleges, enjoyed by its parent. We present two implementations of this idea, using different policy engines.

4.1.1 Systrace-KN. In our first attempt to introduce hierarchical policies to Systrace, we modified the user-level daemon to use KeyNote as its policy engine. Making use of KeyNote's native support for delegation of trust, we built an implicit graph of processes by making the parent process the 'Authorizer' of its offsprings' actions. This construction automatically restricts the privileges of the child process to those of its parent. Figure 1 shows a sample policy generated by Systrace-KN.

At run time, the 'Licensees' and 'Authorizer' fields are filled with the PIDs of the monitored process and its parent respectively. When a system call needs to be evaluated, its description is translated from the native Systrace format into a KeyNote request. The request is then evaluated by KeyNote, returning one of

```
KeyNote-version:  2
Authorizer:  "POLICY"
Licensees:  "systrace"
Conditions:
   (bin=="/bin/foo")->{
    (state=="default")->{
     (syscall=="fswrite")->{
      ((filename == "/usr/libexec/ld.so"));
      ((filename == "/var/run/ld.so.hints"));
      ...
     };
     ((syscall == "munmap"));
     ((syscall == "fsread") && (filename ==
"/<non-existent filename>:  /etc/malloc.conf"));
     ...
     ((syscall == "exit"));
    };
   };
```

Figure 1. Example of a policy, automatically generated by Systrace-KN.

the following values: {*ask,deny,permit*}. Stepping away from the recommendation in the KeyNote RFC, the first result - *ask* - is treated not as having the fewest permissions, but rather as a default one, which is returned if the request does not match the Action Attribute Set. If *ask* is returned in interactive mode, the user is asked for input. If the mode is automatic, the result is downgraded to *deny* just as in the original Systrace.

4.1.2 Systrace-H. Unsatisfactory performance of Systrace-KN (see Section 5) led us to extending the original Systrace policy engine to emulate support for nested policies. Following the idea of KeyNote, we define a notion of `compolicy` (complex policy). Each `compolicy` has a single `policy` (as defined in the original Systrace) as well as a link to an authorizing `compolicy`. Compliance with a `compolicy` is achieved by satisfying restrictions imposed by the associated policy and by all policies associated with each compolicy in the chain of authorizers. Effectively, this means that the system call, issued by a process, has to satisfy the policies of all of its ancestor processes.

A seemingly simpler solution of adding to the original Systrace policy a link to the authorizer proves not to be sufficient, since processes, associated with the same binary (and thus the same policy) might have a different set of ancestors. For example, consider program binaries `/bin/foo`, `/bin/bar`, and

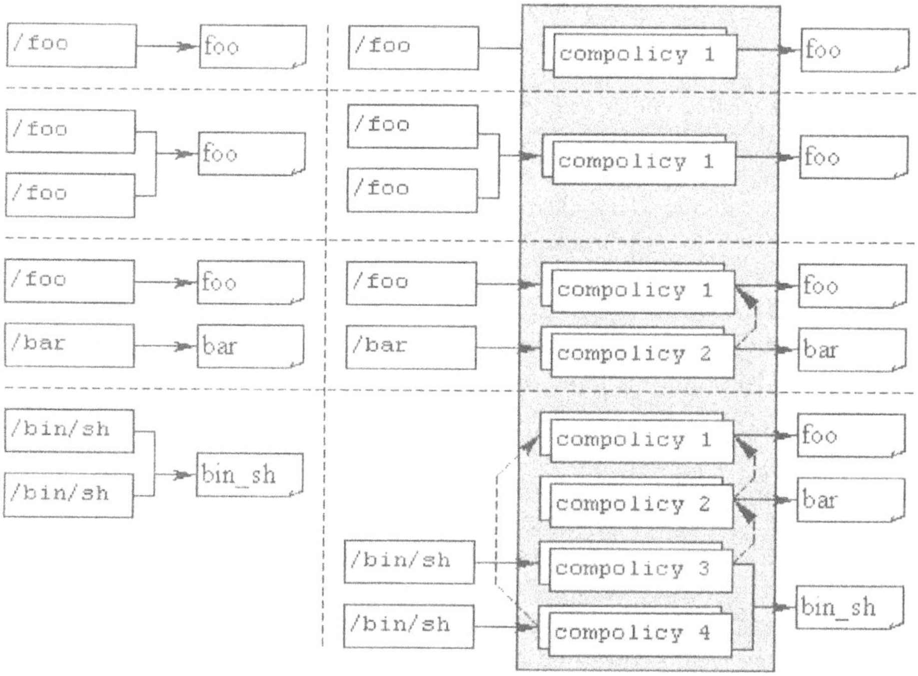

Figure 2. Side-by-side comparison of policy objects, created by original Systrace and those, created by Systrace-H.

`/bin/sh`, and the policies associated with them. The behavior of `/bin/foo` and `/bin/bar` is as follows:

/bin/foo:
```
fork()
child:  exec("/bin/bar")
parent: exec("/bin/sh")
```
/bin/bar:
```
exec("/bin/sh")
```

Running `/bin/foo` under Systrace will create the following policy chains: `foo`←`bar`←`sh` and `foo`←`sh`. Since the two processes are running `/bin/sh` , the behavior of the original systrace would be to use the same policy (bin_sh) for both of them. With hierarchical policies, this approach is incorrect since the privileges of one of the processes are restricted by policies of `/bin/foo` and `/bin/sh`, and the privileges of the other are further limited by policy `/bin/bar`.

Figure 2 illustrates the difference between the old approach and the one taken by our extension.

4.2 Run-Time Policy Modification

In an effort to improve security, a Systrace-aware application might want to restrict its set of permissions to the minimum required by a particular code fragment. We present two approaches to run-time policy modification. The first one uses pre-generated policies, allowing the program to switch between them. The second one allows the program to modify (further restrict) its current Systrace policy with a dynamically generated one.

4.2.1 States. It is often convenient to view an application as a state machine. Being in different states, a program might naturally require different sets of privileges. For example, consider *sendmail.* While being in one state might require permission to establish network connections, these privileges are extraneous if the task at hand is to deliver a message to a local user's mailbox. The idea of narrowing the range of program's privileges is especially useful for further restricting the privileges of the children processes. Execution of a program always starts in *default* state.

State transitions. Naturally, not all state transitions will be valid. A subverted process should not be able to switch to an arbitrary state, aquiring excessive privileges. A separate policy is kept for rules associated with state switching. Every time an application requests to set a different state, this policy is queried. Instead of inventing a new policy engine for state transitions, we used the original Systrace policy engine. Each 'set state' operation is viewed as a system call *set* in emulation *state,* which has two possible arguments: *newstate* and *oldstate*.

File format. Using states required small changes to be made to the Systrace policy file name convention and to the file format itself. The state of the policy is used as an extension to the policy file name. For example, the policy file for `/bin/sh` in its default state should be named `bin_sh.default.` State is also reflected in the first line of the policy file as in the following example:

```
[ bin_sh.default ]
Policy:  /bin/sh, Emulation:  native, State:  default
```

The state policies are written in the same language used for creating regular Systrace policies, and are stored in files with extension *state*. Thus, *state* SHOULD NOT be used as a state identifier. Also, state names that start with an underscore are reserved for use by Systrace as explained in Section 4.2.2. Consider the following code snippet in an application:

```
fork();
if (child) {
    systrace_setstate("risky");
    do_risky();
    ...
}
else if (parent) {
    ...
}
```

Here, the state of the child process is switched to *risky,* possibly restricting the range of actions the process can perform. Note also that since the state transition policy below does not explicitly allow a `risky`→`default` transition, a rogue process will not be able to change its state back to *default* once it enters the *risky* state.

```
[ bin_foo.state ]
Policy:  /bin/foo, Emulation:  state, State:  state
state-set:  oldstate eq "default" and newstate eq "risky"
            then permit
```

Note that when switching states, the policy, associated with the new state does not restrict the old one, but replaces it, as shown in Figure 3. The mechanism for restricting the existing policy is discussed in the next section.

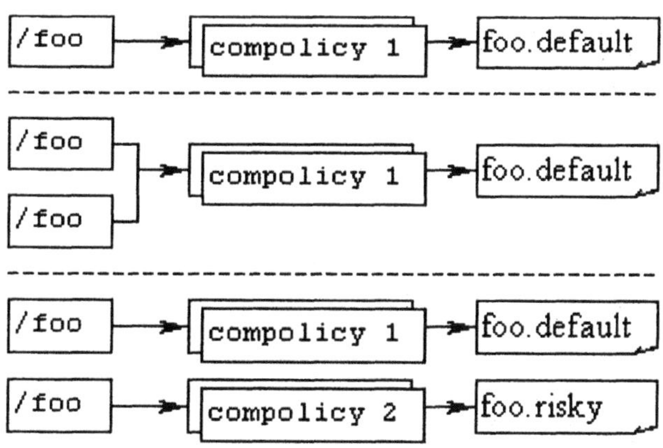

Figure 3. Policies after running the code snippet changing state to `risky`.

4.2.2 Dynamic Policy Creation. While state switching allows applications to choose the most suitable (but static) policy, one might want to have a policy based on information acquired during the program's execution. Dynamic policy creation allows one to refine an application's privileges by providing run-time generated policies.

In order to facilitate creation of dynamic policies, a (pseudo) system call *all* was introduced, which allows a policy to specify restrictions, applying the action, associated with *all*, to all system calls not explicitly described. If a filter for *all* is not specified explicitly, it is taken to be *deny* by default. Consider, for example, how sendmail would make use of this functionality when delivering local mail to user `jdoe`. Let variable `restricted_policy` contain the following policy, generated at run time:

```
native-fsread:  filename sub "/var/mqueue/jdoe/" then permit
native-fsread:  filename sub "/var/mqueue/" then deny
...
native-all:  permit
```

Since policy filters are matched in top-down order, reading from directory `jdoe` is allowed, while an attempt to open a file in someone else's directory will be thwarted. The following code will run the local delivery agent with restricted privileges, allowing it to access specific user's mail queue directory, but not that of any other user:

```
systrace_setpolicy(restricted_policy);
exec(local_delivery_agent,...);
```

Once the Systrace daemon receives a set_policy request, it creates a new `policy` object. This policy's state starts with an underscore, indicating that this is a volatile policy. Such policies do not get saved to disk. Thus, user defined state identifiers should not start with an underscore.

4.2.3 Communication. In order for run-time policy modification to work, there needs to be a way for the controlled application to communicate with the Systrace daemon. We briefly describe how this interaction is accomplished.

To send a message to the user-level Systrace daemon, a Systrace-aware application sends a message to */dev/systrace*, which in turn relays the message to the appropriate Systrace daemon process. The messages are encoded with the following data structure:

```
struct systrace_message {
   unsigned int processed;
   unsigned int code;
   size_t       length;
```

```
    char           message[1];
};
```

Variable `length` specifies the length of the `message`, and `code` identifies the action requested. So far, the only relay codes defined are `SYSTRACE_RELAY_SETSTATE` and `SYSTRACE_RELAY_SETPOLICY`. Structure `systrace_relay` is then populated and sent to `/dev/systrace` via the `SYSTR_RELAY ioctl()`.

```
struct systrace_relay {
    pid_t    pid;
    int      flag;
    key_t    key;
    size_t   size;
};
```

In the `systrace_relay` structure, `pid` is the process ID of the monitored process, `flag` identifies the requested action, `key` is the key of the shared memory region of size `size`.

Having sent the message, the application waits for flag `processed` to be set. The response from the daemon is stored in `code`. The following responses are currently defined:

`SYSTRACE_RELAY_SUCCESS` - Requested operation completed successfully
`SYSTRACE_RELAY_FAILURE` - Unexpected failure
`SYSTRACE_RELAY_PERMISSION` - Application is unauthorized to perform this operation. This error may occur if, for example, an invalid state change is requested.
`SYSTRACE_RELAY_ARGUMENT` - Invalid argument specified. An invalid Systrace policy passed to `systrace_setpolicy()` might cause this error message.

The kernel, upon receipt of the message, signals the Systrace daemon that a message is waiting for it. The daemon picks up the pending relay via a `STRIOCRELAY ioctl()`. When the message is processed, daemon sets `code` to one of the above responses, and the `processed` flag to 1.

5. PERFORMANCE EVALUATION

The policy enforcement agent communicates with the policy engine via three callback functions. These functions are responsible for creating state for a new process (EXEC), evaluating a system call with known (translated) arguments (TRANS), and general calls with no (or not translated) arguments (GEN). In order to measure performance of different versions of Systrace, we timed these three functions. As a test application, we used the `sendmail` daemon, making necessary modifications for it to use policy modification features.

The following table shows the average times (in micro-seconds) needed for each operation. 'Systrace-H (states)' denotes times, observed when running sendmail, modified to have several states. In 'Systrace-H (dynamic)', sendmail was changed to generate a dynamic policy for delivering local mail (as in the example in Section 4.2.2).

	GEN	TRANS	EXEC
Original Systrace	2	37	54
Systrace-KN	902	866	173
Systrace-H	20	37	81
Systrace-H (states)	20	37	89
Systrace-H (dynamic)	20	37	121

One obvious result of these performance measurements is that Systrace-KN is unacceptably slow. Further investigation proved that over 99% of the time was spent in $kn\_query()$ function — the KeyNote function that evaluates a request. This moved us to turn away from a KeyNote-based implementation, but use the ideas of KeyNote to enhance the policy engine of the original Systrace.

In all variations of Systrace-H, the time needed to evaluate a translated system call remained the same as in the original Systrace. The original Systrace policy engine is very lightweight, and more precise measurements would be needed to detect the difference. The time to evaluate a call with no arguments, however, has risen, almost reaching that of TRANS.

The reason for such a slowdown is as follows. In the original Systrace, upon creation of a new process, all general system calls were entered in the kernel policy (pre-loaded) to speed up the execution and to take the responsibility for their evaluation off the user-level Systrace daemon. Thus, all general system calls that needed to be evaluated by the user-level daemon were considered non-compliant with the policy. Such pre-loading is not done in Systrace-H, since it requires all ancestor policies to be queried at the start of the process's execution. Instead, a general system call is evaluated the same way a translated one is (explaining the increase in evaluation time), and the kernel-policy is modified, reflecting the result of the evaluation. Since such evaluation is done once per call per process, the performance degradation is insignificant.

The increase in time when evaluating EXEC is most likely due to the fact that in Systrace-H, there generally exist more `compolicy` objects to search, and the fact that aside from loading the default policy, the state transition policy (if one exists) needs to be loaded as well. Since 'exec' calls are relatively infrequent, this increase results in negligible performance penalty.

6. FUTURE WORK

Having established a mechanism for a controlled application to communicate with the controlling Systrace daemon, we used it to relay requests for policy changes. In the future, the same mechanism may be used to perform a wider variety of tasks, including communication with other processes under Systrace's control.

Lack of tools for programmatically creating Systrace policies might impede the use of dynamically generated policies. Development of tools to facilitate policy generation should greatly decrease the time needed to allow a Systrace-aware application to make use of these features. Such work is in our current work plans.

Finally, we plan to extend the current implementation of Systrace-H to allow pre-loading, per our discussion of possible performance degradation in Section 5.

7. CONCLUSIONS

This paper presents an extension to the Systrace facility for process sandboxing. We argue that using hierarchical policies adds security, while having an acceptable performance overhead. Our mechanism adds the ability to control the behavior of all binaries run by a single aplication by modifying a single policy.

We further showed that an application may enhance its security by using state policies and the ability to dynamically generate policies. The changes necessary to transform an application into a Systrace-aware one are minimal and can be made in minutes.

From our perliminary performance evaluation we conclude that a policy engine as generic and flexible as KeyNote might prove to be too slow for some tasks. A highly specialized policy engine that follows the same evaluation model might be called for instead.

Acknowledgements

This work was supported by NSF under Contract CCR-TC-0208972.

References

[ope,] The OpenBSD Operating System. `http://www.openbsd.org/`.

[Acharya and Raje, 2000] Acharya, Anurag and Raje, Mandar (2000). Mapbox: Using parameterized behavior classes to confine applications. In *Proceedings of the USENIX Security Symposium*, pages 1–17.

[Alexandrov et al., 1998] Alexandrov, A., Kmiec, P., and Schauser, K. (1998). Consh: A confined execution environment for internet computations.

[Balzer and Goldman, 1999] Balzer, Robert and Goldman, Neil (1999). Mediating connectors: A non-bypassable process wrapping technology. In *Proceeding of the 19th IEEE International Conference on Distributed Computing Systems.*

[Berman et al., 1995] Berman, Andrew, Bourassa, Virgil, and Selberg, Erik (1995). TRON: Process-Specific File Protection for the UNIX Operating System. In *Proceedings of the USENIX Technical Conference.*

[Blaze et al., 1999] Blaze, M., Feigenbaum, J., Ioannidis, J., and Keromytis, A. D. (1999). The KeyNote Trust Management System Version 2. RFC 2704.

[Cowan et al., 2000] Cowan, Crispin, Beattie, Steve, Pu, Calton, Wagle, Perry, and Gligor, Virgil (2000). SubDomain: Parsimonious Security for Server Appliances. In *Proceedings of the 14th USENIX System Administration Conference.*

[Custer, 1993] Custer, Helen (1993). *Inside Windows NT.* Microsoft Press.

[Fraser et al., 1999] Fraser, Tim, Badger, Lee, and Feldman, Mark (1999). Hardening COTS Software with Generic Software Wrappers. In *Proceedings of the IEEE Symposium on Security and Privacy.*

[Ghormley et al., 1998] Ghormley, Douglas P., Petrou, David, Rodrigues, Steven H., and Anderson, Thomas E. (1998). SLIC: An Extensibility System for Commodity Operating Systems. In *Proceedings of the USENIX Technical Conference*, pages 39–52.

[Goldberg et al., 1996] Goldberg, Ian, Wagner, David, Thomas, Randi, and Brewer, Eric A. (1996). A Secure Environment for Untrusted Helper Applications. In *Procedings of the USENIX Technical Conference.*

[Hardy, 1985] Hardy, Norman (1985). The KeyKOS. *Operating Systems Review*, 19(4):8–25.

[Hicks et al., 1998] Hicks, M., Kakkar, P., Moore, J. T., Gunter, C. A., and Nettles, S. (1998). PLAN: A Programming Language for Active Networks. Technical Report MS-CIS-98-25, Department of Computer and Information Science, University of Pennsylvania.

[Leroy, 1995] Leroy, X. (1995). Le système Caml Special Light: modules et compilation efficace en Caml. Research report 2721, INRIA.

[Levin et al., 1975] Levin, R., Cohen, E., Corwin, W., and Wulf, W. (1975). Policy/mechanism separation in hydra. In *Proceedings of the 5th ACM Symposium on Operating Systems Principles*, pages 132–140.

[Levy et al., 1998] Levy, Jacob Y., Demailly, Laurent, Ousterhout, John K., and Welch, Brent B. (1998). The Safe-Tcl Security Model. In *Proceedings of the USENIX Technical Conference.*

[McGraw and Felten, 1997] McGraw, Gary and Felten, Edward W. (1997). *Java Security: hostile applets, holes and antidotes.* Wiley, New York, NY.

[Mitchem et al., 1997] Mitchem, T., Lu, R., and O'Brien, R. (1997). Using Kernel Hypervisors to Secure Applications. In *Proceedings of the Annual Computer Security Applications Conference.*

[Provos, 2003] Provos, N. (2003). Improving Host Security with System Call Policies. In *Proceedings of the 12th USENIX Security Symposium.*

[Rajunas et al., 1986] Rajunas, S.A., Hardy, N., Bomberger, A.C., Frantz, W.S., and Landau, C.R. (1986). Security in KeyKOS. In *Proceedings of the IEEE Symposium on Security and Privacy.*

[Shapiro et al., 1999] Shapiro, Jonathan S., Smith, Jonathan M., and Farber, David J. (1999). EROS: a fast capability system. In *Proceedings of the 17th ACM Symposium on Operating Systems Principles*, pages 170–185.

[Spencer et al., 2000] Spencer, R., Smalley, S., Loscocco, P., Hibler, M., Anderson, D., and Lepreau, J. (2000). The flask security architecture: System support for diverse security policies. In *Proceedings of the USENIX Security Symposium*, pages 123–139.

[Tardo and Valente, 1996] Tardo, J. and Valente, L. (1996). Mobile Agent Security and Telescript. In *Proceedings of the 41st IEEE Computer Society Conference (COMPCON)*, pages 58–63.

[Walker et al., 1996] Walker, K. M., Stern, D. F., Badger, L., Oosendorp, K. A., Petkac, M. J., and Sherman, D. L. (1996). Confining root programs with domain and type enforcement. In *Proceedings of the USENIX Security Symposium*, pages 21–36.

FAST DIGITAL CERTIFICATE REVOCATION

An alternative to short lived certificates

Vipul Goyal
Department of Computer Science & Engineering, Institute of Technology, Banaras Hindu University, India. E-mail: vipulg@cpan.org.

Abstract: Digital Certificates are central to the concept of Public Key Infrastructures (PKI) and serve as a cryptographic proof of one's public key. Occasionally, certificates must be declared invalid prior to their due expiration date in case of key compromise or change in identity. Thus all PKIs should provide a mechanism through which an issued certificate may be revoked. The revocation mechanisms are commonly classified into Certificate Revocation Lists (CRLs), trusted dictionaries and online mechanisms. We briefly discuss the existing certificate revocation techniques and then present a new online revocation technique. More precisely, we present an alternative to short lived certificates proposed by Rivest.

Key words: PKIs, certificate revocation, online solutions, short lived certificates

1. INTRODUCTION

A certificate is a digitally signed statement binding the key holder's (principal's) name to a public key and various other attributes. The signer (or the issuer) is commonly called a certificate authority (CA). Certificates act as a mean to provide trusted information about the CA's declaration w. r. t. the principal. The declaration may be of the form-

"We, the Certificate Authority declare that we know Alice. The public key of Alice is ..."

"We further declare that we trust Alice for ..." (optional part)

Certificates are tamperevident (modifying the data makes the signature invalid), unforgeable (only the holder of the secret, signing key can produce the signature). Certificates are the building blocks of a Public Key Infrastructure (PKI). PKI is defined to be "The set of hardware, software people, policies and procedures needed to create, manage, store, distribute, and revoke public key certificates based on public key cryptography" in the IETF PKIX Roadmap [1].

When a certificate is issued, the CA (issuer) declares the period of time for which the certificate is valid. However, there may be some situations when the certificate must abnormally be declared invalid prior to its expiration date. This is called certificate revocation. This can be viewed as "blacklisting" the certificate. This means that the existence of a certificate is a necessary but not sufficient evidence for its validity. A method for revoking certificates and distributing this revocation information to all the involved parties is thus a requirement in PKI. The reasons for revoking a certificate may be: suspected/detected key compromise, change of principal name, change of relationship between a principal and the CA (e.g. Alice may leave or be fired from the company) or end of CA's trust into the principle due to any possible reason.

The revocation mechanism should have an acceptable degree of timeliness, i.e., the interval between when the CA made a record of revocation and when this information became available to the relying parties should be small enough to be acceptable. Further, it is very important for the revocation mechanism to be efficient as the running expenses of a PKI derives mainly from administering revocation [4].

2. AVAILABLE REVOCATION TECHNIQUES

This section briefly outlines a number of available revocation schemes-

2.1 Certificate Revocation Lists (CRLs)

CRLs are the most common and simplest method for certificate revocation. A CRL is a periodically issued list containing the certificate serial number of all the revoked certificates issued by a particular CA. This list is digitally signed by the CRL issuer to avoid tampering. The relying parties willing to validate a certificate issued by a particular CA can then download the most recent CRL of that CA.

Many variants of this "basic" CRL scheme have been designed to improve the performance. These include delta CRLs [2], partitioned CRLs, over-issued CRLs [3], Blacklist CRLs and Redirected Pointers.

CRLs have been criticized for not being able to provide the required service and for being too costly [7, 8, 9, 11, 12]. See [22] for a comparative analysis of CRLs and online revocation techniques.

2.2 Trusted Dictionaries

There are a number of schemes in which the end entities (relying parties) are supplied information in support of validating a single certificate rather than a complete list. Instead of individually signing each revocation reply, trusted dictionary schemes attempt to solve the problem by using one-way hash functions in order to provide lightweight digital signatures.

A notable technique under this category is CRS. Micali introduced the certificate revocation status (CRS) scheme in 1995 and improved it in 1996 [7]. It was designed according to the following strategy: increase the amount of information transmitted by CAs to directories during an update, but design this update information in a way to enable directories to answer certificate revocation queries more succinctly.

CRS introduces two 100-bit values Y and N into a certificate, to represent “valid" and “revoked" status, respectively. These two values are calculated as followings: CA chooses a proper one-way hash function H, determines a lifetime period t and generates two 100-bit random secure values Y_0 and N_0 indicating valid and revoked, respectively. The above $Y = H^t(Y_0)$, $N = H(N_0)$. The scheme works as follows: To make revocation information up-to-date, during each time interval i (where $0 < i < t$), every CA submits the following information to its directories: an authenticated and time stamped 220-bit string L, signed by the CA, containing all serial numbers of issued and not-yet-expired certificates, and 100-bit value V for each certificate indicating whether it has been revoked or not within the current time interval (where $V = H^{t-i}(Y_0)$ for a valid certificate or $V = N_0$ for a revoked one). When a directory receives a certificate validity checking request, it sends proper value V to the verifier, based on which the verifier can check certificate validity.

Micali further revised his CRS scheme in 2002 [23]. The new scheme, called NOVOMODO, has one minor and one major modification: the minor one is that NOVOMODO uses SHA as the hash function, and the major one is that basic NOVOMODO discards directories at all. The author did describe how to build a distributed NOVOMODO, but update cost then is high, which still limits the scheme's scalability.

Aiello et al. [16] improved CRS approach by reducing update costs while maintaining its cheap query costs.

CRS and its variants again cannot provide timely revocation information although they are timelier than CRLs. The recency requirements are determined by the CA rather than the acceptor.

2.3 Online Revocation Mechanisms

As a response to the low timeliness of some periodically updated certificate revocation schemes, protocols for online status checking have been developed. Many certificate based systems cannot tolerate the revocation delay resulting from the periodically updated schemes. With real time revocation checking, any party can confirm/obtain the proof of the certificate validity by performing an online transaction that indicates the current revocation status for a certificate.

We briefly summarize the common online revocation techniques-

2.3.1 On-Line Certificate Status Protocol (OCSP)

OCSP [5] is a protocol developed by IETF in which on-line revocation information is available from an OCSP responder thorough a request/response mechanism. OCSP is designed to check the certificate revocation status exclusively.

The protocol is applied between a client (OCSP requester, acting for the user) and a server (OCSP responder, representing a directory). The client generates a so called OCSP request that primary contains one or even more identifiers of certificates queried for validity check, i.e. their serial number together with other data. Then, the (optionally signed) request is send to the server. The server receiving the OCSP request creates an OCSP response: The response mainly includes a timestamp representing the time when the actual request was generated, furthermore, the identifiers and status values of the requested certificates together with a validity interval. A certificate status value is either set to good, revoked or unknown. Be aware that "good" implies three meanings: firstly, the certificate is not revoked, but secondly, it may also not be issued yet or even thirdly, the time at which the response is produced is not within the validity of the certificate. Status "revoked" stands for a revocation or on hold of the certificate. If the answer is "unknown" the server has no information available about the required certificate. The validity interval specifies the time at which the status being indicated is known to be correct and optional the time at or before newer information will be available about the status of the certificate. The OCSP response should be digitally signed either by the server or by the CA. In case of any error the OCSP response contains an error message. The OCSP response is send to the requesting client of the user who then analyzes the data. A pre-

producing of signed responses is currently optional. OCSP is especially appropriated for attribute certificates where status information always needs to be up-to-date.

2.3.2 OCSP-X, SCVP and DC

There are a number of on-line protocols that are more extensive than OCSP. OCSP-X [19], or OCSP extensions, provide a richer set of semantics for OCSP. With these extensions, an end entity is able to delegate the full task of deciding whether a certificate should be relied upon and whether it is acceptable for a particular operation.

The Simple Certificate Verification Protocol (SCVP) [20] is a separate protocol that is capable of handling (parts of or) the entire certificate validation process. With SCVP, end entities can avoid the overhead involved in processing the certificate validation locally. The protocol may also be used to centrally enforce some validation policy.

The Data Certification Server (DCS) [21] is a trusted third party that can be used as a component in building non-repudiation services. DCS is capable of verifying the correctness of specific data submitted to it. This service may, for example, be used to verify the correctness of a signature, the full certification path, and the revocation status of a certificate. Note that DCS provides more general services than OCSP-X and SCVP.

2.3.3 Obtaining new certificates

Rivest [11] criticizes CRLs and points out several design principles which cannot be fulfilled by CRLs. Rivest proposes an online approach in which if the most recent certificate fails to satisfy the recency requirements of the acceptor, the principal should simply obtain a more recently issued certificate from the CA. Hence, if Alice has a week old certificate indicating employment at the company and Bob is willing to accept at most a day old evidence of employment, Alice should query the online CA and get a new recent certificate created for her. Note that Alice may use this certificate again for other transactions. This technique is also recommended for use in SPKI [25] and SDSI [24].

The approach clearly has advantages i.e. the acceptor is able to set the recency requirements, certificate validation is reduced to just validating the digital signature on the certificate, acceptor need not deal with any revocation mechanism and better load distribution on the sender and the acceptor. A drawback is the increased load on the certificate servers. The certificate servers are now required to sign many more certificates than before.

3. THE PROPOSED SOLUTION

In this section, we present a new online revocation mechanism based on time stamping the certificate serial number. Our technique is based on a variation of [11] and offers significant advantage over [11].

To explain our technique, we first introduce the concept of certificate renewal-

Certificate renewal is the certificate serial number together with timestamp (current date and time) digitally signed by the certificate renewal authority. A certificate renewal with serial number n and timestamp of time t is the proof that certificate having serial number n was not revoked and was still valid at time t.

Now, the mechanism proceeds as follows-

Step 1: The sender makes sure that his latest certificate renewal satisfies the recency criteria of the acceptor. If not, the sender obtains a new certificate renewal as follows

- The sender queries the certificate renewal authority (CRA) by just sending the serial number of his certificate.
- CRA checks the revocation status of the certificate having this serial number. If it is un-revoked, CRA creates the certificate renewal by digitally signing the time-stamped (with current time) serial number. Else, CRA timestamps the serial number with a time before the certificate creation time and signs it. This acts as a proof of *"Certificate Expiry"*.
- The CRA sends the certificate renewal (or certificate expiry) to the sender.

Step 2: The sender sends the certificate renewal and optionally, the certificate to the acceptor. Sending the certificate is not required if the acceptor already has a copy of the sender's certificate in its cache.

Step 3: The acceptor verifies the digital signature on the certificate (in case it was not cached) and the certificate renewal. It also makes sure that the certificate renewal satisfies its recency criteria.

First we present the advantage our scheme offers in general over other revocation mechanisms like CRLs and CRS.

1) Acceptor could simply reject a certificate renewal and ask the sender to obtain a more recent one. Thus, the acceptor is able to set the recency requirements. This is important since acceptor is the party who is running the risk if his decision is wrong, not the CA. Bob may want at most a day old evidence of employment at the bank before granting Alice the access to bank

accounts of the customers. Weekly issued CRLs cannot meet his requirements. CRLs require the verifier to accept a recency guarantee bounded by the rate at which CRLs are generated. Though CRS is better, even it does not satisfy this condition.

2) The bandwidth consumption is quite low contrary to CRLs. This is because individual proof of validity of a certificate is issued by the CRA in the form of certificate renewals rather than long CRLs. Because of the potential size of CRLs, scaling to large communities can be difficult. To verify the certificate of Alice, Bob should download the complete CRL of the Alice's CA. The result of a simulation study [18] indicates that the maximum network load in case of CRLs is about 10 times higher than in case of online approaches.

3) Our method provides real time revocation status information. This is especially important in electronic commerce transactions. Note that CRLs and CRS cannot provide real time revocation status information.

4) The sender supplies all the relevant validity evidence, including recency, to the acceptor. More precisely, this means that the design principle-

"For best load distribution, do work for your certificates yourself"

is fulfilled.

There are several reasons for this principle

- The sender can query the CA as well as the acceptor can.
- The recency information obtained may be useful again to the sender.
- This structure puts any burden on the sender (usually the client) rather than on a possibly overworked acceptor (the server). Even in cases, when the sender is the server (e.g. in https protocol, while establish an SSL connection, server sends its certificate), it is not much work for the server to query the CA and obtain a recent certificate daily (or even hourly). This approach is clearly better than having each client obtain the CRL of the server's CA to verify the server's certificate.
- In many case, this allows the acceptor (server) to be implemented in a stateless manner. For example, Bob can reply to Alice, "Sorry, please make sure that your evidences are at most one week old," and then forget about Alice until she comes back again, rather than having to rummage all over the Internet to see if Alice's certificates are still OK. A stateless server design is less vulnerable to denial-of-service attacks.

Note that CRS also satisfies this design principle.

5) The distribution of requests for the CRA is uniform in our technique. However the distribution of requests for the CRL distribution server is poor. If the weekly CRL is issued by the CA on Monday morning, clearly the request rate for CRLs will be much higher on Mondays and Tuesdays than on Saturdays and Sundays. This high peak request rate shoots up the processing and network bandwidth requirements for the CRL server. Even in the case of CRS, the distribution of requests is not uniform.

6) Sometimes, downloading the CRL may introduce unacceptable latency in certificate validation. Since the acceptor should first download the most recent copy of the CRL of the sender's CA before validation (in case it doesn't have one), the delay introduced in the certificate validation may be significant. Our solution does not require the verifier to contact any third party during the certificate validation process. Thus the delay introduced is limited to the delay due to signature verification.

All the above advantage our technique offers are also possible by using short lived certificates as proposed in [11]. We now proceed to compare our scheme specifically with [11].

1) Network Load

The network load in our revocation scheme is significantly lower than in [11]. The bandwidth consumption is reduced for at least two of the three parties involved (i.e. the sender and the CRA). Additionally, the bandwidth consumption by the acceptor may also be reduced if it can cache the certificates sent to it.

Instead of sending a complete new certificate, the CRA now sends just a certificate renewal to the sender which is significantly smaller in size as compared to a new certificate. Further, in our scheme, the certificate does never change in contrast to [11]. This allows the caching of certificate by the acceptor. Hence the sender does not have to send the certificate now to the acceptor if it is cached.

The average network load with our technique (assuming 1 KB X.509 certificates) will be around 65 % assuming no caching, 40 % assuming 50% cache hit rate and 15 % assuming 99% hit rate as compared to [11].

2) Latency

Since the data to be transferred over the network is reduced for each of the three parties, the latency involved in the communication is also reduced.

3) Computational Requirements for the Parties involved

a) For the sender, the processing load is the same (almost negligible) for our technique as well as [11].

b) For the CRA, the overhead of creating a new certificate is eliminated as well as the data to be signed is reduced.

c) The processing load analysis for the acceptor yields the following-

- When the certificate is not cached:- The acceptor has to verify two digital signatures (certificate signatures and certificate renewal signatures) as compared to one in [11]. Hence, this is a drawback of our approach. However if the signatures schemes like RSA are employed, where signature verification is much faster than signature generation, this drawback may not be very significant.

- When the certificate is cached:- The acceptor has to verify only one digital signature (certificate renewal signature). However, an advantage over [11] is that the data to be verified here is much smaller and the certificate content check and validation is not required. Certificate content check includes verification and validation of various certificate fields like validity period, identity of the holder, public key etc.

4) Storage requirements for the CRA Server

In [11], since the online certificate generation is required, every certificate (or its fields, such as public key and other attributes) should be stored on the server. In our approach, the CRA doesn't need to deal with or even store certificates on its server. Just a database of serial numbers of revoked certificates is sufficient. While signing a certificate renewal for the queried serial number, the CRA just checks for its presence in the database of revoked serial number. If the serial number is not present in the database, it can be inferred that the certificate in question is un-revoked and thus the CRA proceeds to sign the certificate renewal.

This kind of design also offers another interesting advantage. Consider an organization in which certificates are exchanged within the organization and the information specified in the certificates is considered sensitive due to privacy concerns/ some other reasons. Hence this information should be prevented from leakage outside the organization. Storing all the information specified in the certificates on an online server presents an adversary a prime target of attack through which she may obtain all the desired information. However, our technique defends this kind of attack since no certificate information is stored at the server at all.

5) Security of private key of the Certificate Authority

CAs may wish to avoid placing their private key on hosts connected to Internet. However, since [11] required online certificate creation, CAs are forced to do so. This could create mission critical components in the server security design as any compromise of CAs private key may lead to the catastrophic failure of the PKI. Our approach does not require online certificate creation. The keys for certificate creation and certificate renewal could be different. We only require the certificate renewal key to be placed online, which if compromised, would enable the attacker to renew revoked certificates but would not enable her to create new certificates.

6) Clean Separation

With a clean separation between certificate creation and certificate revocation (using certificate renewals), it becomes possible for the CAs to delegate the revocation process. Such a possibility is of great interest in today's world. Organizations nowadays frequently outsource a part of their work to other entities. Hence CAs may choose to stick to the process of certificate creation and outsource the whole revocation process.

Further, with such a clean separation, the failure (key compromise) of revocation system does not imply the failure of certificate creation system and vice versa.

Key compromise Issue-

Rivest [11] proposes that key compromise is different and the issue should be handled differently. It proposes Key Compromise Agents who will form a high speed reliable network among themselves and would issue a certificate of health to each principle. Certificate of health is the proof that the key hasn't been compromised yet. [11] suggests that a key compromise should revoke the certificate of health rather than the ordinary certificate.

To analyze this proposal, we come back to the basic guarantee of a certificate-

"We, the CA, declare that the public key of Alice is ... "

Implied guarantee-

"Any message encrypted with this key will be readable by Alice only. Further, only Alice can produce digital signatures verifiable with this key".

A key compromise obviously renders this certificate guarantee invalid. Further, after the key compromise, the certificate does not seem to be of any use to Alice as she will be probably changing her keys and won't be using public key specified in the certificate anymore. The most obvious solution in this case appears to be the revocation of the certificate. This will eliminate the need of expensive Key Compromise Agents and certificates of health. Hence, we choose to stick with Alice just informing her CA or the certificate renewal authority about the key compromise which would then revoke her certificate and issue a fresh certificate with her new public key on it. This seems to be logically consistent as well as the most efficient solution.

4. CONCLUSION

We proposed a new online revocation scheme based on certificate renewals. Our scheme presents an alternative to short lived certificates. Our scheme offers advantages in terms of network load, communication latency, computational requirements, CRA server storage requirement and CA key security. A clean separation between certificate creation and certificate revocation permits CA's delegation/outsourcing of the revocation process.

We propose that the current notion of digital certificates (e.g. X.509) be extended to include the certificate renewal in an extension field of the certificate itself. This field will keep changing and will be excluded while signing the certificate. This will perhaps give a cleaner and easier to think Public Key Infrastructures with all revocation information embedded in the certificate itself.

REFERENCES

[1] A. Arsenault and S. Turner. PKIX Roadmap, Internet Draft, "Work in progress, IETF PKIX working group", October 1999.

[2] Warwick Ford and Michel S. Baum, Secure Electronic Commerce, Prentice Hall PTR, 1997.

[3] David A. Copper, A model of certificate revocation. In proceedings of the Fifteenth Annual Computer Security Application Conference, December 1999.

[4] Stuart Stubblebine. Recent-secure authentication: Enforcing revocation in distributed systems. In Proceedings 1995 IEEE Symposium on Research in Security and Privacy, pages 224-234, May 1995.

[5] A. Malpani S. Galperin M. Myers, R. Ankney and C. Adams. RFC 2560: X.509 internet public key infrastructure online certificate status protocol - OCSP, June 1999.

[6] Patrick McDaniel and Aviel D. Rubin. A response to "can we eliminate certificate revocation lists?". In Financial Cryptography, pages 245-258, 2000.

[7] S. Micali. Eficient certificate revocation. Technical Memo MIT/LCS/TM-542b, Massachusetts Institute of Technology, Laboratory for Computer Science, March 1996.

[8] J. Millen and R. Wright. Certificate revocation the responsible way. In Post-proceedings of Computer Security, Dependability and Assurance: from Needs to Solutions (CSDA'98). IEEE Computer Society.

[9] M. Myers. Revocation: Options and challenges. In Lecture Notes in Computer Science, volume 1465, pages 165-171, 1998.

[10] Moni Naor and Kobbi Nissim. Certificate revocation and certificate update. In Proceedings 7th USENIX Security Symposium (San Antonio, Texas), Jan 1998.

[11] Ronald L. Rivest. Can we eliminate certificate revocations lists? In Financial Cryptography, pages 178-183, 1998.

[12] Fox and LaMacchia. Certificate revocation: Mechanics and meaning. In Financial Cryptography. LNCS, Springer-Verlag, 1998.

[13] Carl A. Gunter and Trevor Jim. Generalized certificate revocation. In Symposium on Principles of Programming Languages, pages 316-329, 2000.

[14] R. Housley, W. Ford, W. Polk, and D. Solo. RFC 2459: Internet X.509 public key infrastructure certificate and CRL profile, January 1999. Status: PROPOSED STANDARD.

[15] P. C. Kocher. On certificate revocation and validation. In International Conference on Financial Cryptography. LNCS, Springer-Verlag, 1998.

[16] William Aiello, Sachin Lodha, and Rafail Ostrovsky. Fast Digital Identity Revocation. In Advances in Cryptology - CRYPTO '98. Springer, 1998.

[17] Paul Kocher. A Quick Introduction to Certificate Revocation Trees (CRTs). Technical report, ValiCert, 1999.

[18] Andre Arnes, Public Key Certificate Revocation Schemes. Master's thesis, Department of Telematics, Norwegian University of Science and Technology, February 2000.

[19] Phillip Hallam-Baker. OCSP Extensions. Internet Draft, "Work in progress, IETF PKIX working group", September 1999.

[20] Ambarish Malpani and Paul Hoffman. Simple Certificate Validation Protocol. Internet Draft, "Work in progress, IETF PKIX working group", April 1999.

[21] Carlisle Adams and Robert Zuccherato, Data Certification Server Protocols. Internet Draft, "Work in progress, IETF PKIX working group", September 1999.

[22] Vipul Goyal, "Certificate Revocation Lists or Online Mechanisms", Workshop on Security in Information Systems, Portugal, April 2004.

[23] Silvio Micali. NOVOMODO: Scalable Certificate Validation and Simplified PKI Management. In 1st Annual PKI Research Workshop - Proceeding, April 2002.

[24] Ronald L. Rivest and Butler Lampson. SDSI - a simple distributed security infrastructure. (see SDSI web page at http://theory.lcs.mit.edu/~cis/sdsi.html).

[25] C. Ellison, B. Frantz, B. Lampson, R. Rivest, B. Thomas, T. Ylonen, "SPKI Certificate Theory", RFC 2693, September 1999.

MASKS: MANAGING ANONYMITY WHILE SHARING KNOWLEDGE TO SERVERS

Robert Pinto,[1] Lucila Ishitani,[2] Virgílio Almeida,[3] Wagner Meira Júnior,[3] Fabiano A. Fonseca,[3] and Fernando D. Castro[3]

[1] *Universidade Federal de Minas Gerais*
Belo Horizonte, Brazil
robert@dcc.ufmg.br

[2] *Pontifícia Universidade Católica de Minas Gerais*
Universidade Federal de Minas Gerais
Belo Horizonte, Brazil
lucila@dcc.ufmg.br

[3] *Universidade Federal de Minas Gerais*
Belo Horizonte, Brazil
{virgilio,meira,fabiano,fernando}@dcc.ufmg.br

Abstract This work presents an architecture that allows users to enhance their privacy control over the computational environment. Web privacy is a topic that is raising, nowadays, many discussions. Usually, people do not know how their privacy can be violated or what can be done to protect it. Among the generated conflicts, we would like to show up the one that happens between privacy and personalization: by one side, users appreciate the idea of receiving personalized services and do not approve the collection, tracing and analysis of their actions; by the other side, personalization services need this type of information in order to profile their users. The architecture presented in this article helps users to understand better how their privacy can be invaded and, at the same time, gives them a better control of their privacy, through anonymity, without preventing them from receiving personalized services.

Keywords: privacy, personalization, anonymity and framework

1. INTRODUCTION

With the growth and spread of Internet services, information systems have the means to collect and build detailed profiles of individuals that use those services. These profiles can be used by Web sites to offer personalized services to their users. These services include customized content and recommendations based on user past behavior and actual interests. Personalization brings benefits for both sides of a Web interaction: users and sites.

According to Kobsa [Kobsa, 2001], "customers need to feel they have a unique personal relationship with the business" and to confirm this idea, he presents the result of a survey that shows that sites that offered personalized services increased in 47% the number of new customers. This result also shows that personalization benefits sites. The increase in the number of customers tends to increase sales and revenues. However, the personalization needs of collecting and analyzing user data (this work is commonly called *data mining*) can characterize privacy invasion. Privacy in the information age could be characterized as the right of individuals to protect their ability to selectively reveal information about themselves [Rao and Rohatgi, 2000]. Many people that use the Web know that information is collected about them, but they are not certain of what and how much it is collected, neither for what it will be used. In light of this notion, it is evident that the current Web infrastructure represents a serious threat to the privacy of users of services in the Web.

So, there is a great conflict to solve: how can data be disclosed for Web sites, so that personalized services could be offered to users, without violating their privacy? In this work we address this problem of protecting users from companies and online services that collect data during Web browsing activities, without preventing them from receiving personalized services. This framework is called MASKS (Managing Anonymity while Sharing Knowledge to Servers) and is based on the idea of minimizing exposure of individual information by putting an "anonymity barrier" between the private data and Web services, filtering the information that flows across the barrier to the service. In order to evaluate this "filtered information" we made use of an important concept of information theory: the entropy. We compared the entropy of the set of requisitions generated without MASKS and with MASKS and discovered that, although MASKS changes the set of requisitions sent to a site server, this new set of requisition will still offer some value about user interests.

This paper is organized in the following way. Section 2 presents related work. Section 3 presents our proposed solution for the conflict between privacy and personalization. Main characteristics of MASKS' components are presented in sections 4 and 5. The experimental evaluation of MASKS and its results are presented in section 6. And section 7 presents our conclusions and future work.

2. RELATED WORK

There are some previous work that also proposes to solve the conflict between privacy and personalization.

Crowds [Reiter and Rubin, 1998], mix networks [Chaum, 1981]and Onion Routing [Reed et al., 1996]provide anonymity without requiring a unique third-party to forward requests. The basic idea is to hide the real originator of a request, with the cooperation of the members of a group. In Onion, the path of message through the members of a group is pre-determined whereas in Crowds, a path is configured during transmission of a request. As all anonymity tools, both mechanisms do not allow personalization, due to the complete absence of identifiable information available to the web sites. But, there is a special case of anonymity that creates conditions to solve this problem: the persistent anonymity [Goldberg et al., 1997]or pseudonymity.

Pseudonyms are persistent user identification that, logically, can not be connected to the user real identification, but allows the association of a set of requests to a unique user. Lucent Personalized Web Assistant (LPWA)[1] is the most known tool based on the use of pseudonyms. LPWA users can use the same pseudonym each time they return to a site, but they will have different pseudonyms for different sites. It allows Web sites to offer any kind of service that relies on user identification, without linking users to their actual identities. But, just like in real life, if someone discovers the real identity of a user behind a pseudonym, all user's past actions will be automatically exposed.

Arlein et al. [Arlein et al., 2000]proposed an architecture of shared user profiles, so that service providers could customize contents based even on user activities on other sites. This architecture preserves privacy because users can specify which information could be shared and grouped to form a profile. But this proposal is static or, in other words, it does not automatically adapt to changes on user interests.

Another option people have is to hide themselves behind a mask or *personae*. A digital *persona* is a model of the public personality of a person. A person can have many masks. Each of them reflects a specific role or interest that a person can have [Clarke, 1994]. *Persona* [Dunn et al., 1997]is a mechanism that is based on this idea. Each *Persona* is a personal data collection that a client can distribute to sites.

It protects privacy, because users controls what they accept to disclose to sites. The disadvantage of the mechanism is that it requires that users previously specify which information they accept to disclose. Besides

[1] http://www.bell-labs.com/projects/lpwa

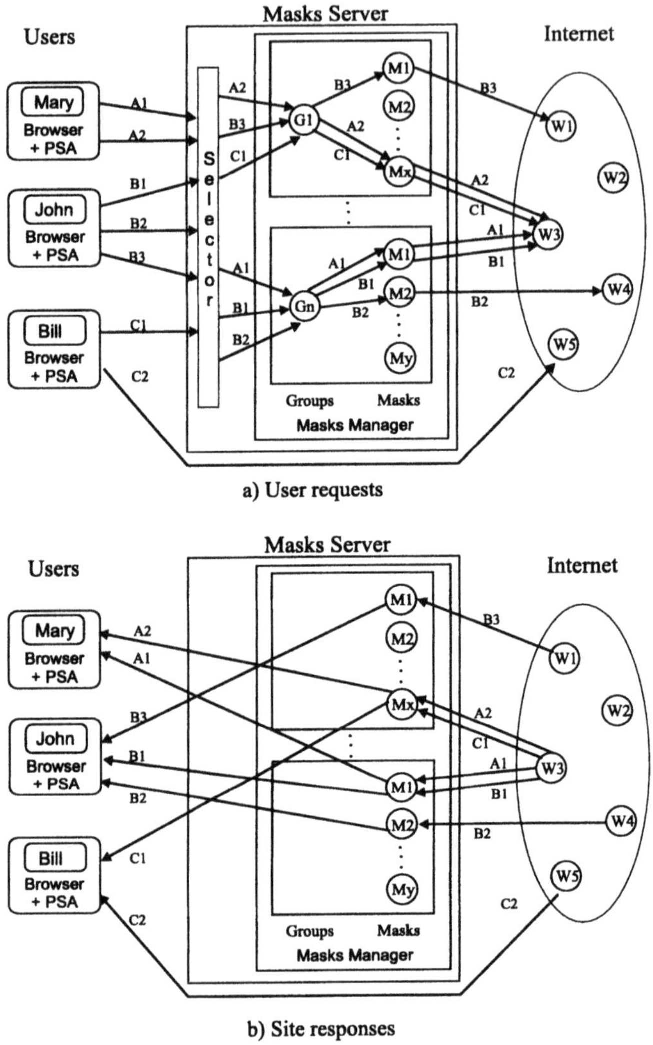

Figure 1. Simplified architecture of MASKS

that, it does not adapt to dynamic aspects of user navigation. The profile it considers is restricted to static data, like name, birth date, address, and so on.

MASKS project proposes to solve the disadvantages of the architectures and technologies presented above.

3. MASKS

This section describes MASKS (Managing Anonymity while Sharing Knowledge to Servers) [Ishitani et al., 2003](Figure 1), a Web-based framework that offers users a privacy-preserving mechanism, through anonymity. Unlike other anonymity tools, it allows some degree of personalization, for it discloses data that can be used by Web sites to provide personalized services, without profiling each individual user.

3.1 Design characteristics

MASKS's process is safe for users, because as it depends only on the last user request to function, it does not need to keep user data. The rationale behind this design decision is that the larger the amount of user information kept, the higher the probability that the server becomes a target of an attack.

Besides protecting user privacy, MASKS offers other benefits to users:

- Partial personalization compatibility - although being an anonymity mechanism, MASKS changes user request in a way that lets Web sites personalize services to users.
- Efficiency - the algorithms implemented are efficient in response time, because although the data structures used are complex, the applied mechanism is very simple. So, we expect that users will not perceive any latency when using MASKS;
- Dynamic - MASKS easily adapts to changes on user interests;
- Easiness of use - users do not need to provide any previous information to MASKS. No special protocol or proprietary technology is required, as MASKS employs standard HTTP and TCP protocols and work with usual identification mechanisms, such as cookies;
- Cover of a broader view of privacy protection - considering the six layers of privacy protection (awareness, control, privacy-enhancing tools, privacy policy, privacy and trust certification and privacy protection laws) [Ishitani et al., 2003], MASKS is the only architecture we know that implements the first three layers.

3.2 MASKS Architecture

The architecture protects user privacy through the use of *masks*. A mask is a temporary identification that a user may adopt while interacting with a Web site. This identification is associated with a user interest in a given topic or specific Web site. So, as users may present distinct interests while navigating in the Web, different masks can be associated with the same user even when they interact with the same site. For example, suppose that Web site W3, in Figure 1, is a portal that offers different classes of information such as travel and sports. Suppose also that Mary requests travel-related services (A1) and later requests sport information (A2). W3 would view the two requests from Mary (A1 and A2) as coming from two distinct users.

The assignment of masks to users is based on the concept of *group*. A group represents a topic of interest, and a user request is assigned to a group according to the semantics of the requested object. As an example, in Figure 1, requests A2 of Mary and C1 of Bill are assigned to the same group. Like with Crowds [Reiter and Rubin, 1998]and mix networks [Chaum, 1981], where web requests are not submitted by the individual user but by a user group (the Crowd) or by a mix network, user privacy is preserved. But differently from Crowds and Mix networks, MASKS provides data about the users' interests. These data can be used by Web servers to personalize the interaction.

Each group may have several masks associated with it, one for each site that offers the type of information associated with the group. For example, there are many sites that provide information about travel. The group associated with the subject travel will have one mask for each of the known sites related to travel. In Figure 1, this situation is represented by the requests B1 and B2 of John.

MASKS architecture has two major components: the Privacy and Security Agent (PSA) and the *Masks Server*. PSA is a program that each user runs in conjunction with the browser, keeping him/her informed about his/her potential privacy risks. PSA also allows users to turn off the masking process, if they want to interact directly with sites, without anonymity (request C2 in Figure 1).

The second component is the *Masks Server*, which is an intermediary that sits between users and Web sites, working as a proxy. It is responsible for managing masks and assigning them to the users.

4. PRIVACY AND SECURITY AGENT

Privacy and Security Agent (PSA) is a *plugin*, acting as an intermediary between users, *Masks Server* and Web sites. PSA receives user requisitions and resend them to *Masks Server* or to Web sites, according to user wishes of being masked or not. It is also a PSA function to receive sites' responses and to transmit them to users, with a privacy risk evaluation.

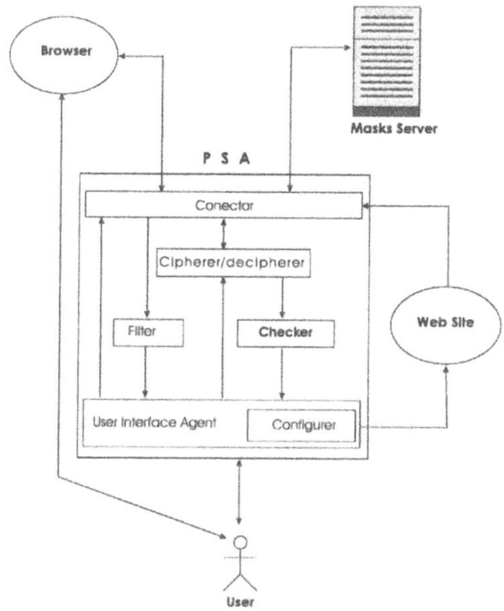

Figure 2. Simplified architecture of PSA

4.1 PSA Architecture

The main modules which compose PSA architecture are (Figure 2):

Conector: Module that acts as an intermediary between the user's browser and *Masks Server.* This module intercepts requests sent by the users to a Web site. It is also responsible to transmit responses of *Masks Server* to the user's browser, after their processing by the other modules of PSA.

Cipherer/decipherer: Module that ciphers and deciphers URLs sent to and received from a *Masks Server.* It aims to prevent third-parties from having access to user private information, such as: accessed sites, searched topics, the IP address of a set of transmitted data.

Filter: Module that filters HTTP headers. Examples of information that can be filtered: browser identification and the page that the user was visiting when he/she sent the last request (*referer*).

Checker: Module that blocks known mechanisms of information collection, like *Web Bugs.*

User Interface Agent: This module keeps users informed about their privacy risks. It includes another module: the Configurer.

Configurer: The function of this module is to allow users to choose the type of interface they want to use with PSA (pop-up windows or icons) and if they want to interact with Web sites behind masks or not.

The PSA processing of a user request is as following: first of all, the request is filtered. If any problem is detected, the user is informed about that and he/she can opt to continue his/her navigation masked or not. Before sending a filtered request to a *Masks Server*, PSA ciphers it.

The inverse process, or the PSA processing of a response sent by a *Masks Server*, starts with deciphering of the response. After that, the deciphered response is checked and "cleaned" by the Checker module. If any problem is detected, the user is informed. The response will only be transmitted to the browser, after satisfying PSA and user requirements.

4.2 Implementation

The first version of PSA prototype was implemented for the *Mozilla* browser. This decision was based on the fact that this browser makes easier the implementation of *plugins* and user interface, through the use of a language known as *XUL* (*XML-based User Interface Language*).

The following components were implemented through a *XPCOM* architecture: the conector, the filter and, partially, the checker and configurer. Those components actions are activated by a user interface available on the browser interface.

5. MASKS SERVER

Masks Server is the main component of the MASKS architecture, as it is the responsible for the masks managing. It is the intermediary between PSA and Web sites, acting as a proxy.

Selector is the Masks Server component, responsible for selecting the interest group of each user requisition. Since the masks are divided into groups, one key issue for the effectiveness of the server is the definition of how objects are assigned to groups. One strategy we evaluated is the use of a semantic tree, more specifically the tree defined by the Open Directory Project[2]. This tree organizes Web sites based on a human-based classification procedure. It is costless and completely available, representing a start point to define the groups and their relationships to each other. Each node of the tree represents a semantic category, or

[2]http://dmoz.org

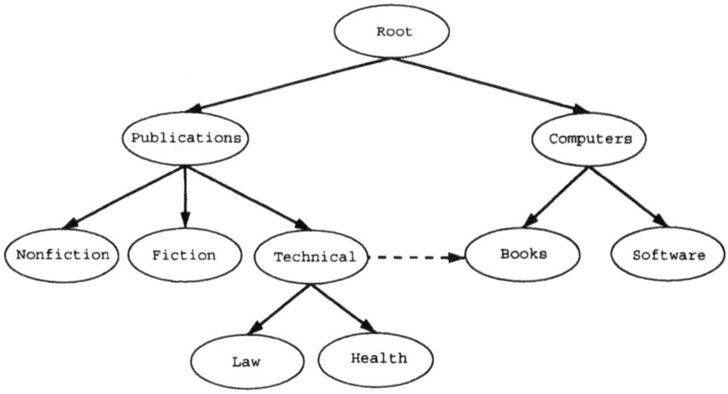

Figure 3. Example of a semantic tree

a *group* in our approach. Figure 3 illustrates an example of a semantic tree.

5.1 Implementation

For the implementation of a prototype of *Masks Server* we chose *Squid*[3]. *Squid* is the result of the work of many people from the Internet Community, coordinated by *Duane Wessels*, of the *National Laboratory for Applied Network Research.* This architecture has the following advantages: its code is free and open, it works as an HTPP proxy, its scalability level is satisfactory. Besides these, *Squid* easily allows the capture of HTTP headers, which facilitate the processing of cookies sent by Web servers.

5.1.1 Cookie Processing.

Cookie processing is shown in Figure 4. Figure 4-(a) represents a request sequence directly sent to a Web server, by the users John and Mary. Figure 4-(b) represents the same request sequence, with a *Masks Server* as an intermediary. In both situations, the Web server returns, with the response, a cookie that can be used for personalization services.

As can be seen, without using a Masks Server, users are directly receiving cookies. The Web Server asks for these cookies in the following requests. These cookies can be changed or not by the Web server. For example, in Figure 4-(a) the user John sends a new request along with the *cookie* **ABC**. The response he receives contains the new *cookie* **ABD**.

[3]http://www.squid-cache.org

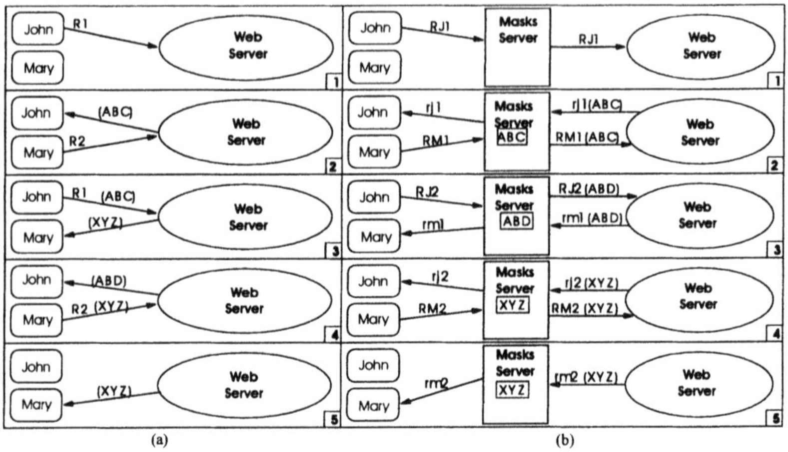

Figure 4. Cookie processing without/with a MASKS Server

By the other side, the Web server does not change the cookie **XYZ** associated to Mary's request.

When a Web user is using a *Masks Server*, this situation becomes different. The Web server will continue to do the same operations with the *cookies*, but the users will not receive the cookies. For example, suppose that in Figure 4-(b), John and Mary are interested on the same subject. According to the MASKS architecture, they both will be associated to the same interest group/mask. So, *Masks Server* resends user requests as if they belonged to a unique user. As can be seen in Figure 4-(b), the response to John's request **RJ1**, will be **rj1** along with the *cookie* **ABC**. When Mary sends her first request **RM1**, as it belongs to the same group of John's request, it will be resent to the Web Server, along with the *cookie* **ABC**. The response will be **rm1**, along with the *cookie* **ABD**. When John sends his second request **RJ2**, the *Masks Server* will resend it with the *cookie* **ABD**.

6. EXPERIMENTAL EVALUATION

In this section we present the methodology used to evaluate the quality of the data disclosed by MASKS to the Web sites. We also present the results of our experiments.

6.1 Methodology

It is hard to evaluate at what extent the data disclosed to Web sites will affect the personalization strategies adopted by the site, since there is a great number of different techniques that may be used. But it is feasible to estimate the worth of the *disclosed information*, which will always be the same regardless of the personalization strategy used by each Web site. According to the Information Theory the measure of the amount of information one variable contains is given by the *entropy* [Shannon, 1948].

In our case, as we want to discover if a set of masked requests will disclose valuable information to Web sites, the entropy will be related to the amount of information of each set. The smaller the occurrence probability of a request set, the bigger the entropy. And the bigger the occurrence probability of a request set, the smaller the entropy.

For calculating the entropies, we represented user navigation by a Markov model in which each state will represent a page and the links will represent page transitions, and their number of occurrences. Figure 5 illustrates the model adopted in our experiment and the entropy calculation.

In Figure 5-a), we present the basic subjects of seven pages of a fictitious site W. For each subject, MASKS will associate a different mask. Each page will have a occurrence probability, computed in the following way: *p(page)= number of page requests / total number of requests to all pages of the site.* In Figure 5-b), we can see that the total number of requests is 17 and that the number of requests to P1 is 3. Thus, $p(P1) = 3/17 \approx 0,176$.

In Figure 5-b), we present a request sequence that arrives to the W site server, in two distinct days, including the user that sent the requests. In Figure 5-c), we present the same sequence, with masked users. In Figures 5-d) and 5-e), we present tables that detail the requests that compose sessions. In order to calculate the entropies, we build the models shown in Figures 5-f) and 5-g). These models were created based on the data of the Figures 5-d) and 5-e). Each node of the model represents one page and contains the number of times the page was accessed. Each link represents a page transition and has, associated to it, the number of times the transition has occurred. As the masking process changes the sessions, the models will have different entropies values.

The entropy calculation was based on the theory proposed in [Levene and Loizou, 1999]. The entropy value $H(M)$ of the model M can be calculated by:

$$H(M) \approx -\sum_{i=1}^{n}\sum_{j=1}^{n}\frac{m_{i,j}}{n}log\frac{m_{i,j}}{m_i},$$

where n is the total number of pages that compose the model, $m_{i,j}$ is the total page transition number from page i to page j and m_i is the total number of requsitions of page i.

Pages	Subject	Masks
P1	Sports	M1
P2	Sports	M1
P3	Finances	M2
P4	Finances	M2
P5	Finances	M2
P6	Tourism	M3
P7	Tourism	M3

a) Pages and their masks

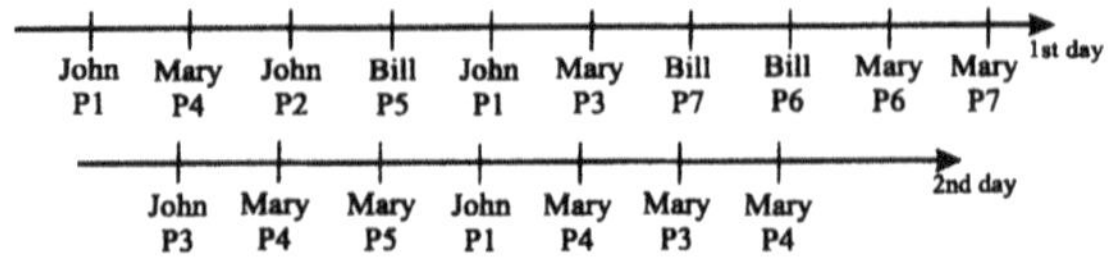

b) Sequence that original requests arrive at site server

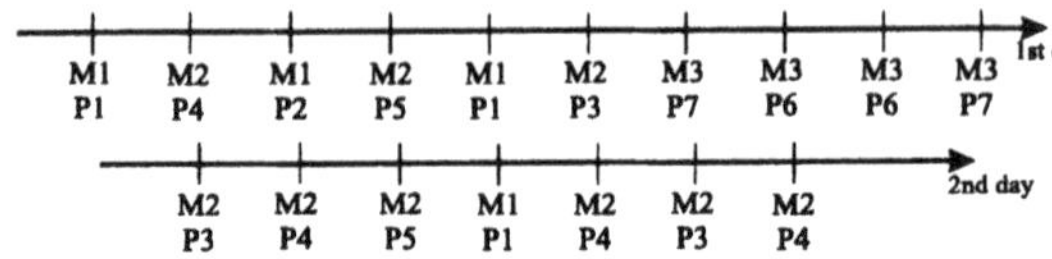

c) Sequence that masked requests arrive at site server

User	Session
John	P1, P2, P1
John	P3, P1
Mary	P4, P3, P6, P7
Mary	P4, P5, P4, P3, P4
Bill	P5, P7, P6

d) Original sessions

User	Session
M1	P1, P2, P1
M1	P1
M2	P4, P5, P3
M2	P3, P4, P5, P4, P3, P4
M3	P7, P6, P6, P7

e) Masked requests sessions

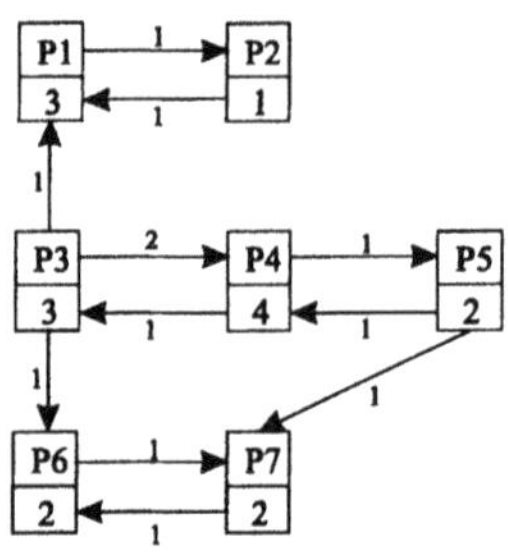

f) Navigation model for original requests
(Model entropy = 1,37)

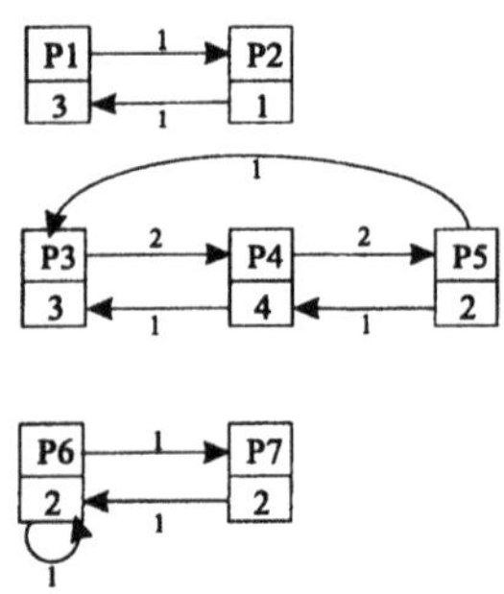

g) Navigation model for masked requests
(Model entropy = 1,17)

Figure 5. Original and masked models and sessions

6.2 Results

The simulation was done using real logs of a virtual bookstore, collected during seven days. We have chosen the logs of a unique Web site and not a generalized log file, such as the one stored by a proxy, because we had to evaluate the value of the data disclosed for personalization from the point of view of a Web site. In other words, for offering per-

Model	Entropy	IDs	Sessions	Avg. length
Original	8.01	227,047	242,990	1.52
Masked (complete tree)	4.72	12,105	166,345	2.22
Masked (five-level tree)	5.31	9,996	157,436	2.35
Masked (four-level tree)	5.97	8,780	147,459	2.50
Masked (three-level tree)	7.20	7,912	138,426	2.67
Masked (two-level tree)	7.79	7,580	131,858	2.80
Masked (one-level tree)	8.47	7,517	127,325	2,90
Typically anonymized	–	369,832	369,832	1

Table 1. Results Comparison

sonalized services, Web sites must have access to user data (requests, profiles and navigation pattern). So, to discover if the personalization process will be affected, we have to study the difference between the original sessions of a site and the new masked one.

We simulated the requests for the following cases:

1 original user sessions;

2 masked sessions, considering different levels of the semantic tree - sessions composed by requests whose client id field was replaced by the group mask. For the group assignment, we considered from one to five levels of the semantic tree and the complete semantic tree;

3 sessions composed by only one user request - this simulation was done in order to evaluate the result of other anonymizing mechanisms: each request is considered as being sent by a different user and because of that, we have sessions of only one request.

For each case we generated a distinct model. For each generated model, Table 1 presents some information about it: column **Entropy** shows the entropy of the model; column **IDs**, the number of different user identifications; column **Sessions**, the total number of sessions generated; **Avg. length**, the average session length. As expected, the entropies of the masked models are different from the entropy of the original model, but even so, the masked models offer more informational value than the typical anonymization process. It is not possible to calculate the entropy of the model generated by a sequence of typically anonymized requisitions. That is because the trails are composed of only one requisition and, as so, there are no transitions available to compute the entropy of user navigation.

We can observe from Table 1, that the number of masked sessions and user IDs are smaller than those from the original model, because according to the masking process, the original requests are semantically grouped. This is a good result, because the more users mixed in a group, the more protected they are, as it will become more difficult to define the real profile of a specific user.

Table 1 also show us that there is no great difference between the masked models using some or all the levels of the semantic tree. It is clear that a greater number of levels contains a greater number of semantic groups. And, if there are more semantic groups, more different masks will be generated. This leads to a greater number of sessions, and a reduction on the average length of the sessions. But, as the semantic tree is very wide, even making use of only the first two of its levels, there will be a great variety of semantic groups. This result deserves attention, because it shows that there is no need to consider all the semantic tree to group user requests. Thus, the Masks Server can save some memory to store the groups and as it will have less semantic groups to analyze, the response time to resend a user request will be also smaller.

7. CONCLUSIONS AND FUTURE WORK

In this work, we show that, using MASKS, it is possible to disclose data that can be used to offer personalized services adapted to user interest without privacy invasion. The methodology used to prove it was based on the modeling of user navigation through a Markov chain model and the computation of the model entropy.

The results obtained emphasizes the usefulness of MASKS to the Web user activities of browsing and searching for specific information, which are the main points of privacy invasion [CyberDialogue, 1999]and the activities that spend most of the user time [Menascé et al., 1999].

As future work, it will be interesting to improve the masks association code and to evaluate other site classification strategies. There is also the need to develop a complete prototype of PSA.

Another future project is to implement a tree of *Masks Servers* that act cooperatively, exchanging information about semantic groups already known. This strategy will also provide the advantage of avoiding a single point of trust, as users will have more options of servers to provide them anonymity and their data can be distributed among different *Masks Servers*.

References

[Arlein et al., 2000] Arlein, R. M., Jai, B., Jakobsson, M., Monrose, F., and Reiter, M. K. (2000). Privacy-preserving global customization. *Proc. of Conference on Electronic Commerce (EC'00)*, pages 176–184.

[Chaum, 1981] Chaum, D. (1981). Untraceable electronic mail, return addresses, and digital pseudonyms. *Communications of the ACM*, 24(2):84–88.

[Clarke, 1994] Clarke, R. (1994). The digital persona and its application to data surveillance. *The Information Society*, 10(2).

[CyberDialogue, 1999] CyberDialogue (1999). American internet user survey: Privacy x personalization - part I. http://www.cybersitter.com.

[Dunn et al., 1997] Dunn, M., Gwertzman, J., Layman, A., and Partovi, H. (1997). Privacy and profiling on the web. Technical note, World Wide Web Consortium. http://www.w3.org/TR/NOTE-Web-privacy.html.

[Goldberg et al., 1997] Goldberg, I., Wagner, D., and Brewer, E. (1997). Privacy-enhancing technologies for the internet. *Proc. of IEEE Spring COMPCON.* http://citeseer.nj.nec.com/54687.html.

[Ishitani et al., 2003] Ishitani, L., Almeida, V., and Meira Jr., W. (2003). Masks: Bringing anonymity and personalization together. *IEEE Security & Privacy Magazine*, 1(3):18–23.

[Kobsa, 2001] Kobsa, A. (2001). Tailoring privacy to users' needs. *Proc. of 8th International Conference on User Modeling.* http://www.ics.uci.dcu/ kobsa/papers/2001-UM01-kobsa.pdf.

[Levene and Loizou, 1999] Levene, M. and Loizou, G. (1999). Computing the entropy of user navigation in the web. Research Note RN/99/42, Department of Computer Science, University College London.

[Menascé et al., 1999] Menascé, D. A., Almeida, V., Fonseca, R. C., and Mendes, M. (1999). A methodology for workload characterization for e-commerce servers. In *1st ACM Conference in Electronic Commerce (EC-99)*, pages 119–128.

[Rao and Rohatgi, 2000] Rao, J. R. and Rohatgi, P. (2000). Can pseudonymity really guarantee privacy? *9th USENIX Security Symposium.*

[Reed et al., 1996] Reed, M. G., Syverson, P. F., and Goldschlag, D. M. (1996). Proxies for anonymous routing. In *12th Annual Computer Security Applications Conference*, San Diego.

[Reiter and Rubin, 1998] Reiter, M. K. and Rubin, A. D. (1998). Crowds: Anonymity for web transactions. *ACM Trans. on Information and System Security*, 1(1):66–92.

[Shannon, 1948] Shannon, C. E. (1948). A mathematical theory of communication. *The Bell System Technical Journal*, 27:379–423, 623–656.

SECURITY AND DIFFERENTIATED HOTSPOT SERVICES THROUGH POLICY-BASED MANAGEMENT ARCHITECTURE

Idir FODIL and Vincent JARDIN
6WIND, Research and Development,
{Idir.fodil, Vincent.Jardin}@6wind.com

Abstract: We have studied the case of deploying services in public wireless networks based on IEEE802.11 standard. Due to low cost, easy deployment, cost effectiveness and high performance, this technology appears as a very attractive solution for providing internet access and services in public places called hotspot like airports, hotels, train stations... etc Actually, there are numerous solutions that allow user management in WLAN networks. However, most of them do not support multiple service providers and provide all users with the same level of services to Internet access. In our paper, we propose a new software management architecture for hotspot networks, which is based on policy-based management principles introduced as a result of collaboration with the IETF. Our solution enables multiple service provider support and it allows user and service differentiation in hotspot networks. It provides efficient, flexible and scalable user management solution by implementing coherent combination of AAA functions, quality of service guarantee and security assurance for hotspot operators and service providers. For policy configuration, XML schemes have been defined, offering open, easy and customizable management architecture. Moreover, since our solution is layer 2 agnostic, it can be extended to different access technologies such as DSL, PLC... This management architecture has been implemented, tested and validated on the 6WINDGate™ routers and it can easily be ported onto other software architectures and open standard platforms.

Key words: Hotspot, security, AAA, WLAN, 802.11, Policy Management, services, SLA, multiple service providers

1. INTRODUCTION

Today, service management remains a strong concern for both service providers and customers. Service providers are not only under pressure to sell services that are guaranteed and differentiated but also to provide "always-on" connectivity for their customers. At the same time, customers are demanding for more discerning services such as security, mobility, and quality of service (QoS). Achieving these services separately is easy as a lot of standards exist and implementations are widely available. However, combining these services at the same time remains a challenging task since there are a lot of dependencies between them, leading to instable and non performing networks. For those reasons, a higher layer of service management is strongly needed to provide both users and service providers with guaranteed consistency and efficient service deployment. To this end, we have studied the case of deploying services in public wireless networks based on IEEE802.11 standard. Due to low cost, easy deployment, cost effectiveness and high performance, this technology appears as a very attractive solution in order to provide internet access and services in public places called hotspot like airports, hotels, train stations… etc

The Wi-Fi hotspot usage is actually quite inconvenient since users have to buy a new account with each hotspot provider, and has no security assurance from the hotspot provider. It means that the actual hotspot cannot support efficiently roaming. Such inconvenience would reduce the user's interest in using the hotspot services. One solution to this problem is to push the user's existing service provider's contracts into the hotspot. This service provider can be any type of entity that offers the users certain types of services and maintains the user's accounts. These service providers may include Internet service providers, content providers, or cellular operators.

This convenience and security assurance from the existing service providers give greater interest and confidence in using the hotspot services. In such environments, where hotspot operators can have contracts with several service providers (SP), user access management appears as the most important issue to resolve. Hotspot operators must guarantee the SP a subscribed contract that includes dynamic or static bandwidth reservation and security insurance. Moreover, they must be able to provide different levels of services for both service providers and their own users. Concerning service providers, they must be able to manage their mobile customers by providing them with their subscribed contract, and by achieving service differentiation between users in the hotspot network.

This management requires a solution that allows supporting multiple service providers, securing authentication and authorization of users according to their service provider, provisioning hotspot network according to the user service level agreement (SLA), and that allows different billing models. User management implies hotspot access equipment (router, switch...) adaptation for each new user according to his SLA. Concerning multiple service provider support, a new architecture must be built on top of the access equipment in order to provide virtual dedicated equipment for the hotspot operator and for each service provider. Actually, there are numerous solutions that allow this user management in WLAN networks. However, most of them do not support multiple service providers and provide all users with the same level of services to Internet access. This is due to the fact that these solutions are layer 2 based, and providing a scalable service level differentiation can only be achieved at IP level, which is the layer 3. For these reasons, a new management approach that allows access equipment virtualization, dynamic adaptation and reconfiguration is strongly needed.

In our paper, we propose a new software management architecture for hotspot networks, which is based on policy-based management principles introduced as a result of collaboration with the IETF. Our solution enables multiple service provider support and it allows user and service differentiation in hotspot networks. It provides efficient, flexible and scalable user management solution by implementing coherent combination of AAA functions, quality of service guarantee and security assurance for hotspot operators and service providers. For policy configuration, some XML schemes have been defined, offering open, easy and customizable management architecture. Moreover, since our solution is layer 2 agnostic, it can be extended to different access technologies such as DSL, PLC... This management architecture has been implemented, tested and validated on the 6WINDGate™ routers and it can easily be ported onto other software architectures and open standard platforms.

The paper is organized as follows: in section 2, we describe hotspot networks, services and their requirements; in section 3, we list and discuss the existing solutions; in section 4, we describe our policy based solution and its implementation on 6WINDGate routers; in section 5, we describe a use-case scenario of access management in hotspot network with multiple service providers; finally, we conclude this paper in section 6 and some future works are presented.

2. HOTSPOTS OVERVIEW

Initially, WLAN was seen as a potential threat to existing network services for both wireless operators and Internet service providers[24]. But, with the growing deployment of hotspot networks, both have taken more positive view and created new business opportunities by including hotspot networks as complement to their existing offerings. To cater to this demand and to capitalize on their customer relationship, a number of fixed network operators are investing in public hotspots to which they give access to their own DSL subscribers. Now mobile operators are moving to incorporate Wi-Fi access into their service offers.

To provide attractive hotspot services with user security assurance, convenience and always the same level of services, it is mandatory that the customers use the same login ID in all these places. To achieve this goal, a customer relationship must be established between hotspot operators and service providers, and between service providers themselves. This includes marketing, customer service, billing and collection, providing secure access, and account management.

Stronger encryption of the wireless traffic is required to prevent eavesdropping and to secure authentication, especially when business users are accessing company networks from a hotspot. Moreover, business users in particular will require more guarantees about their quality of service to justify higher prices. Last but not least, more varied billing strategies must been supported like free access, prepaid access for a certain amount of time or volume, pay per use period and differential fees for higher bandwidth. Finally, roaming and multiple service providers must be supported in all hotspot networks, which always achieve the best services and the required connections.

To get benefit from this new business model, networks services and architectures, the following requirements must been satisfied:

- Multiple service provider support.
- Secured Authentication, Authorization and Accounting (billing models).
- Quality of service assurance
  - Static: some service providers may pay for fixed bandwidth in some hotspot networks, because they are deployed in places that attract lot of people. In these cases, the quality of service is guaranteed per service and it is the service provider responsibility to ensure per user quality of service guarantees.
  - Dynamic: in other places, which are less frequented, service providers may pay for bandwidth only when their customers use the hotspot. In this case, the hotspot operator must be able to ensure the quality of service for the user.
- Dynamic User management according to profiles, authorized times, and network resources.
  - Achieving user's SLA.
  - Service differentiation between users.
  - Rapidly solving problems when they occur.

In the next section, we will detail the existing solutions for hotspot networks, and we will discuss their advantages and drawbacks concerning the above requirements.

3. EXISTING SOLUTIONS

In the hotspot networks, users are mobile. They come from different ISPs, have different SLA, and execute different applications. For those reasons, one of the important management issue is related to user access management (arrival and departure of new users) in the WLAN since a complex process is required. This process includes authentication, authorization, accounting and SLA provisioning. We detail a non exhaustive list of solutions that allows user management.

3.1 IEEE 802.1x:

The first one and the most used is the IEEE 802.1x ([7], [8]) standard which defines a method for executing EAP protocol over Ethernet frames [17]. EAP has been defined as an extension to the PPP protocol, and can carry any authentication mechanism. EAP messages are exchanged between an EAP client (mobile user) and EAP server (remote authentication server), and are completely transparent to the access point (AP). The AP has only to maintain a trust relationship with the remote authentication server. Initially, only EAP messages can go through the AP, but when the user is successfully authenticated, the associated MAC addressed is authorized on the access point. The advantage of IEEE 802.1x is the use of EAP protocol which allows mutual authentication between access points and users, as EAP-Key establishment between them.

The first drawback of the 802.1x standard is that users have to re-authenticate when they change access point, since EAP is used between users and access points. This can be avoided by doing context transfer between access points, but generating a big handoff delay. Another drawback of 802.1x solution is that users (different SLA) cannot be differentiated because the authentication is done on the access points based on the MAC layer. Moreover, access points can't dynamically select the authentication server based on user request but rather are configured to communicate with a fixed authentication server.

3.2 PANA (Protocol for carrying Authentication for Access Networks)

Currently under design in the PANA working group of the IETF [19], its main purpose is to implement Network layer access authentication protocol by defining solutions that are layer 2 agnostic and IPv4/IPv6 compliant. These solutions define a client-server messaging protocol that will allow authentication payload to be carried between the host/client (PaC) and an agent/server (PAA) in the access network for authentication and authorization purposes regardless of the AAA infrastructure that may (or may not) reside on the network. Since its goal is to provide Network layer secure access control by carrying authentication methods, PANA will reuse EAP protocol and its extensions.

PANA protocol brings the advantage of using EAP between users and authentication agent that can be access router or switch, avoiding user re-authentication problem of 802.1x. Service differentiation between users and multiple provider support in hotspot networks cannot be achieved through PANA protocol, because of the lack of EP (enforcement point) provisioning specification in the PANA architecture and the use of the EAP protocol. Some works are currently in progress in PANA working group for definition of EP provisioning (draft) methods using SNMP, COPS-PR, Diameter or Forces. Moreover, PANA deployment may suffer from 802.1x deployment that is actually used in most of the APs (access points).

3.3 LWAPP (LightWeight Access Point Protocol)

LWAPP is an IETF draft standard [20], which was primarily designed by a company named Airespace (Wi-Fi Network Management Company). The main goal of LWAPP is to be a protocol that provides centralized management for access points in 802.11 Networks. LWAPP idea is the following: since an access point has its own IP address and works as an access server, it would be more benefitable if the access point worked as a layer 3 device instead of working only as layer 2 device. Thus all the access points can be controlled (managed) through a level 3 router or console, reducing filtering, policy processing, traffic management, authentication, and encryption needed in an access point. A generic encapsulation and transport mechanism is also provided by LWAPP to enable interoperability between LWAPP management console and the LWAPP access points.

Currently LWAPP is available in an Airespace product called AireWave Director Software, but it is not a real IETF standard yet because it has not reached a state of consensus.

3.4 IPSec VPN Solution

The second solution completely based on IP is currently used by multiple hotspot service providers, which use existing hotspot operator to provide wireless services to their subscribers (e.g. Boingo). With intend to provide this scalable level 3 service, the hotspot operators have to join some service providers by installing "hotspot in a box" package, and mobile users have to install client software that helps them to locate service provider compatible access points. After successful authentication and authorization based on login/password, an IPSec tunnel [21] is created between client and a specific IPSec server.

The advantage of this layer 3 solution for the service provider it that it controls the traffic of its customers since all its traffic goes through its network infrastructure. However the main drawback of this solution is that only the services of a single service provider are supported (not multiple service providers), due to the routes through the IPSec tunnel and due to the packages installed on access points.

3.5 Discussion

In all the above solutions, several service providers in the same hotspot are not supported, there is no guarantee of the quality of service and the mobile users have access to the same level of services. Since service providers sell different types of contracts to their users, differentiating them is a crucial aspect in the hotspots. Service differentiation can be done in IP level (layer 3) by provisioning the SLAs of the users and by configuring dynamically the access routers. SLA provisioning can be done during the authentication and authorization process. But dynamic router configuration is more complex because nowadays routers are provisioned using their CLI (Command Line Interface) and they are monitored using SNMP. These tools are not suitable for the hotspot networks where the users roam frequently, because the associated configurations of the users need to be installed into its access router when a user arrives and they need to be removed when he leaves. For that reason, a new approach is needed to allow automated configuration of the access routers according to the service provider needs, the hotspot configuration and the SLA of the roaming users. This approach is based on policies that are some sets of events, some conditions and their related actions. The new events launch the evaluation of some conditions that entail the execution of the actions.

4. POLICY BASED SOLUTION

The requirements, which are related to user and network management in hotspot with multiple service provider environments, can be easily achieved separately. But providing a solution, which meets all the requirements, remains very difficult, since it involves some new network management architectures. For these reasons, we investigate the use of policy based management approach in hotspot networks in order to offer the hotspot operators and service providers a solution that allows simple, flexible and scalable user management. Based on the use of policies installed on the access router by the service provider and according to user SLA containing allowed services and QoS parameters, the access router configures itself dynamically to ensure the contracted service. In our solution, the entire authentication, authorization and service level provisioning is achieved at the IP layer, by using web-based approach associated to a Radius authentication server [18]. About policies, hotspot operators and service providers can implement their own policies on the hotspot access router according to their contracts. Policies are separated and we assume that no conflict can happen between them since the access router appears as a dedicated router for each service provider. We will first detail the policies implemented on the router. Then we will detail the design of the architecture of the router for multiple policies of many service providers and we will finish with a description of the management of the user accesses into the hotspot networks.

One important point to note is that hotspot operators implement their own policies for managing their user services. Therefore hotspot operators can be seen as a service provider.

4.1 Policy Specification

All policies defined in our solution are described and validated using XML schemas and installed using the two following: CLI (Command Line Interface), Web interface directly or from a remote administration machine.

Policies from our solution are grouped in 2 families: operator policies and customer policies. All policies are defined using an XML scheme.

4.1.1 Operator policies

They constitute the set of hotspot operator and service provider requirements. These policies are handled only by the hotspot operator. There are 2 types of operator policies.

- **Contract policies:** point to the subscribed contract between service providers and hotspot operator. These policies contain parameters related to service provider name, associated bandwidth on the Hotspot, and AAA information.

- **bandwidth policies** : There are two types of bandwidth policies, depending on the contract between service provider and hotspot operator.
  - **Static bandwidth policies :** These policies enable service providers and hotspot operator to specify services and their associated bandwidth. The hotspot operator and the service providers are able to manage their own bandwidth by dividing it between their services and by allowing borrowing between them.
  - **Dynamic bandwidth policies:** a service provider can subscribe a contract with a hotspot operator specifying dynamic bandwidth reservation. This reservation is done according to parameters pushed by the SP when a new user connects to the hotspot. These parameters are carried by the authentication protocol.

4.1.2 Client Policies

They define a set of rules chosen by the service provider administrator

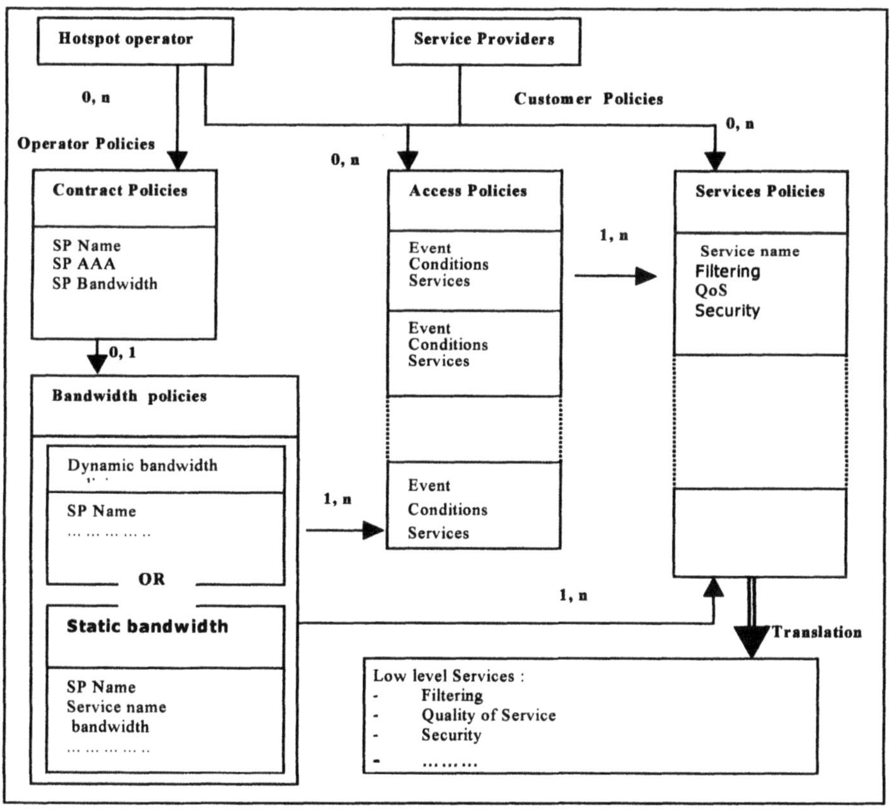

Fig1. Hotspot Policy Specification

and the hotspot administrator in order to manage the client's accesses to the network and to provide them with their subscribed SLA. Two subtypes of policies exist:

- **Access-policies:** materialize the added value that a SP and hotspot operator may offer to their customers. These policies allow dynamic service deployment by providing an admission control mechanism according to the profiles of the users, the number of users on the network, the time, the date and many other parameters. For example, an access control based on the number of users into the network, can be installed in the access equipment using these policies.
- **Service policies:** provide low level service specifications that need to be installed on the access equipment (in our case : the router). This specification gathers quality of service, filtering, security and other parameters.

4.2 Policy Implementation

When a service provider subscribes a contract with a hotspot operator, a new module is instantiated into the router.

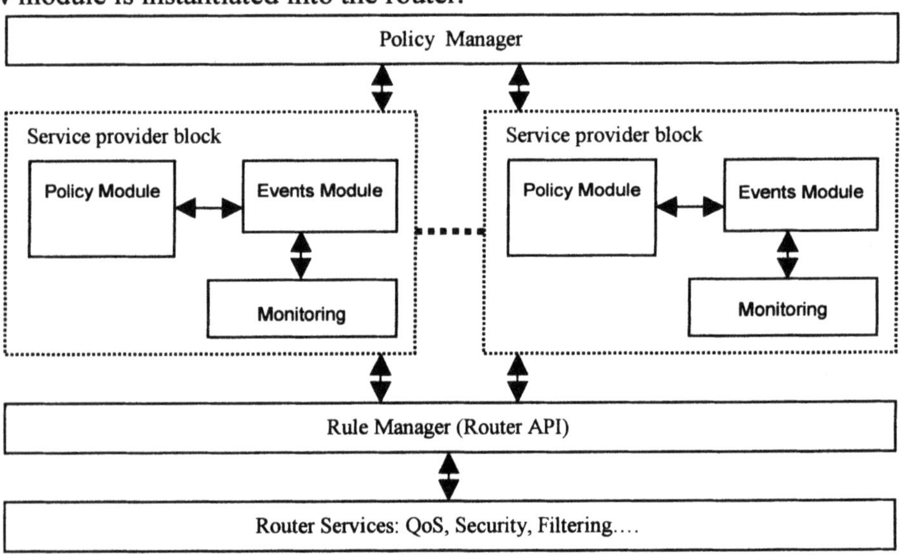

Fig2. Hotspot Policy Architecture

This module is called service provider block. After, the service provider installs its policies on the access router. These policies are received by a module named policy manager which forwards them to the associated service provider block. Policy module stores these policies in tree structure. Policy module checks if the rule can be directly applied. If it's not the case, the policy module notifies the event module that it is waiting for specified event.

Events can be external (e.g. new user) or internal (e.g. QoS parameter) and when they occurred the events module notifies the policy module which will apply the associated policy. To apply a policy, the policy module must send it to the Rule manager. This module will apply the policy by translating it to router rules using router API.

- Policy Manager: ensure policies reception and forwarding to the appropriate service provider block.
- Service provider block: ensure policy storage and enforcement on the router. It is composed of three parts:
  - Policy Module: ensure policies reception, storage and enforcement. It communicates with the events module to get notifications of new events.
  - Events Module: responsible of events management. It notify Policy module when new event occurred and communicate with monitoring module to supervise internal router parameters (security, QoS, filtering….)
  - Monitoring: responsible of monitoring internal router parameters (QoS, security...).
- Rule Manager: apply policies sent by the policy module. It translates them into router rules using router API.
- Router Services: gather all services provided by the router such as security, filtering, quality of service, mobility …

This architecture has been used for implementing policies in case of access control in Hot Spot networks

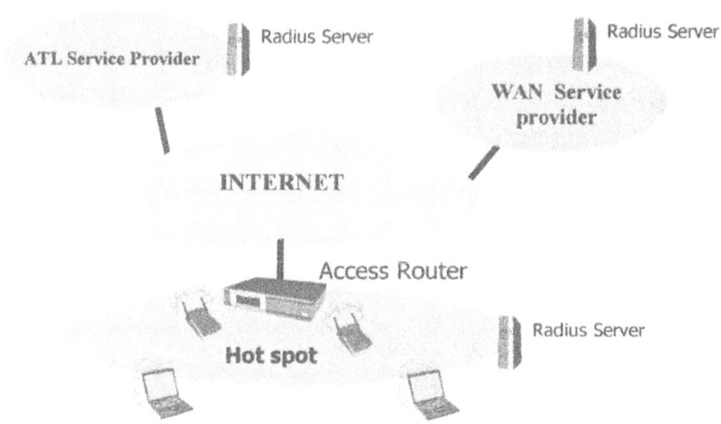

Fig3. Hotspot user management platform

5. ACCESS CONTROL SCENARIO

To illustrate the use of our policy-based solution for user access management in hotspot networks with multiple service providers, we describe a usage scenario. This scenario may be summarized as follows: A hotspot operator called "HSP" have an Internet access of 2Mbps. HSP wants to provide internet access to its own customers and reserve for these 500Kps. Two service providers subscribe contracts with HSP. The first one "ATL" buys a static bandwidth of 500Kbps. The second one "WAN" buys a dynamic bandwidth.

5.1 Access router Configuration with Policies

The policies are installed in the 6WINDGate Access router in order to dynamically react when new users arrive on the Hotspot. The router is basically closed and only flows authorized by the service providers or by the hotspot operator can pass through after authentication. The following policies are installed into the router.

- **Operator Policies:** HSP configure the access router with the following policies
  - Contract policies

```
<sp-information> <sp-name name ="HSP"/>
<sp-aaa   server="10.16.0.142" secret="testing123" port="1812"/>
<sp-bandwidth rate="500kbps"/> </sp-information>
<sp-information><sp-name name ="ATL"/>
<sp-aaa   server="192.17.52.36" secret="testing123" port="1812"/>
<sp-bandwidth rate="500kbps"/> </sp-information>
<sp-information><sp-name name ="WAN"/>
<sp-aaa   server="192.36.96.52" secret="testing123" port="1812"/>
</sp-information>
```

  - Bandwidth Policies
    - Static bandwidth

```
<sp-bandwidth><sp-name name ="ATL"/> <service name ="web"  bandwidth= "380kbps"/>
<service name ="ftp"  bandwidth= "120kbps"/>  </sp-bandwidth>
```

    - Dynamic bandwidth

```
<sp-bandwidth><sp-name name="WAN"/>
<sp-reservation = "dynamic"/> <sp-bandwidth/>
```

- Customer Policies
- Hotspot operator: HSP want to provide only Internet access for a maximum number of 15 users. All users have the same profile which is "guest". The following access and services policies are installed in the access router:

```
<access-policy> <sp-name name ="HSP"/>
 <policy-ident number ="1"/>
 <event type ="new-user"/>
 <conditions>
  <and>
   <type-condition          profile = "guest"/>
   <usernumber profile ="guest" maximum="15"/>
  </and>
  <then>
  <service name = "web">
  </then>
</access-policy>
```

```
<service>
 <service-name name ="web"/>
   <actions>
     <action-ident       number ="1"/>
     <action-type        type ="filtering"/>
     <action-todo        do = "allow"/>
     <action-ipversion   ver ="4"/>
     <action-protocol    pro ="6"/>
     <ipsource           ipsrc ="host"/>
     <portsource      portsrc ="any"/>
     <ipdesttination   ipdest="any"/>
     <portdestination portdest ="80"/>
 </actions>
</service>
```

- ATL: ATL have two types of users: gold, and silver. Gold users can access to web, and FTP. Silver users can access only to web services. Moreover, only 5 gold users have access simultaneously to the WEB and FTP services.

```
<access-policy>
   <sp-name name ="ATL"/>
   <policy-ident number ="1"/>
   <event type ="new-user"/>
   <conditions>
   <and>
      <type-condition profile = "gold"/>
      <usernumber profile="gold" maximum="5"/>
   </and>
   <then> <service name = "web">
      <service name = "ftp"> </then>
   <else>
      <service name = "web">
   </else>
</access-policy>
```

```
<service>
   <service-name name ="web"/>
   <actions>
      <action-ident number ="1"/>
      <action-type       type ="filtering"/>
      <action-todo       do = "allow"/>
      <action-ipversion  ver ="4"/>
      <action-protocol   pro ="6"/>
      <ipsource          ipsrc ="host"/>
      <portsource        portsrc ="any"/>
      <ipdesttination    ipdest="any"/>
      <portdestination portdest ="80"/>

      <action-ident      number ="2"/>
      <action-type    type ="filtering"/>
      <action-todo   do = "allow"/>
      <action-ipversion  ver ="4"/>
      <action-protocol   pro ="6"/>
      <ipsource          ipsrc ="host"/>
      <portsource        portsrc ="any"/>
      <ipdesttination    ipdest="any"/>
      <portdestination portdest ="21"/>
   </actions>
</service>
```

- WAN: WAN also have two types of users: gold, and silver. Gold users can access to WEB, and FTP. Silver users can access only to web services.

5.2 Radius Server Configuration

In the radius server configurations of the hotspot operator and the two service providers, we have added attributes in the parameters related to users. In the three radius servers, we have added the two following attributes for each user:

- POL_PROFILE: specify the user profile (guest: for HSP, gold or silver for ATL, and gold or silver for WAN).
- POL_TIME: for each user we have added an authorized time connection (Guest users are authorized for 20 minutes, ATL: gold users for infinite time and silver users for 3 hours, WAN: gold users for infinite time and Silver users for 1 hours).

Since WAN has a dynamic bandwidth subscription, we have added bandwidth attributes for each user in the radius server of WAN

- POL_SERVICE: WEB or FTP
- POL_BANDWIDTH: 20Kbps for Web and 15kbps for FTP.

5.3 How does it work?

- **Bandwidth reservation:** in the access router, the service provider ATL has a class of service of 500kbps which is divided into two classes: web class with 380kbps and ftp class with 120 kbps. HSP a class of service of 500kbps which is reserved for internet access for HSP customers. The 1MBps remaining bandwidth is used by HSP for its own services, its contract with WAN and for possible new contracts with other service providers.
- **Filtering:** Initially, only DNS request can go through the access router and all other flows are forbidden. This is done by setting firewall rules in the 6WINDGate router.
- **Authentication and authorization of users:** Authentication is done using the https protocol combined with radius protocol. This is achieved thanks to web portal and radius client embedded in the access router. The benefits of this solution are that no specific configuration is required on the machines of the users. Moreover the web browsers, which support the HTTPS protocol, are universally available. In this web portal, HSP, WAN and ATL appears offering the user possibility to choice its service provider.
  - When new user arrives at the hotspot, he/she obtains an IPv4/IPv6 address using stateless or statefull configuration mechanism. Statefull mechanism is achieved through DHCPv4 server, and DHCPv6 server located in the access router. IPv6 stateless mechanism is realized thanks to router advertisement messages sent by access router.
  - When the user activates Internet browser, automatically the web portal embarked in the router is displayed. The user must then insert its login, password and choice its service provider. Once this information is validated, the router sends a request to the radius server to authenticate him. The radius server responds with accept or reject.

- **Service deployment:** according to service provider, and radius response containing the user's SLA, the policies installed in the router are dynamically translated into router rules allowing users to access its contracted services. These rules are based on the policies and the IP address of the user
- **Time management:** Users are automatically disconnected from the network when their authorized time duration expired. Associated filtering and quality of service rules are dynamically deleted.
- **Data volume management:** If users are allowed to certain amount of traffic volume, we install with the filtering rules traffic conditioners. These conditioners are automatically removed when the data volume is reached and dynamically filtering rules are also removed.
- **Accounting:** Once the user is disconnected by itself or by the access router, an accounting message is sent to the radius server containing time connection duration and also data volume.

6. CONCLUSION AND FUTURE WORKS

In this paper, a new network management architecture for hotspot network has been overviewed. The lack of solutions that allow multiple service provider support, service guarantee per user and per SP, and service differentiation led us to propose this architecture. Our solution allows service providers and hotspot operators to get benefit from the large deployment of public Wireless LAN. This solution is based on the use of policies in access router, which provide high level configuration tool and dynamic router behavior according to service providers criteria and users contracted services. Secured Authentication and authorization, dynamic service deployment, quality of service guarantee and different billing schemas are managed in one way offering thus simple, flexible and scalable tool for both hotspot operators and service providers. Because of the IP based, our solution can work over different air interfaces, across wireless LAN cards from different vendors and it does not require any modification to layer2. Moreover, it can be extended to different access technologies such as PLC, DSL...

We are currently working on the deployment of this solution in the context of INFRADIO project [22]. It is a RNRT project that aims to deploy a large IPv6 WLAN into the Paris6 University and the ENST Paris. Another work we will investigate is related to the study of combining IPSec security protocol with our solution in order to provide hotspot Security.

7. REFERENCES

[1]IEEE. 802.11b/d3.0 Wireless LAN Medium Access Control (MAC) and Physical Layer (PHY) Specification, August 1999.

[2]Junbiao Zhang and al, "Virtual Operator based AAA in Wireless LAN Hot Spots with Ad-hoc Networking Support", Mobile Computing and Communications Review, Volume 6, Number3.

[3]Terry Schmidt and Anthony Townsend, "Why WI-FI Wants to be free", Communications of the ACM, Vol. 46, N° 5, May 2003.

[4]Joseph W. Graham II, *"Authenticating Public Access networking"*, SIGUCCS'02, November 20-23, 2002, Providence, Rhode Island, USA.

[5]Upkar Varshney and Ron Vetter, "Emerging Mobile and Wireless Networks", Communications of the ACM, Vol. 43, N°. 6, June 2000.

[6]Rajeswari Malladi and Dharma P. Agrawal, "Current and Future Applications of Mobile and Wireless Networks", Communications of the ACM, Vol. 45, N°. 10, October 2002.

[7]IEEE Daft P802.1X/D11: Standard for Port based Network Access Control, LAN MAN Standards Committee of the IEEE Computer Society, March 27, 2001.

[8]Pekka Nikander, "Authorization and charging in public WLANs using FreeBSD and 802.1x", USENIX annual technical conference, June 10-15 2002.

[9]Jim Martin, and Arne Nilson, "On Service Level Agreements for IP Networks", IEEE Infocom Conference, June 2002.

[10]S. Salsano et al., "Definition and usage of SLS in the AQUILA Consortium", Internet Draft, November 2000.

[11]Bob Moore, Ed Ellesson, John Strassner, and Andrea Westerinen, "RFC 3060: Policy Core Information Model – version 1 Specification". IETF, February 2001.

[12]J Jason, L Rafalow, and E Vyncke, "IPsec Configuration Policy Model", Internet draft, November 2001.

[13]Y Snir, Y Ramberg, J Strassner, R Cohen, and B Moore, "Policy QoS Information Model", Internet draft, November 2001.

[14]Raj Yvatkar, Dimitrios Pendarakis, and Rocj Guerin, " RFC 2753: A Framework for Policy-Based Admission Control". IETF, Informational, January 2000.

[15]A.Westrinen and al, "RFC 3198: Terminology for Policy Based Management ", IETF, November 2001.

[16]David Kosiur, "Understanding Policy-Based Networking". Wiley Computer Publishing, 2001.

[17]L. Blunk and J. Vollbrecht, "RFC 2284: PPP Extensible Authentication Protocol (EAP)". IETF,March 1998.

[18]C. Rigney, S. Willens, A.Rubens, and W. Simpson, "RFC 2865: Remote Authentic ation Dial in User Service (Radius)", IETF, June 2000.

[19]Alper E. Yegin, Yoshihiro Ohba, Reinaldo Penno, George Tsirtsis ,and Cliff Wang, " Protocol for Carrying Authentication for Network Access (PANA) Requirements", Internet Draft , June 2003.

[20]P. Kalhoun and al., "Light Weight Access Point Protocol", Internet Draft, June 2003.

[21]S.Kent, and R. Atkinson, " RFC 2401: Security Architecture for the Internet Protocol",IETF, November 1998.

[22]INFRADIO Project : http://rp.lip6.fr/infradio/

[23]Charny, B. (2002c). Want Wi-Fi? Verizon takes it home. *CNET News.com*, October 9.

[24]A.Mahler and C.Steinfield The Evolving Hot Spot Market for Broadband Access "ITU Telecom World 2003 Forum panel on Technologies for Broadband, Geneva, October 2003"

KEY MANAGEMENT FOR SECURE MULTICAST IN HYBRID SATELLITE NETWORKS

Ayan Roy-Chowdhury and John S. Baras
Department of Electrical and Computer Engineering and
Institute for Systems Research
University of Maryland College Park, MD 20742, USA
ayan@umd.edu; baras@isr.umd.edu

Abstract This paper proposes a design for key management for secure multicast in hybrid satellite networks. Communication satellites offer an efficient way to extend IP multicast services for groups in wide-area networks. In order to be commercially viable, the multicast traffic should be accessible only to paying subscribers. Access control can be achieved by data encryption. This requires secure and efficient methods to generate, distribute and update the keys. Most current key management protocols do not scale well when applied to large dynamic groups in wide-area networks. This paper attempts to solve the above problem for groups in a hybrid network that is composed of terrestrial Ethernet LANs interconnected by ATM-based satellite channels. We investigate current group key management protocols, and design a framework for secure and scalable key management for the multicast routing architecture in the satellite network. The proposed framework is presented in detail, alongwith analysis and simulation results.

Keywords: Satellite network, secure multicast, group key management.

1. INTRODUCTION

Multicasting is a network-layer mechanism for one-to-many or many-to-many communication that is efficient in terms of usage of network resources. With the growth of the Internet, web applications using Internet Protocol (IP) based multicast routing protocols are becoming increasingly popular. Examples are webcasts, video and voice conferencing and Internet gaming.

Satellite networks offer a natural method to extend the multicast services in wide-area networks where the sources and recipients are widely separated from one another. There is, however, little support today for IP multicast services over satellites. Apart from the problems involved in creating an efficient routing mechanism for satellite multicast, another major challenge is to secure the multicast data in the satellite network. The IP multicast paradigm allows free access to the multicast data to anyone interested in receiving it. However, in order for a multicast service to be commercially viable, access to the multicast data should be restricted to paying or authorized receivers. Access control can be achieved by means of encryption - the source encrypts the application content using a key; the decryption key is distributed to all authorized receivers. The mechanism of key distribution is challenging when the set of authorized receivers changes dynamically with time. The design problem becomes more complex when we consider large groups of the order of thousands of members, spread over a wide geographical area, as might be the case for satellite networks.

In [Roy-Chowdhury, 2003] we have proposed a multicast routing architecture that scales to large groups in a wide-area hybrid satellite network. In this paper we address the problem of group key management for secure multicast in the above network. We propose a framework for secure key management for groups operating in this network, with the primary objective of minimizing the communication over the satellite links, and present simulation results to demonstrate the feasibility of the proposed framework.

The rest of the paper is organized as follows. We review current proposals for group key management in section 2. Section 3 describes in brief the network architecture. The design of the key management framework is given in section 4. Various analyses of the framework are in sections 5, 6. We describe our simulation and results in section 7. We conclude in section 8, highlighting future research directions.

2. REVIEW OF KEY MANAGEMENT PROTOCOLS

We describe in brief some of the fundamental ideas presented in group key management. A more detailed analysis of the protocols presented here can be found in [Roy-Chowdhury, 2003].

Most of the protocols proposed to date fall in two categories: centralized key distribution schemes and distributed key generation schemes. In centralized key distribution, there is a centralized key controller to whom all members send join and leave requests. The key controller is fully trusted and is responsible for key generation and distribution to the group members, and for key updates, triggered periodically or on membership changes. The centralized schemes provide a higher degree of security and are more efficient. Their major weakness is the dependence on a central entity, which can be a single point of failure. The Key Predistribution System, proposed in [Matsumoto and Imai, 1988], and the Broadcast Encryption scheme proposed in [Fiat and Naor, 1994] are examples in which the key controller pre-computes the group keys for all possible groups. In these schemes, the memory requirements can become prohibitively high for large groups. Another category of centralized schemes are the *threshold encryption* protocols, such as [Berkovits, 1991]. They require collaboration between participants (who might not know each other, as in IP multicast), and might have high storage requirements for large groups. Secure Lock is a secure broadcasting scheme proposed in [Chiou and Chen, 1989]. Here the number of key encryptions done at the centralized controller increases linearly with the number of group members. The system is one-to-many, and cannot be used if there are multiple sources. Another one-to-many system is the Conditional Access System (CAS) [Kim et al., 1996], which is popular for data confidentiality in satellite broadcasts. Group Key Management Protocol (GKMP) [Harney and Muckenhirn, 1997] has been proposed for groups with multiple sources and receivers. In GKMP, the communication overhead in sending messages for the initial system setup, and key update messages on member leaves, is high for large groups.

In distributed key generation schemes all the group members (or a chosen subset), contribute shares in a round of message exchanges to generate a common group key. Key agreement for secure multicast using hidden fractional keys (HFK), proposed in [Poovendran, 1999], is an example. The scheme does not handle membership changes well. A suite of protocols have been proposed in [Steiner et al., 2000] for fully distributed group key agreement. In this system, the computational burden on each entity for generating the group key, can be prohibitively high for large groups. Both the preceding protocols, and distributed schemes in general, require that all members participating in the key setup are aware of one another, and can send messages in order to the others. This is not necessarily true in several group situations, such as IP multicast. Also, the protocols in general suffer from a high overhead in communication for key generation when applied to large groups. Several other distributed protocols have been proposed in [Burmester and Desmedt, 1994; Steer et al., 1990]; all are susceptible to similar inefficiency problems in large groups.

A family of protocols have been proposed for key management based on logical trees, originally in [Wong et al., 2000; Wallner et al., 1999]. The original protocol is called the Logical Key Hierarchy (LKH). The centralized logical tree-based protocols have a group controller (GC), who constructs a logical key tree with the group members at the leaf nodes of the tree. The internal nodes of the tree are usually logical nodes and correspond to the key encrypting keys (KEK) which are used to securely transport key updates to the group. The root of the tree is the session key or traffic encrypting key (TEK). The key tree protocols have logarithmic communication, storage and computation complexity. These protocols scale very well to large groups. Their primary drawback is the use of a centralized GC; [Rodeh et al., 2000] has suggested a distributed version that does not have a GC. Various modifications to the original LKH protocol have been made that try to reduce the communication and computational complexity, for example, [Canetti et al., 1999; Perrig et al., 2001].

3. NETWORK ARCHITECTURE

The network architecture is given in figure 1. We consider a group of terrestrial Ethernet-based networks geographically separated and spread over a wide area. We term them the "subnetworks" in our overall network. Each subnetwork has one or more satellite gateways, which interconnect the subnetworks via ATM-based satellite links using a geostationary satellite.

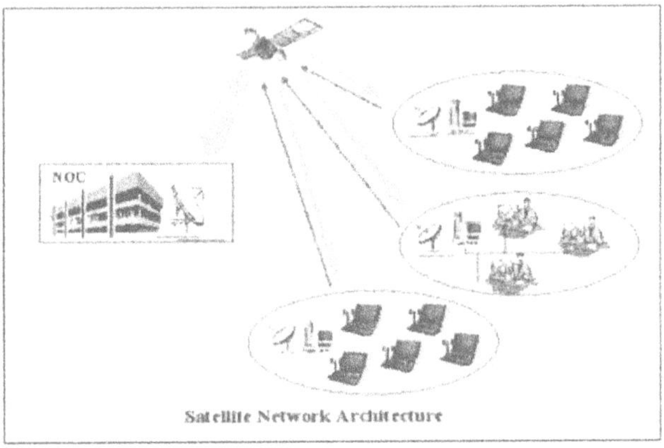

Figure 1. The Network Architecture

In [Roy-Chowdhury, 2003], we have described a routing architecture whereby sources and receivers spread across different subnetworks can form an IP multicast group. The IP multicast framework has two components: the multicast routing within a subnetwork is based on *Protocol Independent Multicast - Sparse Mode* [Deering et al., 1996], while the multicast routing between the subnetworks over the ATM satellite links uses the ATM *Multicast Address Resolution Server* (MARS) with VC mesh architecture [Armitage, 1997]. A satellite gateway router in each subnetwork acts as the root of the multicast tree within its subnetwork. This router is known as the Rendezvous Point (RP). The MARS is used for address mapping for IP multicast over the ATM links. It is located at the Network Operations Center (NOC). The key management framework proposed in this paper builds on the multicast routing architecture.

4. TIERED KEY MANAGEMENT IN SATELLITE ATM NETWORK

The primary metric that we consider for our design is the communication overhead in the network. The propagation delay in the geostationary satellite links is high, of the order of 250ms in one hop. The uplink bandwidth is limited to 1.5Mbps. Also, geostationary satellites operating in the Ka-band are prone to channel errors due to atmospheric conditions such as rain fade. We therefore need a key management scheme that minimizes the communication over the satellite links, to reduce the delay in group initialization or key updates, and also to minimize the possibility of error conditions where the group keys do not reach all the members due to channel conditions. We assume that the hosts in each terrestrial network have significant processing power and memory capacity, similar to the workstations and personal computers prevalent today. Hence computation and storage are not critical issues.

The hierarchical structure of the network creates two distinct levels in the network - the terrestrial subnetworks, and the satellite connections between the subnetworks forming an "overlay", as shown in figure 2. We consider the different subnetworks to be independent domains, such as company networks, which might follow different security policies. Reconciling the security policies across the subnetworks to build a single key management framework would be a difficult task. Also, a single key framework would suffer from the *1-affects-n* scalability problem [Mittra, 1997] and the probability of updates in the keys stored at a member would be much higher due to the dynamics of member joins and leaves overall. Join or leave of a member in any subnetwork would trigger updates even in remote subnetworks, which is highly inefficient. The key management communication over the satellite links would hence be frequent, which is undesirable from our perspective due to the reasons mentioned above.

We therefore divide the key management into two tiers - one at the subnetwork level, while the other at the level of the satellite overlay (the concept of dividing a system into subgroups for scalable key management was originally proposed in Iolus [Mittra, 1997]). The key generation and distribution in each subnetwork is independent of one another, and also of the key generation and distribution in the overlay; we add mechanisms so that the encrypted data can be transferred securely across the different key management areas. The key management in each logical group is based on centralized key trees, which is selected due to its scalability to large groups, in terms of the communication required to initialize and update the group keys. The framework therefore has two tree levels: a global *RP Tree* for managing the keys between the subnet RPs in the overlay; and the local *SN Tree* for managing the keys amongst the hosts in each subnet. Each subnetwork has its own SN Tree. We term this framework, *Tiered Tree Based Key Management*.

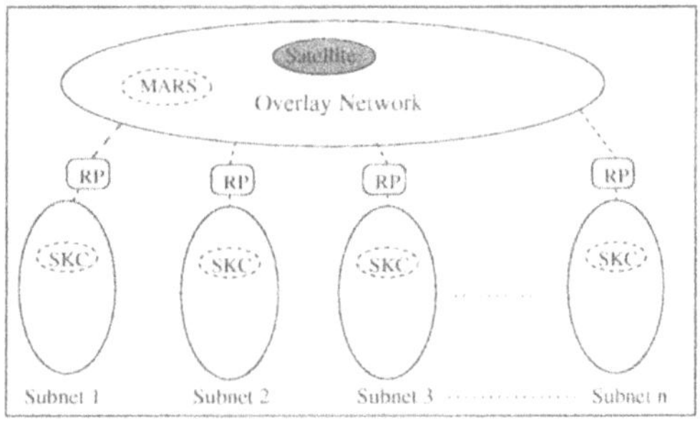

Figure 2. Logical Grouping in the Satellite Network

4.1 Trust Model and Security Assumptions

The network entities that are relevant in the security framework are the ATM multicast server (MARS), the Rendezvous Points and Key Server in each subnetwork and the end-hosts.

In the routing framework, the MARS maintains the database of multicast group membership at the subnetwork level. It periodically sends the group membership information to all the RPs that are subscribed to the group. We envision the MARS, located at the NOC, to be owned by the network service provider, and it keeps track of the different customers (which can be each subnetwork) who are using the network services provided. The customers would prefer to keep their traffic confidential and not allow the network provider read the transmissions. Therefore in the security framework, we model the MARS as the trusted third party for performing access control, based on group policy, for different subnetworks that want to join or send to a given multicast group. In addition, the MARS acts as a Certificate Authority (CA) for verifying the public keys of the RPs when needed. However, the MARS is not trusted with the multicast traffic. The MARS should not receive the application data (unless it explicitly subscribes as a member to the multicast group).

The RP in each subnetwork is located at the satellite terminal, which is the communication gateway to/from all entities outside the subnetwork. In the security design, the RP is trusted to securely transmit multicast data from local sources to remote subnetworks over the satellite links, and to receive multicast data from remote sources in case there are local receivers. However, the RP is not trusted to read the multicast traffic. We place this limitation since the satellite gateway in each subnetwork would usually be owned by the network provider, who might not be authorized to read group information as discussed previously.

The end-hosts are trusted to securely encrypt or decrypt the multicast traffic.

Since we choose centralized key management schemes, we need a group controller/key server for the hosts in a subnetwork. The RP would have been a good candidate for the key server since it is the root of the multicast tree in its subnetwork. However, as stated above, it might be undesirable to allow the RP to read the group traffic, which it can easily do if it generates the encryption/decryption keys for the hosts. Therefore the security framework introduces a key server in each subnetwork, distinct from the RP, and responsible for managing group keys in its subnet. It is termed the *Subnetwork Key Controller* (SKC). The SKC does access control operations on local group members, and performs key generation, distribution and periodic key updates for all groups that have members in its local subnet. Each end-host and the RP is assumed to *apriori* establish a *secure channel* to the SKC for receiving the key information.

In addition to the above, we make the assumption that the IP/ATM multicast routing is secure.

4.2 Key Management in the Overlay: RP Tree

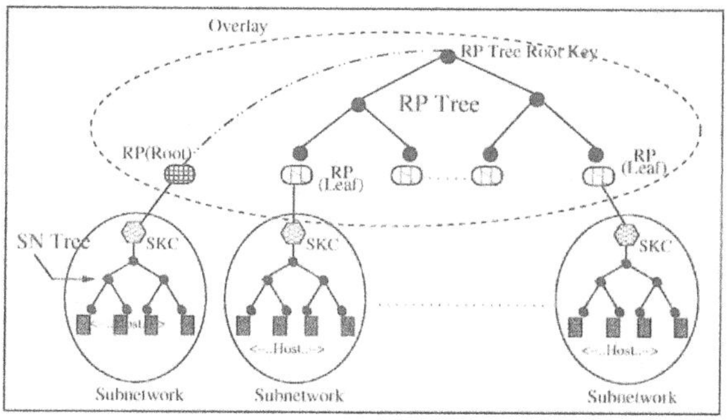

Figure 3. RP Tree and SN Tree

Figure 3 illustrates the key trees for the overlay and each subnetwork in our framework. The logical key tree in the overlay is the RP Tree, that in each subnetwork is the SN Tree. The RPs in different subnetworks are located at the leaves of the RP tree. The root of the RP tree for any group is one of the RPs in the group, while the intermediate nodes as shown in figure 3 are logical nodes which correspond to the KEKs in the key tree. Likewise, in the SN Tree, the SKC is the root, while the hosts are at the leaf nodes of the tree. The RP Tree is described below, and the SN Tree is described in the next section.

RP Tree Setup

The RP tree is constructed by making additions to the MARS message exchange protocol, which is described in [Armitage, 1997].

Sender RP Request: When a RP has hosts downstream who want to send to group G, the RP sends a request message to the MARS for the list of group members. If the MARS database has a non-empty entry of RPs subscribed to

G, the MARS adds the requesting RP to the entry, and returns the details of the entry to the requesting RP in a reply message. The reply message is broadcast to all RPs in G present in the MARS entry at that time. The message has the IP address and public key of each valid RP, and the address of the RP Tree root. If MARS has no entry for G (i.e., the requesting RP is the first to join G at the MARS), then MARS creates a new entry for G, adds the requesting RP ATM address to the entry, and sends a negative acknowledgment in reply.

Receiver RP Join: When a RP has hosts in its local subnetwork requesting to join a group G as receivers, the RP sends a join request to the MARS. The MARS adds the joining RP's address, public key to the database entry at MARS for group G. If the entry does not exist, then a new entry is created. Subsequently the MARS broadcasts the list of RP group members in a regular membership update message to all the RPs subscribed to G. The public key of the RPs are needed to bootstrap the RP Tree, since we do not assume that the different RPs have secure associations between one another established previously. Once the multicast tree and RP key tree state is created in local memory, for subsequent join or send requests from downstream nodes, an RP does not send MARS requests.

Selection of the RP Tree Root: The root of the RP tree is selected to be the sender RP that is the *earliest to join* the group amongst the sender RPs in the MARS database entry. The selection is done by the MARS based on the join time in the requests it receives. The address and public key information of the root RP becomes known to all the group RPs from the MARS message they receive. In case the root RP leaves the group, the MARS checks the joining times of the remaining sender RPs, selects the earliest-to-join, and broadcasts a new message to the group. The RPs update their local group security information upon receiving the MARS message.

Tree Setup at the Root: When a sender RPs receives the MARS message, it checks whether it is the root. If so, it proceeds to set up the logical key tree in its local node. The information about the leaves of the key tree are obtained from the MARS message.

Key Information Transmission: Once the RP tree has been setup at the root, the root creates one message containing all the keys of the RP tree, encrypted as appropriate, and broadcasts the message over the satellite links to all the other RPs in the group. We assume that the root RP has no prior secure association with the leaf RPs. So when the RP tree is created, the initial communication from the root to the leaf RPs are encrypted with the public keys of the leaf RPs. Upon reception, each leaf RP decrypts its relevant key information using its private key, and obtains all the keys on the path from its leaf to the root of the tree. The key corresponding to the tree root is now used as the session key. Subsequent communication from the root RP to the leaf RPs uses the shared long-term secret, corresponding to the leaf node of each receiver RP, that is sent by the root in the initial communication.

RP Tree Update on Member Join, Leave
When a RP wants to join an existing group as a member, it sends a join request to the MARS. The MARS adds the RP to the group entry. When a *leaf* RP leaves a group it sends a leave request to the MARS for the group. The MARS ensures that the leaving RP is not the RP tree root and removes the RP information from the group entry. The join or leave message is retransmitted to the existing group members to update them about change in the group membership. On getting the join/leave message, the root RP updates the keys in the RP tree as required, and sends the updated keys to the affected group members. When the root RP sends a leave message, the MARS removes the root from the group entry; runs the algorithm to select a new root RP; creates a new update message and immediately sends the update to the remaining group members. The new root, upon receiving the update message, proceeds to create a new RP tree. Till the new tree is created, the group information is secured using the existing session key. The drawback is that the old root RP can still receive all the information, but it prevents "blackout periods".

In case the multicast group has only one sender RP (the root) (in situations where there is only one source host, or all the sources are concentrated in the same subnet), the root RP leaving implies there are no sources left. The MARS on getting the leave message cannot locate a new root and hence does not send out a new update message. The group entry will be erased from the MARS database on a timeout.

4.3 Key Management in the Subnetwork: SN Tree

The key server in each subnet, known as the Subnetwork Key Controller (SKC), manages the subnetwork key tree (SN tree). We assume that the security module in all hosts and the RP are aware of the address of the SKC.

SN Tree Setup: When an end-host wants to join a multicast group G as a receiver, or intends to send to a multicast group as a sender, it first sends a *join request* message to the SKC specifying the IP address of G. In the subnet, the SKC does not differentiate between a sending host and a receiving host.

When the SKC receives a join request, it checks its local database for an entry for the group. If none exists, the SKC creates an entry and the corresponding key tree. The SKC also generates a *datahiding key* for the group. The datahiding key for group G has to be identical across subnetworks; the SKC in a subnetwork has to contact the SKCs in other subnetworks (that have members in G) to agree on the datahiding key for G. The datahiding key is long-term; once created, it does not change for the lifetime of group G, despite member joins and leaves. The SKC assigns the joining host to a leaf in the tree.

It then encrypts all the keys in the path from the leaf node to the root and the datahiding key using the long-term secret it shares with the joining host; it also encrypts only the session key for the RP. The SKC then forms a *key information* message containing the encrypted keys, and transmits the key information message to the host and the local RP. The host decrypts the tree keys and group datahiding key and stores them in local memory. The RP decrypts the session key, creates an entry for the group in local memory, and stores the session key in the entry.

When there are existing group members, or multiple members joining simultaneously, the message will contain all the relevant tree keys encrypted for all affected members.

SN Tree Update on Member Join: When one host sends a join request for group G to the SKC, the controller adds the host to the key tree following the standard procedure for adding group members in LKH, and sends the updated group keys to all the members. The local RP is also informed about the update in the session key. The new member gets all the keys in the path from its root to the leaf in the SN Tree, and also the datahiding key. For multiple members joining simultaneously, the sequence is similar, with the added processing at the SKC to find the minimum number of valid KEKs to send the update information.

SN Tree Update on Member Leave: When a member leaves, all the keys on the path from the member leaf to the root are invalidated. The SKC generates new keys in replacement, and sends the fresh keys to all affected members, and the RP. For bulk member revocation, the SKC has to identify all the invalid keys, and find the minimal number of valid keys that are required to transmit the updated keys.

Synchronization of Group Information at the RP: At all times, the RP maintains integrated state information for a group. When the RP is a leaf of the RP tree, the group entry in its local memory specifies it is a leaf, and contains the path keys to the root of the RP tree, and also the local subnetwork session key. If a leaf RP becomes a root (in situations where the previous root RP has left the group), then a root entry is created. The subnetwork session key is transferred from the leaf entry to the root entry. Note however, a root RP for group G does not become a leaf RP for G at any time when it is continuously subscribed to G.

4.4 Secure Data Transmission in a Group

Multicast traffic can be transmitted securely when the SN trees and the RP tree have been established. The sequence is described here.

1 Source host j in subnetwork i, a_{ij}, encrypts the data m for group G twice: first using the datahiding key DK_G to produce ciphertext $C =$

$E_{DK_G}(m)$. The encrypted data is re-encrypted using the subnetwork session key SK_{G_i} to produce ciphertext $\hat{C} = E_{SK_{G_i}}(C)$.

2 a_{ij} sends the doubly-encrypted data to the local multicast tree and the RP.

3 The group members a_{ik} in the local multicast tree decrypt $\hat{C}$ to retrieve the multicast traffic: $C = D_{SK_{G_i}}\left(\hat{C}\right)$, $m = D_{DK_G}(C)$.

4 The RP decrypts $\hat{C}$ to obtain C. It cannot decrypt C to get m, since it does not know DK_G. The RP re-encrypts C with the RP tree session key $SK_{G_{RP}}$ and transmits the ciphertext $\hat{C}' = E_{SK_{G_{RP}}}(C)$ to the other subnetworks over the satellite link.

5 RP_j in subnetwork j receives the encrypted transmission. It decrypts $\hat{C}'$ to obtain $C = D_{SK_{G_{RP}}}\left(\hat{C}'\right)$. RP_j cannot decrypt C since it does not know DK_G. It re-encrypts C using the local subnetwork session key SK_{G_j} for G to generate ciphertext $\hat{C}'' = E_{SK_{G_j}}(C)$; RP_j sends $\hat{C}''$ along the multicast tree in its subnet.

6 Each host a_{jk} in subnetwork j subscribed to G receives $\hat{C}''$. It decrypts the ciphertext using SK_{G_j} to obtain C. a_{jk} decrypts C using the datahiding key DK_G to obtain m: $m = D_{DK_G}(C)$.

5. SECURITY ANALYSIS

5.1 Passive Adversary

SN Tree: Let A be a passive adversary, who is never a group member. We assume A eavesdrops on all traffic in an arbitrary subnetwork and receives all the encrypted key information and data packets. A cannot decrypt the data packets, since it does not know either the subnetwork session key or the datahiding key. A brute-force attack to find the group key takes $\Omega\left(2^k\right)$ operations where k is the length of the group key. A cannot do better than this, since it does not know any of the KEKs in the tree.

RP Tree: We assume A has the capability of listening to the satellite traffic and receives all the traffic in a complete session, that is, A can be a passive eavesdropping RP. A still cannot decrypt the encrypted traffic, since it does not know the RP session key. It cannot obtain the session key from the RP tree key messages, because it does not have any of the keys used to decrypt the key messages.

MARS: If the MARS is a passive adversary, then under normal operation of the network, the multicast traffic will not reach it at all, since the routing path from a source RP to the set of receiver RPs will not include the MARS.

5.2 Active Adversary

SN Tree: Let B be an active adversary, who has been a group member during some previous time period. In the key management protocol, when B joins the group in any subnet, it cannot derive any previous group key by doing better than exhaustive search, i.e., $\Omega\left(2^k\right)$ operations. Even if B has listened to and stored past group traffic, it cannot obtain any of the decryption keys for the previous enciphered messages. The only keys it gets are the updated keys that are sent to it by the SKC.

Assume B leaves the group and tries to read the group traffic after it has left. B has with it the set of keys on its key path, and the datahiding key. However, it cannot read the group traffic at a later time, since the key server updates all the keys on the key path that B knows, including the session key, and securely transmits the updated keys to the other members using long-term keys that B does not know. The datahiding key does not change. But this does not help B since it first needs to decrypt using the current session key, which B does not possess.

RP Tree: Let B be an RP who was a member of the group at some previous time. Before B had joined the RP tree, it could not decrypt the data traffic since it did not know the group key at a previous time instant. After B joins the RP tree and collects the keys in its key path, it leaves. Once B has left, the root of the tree (assuming B was not the root), updates all the keys in the RP tree known to B, including the RP session key. B cannot obtain the updated keys from the key message since it does not know the decryption keys used to send the updated keys to the other RPs.

The only time when B, as an RP, could read the data after leaving, is if B was the immediate previous root of the RP tree. Then for the interval of time it takes the new root to create a new tree, the group traffic would continue to be encrypted using the old RP Tree session key, allowing B access to the data. However, B can obtain only the ciphertext of the data, encrypted with the datahiding key, which B does not know.

MARS: The MARS can join a multicast group by adding its ATM address to the list of addresses for the multicast group, and sending the list to the source RPs. The routing paths created by the source RPs will then include a branch to the MARS. Subsequently the MARS will receive all the key traffic on the RP tree, and all the encrypted multicast traffic. But even in this situation, the MARS will not be able to read the multicast data, because the multicast traffic is first encrypted with the datahiding key, to which the MARS does not have access.

	RP Root	SKC
Tree setup	$(n_1 - 1)\, k_p + \frac{d_1(n_1-1)}{d_1-1} k_s$	$\left(n_2 + \frac{d_2(n_2-1)}{d_2-1} + 1\right) k_s$
Member join to existing group in subnet	0	$(d_2 h_2 + 1)\, k_s + k_s$
Adding a subnet to existing group	$(d_1 h_1 + 1)\, k_s + k_p$	$\lceil \left(n_2 + \frac{d_2(n_2-1)}{d_2-1} + 1\right) k_s \rceil$
Evicting a member from subnet	0	$(d_2 h_2 - 1)\, k_s$
Evicting a subnet	$(d_1 h_1 - 1)\, k_s$	0

Table 1. Communication Cost in Tiered Tree Based Key Management with LKH algorithm.

RP root	SKC	RP	Member
$\lceil \frac{(d_1 n_1 - 1)}{d_1-1} k_s + n_1 k_p \rceil$	$\lceil \frac{(d_2 n_2 - 1)}{d_2-1} k_s + 2 \rceil$	$\lceil h_1 + 2 \rceil$	$\lceil h_2 + 2 \rceil$

Table 2. Storage Cost in Tiered Tree Based Key Management with LKH algorithm.

6. COST ANALYSIS

We compute the cost for communication and storage for the basic key tree scheme: LKH in the overlay and in each subnet. The results are derived by applying the cost metrics of the basic LKH to the RP tree and the SN tree, and by aggregating the two. Table 1 shows the communication overhead for the RP tree and SN tree individually, while Table 2 gives the total storage cost in the framework, using basic LKH algorithm. n is the total number of members in the group; n_1 is the number of RPs, n_2 is the number of members in each subnet; d_1, h_1 are respectively the degree and height of the RP tree; d_2, h_2 are respectively the degree and height of the SN tree; k_p is the length of a public key and k_s is the length of a symmetric key. The figures for the communication cost are only approximate; we do not rigorously consider the fact that the root of the RP tree itself is a group member. The storage costs consider that the RP root stores the public keys of all subscribed RPs, though the public keys are not needed except for the initial setup.

A comparison of the cost in the proposed framework, with respect to the protocols mentioned in section 2, can be found in [Roy-Chowdhury, 2003]. The comparison shows that in most of the cases the proposed framework fares at least as well or better than the other protocols for various metrics.

7. SIMULATION

We have verified the validity and feasibility of our framework through simulations using OPNET Modeler 9.0 [Opnet, 2002]. We used the multicast simulation setup from [Roy-Chowdhury, 2003] and added our security features to it.

We consider three multicast groups in the network, each spread across 31 subnetworks. Each group has 10 sources in 10 subnetworks, and 1075 receivers spread across all the subnetworks. For encryption, we simulate use of 64 bit symmetric keys and 1024 bit public keys. The simulation was run for 300 seconds.

MARS selected 3 different RPs as the root of the RP trees for the three groups. These RPs are leaves in the RP trees for the groups for which they are not the RP tree root. Thus in our framework, the key management in the overlay can be distributed among different RPs for different groups.

The savings in terms of bytes of key information sent per second is illustrated in figure 4, which compares the total key information sent for all the groups in the RP trees and all the SN trees, to the total key information sent on the RP trees only. As the graph shows, the resource savings on the satellite links is substantial using the tiered tree scheme. Even though the group dynamics are high, the amount of message exchanges are very few in the RP tree. This is because the RPs remain subscribed to the RP tree as long as there is at least one member in its local subnetwork sending to or receiving from the group; the frequency of joins and leaves in the subnetwork is transparent to the RP tree. This is precisely our intention, to minimize the cost of message exchanges over the satellite links. The figure also illustrates another important point of our key management scheme, namely, *1-affects-n* scalability. The effect of frequent member joins and leaves in one subnetwork remains localized within the subnetwork, and does not affect the group dynamics in other subnetworks.

8. CONCLUSION

In this paper we have proposed a framework for key management for secure multicast in a wide-area hybrid satellite network. Our design is scalable and efficient and well suited for the unique network architecture that we consider. The framework is essentially a generic design; different types of key management algorithms can be applied in each logical grouping. We considered tree based algorithms due to their scalability and robustness for large groups. However, if the subnetworks in a group are limited and remain static, then GKMP might be a good candidate for the overlay. Likewise, if total members within a subnetwork are small, then we can use GKMP or HFK in a subnet, for example. We intend to analyze the costs and tradeoffs involved in using different key generation schemes in the proposed framework.

The generation of the datahiding key for a group requires the SKCs of all subnetworks in the group to be in agreement about the datahiding key. We

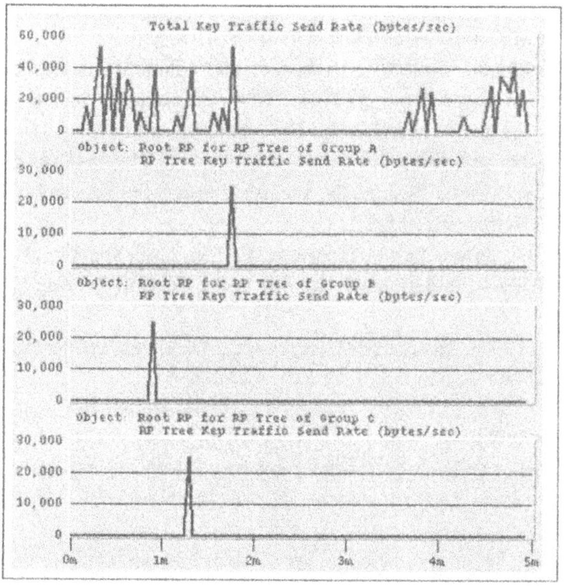

Figure 4. Tiered Tree Framework - Many-to-Many: Total Key Traffic vs. RP Tree Traffic for 3 Groups (Y-axis shows the traffic in bytes/sec; X-axis is the simulation duration in minutes).

have not considered the key management for the datahiding key, since that is a one time message exchange. A simple mechanism to do this is for the SKC in the root RP subnetwork to generate the key and send it to the SKCs in the other subscribed subnetworks; the generating SKC can know of the other subnetworks in a message from the root RP.

In other future work, we plan to investigate mechanisms for source authentication, with suitable modifications for groups operating in the hybrid broadcast network.

9. ACKNOWLEDGMENTS

The first author thanks Dr. Virgil Gligor and Vijay Bharadwaj for helpful discussions on the security framework, and Nalini Bharatula for help with the simulations. Thanks are also due to the anonymous reviewers for their valuable feedback.

The research work reported here was supported by a grant from Lockheed Martin Global Telecommunications, through Maryland Industrial Partnerships under contract number 251715, and by NASA under award number NCC 8235.

References

Armitage, G. (1997). "IP Multicasting over ATM Networks". *IEEE Journal on Selected Areas in Communications*, 15(3):445–457.

Berkovits, S. (1991). "How To Broadcast A Secret". *Advances in Cryptology - EUROCRYPT '91, Lecture Notes in Computer Science, LNCS*, 547:535–541.

Burmester, M. and Desmedt, Y. (1994). "A Secure and Efficient Conference Key Distribution System". *Advances in Cryptology - EUROCRYPT '94, Lecture Notes in Computer Science.*

Canetti, R., Garay, J., Itkis, G., Micciancio, D., Naor, M., and Pinkas, B. (1999). "Multicast Security: A Taxonomy and Some Efficient Constructions". *Proceedings of INFOCOMM '99.*

Chiou, G. and Chen, W. (1989). "Secure Broadcasting Using the Secure Lock". *IEEE Transactions on Software Engineering*, 15(8).

Deering, S.E., Estrin, D., Farinacci, D., Jacobson, V., Liu, C-G, and Wei, L. (1996). "The PIM Architecture for Wide-Area Multicast Routing". *IEEE/ACM Transactions on Networking*, 4(2):153–162.

Fiat, A. and Naor, M. (1994). "Broadcast Encryption". *Advances in Cryptology - CRYPTO '93, Lecture Notes in Computer Science, LNCS*, 773:480–491.

Harney, H. and Muckenhirn, C. (1997). *"Group Key Management Protocol (GKMP) Architecture"*. Internet RFC 2094.

Kim, K. S., Kim, S. J., and Won, D. H. (1996). "Conditional Access System Using Smart Card". *Proc. of JCCI'96, The 6th Joint Conference on Communication and Information*, pages 180–183.

Matsumoto, T. and Imai, H. (1988). "On the KEY PREDISTRIBUTION SYSTEM: A Partical Solution to the Key Distribution Problem". *Advances in Cryptology - CRYPTO '87, Lecture Notes in Computer Science, LNCS*, 293:185–193.

Mittra, S. (1997). "Iolus: A Framework for Scalable Secure Multicasting". In *Proceedings of ACM SIGCOMM'97*, pages 277–288.

Opnet (2002). Opnet Modeler 9.0. http://www.opnet.com/products/modeler/home.html.

Perrig, A., Song, D., and Tygar, J.D. (2001). "ELK, a New Protocol for Efficient Large-Group Key Distribution". *Proceedings of IEEE Security and Privacy Symposium S&P2001.*

Poovendran, R. (1999). *"Key Management for Secure Multicast Communication"*. PhD thesis, University of Maryland College Park.

Rodeh, O., Birman, K., and Dolev, D. (2000). "Optimized Group Rekey for Group Communication Systems". *Proceedings of Network and Distributed System Security Symposium (NDSS'00).*

Roy-Chowdhury, A. (2003). "IP Routing and Key Management for Secure Multicast in Satellite ATM Networks". Master's thesis, Univeristy of Maryland College Park.

Steer, D., Strawczynski, L., Diffie, W., and Wiener, M. (1990). "A Secure Audio Teleconference System". *Advances in Cryptology - CRYPTO '88, Lecture Notes in Computer Science, LNCS*, 403:520–528.

Steiner, M., Tsudik, G., and Waidner, M. (2000). "Key Agreement in Dynamic Peer Groups". *IEEE Transactions on Parallel and Distributed Systems*, 11(8):769–780.

Wallner, D., Harder, E., and Agee, R. (1999). *"Key Management for Multicast: Issues and Architectures"*. Internet RFC 2627.

Wong, C. K., Gouda, M., and Lam, S. S. (2000). "Secure Group Communications Using Key Graphs". *IEEE/ACM Transactions on Networking*, 8(1):16–30.

GPSR Compliance
The European Union's (EU) General Product Safety Regulation (GPSR) is a set of rules that requires consumer products to be safe and our obligations to ensure this.

If you have any concerns about our products, you can contact us on

ProductSafety@springernature.com

In case Publisher is established outside the EU, the EU authorized representative is:

Springer Nature Customer Service Center GmbH
Europaplatz 3
69115 Heidelberg, Germany

www.ingramcontent.com/pod-product-compliance
Ingram Content Group UK Ltd.
Pitfield, Milton Keynes, MK11 3LW, UK
UKHW020130290926
15471UKWH00003B/384
* 9 7 8 1 4 7 5 7 8 0 1 5 4 *